Ion Exchange Technology II

Inamuddin • Mohammad Luqman
Editors

Ion Exchange Technology II

Applications

Springer

Editors
Inamuddin
Assistant Professor
Department of Applied Chemistry
Faculty of Engineering & Technology
Aligarh Muslim University
Aligarh, India

Mohammad Luqman
Assistant Professor
Chemical Engineering Department
College of Engineering
King Saud University
Riyadh, Kingdom of Saudi Arabia

ISBN 978-94-007-4025-9 ISBN 978-94-007-4026-6 (eBook)
DOI 10.1007/978-94-007-4026-6
Springer Dordrecht Heidelberg New York London

Library of Congress Control Number: 2012938478

Printed on acid-free paper

Springer is part of Springer Science+Business Media (www.springer.com)

Preface

Ion exchange is a process of exchanging ions between stationary and mobile phases. It is a natural process that has been in practice for ages. Since commercial development took place in the last century, both academic and industrial research have been improving technology to find new applications.

This edition covers applications of ion exchange materials and technology in various fields, including fuel cells, catalysis, wastewater treatment and water purification, selective separation, concentration and recovery of toxic metal cations and anions, chemical and biochemical separations, pharmaceuticals, therapeutic applications, and electronics.

Chapter 1 covers the basic principles and modern applications of ion exchange technology in the separation of amino acids, peptides, and proteins, which are very useful biomaterials. Chapter 2 deals with whey disposal, which presents a huge obstacle for the dairy industry, as it is an expensive and problematic process. Chapter 3 presents the application of ion exchangers in the speciation and fractionation of elements in food and beverages. The main method used in this area is solid-phase extraction, which includes procedures with single sorbents or two coupled sorbents. Chapter 4 covers the applications of ion exchangers in the alcohol manufacturing industry. Ion exchangers contribute to remove potassium and to stabilize wine against the sediments of potassium bitartrate crystals in the acid cycle. Furthermore they help treating wine for cation and removing anion from grape in grape sugar production. The use of ion exchange resins in continuous sugar process industry is reviewed in Chap. 5. a particular focus is given to chromatographic methods and, specifically to continuous annular chromatography and simulated moving bed (SMB) apparatus.

Chapter 6 deals with the study of the application of ion exchange resins as catalysts in the synthesis of isobutyl acetate, one of the industrially important reactions. The ability of ion exchange resins to bind to drugs, which helps in masking the bitter taste of drugs, drug release, and drug stabilization, has made it a popular choice with pharmaceutical companies. Chapter 7 focuses on the therapeutic applications of ion exchange resins. Chapter 8 adds to chapter 7 by reviewing the application of ion exchange resins in kidney dialysis. The development and use of synthetic ion exchange resins for kidney dialysis is a relatively recent achievement.

An overview of zeolites as inorganic ion exchangers for environmental applications is presented in Chap. 9. The presence of fluoride, pesticides, radionuclides, organic wastes, and heavy toxic metal ions in the aquatic environment has been of great concern to engineers, environmentalists, and scientists because of their increased discharge, toxic nature, and adverse effects on receiving waters. Chapter 10 addresses these points using ion exchange materials and technology. Chapter 11 presents an overview of the ion exchange of metal ions, focusing on their recovery, separation, and pre-concentration. In Chap. 12, a study for a chelating resin containing iminodiacetic acid group (Diaion CR11) is presented to separate trivalent chromium, copper, and iron from synthetic and industrial effluents. In Chap. 13, a similar study to determine and model the effect of cadmium and zinc ions in solution for the removal of Cr(VI) via ion exchange with hydrotalcite, a clay mineral media, is discussed. Chapter 14 presents an overview of the role of the absorbing materials, especially phenol-based resins, for "3d" and "4f" metals ions from the discharge of the related industries.

Planar chromatography, i.e., paper and thin-layer chromatography, is a simple and fast method for the separation of various classes of organic and inorganic compounds. An overview related to the usage of inorganic ion exchangers in planar chromatographic separations is presented in Chap. 15. Chapter 16 deals with a similar topic, i.e., an overview of cation-exchanged silica gel–based thin layer chromatography of organic and inorganic compounds. Finally, Chap. 17 discusses different ion exchange materials used for the removal of various anions (nitrate, fluoride, perchlorate, arsenate, chromate, phosphate, thiocynate, etc.) from water.

These chapters provide an in-depth practical knowledge of ion exchange materials suitable for postgraduate students, researchers, and R&D specialists in the chemical and biochemical industries.

Inamuddin
Mohammad Luqman
Editors

Acknowledgments

We are most indebted to the grace of the Almighty "One Universal Being," who inspires entire Humanity to knowledge, and who blessed us with the needed favor to complete this work.

This edition is the outcome of remarkable contributions of experts in the field of ion exchange technology with up-to-date and comprehensive reviews and research work. We are thankful to all the authors for their esteemed contribution to this book. We would also like to thank all the publishers and authors who granted us permission to use their copyright material. Although sincere efforts were made to obtain copyright permissions from the respective owners and to include citations with the reproduced materials, we would like to offer our sincere apologies to any copyright holder whose rights may have been unknowingly infringed.

Dr. Inamuddin would like to express his deep sense of gratitude to Profs. Syed Ashfaq Nabi, Ishtiyaq Ahmad, Rakesh Kumar Mahaja, Seon Jeon Kim, Kenneth I. Ozoemena, Saleem-ur-Rahman, S.M.J. Zaidi, Gaber E. Eldesoky, Zeid A. Al-Othman, Sheikh Raisuddin, Byong-Hun Jeon, and A.I. Yahya, and to Drs. B.D. Malhotra and Raju Khan for their valuable suggestions, guidance, and constant inspiration.

He would also like to thank his departmental colleagues, Profs. M. Mobin, Asif Ali Khan, and R.A.K. Rao, and Drs. M.Z.A. Rafiqui, Abu Nasar, Rais Ahmad, and Yasser Azim. Without their continuous encouragement, this book would have not been brought to its final form. Dr. Inamuddin is at a loss of words to express his gratitude to his friends and colleagues, Drs. M.M. Alam, Amir-Al-Ahmad, Zafar Alam, Mu. Naushad, Salabh Jain, Hemendra Kumar Tiwari, Adesh Bhadana, Shakeel Ahmad Khan, Satish Singh, and others, for their timely help, good wishes, encouragement, and affection.

Dr. Luqman would like to acknowledge the lively environment provided by Samsung Cheil Industries during time with them. He is also very grateful to his colleagues, more specifically to the Chairman of the Chemical Engineering

Department, King Saud University, and other departmental staff for their help, support, and encouragement throughout the preparation of this book.

Last but not least, we would like to express our heartfelt gratitude to our family members for their constant inspiration and gracious support.

Inamuddin
Mohammad Luqman
Editors

Editors' Bios

Dr. Inamuddin is currently working as Assistant Professor in the Department of Applied Chemistry, Aligarh Muslim University (AMU), India. He received his Master of Science degree in Organic Chemistry from Chaudhry Charan Singh (CCS) University, Meerut, India, in 2002 and his Master of Philosophy and Doctor of Philosophy degrees in Applied Chemistry from AMU in 2004 and 2007, respectively. He has extensive research experience in the multidisciplinary fields of Analytical Chemistry, Material Chemistry, and Electrochemistry and, more specifically, Renewable Energy and Environment. He has worked in different projects funded by the University Grants Commission (UGC), Government of India, and Council of Scientific and Industrial Research (CSIR), Govt. of India. He has received Fast Track Young Scientist Award of Science and Technology, Government of India, to work in the area of bending actuators and artificial muscles. He has published 30 research articles and 4 book chapters of international repute. Recently, he has edited the book entitled *Advanced Organic–inorganic Composites: Materials, Devices and Allied Applications* and co-edited the books entitled *Green Solvents I: Properties and Applications in Chemistry and Green Solvents II: Properties and Applications of Ionic Liquids* published by Nova Science Publishers, Inc, and Springer, United Kingdom, respectively. He is presently working as editor in chief of the Journal of Chemical and Environmental Research published from The Muslim Association for the Advancement of Science (MAAS), India. He has worked as a Postdoctoral Fellow, leading a research team at Creative Research Initiative Center for Bio-Artificial Muscle, Hanyang University, South Korea, in the field of renewable energy, especially biofuel cells. He has also worked as Postdoctoral Fellow at Center of Research Excellence in Renewable Energy, King Fahd University of Petroleum and Minerals, Saudi Arabia, in the field of polymer electrolyte membrane fuel cells and computer fluid dynamics of polymer electrolyte membrane fuel cells. He is a life member of the *Journal of the Indian Chemical Society*.

Mohammad Luqman, Ph.D. is serving as an Assistant Professor in Chemical Engineering Department, King Saud University (KSU), Saudi Arabia. He teaches Polymer/Material Science and Engineering courses. Before joining the KSU, he served as an Assistant Manager in Samsung Cheil Industries, Korea. He worked on the development of heat–resistant polymers, organic glass, and block copolymers as impact modifiers and compatibilizers for engineering alloys. He served as a post-doctoral fellow at Artificial Muscle Research Center, Konkuk University, Korea, in the field of Ionic Polymer Metal Composites. He was awarded the Ph.D. degree in 2007, in the field of Ionomers, by Chosun University, Korea. Dr. Luqman has published numerous papers and book chapters in the field of Ionomers, Ion exchange Polymers, and Polymer Nanocomposites. One of his papers, published in *Polymer*, Elsevier, was Ranked the "first" among the "Top 25 Hottest Articles" in "all" chemistry journals during April–June 2008. Dr. Luqman is acting as a Regional (Middle East) Editor of the International Journal *Chemical and Environmental Research*," Published from MAAS, India. He also serves as a Section Editor of *Journal of Industrial Research & Technology*, by HATAM Publishers, Malaysia. He has delivered few invited lectures at International podiums and very recently has served as an Invited Editor of a reference book entitled *Recent Advances in Plasticizers*.

Contents

Contributors

Zeid A. AL-Othman Department of Chemistry, College of Science, King Saud University, Riyadh, Saudi Arabia

Kendra Axness Wisonsin Department of Natural Resources, Madison, WI, USA

Amit Bhatnagar Chemical Engineering Department, LSRE – Laboratory of Separation and Reaction Engineering, University of Porto (FEUP), Porto, Portugal

Simão Pedro Cardoso CICECO/Department of Chemistry, University of Aveiro, Aveiro, Portugal

Ildefonso Caro Department of Chemical Engineering and Food Technology, University of Cádiz, Puerto Real, Spain

Alime Çıtak Department of Chemical Engineering, Eskişehir Osmangazi University, Eskişehir, Turkey

David M. Dolan Department of Natural and Applied Sciences, University of Wisconsin-Green Bay, Green Bay, WI, USA

Armando Costa Duarte CESAM, University of Aveiro, Aveiro, Portugal

Fazal-Ur-Rehman Department of Anatomy, Jawaharlal Nehru Medical College, Aligarh Muslim University, Aligarh, India

Licínio M. Gando-Ferreira Department of Chemical Engineering, Centre for Chemical Processes Engineering and Forest Products (CIEPQPF), University of Coimbra, Pólo II, Coimbra, Portugal

Pedro Sá Gomes CAT Catalytic Center - Institute for Technical and Macromolecular Chemistry Rheinisch-Westfälische Technische Hochschule – RWTH Aachen University, Aachen, Germany

Juan Gómez Department of Chemical Engineering and Food Technology, University of Cádiz, Puerto Real, Spain

Piotr Jamroz Division of Analytical Chemistry, Wroclaw University of Technology, Wroclaw, Poland

J.D. Joshi Sardar Patel University, Dist. Anand, Gujarat, India

Sheeba Nuzhat Khan Department of Biochemistry, Aligarh Muslim University, Aligarh, India

Eva Kumar Chemical Engineering Department, LSRE – Laboratory of Separation and Reaction Engineering, University of Porto (FEUP), Porto, Portugal

Cristina Lasanta Department of Chemical Engineering and Food Technology, University of Cádiz, Puerto Real, Spain

Patrícia Ferreira Lito CICECO/Department of Chemistry, University of Aveiro, Aveiro, Portugal

Cláudia Batista Lopes CESAM, University of Aveiro, Aveiro, Portugal

Luka Mihajlovic University of Belgrade, Belgrade, Serbia

Dušanka Milojković-Opsenica Faculty of Chemistry, University of Belgrade, Belgrade, Serbia

Ali Mohammad Department of Applied Chemistry, Aligarh Muslim University, Aligarh, India

Abdul Moheman Department of Chemistry, Aligarh Muslim University, Aligarh, India

Gaber E. El-Desoky Department of Chemistry, College of Science, King Saud University, Riyadh Saudi Arabia

Mu. Naushad Department of Chemistry, College of Science, King Saud University, Riyadh, Saudi Arabia

Jana Ognjenovic University of Belgrade, Belgrade, Serbia

Eduarda Pereira CESAM, University of Aveiro, Aveiro, Portugal

Pawel Pohl Division of Analytical Chemistry, Wroclaw University of Technology, Wroclaw, Poland

Jelena Radosavljevic University of Belgrade, Belgrade, Serbia

Alírio E. Rodrigues Laboratory of Separation and Reaction Engineering, Associate Laboratory LSRE/LCM, Department of Chemical Engineering, University of Porto, Porto, Portugal

Carlos Manuel Silva CICECO/Department of Chemistry, University of Aveiro, Aveiro, Portugal

Viviana M.T.M. Silva Laboratory of Separation and Reaction Engineering, Associate Laboratory LSRE/LCM, Department of Chemical Engineering, University of Porto, Porto, Portugal

Dragana Stanic University of Belgrade, Belgrade, Serbia

Helena Stecka Division of Analytical Chemistry, Wroclaw University of Technology, Wroclaw, Poland

Marija Stojadinovic University of Belgrade, Belgrade, Serbia

Patricia A. Terry Department of Natural and Applied Sciences, University of Wisconsin-Green Bay, Green Bay, WI, USA

Živoslav Tešić Faculty of Chemistry, University of Belgrade, Belgrade, Serbia

Tanja Cirkovic Velickovic University of Belgrade, Belgrade, Serbia

Sadaf Zaidi Department of Chemical Engineering, Aligarh Muslim University, Aligarh, UP, India

List of Abbreviations

z	Axial coordinate (m)
D_b	Axial dispersion (m^2/s)
d	Diameter (m)
q^{eq}	Equilibrium adsorbed concentration ($kg/m_{solid}{}^3$)
u	Interstitial velocity (m/s)
L	Length (m)
	Liquid flow rate (m^3/s)
C	Liquid phase concentration (kg/m^3)
	Mass transfer coefficient (s−1)
α_B^A	Separation factor
	Solid flow rate (m^3/s)
u_S	Solid velocity (m/s)
t_s	Switching time (s)
t	Time variable (s)
$(C_2H_5)_2O$	Diethyl ether
12MR	Twelve-membered ring
1M2P	1-methoxy-2-propanol
2M1P	2-methoxy-1-propanol
4MR	Four-membered ring
4-VP	4-Vinylpyridine
6MR	Six-membered ring
8MR	Eight-membered ring
abAz, aaBs	Activities
ABEC	Aqueous biphasic extraction chromatographic
AC	Activated carbon
AcOH	Acetic acid
AE	Aminoethyl
$AgNO_3$	Silver nitrate
Al(III)	Aluminum(III)
$AlCl_3$	Aluminium chloride
Amb200CT	Amberlite-type cation exchange resin

APAS	Aminophosphonic acid silica
As(III)	Trivalent arsenite
As(V)	Pentavalent arsenate
ASV	Anodic stripping voltammetry
B.C.E	Before the common era
BAT	Best available technology
BEA	Zeolite beta
BS	Banana stem
BT	Breakthrough
BuOAc	Butyl acetate
BuOH	Butanol
BV	Bed volume
C	Cholic acid
C.E	Common Era
C_6H_{14}	Hexane
$C_6H_5CH_3$	Toluene
CBUs	Composite building units
CCl_4	Carbon tetra chloride
CCP	Colloidal calcium phosphate
CD	Cyclodextrin
CDC	Chenodeoxycholic acid
$CdCl_2$	Cadmium chloride
CE	Cation-exchange, cation-exchanger
Ce(III)	Cerium(III)
CEC	Cation-exchange capacity
CER	Ceralite IRA 400
CF	Feed concentration, mg/L
CH_2Cl_2	Dichloromethane
$CHCl_3$	Chloroform
Clin-Fe	Clinoptilolite-iron
ClO_4^-	Perchlorate
CM	Carboxymethyl
CMC	Ceramic matrix composites
CMX	Cation exchange membranes
COD	Chemical oxygen demand
CP	Coconut coir pith
CPE	Carbon paste electrode
CRM	Certified reference material
CSV	Cathodic stripping voltammetry
$CuSO_4$	Copper sulfate
CV	Crystal violet
Cyt *c*	Cytochrome *c*
D	Diffusion coefficient
D4R	Double four-membered ring
D6R	Double six-membered ring

D_{app}	Apparent diffusion coefficient
Dax	Axial dispersion coefficient, cm^2/s
DC	Defluoridation capacity
DD-MSWV	Double differential multiple square wave voltammetry
DDT	Dichlorodiphenyltrichloroethane
DE	Decolorization efficiency
DEA	Diethylamine
DEAE	Diethylaminoethyl
DEAPA	Diethylaminopropylamine
DETA	Diethylenetriamine
DFs	Decontamination factors
DF-STEM	Dark field scanning transmission electron microscopy
DL	Detection limit
Dm	Molecular diffusivity, cm^2/s
DMAHP	Dimethylaminohydroxypropyl
DMAPA	Dimethylaminopropylamine
DMFC	Direct methanol fuel cell
DMG	Dimethylglyoxime
DMSO	Dimethyl sulphoxide
DNA	Deoxyribonucleic acid
Dp	Intraparticular diffusion coefficient, cm^2/s
dp	Particle diameter, cm
DP-ASV	Differential pulse anodic stripping voltammetry
DPCSG	1,5-diphenylcarbazide doped sol–gel silica
DPGME	Di-propylglycol-methylether
DPV	Differential pulse voltammetry
D-R	Dubinin-Radushkevich
DR	Dubinin-Raduskevish isotherm
DS	Degree of sulfonation
D_s	Diffusion coefficient in the solution phase
DVB	Di-vinyl benzene
E(OC)	Open circuit electrochemical potential
$E(OC)_{CE}$	Open circuit electrochemical potential
$E_{1/2}$	Half-wave potential
E_{appl}	Applied potential
ECL	Electrochemiluminescence
ECTEOLA	The product of reaction of epichlorohydrin, triethanolamine, and alkali cellulose
ED	Electrodialysis
EDA	Ethylenediamine
EDR	Electrodialysis reversal
EDS	Energy dispersive X-ray spectroscopy
EDTA	Ethylenediaminetetraacetate
EDX	Energy dispersive X-ray spectroscopy
E_i	Initial potential

EIS	Electrochemical impedance spectroscopy
ENM	Electrospin nano-fiber membrane
EPA	Environmental protection agency
Ep_b	Backward peak potential
EPBI(DMG)	Epoxidized polybenzimidazole(Dimethylglyoxime)
Ep_f	Forward peak potential
ETSS	Ethyl styrene sulfonate
EU	European union
EW	Equivalent weight
F	Faraday constant
F-AAS	Flame atomic absorption spectrometry
FAU	Faujasite
Fc	Ferrocene
Fc^+	Ferricinium cation
FCC	Fluid catalytic cracking
Fe(III)	Iron(III)
$FeCl_3$	Ferric chloride
$FeCl_4^-$	Tetrachloroferrate
Fe-LLT	$(La,Fe)TiO_3$
$FeSO_4$	Ferrous sulfate
FS	Full scale
FTIR	Fourier transform infrared
FVA	Fibrous anion exchanger
GC	Gas chromatography
GCE	Glassy carbon electrode
GDC	Glycodeoxycholic acid
GF-AAS	Graphite furnace atomic absorption spectrometry
GIS	Gismondine
GLC	Glycolithocholic acid
GMA/MBA	Glycidyl methacrylate/N,N´-methylene bis-acrylamide
GME	Gmelinite
$H_2AsO_4^-$	Dihydrogen arsenate
H_3AsO_3	Arsenious acid
H_3Cyt	Cytric acid
HA	Humic acid
HAIX-F	Hybrid anion exchange fiber
HASB	Hard soft acid base
$HAsO_4^{2-}$	Hydrogen arsenate
HA-Zr-PILC	Humic acid-immobilized zirconium-pillared clay
HBPEI	Hyperbranched polyethylenimine
HFO	Hydrated Fe(III) oxide
HIX	Hybrid ion exchange
HIX-NF	Hybrid ion exchange-nanofiltration
HMW	High molecular weight
HPA	Hydrated tungstophosphoric acid

HPCIC	High performance chelation ion chromatography
HPLC	High performance liquid chromatography
HPO_4^{2-}	Mono hydrogen phosphate
HPTLC	High performance thin layer chromatography
HREM	High resolution electron microscopy
hRF	RF ×100
HTC	Hydrotalcite
HZO	Hydrous zirconium oxide
IAEA	International atomic energy agency
ICP-MS	Inductively coupled plasma mass spectrometry
ICP-OES	Inductively coupled plasma optical emission spectrometry
IDA	Iminodiacetic acid
IE fiber	Ion exchange fiber
IE	Ion exchange
IEC	Ion exchange capacity
IEV	Ion exchange voltammetry
IND	Indion FR 10
IO	Iontosorb oxin
Ip_p	Peak current for analytes in the polymer phase
Ip_s	Peak current for analytes in the solution phase
IR	Infrared
IS	Iontosorb salicyl
ITO	Indium tin oxide
IZA	International zeolite association
K_2CrO_4	Potassium dichromate
Ka	Equilibrium constant
KBr	Potassium bromide
k_D	Distribution coefficient
Kd	Distribution coefficient
k_{ex}	Site to site electron exchange rate constant
kf	Film mass transfer coefficient, cm/s
KFI	KFI framework type zeolite
KI	Potassium iodide
kL	Langmuir parameter, L/mg
K_X^M	Selectivity coefficient
Ks	Selectivity coefficient
L	Bed height, cm
La(III)	Lanthanu(III)m
La_2O_3	Lantanum oxide
LB	Langmuir-Blodgett
LBL	Layer-by-layer
LC	Lithocholic acid
LDH	Layered double hydroxides
LDPE	Low density polyethylene
L-His	L-Histidine

LiBr	Lithium bromide
LMW	Low molecular weight
LOD	Limit of detection
LOQ	Limit of quantitation
L-Phe	L-Phenylalanine
L-Pro	L-Proline
LRW	Liquid radioactive waste
LS	Lab scale
LTA	Linde type A framework type zeolite
L-Trp	L-Trptophan
MAP	Maximum aluminium P, GIS framework type zeolite
MAS	Magic angle spinning
MAs, MBs	Molarities
mAz, mBz	Molalities
MB	Methylene blue
MCM-41	Mobil mesoporous material
MEL	Maximum exchange level
MELs	Maximum exchange levels
MFI	ZSM-5 (five)
MG	Malachite green
MHL	Metal proton ligand
MMC	Metal matrix composites
MMT	Montmorillonite
MMW	Moderate molecular weight
$MnSO_4$	Manganese sulfate
MOR	Mordenite
MS	Mass spectrometry
MSWV	Multiple square wave voltammetry
MTA	Methylthriamyl ether
MTB	Methylthributhyl ether
MV	Methyl viologen
MW	Molecular weight
MWW	MCM-22 (twenty-two)
N	Noise
n	Number of columns (–)
N,N-Me2-L-Phe	N,N-dimthyl-phenylalanine
NaOH	Sodium hydroxide
ND	Number of mass transfer units by pore diffusion (= τεP DP /R2)
N-DC	N, N′ di(caroxymethyl)dithiocarbamate
Nf	Number of film mass transfer units (= 3 (1 – ε) kf τ/(ε R))
NH_3	Ammonia
NH_4Cl	Ammonium chloride
$NiSO_4$	Nickel sulfate
nm	Nanometer
NMP	N-methyl-2-pyrrolidinon

NMR	Nuclear magnetic resonance
NORM	Naturally occurring radioactive materials
NP	Nernst-Planck
OIV	International organisation of vine and wine
P	Phosphate
PA	Polyamide
PAB	4-Aminobenzyl
PAMAM	Polyamidoamine
PAN	Polyacrylonitrile
PAn/SD	Polyaniline/sawdust
PANF	Polyacrylonitrile fiber
PANI	Polyaniline resin
PANI/HA	Polyaniline/humic acid nanocomposite
PBI	Poly(benzimidazole)
PC	Polycarbonate
PC	Paper chromatography
PDDMACl	Poly(diallyldimethylammonium chloride)
PDDPCl	Poly(1,1-dimethyl-3,5-dimethylenepiperidinium chloride)
Pe	Axial Peclet number (= uL/εDax)
Pe	Peclet number (–)
PE	Polyethylene
PEEK	Poly(ether ether ketone)
PEG	Polyethylene glycol
PEI	Poly(ether imine)
PEIMPA	Polyethylenimine methylenephosphonic acid
PEK	Poly ether ketone
PEKEKK	Poly(ether ketone ether ketone ketone)
PEM	Poly(ethylene mercaptoacetimide)
PEMFC	Polymer electrolyte membrane fuel cell
PES	Polyether sulphone
PET	Poly(ethyleneterphthalate)
PFSA	Perfluorosulfonic acid
PGCP	Coconut coir pith grafted with polyhydroxyethylmethacrylate
PGHyFeO–COOH	Carboxylate functional group into polyacrylamide-grafted hydrous iron(III) oxide
PI	Polyimides
PILC	Pillared clay
PLE	Polymeric ligand exchanger
PLE's	Polymeric ligand exchangers
PMA	Poly mtharcylate
PMC	Polymer matrix composites
PMeT	Poly(3-methylthiophene)
POPs	Persistent organochlorine pollutants
PP	Polypropylene
PPO	Poly(phenylene oxide)

PP-ST-DVB	Polypropylene with grafted polystyrene with divinylbenzene
PPy	Polypyrrole
PPy/SD	Polypyrrole/sawdust
PrO	Propylene oxide
PrOH	Propanol
PSDC	PS-DVB copolymer
PS-DVB	Polystyrene divinylbenzene
PSS	Poly(4-styrene sulfonate)
PSSH	Poly(styrenesulfonic acid)
PSSNa	Poly(sodium styrenesulfonate)
PSU	Polysulphone Udel
PSU-NH_2	Aminated polysulfone Udel
PTFE	Poly(tetrafluoroethylene)
PV	Pervaporation
PVA	Poly (vinyl alcohol)
PVA/EDTA	Poly(vinyl alcohol)/EDTA
PVA/P4VP	Poly(vinyl alcohol)/poly(4-vinylpyridine)
PVC	Poly (vinyl chloride)
PVI/PVP	Polyvinylimidazole / Polyvinylpyrrolidone copolymers
PVP	Polyvinyl pyrollidone
PVPP	Polyvinylpyrrolidone
Q	Adsorbed concentration in equilibrium conditions, mg/g wet resin
QA	Quality assurance
QC	Quality control
QCS	Quaternary chitosan salt
qmax	Maximum adsorption capacity (Langmuir parameter), mg/g wet resin
R	Particle radius, cm
R2	Coefficient of determination
Re	Reynolds νυμβερ (= υ ρ dπ /ε μ)
REC	Real exchange level
RF	Retardation factor
RO	Reverse osmosis
ROS	Reactive oxygen species
RP	Reversed-phase
RP-18 silica gel	Octadecyl silica gel
RPA	Amino pentamine resin
RQA	Quaternary amine resin
Rv	Product of ionic radius of cation and its valence
S	Signal
SAED	Selected area electron diffraction
SBA	Strong basic anion-exchange
SDB	Styrene-divinylbenzene
SEC	Size exclusion chromatography

SEE	Standard error of estimate
SEM	Scanning electron microscopy
s-IPNs	Semi-interpenetrating polymer networks
SIR	Solvent impregnated resins
SMM	Surface modifying macromolecules
SPE	Screen printed electrode
SPEEK	Sulfonated poly(ether ether ketone)
SPI	Sulfonated polyimide
SPME	Solid phase microextraction
SPSU	Ortho-sulfonesulfonated poly (ethersulfone)
$SrCl_2$	Strontium chloride
ST	Polystyrene
ST-DVB	Styrene–divinylbenzene
STEM	Scanning transmission electron microscopy
STPP	Sodium tripolyphosphate
SWV	Square wave voltammetry
T	Temperature
t	Time
TAG	Triacylglycerol
TCB	Phenol-trichlorobenzene
TDS	Total dissolved solids
TEC	Theoretical exchange level
TEM	Transmission electron microscopy
TETA	Triethylenetetraamine
Tg	Glass transition temperature
THF	Tetrahydrofuran
TiOx(OH)y	Ti-oxohydroxide
TLC	Thin layer chromatography
TMFE	Thin mercury film electrode
TPA	Tripropylamine
TPABr	Tetrapropylammonium bromide
TPIR	Total polyphenol index removed.
U	Bed superficial velocity, cm/s
u*	Normalized radial coordinate (r/R)
UF	Ultra filtration
UHT	Ultra high temperature
USEPA	United States environmental protect agency
UV	Ultraviolet light
V	Scan rate
Vc	Estimated volume occupied by cations in FAU supercavity
VOCs	Volatile organic chemicals
WBA	Weak base anion
WHO	World health organization
WNF	Waste nuclear fuel
WWTG	World wine trade group

x	Dimensionless axial coordinate (–)
X	Dimensionless solute concentration in the bulk liquid phase (C/CE)
XAD	Commercial polystrene divinylbenzene resin
Xp	Dimensionless solute concentration in the liquid inside pores (Cp/CE)
Y(III)	Yttrium(III)
Z	Bed axial coordinate, cm
z*	Normalized axial coordinate (z/L)
ZA, ZB, SA, SB	Equivalent fractions
ZK-5	KFI framework type zeolite
Zn-LLT	$(La,Zn)TiO_3$
ZP	Zirconium phosphate
ZrP	Zirconium phosphate
ZSM-5 MFI	Framework type zeolite
ΔG0	Free energy change
ΔH	Enthalpy change
ΔS	Entropy change
δ	Thickness of the diffusion layer
η	Interaction term
μ	Viscosity of the fluid, g/cm.s
Φ	Thickness of the polymer layer
ε	Bed porosity
ε_p	Particle porosity
θ	Normalized time (t/τ)
ρ	Density of the fluid, g/cm^3
ρ_h	Wet density of the adsorbent, g/cm^3
τ	Space time, s

Subscripts and Superscripts

E	Eluent stream
X	Extract stream
k	Number of column (k =1,2,…12)
j	Number of section (j =I,II,III,IV)
E	Raffinate stream
i	Species in binary mixture
*	Operating conditions in SMB
AER	Anion-exchange resin
AEX	Anion-exchange
ALA	Alpha-lactalbumin
API	Active pharmaceutical ingredient
ARF	Acute renal failure

b	Bulk
BET	Surface area and porosity measurement equipment
BLG	Beta-lactoglobulin
BSA	Bovine serum albumin
c	Column
CAC	Continuous annular chromatography
CEC	Capillary electrochromatography
CER	Cation-exchange resin
CKD	Chronic kidney disease
CM	Carboxymethyl
CMP	Caseinomacropeptide
CsEBS	Cesium salts of ethylbenzenesulfoacid
CSS	Cyclic steady state
DB-WAX	GC column
DEAE	Diethylaminoethyl
DVB	Divinylbenzene
ED	Electrodialysis
ESRD	End stage renal disease
ETBE	Ethyl *tert*-butyl ether
EW	Egg white
F	Feed stream
FDA's	Food and drug administration's
FID	Flame ionization detector
GFR	Glomerular filtration rate
GIT	Gastrointestinal tract
Hb	Haemoglobin
HD	Hemodialysis
HGMF	High-gradient magnetic fishing
HPLC	High performance liquid chromatography
HSA	Human serum albumin
IE	Ion exchange
IEC	Ion exchange chromatography
IER	Ion exchange resin
Ig	Immunoglobulin
IO	Integrated-optic
It	Intermediary (Pseudo-SMB process) stream
LDF	Linear driving force
LDL	Low density lipoprotein
LDL-C	Low density lipoprotein cholestrol
LF	Lactoferrin
LP	Lactoperoxidase
Lys	Lysosyme
m- DVB	meta-Divinylbenzene
MA	Membrane adsorber

MAC	Membrane adsorption chromatography
MANAE	Monoaminoethyl-n-aminoethyl
MCGs	Medicated chewing gums
MCLC	Membrane convective liquid chromatography
MDDS	Mobile drug delivary system
MINI, MIDI(d), 3-21G*	Basis sets for non-empirical calculations
MMM	Mixed matrix membrane
MP2/3-21G* and MP2/MIDI(d)	Level of theory of non-empirical calculations with using basis sets 3-21G* and MIDI(d) and with accounting for electronic correlation in the frame of the second order Moeller-Plesset perturbation theory
MS	Mass spectrometry
MTBE	Methyl *tert*-butyl ether
NCBI	National center for biotechnology information
p	Particle
PAD	Pulsed amperometric detection
PD	Peritoneal dialysis
PU	Purity
QPI	Quaternized polyethyleneimine
RE	Recovery
RESD	Renal end stage disease
RHF	Restricted Hartree-Fock method for closed shalls
Rnase	Ribonuclease
ROHF	Restricted open shall Hartree-Fock method
RPC	Reverse phase chromatography
SCF MO LCAO	Model, in which a molecular orbital (MO) is represented as a linear combination of atomic orbitals (LCAO), are examined in light of ab initio self-consistent field (SCF) computations with bases of various sizes
SCX	Strong cation-exchange
SMB	Simulated moving bed
SMCGs	Synthetic medicated chewing gums
SPS	Sodium polystyrene sulfonate
ST-DVB Matrix	Styrene – divinylbenzene matrix
TAME	*Tert*-amyl-methyl ether
TMB	True moving bed
UDCA	Ursodeoxycholic acid
UF	Ultrafiltration
WKB method	Wentzel-Kramers-Brillouin method
WPC	Whey protein concentrate
WPI	Whey protein isolate

Nomenclature

a	Minimum approximation distance between ions
A	External particle surface area
A_γ	Debye-Huckel constant
A'_{ij}, A'_{ji}	Margules parameters
A^{z_A}, B^{z_B}, C^{z_C}	Counter ions with valences z_A, z_B, z_C
$\overline{A}^{z_A}, \overline{B}^{z_B}$	Counter ions with valences z_A, z_B inside the exchanger
a_i	Activity of species i in solution
$\overline{a}_i$	Activity of species i in exchanger
$A_i^{z_i}$	Generic counter ion i with valence z_i
a_p	External surface area per unit particle volume
B	Second Virial coefficient
$B_{j,i}$	Langmuir constant
C_b	Solute concentration at breakthrough time
$C_{ef,i}$	Concentration of sorbate in the effluent
$C_{F,i}$	Concentration of species i in the feed
C_i	Molar concentration of species i in solution
C_i^*	Molar concentration of species i at the exchanger/film interface
$C_{N,i}$	Normality of species i
$C_{N,t}$	Total normality of solution
$C_{p,i}$	Molar concentration of species i inside the pores
$\overline{C}_{p,i}$	Average concentration of species i inside the pores
C_{sat}	Saturation concentration
C_t	Total molar concentration of ionic species in solution
d	Particle diameter
D_A, D_B	Self-diffusion coefficients of species A and B
D_{AB}	Interdiffusion coefficient
$D_{eff,p,i}$	Effective diffusion coefficient of species i in macropores
$D_{eff,s,i}$	Effective diffusion coefficient of species i in micropores
D_f	Diffusion coefficient in the film
D_i	Diffusion coefficient of species i
$Đ_{ij}$	MS surface diffusivity of the pair $i-j$
$Đ_{is}$	MS surface diffusivity corresponding to the interaction between i and the fixed ionic charges
D_L	Axial dispersion coefficient
e	Electron charge
$E_{i,j}$	Energy of adsorption of ion i on site j
$\overline{E}_i$	Average adsorption of ion i
F	Faraday constant
F_i	Fractional attainment of equilibrium of species i
g_{ij}	Energy parameter characteristic of the $i-j$ interaction
I	Ionic strength

J_i	Diffusion flux of species i
k	Boltzmann's constant
k_1	Rate constant of the first order sorption
k_2	Rate constant of the second order sorption
K_{aB}^{A}	Corrected selectivity coefficient
k_{AB}	Bohart and Adams rate constant
K_{B}^{A}	Thermodynamic (equilibrium) constant
K_C	Selectivity coefficient
K_D	Distribution coefficient
k_f	Convective mass transfer coefficient
K_{LDF}	Linear driving force coefficient
$K_S^{M_xM_m}$	Stability constant
k_{Th}	Thomas rate constant
k_{YN}	Yoon-Nelson rate constant
L	Column length
M^{m+}	Cation
m_i	Molality of species i
m_t	Total molality of ionic species
n	Freundlich constant, number of ionic species in solution
N_0	Avogadro's constant
n_c	Number of counter ions
n_f	Number of functional groups
N_i	Molar flux of species i
$N_{p,i}$	Diffusion fluxes of species i through the macropores
$N_{s,i}$	Diffusion fluxes of species i through the micropores
n_w	Number of water molecules in the zeolite
$n_x + n_y$	Total number of tetrahedral in the unit cell of zeolite
q_i	Molar concentration of ionic species i in exchanger
p	Parameter in binomial distribution
p_j	Equivalent fraction of exchanger site of type j
$\overline{q}_i$	Average loading of ionic species i in exchanger
q_i^*	Resinate concentration in equilibrium with the fluid concentration
Q_i	Equivalent ionic concentration of species i in exchanger
Q_i^e	Surface excess of ion i
$Q_{j,i}$	Equivalent ionic concentration of species i on exchanger site j
q_M	Kusik-Meissner parameter
q_{max}	Maximum sorbate concentration in the solid phase
q_s	Molar concentration of ionic fixed groups in exchanger
q_t	Total molar concentration of ionic species in exchanger
Q_t	Ion exchange capacity (in equivalents)
r	Radial position
R	Particle radius
$\Re$	Universal gas constant
t	Time
T	Absolute temperature

$t_{1/2}$	Time required for 50% sorbate breakthrough; stoichiometric time
t_b	Breakthrough time
U_0	Superficial velocity
u_i	Electrochemical mobility of species i, velocity of diffusing species i
V_{ef}	Volume of effluent
V_L	Volume of fluid phase
V_s	Volume of solid phase
V_{ZLC}	Volume of the ZLC column
W_{exch}	Mass of exchanger
W_{ij}	Weighting factor
x_i	Ionic fraction of species i in solution
X_i	Equivalent ionic fraction of species i in solution
X^{x-}	Anion
y_i	Ionic fraction of species i in exchanger
Y_i	Equivalent ionic fraction of species i in exchanger
y_s	Mole fraction of ionic fixed groups in exchanger
z_i	Valence of ionic species i

Subscripts

e	Equilibrium
f	Free
0	Initial condition
s	Solid, fixed ionic groups of the exchanger
t	Total
intra	Intraparticle

Greek Letters

$\overline{\alpha}_j^i$	Average separation factor
δ	Film thickness
α_B^A	Separation factor
$\overline{\gamma}_i$	Activity coefficient of species i in exchanger
γ_i	Activity coefficient of species i in solution
ε_b	Bed porosity
μ_i	Chemical potential of species i in solution
ρ_w	Density of pure solvent
ε	Dielectric constant
θ	Dimensionless time (–)
λ_i	Distribution coefficient of species i
φ	Electric potential
ε	External porosity (–)
$\tau_{d,m}$	Maximum value of τ_d
$\tau_{d,i}$	Minimum value of τ_d

ν_a	Number of anions per electrolyte
ν_c	Number of cations per electrolyte
ε_p	Particle porosity
v_i	Pure-component molar volume
γ	Ratio of fluid and solid velocity in section j (–)
Γ	Reduced activity coefficient of Meissner and Kusik
σ_i	Standard deviation of energy distribution
$\bar{\mu}_i$	Surface chemical potential of species i
Γ_{ij}	Thermodynamic factor
τ_d	Time constant for intraparticle diffusion
ν	Volumetric flow rate, number of site types
$\Lambda_{ij}, \Lambda_{ji}$	Wilson parameters

Chapter 1
Separation of Amino Acids, Peptides, and Proteins by Ion Exchange Chromatography

Tanja Cirkovic Velickovic, Jana Ognjenovic, and Luka Mihajlovic

Abstract Separation of amino acids, peptides, and proteins (bioanalytes) via ion exchange (IE) has widespread usage because it is usually very simple to design and it has high capacity and easily achievable control of the separation process. Amino acids, as principal constituents of proteins and having a plethora of biological functions of their own, are always in focus when developing novel methods. Separation and quantification of amino acids is essential in food science, medicine, agricultural science, etc. Peptides exist in nature and have diverse functions. Digestion of proteins by enzymes also gives complex mixtures of peptides and IE finds its application in peptide separation. There are lots of reasons for the popularity of IE in protein isolation and purification. It is used in research, analysis, and large-scale purification of proteins. Ion exchange is ideal for the initial capture of proteins because of its high capacity, relatively low cost, and its ability to survive rigorous cleaning regimes. This chapter covers basic principles and modern applications of IE in separation of amino acids, peptides, and proteins.

1.1 Introduction

The technique of ion exchange (IE) is based on interactions between charged moieties. Amino acids, peptides, and proteins (bioanalytes) are water-soluble, charged amphoteric molecules and as such, able to bind via Coulomb's interactions to oppositely charged moieties covalently linked to an insoluble carrier (matrix, or a stationary phase). Stationary phases for ion exchange separations are characterized by the nature and strength of the acidic or basic functions on their surfaces and the

---

T. Cirkovic Velickovic (✉) • J. Ognjenovic • L. Mihajlovic
University of Belgrade, Studentski Trg 12-16, 11000 Belgrade, Serbia
e-mail: tcirkov@chem.bg.ac.rs

Inamuddin and M. Luqman (eds.), *Ion Exchange Technology II: Applications*,
DOI 10.1007/978-94-007-4026-6_1, © Springer Science+Business Media B.V. 2012

types of ions that they attract. Cation exchangers retain and separate positively charged ions (analytes) on a negative surface, while anion exchangers retain and separate negatively charged ions on a positive surface. There are at least two general approaches for separation and elution with each type of ion exchanger [1]. Strength of the interaction between stationary phase and the analyte can be controlled by changing the ionic strength of the solution. Also, ionic properties of ionizable functional groups by pH titration of the stationary phase or the analyte can be controlled, allowing great variability in the design of IE experiments.

Strong ion exchangers bear functional groups that are always ionized, over a wide range of pH values. They are typically used to retain and separate weak acids or bases. These weak ions may be eluted by displacement with ions that are more strongly attracted to the stationary phase exchange sites (so-called salt elution). Alternately, weak ions may be retained on the column and subsequently neutralized by changing the pH of the mobile phase, causing them to elute (so-called pH elution). Combined salt/pH elution is possible, as well as isocratic elution.

Weak ion exchangers bear functional groups that above or below a certain pH value may lose their ability to retain ions by charge, i.e., those groups are titrable over a narrower pH range comparing to the strong ion exchanger. When charged, they are used to retain and separate strong ions. Strong ions are ionized over a wide pH range and often bind strongly to the exchange sites of a matrix. It is often difficult to elute strong ions by displacement; therefore, the matrix exchange sites may be neutralized by permitting elution of the charged analytes with a mobile phase [1, 2].

Separation of bioanalytes via ion exchange has widespread usage because it is usually very simple to design and it has high capacity and easily achievable control of the separation process. Both strong and weak ion exchangers found application in the separation of amino acids, peptides, and proteins [3–6]. The binding of proteins to the IE stationary phase predominantly occur via charged amino acid residues at the surface of the protein, although certain posttranslational modifications may contribute to the binding (such as sialyl, sialoglycosyl, phosphoryl groups) [7]. Majority of amino acids, peptides, and proteins are weak ions. Sometimes, it is more convenient to apply methods for separation of strong ions by neutralizing the stationary phase in order to preserve the biological function of the molecules. Although the principles behind separation of amino acids, peptides, and proteins are the same, separation of large molecules, such as proteins by IE, requires matrices of high porosity and moderate density of ionizable groups [8]. Original matrices, developed for separation of small molecules, have been modified and a great variety of stationary phases suitable for separation of proteins is available today [9].

In most cases, maintenance of the functional integrity of a protein during separation is necessary. Mild conditions and short separation time of the ion exchange separation help proteins retain their conformation and thereby the biological function, which is often necessary for many downstream applications.

1.2 General Principles of Ion Exchange Chromatography of Bioanalytes

1.2.1 Theory

For amphoteric compounds, such as amino acids, peptides, and proteins, *the isoelectric point (pI, the pH at which the net charge is zero) of the compound and its stability* at various pH values determine the separation strategy.

At a pH above its pI, the bioanalyte of interest will be negatively charged, and at a pH below its pI, the compound will be positively charged. If the compound is stable at a pH above its pI, an anion exchanger is used (Table 1.1). If the compound is stable at a pH below its pI, a cation exchanger is used (Table 1.2). The operating pH also determines the type of exchanger to use (Tables 1.1 and 1.2).

It should be kept in mind that the pKa value of a residue represents a pH at which ionized and nonionized forms of a functional group are at equilibrium. For a group to be more than 90% titrated, usually 0.5 pH value above the pKa is necessary. To assure an essentially neutral, or a fully charged, analyte or particle surface, the pH must be adjusted to a value at least 2 units beyond the pKa. It should be kept in mind

Table 1.1 Short guidelines for ion exchange of bioanalytes using anionic exchangers

Strength of the ion exchanger	Weak anionic, pKa around 9	Strong anionic
Net charge of the bioanalyte	Negative at pH above its pI	Negative at pH above its Pi
Charge of exchanger	Positive below pH 9	Positive
Running conditions	0.5–1.5 units below the pKa of the exchanger	0.5–1.5 units above the pI of the bioanalyte
Bioanalytes separation	Separation possible for bioanalytes of pI 7.5 and below	Separation of weakly basic and acidic bioanalytes, because of usually better pH stability outside extreme pH values (below 3 and above 10)

Table 1.2 Short guidelines for ion exchange of bioanalytes using cationic exchangers

Strength of the ion exchanger	Weak cationic, pKa around 5	Strong cationic
Net charge of the bioanalyte	Positive at pH below its pI	Positive at pH below its pI
Charge of exchanger	Negative above pH 5	Negative
Running conditions	0.5–1.5 units above the pKa of the exchanger	0.5–1.5 Units below the pI of the bioanalyte
Bioanalytes separation	Separation possible for bioanalytes of pI 6.5 and above	Separation of weakly acidic and basic bioanalytes, because of usually better pH stability outside extreme pH values (below 3 and above 10)

that the pH value and additives of the mobile phase can alter the surface properties of a protein such as the charge density and accessibility or relative location of charged residues, making IE separation of proteins sometimes unpredictable and proving the superiority of practice over theory in IE of bioanalytes. Therefore, the guidelines presented in Tables 1.1 and 1.2 should be regarded as suggestions when optimizing IE separation of bioanalytes.

When using a strong cation exchanger for separation of strong bases, both remain charged, making the base difficult to elute. It can only be removed by a competing base that binds even stronger and displaces the compound of interest at the exchange sites. This approach is not very practical as very strong acids and bases may be corrosive to materials used in IE and may cause structural instability of bioanalytes. A more practical approach is to use a weak cation exchanger for separation of strong bases and neutralize the surface of the exchanger by lowering the pH of the mobile phase.

The basis for IE is competition between ions of interest and other ions for oppositely charged groups on an ion exchanger. The interaction of a protein and an ion exchanger depends not only on the net charge and the ionic strength but also on the surface charge of a protein. It is clear that the more highly charged a protein is, the more strongly it will bind to a given, oppositely charged ion exchanger. Similarly, more highly charged ion exchanger (i.e., those with a higher degree of substitution with charged groups) usually bind proteins more effectively than weakly charged ones. Conditions, for example, pH, that alter the charge on either the protein or ion exchanger will affect their interactions and may be used to influence the ion exchange process [10].

The pH is one of the most important parameters in determining protein binding as it determines the charge on both the protein and the ion exchanger. Although proteins are complex ampholytes that have both positive and negative charges over a wide pH range, as a rule, binding to an ion exchanger occurs only when there is a net charge on the protein of opposite sign to that found on the ion exchanger. The isoelectric point of a bioanalyte depends on ionizable amino acid residues in its structure. Positive charges are usually provided by histidines, arginines, and lysines, depending on the pH of the surrounding buffer. Any free N-terminal amine will also contribute with a positive charge below pH 8. Negative charges are principally provided by aspartate and glutamate residues and the C-terminal carboxyl group. Virtually all these residues are ionized above pH 6. At higher pH values (above 8), cysteines may become ionized too. The charged groups nearly always reside on the protein surface. Exceptions are mainly metalloproteins, where an internal metal ion is often coordinated by charged residues. Posttranslational modifications can also influence charge properties of side residues, and those often contribute to the so-called microheterogeneity often noticed in IE separation of proteins.

In addition, influences from neighboring groups and the position in the tertiary structure will affect the pKa for the side-chain groups [11]. The combined influence of all of the charged side chains will give the protein a varying net charge depending on the pH of the solute. Therefore, it is possible to separate proteins using either fixed positive charges on the stationary phase, anion exchanger, or fixed negative charges, cation exchanger.

A protein must displace the counterions to become attached, and consequently, the net charge on the protein will be the same as that of the counterions displaced, thereof the term "ion exchange." At pH values far away from pI, proteins bind strongly and, in practice, do not desorb at low ionic strength. Near its pI, the net charge of a protein is decreased, and consequently, it binds less strongly. Because the charges on the protein surface are distributed asymmetrically, binding can also occur close to the pI even when overall charge is the same as that on the ion exchanger. Due to Donnan effects, protons are attracted and hydroxyl ions are expelled from the microenvironment of cation-exchange groups. The pH in the matrix is therefore usually one unit lower than in the surrounding buffer. Quite similarly, for anion exchangers, an increase in pH of about one unit occurs. For instance, if a protein is adsorbed at pH 5, it will be exposed to pH 4, which may lead to denaturation if the stability of the protein is poor.

1.2.2 The Matrix

The matrix may be based on inorganic compounds, synthetic resins, or polysaccharides and can be porous and nonporous. The standard supports are beads of polystyrene, silica, or agarose. The matrix characteristics are important for chromatographic properties such as efficiency, capacity, recovery, and chemical and mechanical stability. A wide range of support matrices are utilized in the manufacture of ion exchange media.

The porosity of a stationary phase refers to the total pore volume within the matrix of the support. A very porous support may have either many small pores or a few large pores. The largest compound able to enter the pores under a given set of conditions determines the exclusion limit of a support. Porous media with high exclusion limits are recommended for high molecular weight compounds such as protein and other biomolecules. Low- or high-porosity media with low exclusion limits are recommended for the separation of low molecular weight compounds such as inorganic ions, organic acids, and amino acids.

The first ion exchangers were based on synthetic resins of high hydrophobicity and low porosity, suitable for exchange of small molecules, such as inorganic ions, organic acids, and amino acids. Due to their hydrophobicity and high charge density, these matrices were not suitable for separation of biological samples.

Ionic interactions as the basis for separation and purification of proteins by IEC (ion exchange chromatography) have been applied successfully since the late 1940s. Originally limited to the separation of basic proteins [12], with the introduction of cellulose-based matrices by Sober and Petersen (e.g., carboxymethyl [CM] and diethylaminoethyl [DEAE] derivatives of cellulose), the utility of IEC was extended to include a wide range of proteins. The ideal matrix should be inert, highly porous, and structurally rigid. Rigidity is essential for scale-up and high-flow-rate processes [13]. Particle diameter and porosity of the matrices greatly influence the medium's capacity and resolution. A medium composed of small-diameter particles reaches equilibrium faster (due to short distances for proteins to diffuse) and results in higher resolution than larger-diameter particles. Porosity can influence the binding capacity because particles with larger pores allow larger proteins to enter the pores, thus increasing a number of available sites for protein to bind. The charged moiety bound to the matrix determines the useful pH range and the type of ion exchanger. There is much accumulated experience on this type of chromatography documented in the literature.

Ion exchangers based on dextran (Sephadex) and agarose (Sepharose CL-6B) and cross-linked cellulose (DEAE Sephacel) were the first ion exchange matrices that combined a spherical form with high porosity. The result was an improvement in flow properties and capacity for biological macromolecules.

Ion exchange chromatography matrices are available as dry, granular material or as preswollen loose beads, but prepacked columns are nowadays common, especially for small-scale analytical work [14, 15]. Although the method is essentially the same, now IE can also be carried out on monolithic columns, on IE membranes [16], and on ion exchange high-performance liquid chromatography (HPLC) columns [17].

Besides porosity, particle size is of particular importance in ion exchange chromatography of bioanalytes. Particle size is measured in micrometers, with dry mesh or wet mesh designations. Smaller particle sizes provide higher resolution and typically require lower operational flow rates; larger particle sizes yield lower resolution but can be operated at higher flow rates.

1.2.3 Charged Groups

The charged moiety bound to the matrix determines the useful pH range and the type of ion exchanger. Their total number and availability determines the capacity of the ion exchanger.

The ion exchangers are usually classified as weak and strong. The names refer to the pKa values of their charged groups, and it does not say anything about the strength with which they bind proteins [18]. Strong ion exchangers are chemically sulfonate and quaternary amino groups (Table 1.3). Most commonly applied ion exchangers are those with immobilized DEAE and CM groups (weak ion exchangers). Phospho-group-immobilized matrices can be used both as ion exchangers and as affinity media for separation of enzymes recognizing phosphoryl esters. The synthesis of an ion exchanger usually takes place by one-step chemical reaction in which a charged molecule containing a reactive group (often a halogenide) is allowed to react with the hydroxyl group–containing matrix (such as a polysaccharides) under strongly alkaline conditions. Most DEAE and CM ion exchangers are produced this way [1].

Table 1.3 Most commonly used charged groups of ion exchange stationary phases

Ion exchanger	Chemical structure of the functional group	Abbreviation
Cation, weak	Carboxymethyl-	CM
Cation, strong	Sulfopropyl-	SP
Cation, strong	Sulfoethyl-	SE
Cation, intermediate	Phospho-	P
Anion, weak	Diethylaminoethyl-	DEAE
Anion, weak	Dimethylaminoethyl-	DMAE
Anion, weak	Polyethyleneimine	PI
Anion, strong	Trimethylaminoethyl-	TMAE
Anion, strong	Trimethylaminomethyl-	Q
Anion, strong	Trimethylaminohydroxypropyl-	QA
Anion, strong	Diethyl-(2-hydroxypropyl)-aminoethyl	QAE
Anion, strong	Quaternized polyethyleneimine	QPI

1.2.4 Commercially Available Exchangers

A great range of commercially available exchangers is available today, having their characteristics clearly described. The capacity of the exchanger is usually defined toward a standard protein, such as HSA (human serum albumin) or lysozyme. As porosity and particle size are of great importance in separation of bioanalytes, the proper support has to be chosen depending on:

- Size of bioanalyte: The higher the size of the analyte, the greater the porosity of the support.
- Its pI value: anion versus cation exchangers, weak versus strong, depending on the separation strategy.
- Separation requirements: analytical, semipreparative, or preparative scale separation/purification of a bioanalyte of interest.

Great range of commercially available exchangers is shown in Table 1.4.

1.2.5 Significance of Buffers in Ion Exchange

As pH value is one of important parameters in IE separation of peptides and proteins, the pH value has to be controlled by buffers. Buffering components used in the IE are presented in the Table 1.5 and arranged according to their pKa values. Volatile buffer systems are especially useful if concentration of the sample by evaporation is to be performed after the IE separation. Combination of volatile components will result in a volatile buffer system. That is why acetic acid/pyridine or bicarbonate/ammonium buffer systems are appropriate for separation of proteins prior to evaporation.

Table 1.4 Commercially available exchangers. Data have been compiled from the manufacturer's booklets. The protein in capacity measurements were bovine serum albumin (BSA), human serum albumin (HSA), lysozyme (Lys), bovine hemoglobin (Hb), and ribonuclease (RNase)

Name	Matrix	Functional group	Degree of substitution (μmol/mL)	Available capacity (mg/mL)	Company
DE 23	Fibrous cellulose	DEAE	150	60 BSA	Whatman
CM 23	Fibrous cellulose	CM	80	85 Lys	Whatman
DE 52	Microgranular cellulose	DEAE	190	130 BSA	Whatman
CM 52	Microgranular cellulose	CM	190	210 Lys	Whatman
DE 53	Microgranular cellulose	DEAE	400	150 BSA	Whatman
CN32	Microgranular cellulose	CM	180	200 Lys	Whatman
DEAE Sephacel	Beaded cellulose	DEAE	170	160 BSA	GE Healthcare
DEAE Sephadex A-25	Dextran, Sephadex G-25	DEAE	500	70 Hb	GE Healthcare
QAE Sephadex A-25	Dextran, Sephadex G-25	QAE	500	50 Hb	GE Healthcare
CM Sephadex C-25	Dextran, Sephadex G-25	CM	560	50 Hb	GE Healthcare
SP Sephadex C-25	Dextran, Sephadex G-25	SP	300	30 Hb	GE Healthcare
DEAE Sephadex A-50	Dextran, Sephadex G-50	DEAE	175	250 Hb	GE Healthcare
QAE Sephadex A-50	Dextran, Sephadex G-50	QAE	100	200 Hb	GE Healthcare
CM Sephadex C-50	Dextran, Sephadex G-50	CM	170	350 Hb	GE Healthcare
SP Sephadex C-25	Dextran, Sephadex G-50	SP	90	270 Hb	GE Healthcare
DEAE Sepharose CL-6B	Agarose, 6% cross-linked	DEAE	150	100 Hb	GE Healthcare
CM Sepharose CL-6B	Agarose, 6% cross-linked	CM	120	100 Hb	GE Healthcare
DEAE Bio-Gel A	Agarose	DEAE	20	45 Hb	Bio-Rad
CM Bio-Gel A	Agarose	CM	20	45 Hb	Bio-Rad
DEAE-trisacryl M	Trisacrylate polymer	DEAE	300	90 Hb	Sepracor
CM-trisacryl M	Trisacrylate polymer	CM	200	100 Hb	Sepracor
SP-trisacryl M	Trisacrylate polymer	SP	230	150 BSA	Sepracor
Fractogel TMAE 650 (S)	Synthetic organic polymer	TMAE	–	100 BSA	Merck
Fractogel DEAE 650 (S)	Synthetic organic polymer	DEAE	–	100 BSA	Merck
Fractogel DMAE 650 (S)	Synthetic organic polymer	DMAE	–	100 BSA	Merck

(continued)

Table 1.4 (continued)

Name	Matrix	Functional group	Degree of substitution (μmol/mL)	Available capacity (mg/mL)	Company
Fractogel SO_3 650 (S)	Synthetic organic polymer	S	–	100 Lys	Merck
Fractogel COO 650 (S)	Synthetic organic polymer	COO–	–	100 Hb	Merck
High-performance media					
Mono Q	Synthetic organic polymer	Q	270–370	65 HAS	GE Healthcare
Mono S	Synthetic organic polymer	S	140–180	75 RNase	GE Healthcare
SOURCE 15Q	Synthetic organic polymer	Q	–	25	GE Healthcare
SOURCE 15 S	Synthetic organic polymer	S	–	25	GE Healthcare
SOURCE 30Q	Synthetic organic polymer	Q	–	40 BSA	GE Healthcare
SOURCE 30 S	Synthetic organic polymer	S	–	80 Lys	GE Healthcare
Q Sepharose high performance	Cross-linked agarose	Q	200	70 BSA	GE Healthcare
SP Sepharose high performance	Cross-linked agarose	SP	200	70 RNase	GE Healthcare
DEAE-5-PW	Synthetic organic polymer	DEAE	–	1.5–3	Bio-Rad
SP-5-PW	Synthetic organic polymer	SP	–	1.5–3	Bio-Rad
Mini S	Nonporous synthetic polymer	S	20	4–6	GE Healthcare
Mini Q	Nonporous synthetic polymer	Q	75	4–6	GE Healthcare
HRLC MA7P	Nonporous synthetic polymer	PEI	–	0.6–2	Bio-Rad
HRLC MA7C	Nonporous synthetic polymer	CM	–	0.6–2	Bio-Rad
DEAE-3SW	Silica	DEAE	–	120 BSA	Toso Haas
CM-3SW	Silica	CM	–	45 Hb	Toso Haas

When choosing a suitable buffer system, one has to have in mind that the chosen buffer's pKa should be close to the working pH, and there should not be any interference with mobile phase additives or subsequent screening assays [7]. In addition, the chemical nature of the used ions has considerable influence on separation of ions in ion exchange chromatography.

Many ion exchange media are available in several ionic forms and may be converted from one form to another. The ionic form of a support refers to the counterion presently adsorbed to the support's functional group. Counterions will exhibit

Table 1.5 Examples of the most commonly used zwitterionic and nonzwitterionic buffers (Data compiled from references [7, 21–24])

Buffer component	Titrate with	pKa (20°C)	pH range	Zwitterionic
Glycine	HCl	2.4 (carboxylate)	2.0–3.5	No
Citrate	Phosphoric acid	3.14	3–8	No
		4.77		
		6.39		
Acetate	NaOH	4.75	3.7–5.8	No
Maleate	NaOH		5.5–7.2	No
Phosphate	HCl, sodium hydrogen phosphate	2.12	6.2–8.2	No
		7.21		
		12.67		
Tris (N-Tris-(hydroxymethyl)aminomethane)	HCl	8.3	7.3–9.3	No
Glycine	NaOH	9.5 (amino group)	8.5–10.5	No
Mes (2-(N-morpholino)ethanesulfonic acid)	NaOH	6.15	5.5–7	Yes
Mops (3-morpholinopropanesulfonic acid)	NaOH	7.2	6.5–8.2	Yes
Tricine (N-[Tris-(hydroxymethyl)methyl]glycine)	NaOH	8.15	7–9	Yes
Ches (2-(cyclohexylamino)-ethansulfonic acid)	NaOH	9.55	9–10	Yes
Caps (3-(cyclohexylamino)-1-propanesulfonic acid)	NaOH	10.4	9.8–11	Yes

specific selectivity for each support. The lower the selectivity of a counterion toward the support, the more readily it is exchanged for another ion of like charge. Consequently, the appropriate ionic form will depend on the relative selectivity of the sample ion to be adsorbed.

In general, the ionic form should have a lower selectivity for the functional group than the sample ion so that the sample ion will displace the counterion and be adsorbed to the support. The sample ion can then be eluted by a second counterion with a higher selectivity for the support.

The affinities of anions to anion-exchange supports follow the Hofmeister series of ions, arranged by their effects on water structure. Chloride anion, one of the most widely used anions in elution from anion exchangers, has a moderate water structure-making ability, being in the middle of Hofmeister series of anions. It should also be noted that Goods' buffers (a groups of zwitterionic organic compounds, primarily sulfonates and tertiary amines) show excellent compatibility with biological systems [19, 20].

1.3 Separation of Amino Acids: Examples and Applications

Amino acids, as principal constituents of proteins and having a plethora of biological functions of their own, are always in focus when developing novel methods. Separation and quantification of amino acids is essential in food science, medicine, agricultural science, etc.

Although fully automated separation methods for amino acids have been developed long time ago [25, 26], more and more new methods have been described to be suitable for wide range of applications [27, 28]. The amino acid composition of proteins was determined by separation and quantification of constituent amino acids, following acid hydrolysis of a given protein [29]. Recently, analyzing amino acids in food stuffs and various biological samples is of growing importance [27, 30, 31]. Specialized applications exist for detection of nonprotein amino acids, but in most cases, samples of amino acids for further separation and quantification are prepared by sample hydrolysis with hydrochloric acid. The complexity of the environment does not allow the separation of amino acid to become a routine practice.

Amino acids are widely used in biotechnology applications. Since amino acids are natural compounds, they can be safely used in pharmaceutical applications, for example, as a solvent additive for protein purification and as an excipient for protein formulations. At high concentrations, certain amino acids are found to raise intracellular osmotic pressure and adjust to the high salt concentrations of the surrounding medium. They are called "compatible solutes" since they do not affect macromolecular function. Not only are they needed to increase the osmotic pressure, they are known to increase the stability of the proteins. Sucrose, glycerol, and certain amino acids are used to enhance the stability of unstable proteins after isolation from natural environments.

The mechanism of the action of these protein-stabilizing amino acids is relatively well understood. A recent review covers various biotechnology applications of amino acids, in particular arginine [32].

1.3.1 Use of Tandem Mass Spectrometry for Amino Acid Quantification

Analysis of amino acids has been commonly performed by precolumn derivatization with orthophthaldialdehyde and/or 9-fluorenylmethylchloroformate followed by reversed-phase high-performance liquid chromatography (RP-HPLC) and fluorescence detection. These methods impose certain limitations, which are overcome by use of mass spectrometry as the detection system. In a paper by Thiele and coworkers [33], ion exchange chromatography on a strong cation-exchange matrix was used for amino acid separation of barley shoots hydrolysates, which were subsequently quantified by tandem MS. Strong cation matrix was selected due to the fact that at pH 3, all amino acids in the sample are in their ammonium ion form. All 20 amino acids were detected and quantified.

Although superior in its precision and accuracy, analysis of MS results can pose problems. Also, mass spectrometry is inherently more costly than other methods of detection. Because of these reasons, several methods for separation and quantification of amino acids have been developed during the last decade.

1.3.2 Pulsed Amperometric Detection Integrated with Anion-Exchange Chromatography for Separation and Detection of Amino Acids

Pulsed amperometric detection (PAD) integrated with anion-exchange chromatography has recently gained attention as the method of choice for separation and detection of amino acids. The method was pioneered by Jandik et al. [34] and is based on pulsed amperometric detection with optimized waveforms for amino acid detection. Separation is achieved on a strong cation-exchange matrix, and the detection method enables accurate quantification of all amino acids. The method employs AEX (anion-exchange chromatography) for separation of sugars from amino acids in order to simplify detection.

This methodological setup has been used for a wide variety of applications. In a paper by Rombouts et al., wheat gluten protein hydrolysate was analyzed in order to quantify its amino acid composition. In contrast to most conventional methods, the analysis requires neither pre- or postcolumn derivatization nor oxidation of the sample. Samples of wheat gluten hydrolysate were separated with AminoPac PA10 analytical column. The chosen detection method, namely, integrated PAD, makes a derivatization step unnecessary [35]. Genzel et al. managed to quantify amino acids in mammalian cell culture media containing serum and high concentrations of glucose using similar methodology [36].

In a similar paper, tryptophan concentrations in proteins, animal feed, cell cultures, and fermentation broths were directly determined. Tryptophan content of a protein or peptide is difficult to determine due to its decomposition during the acid hydrolysis used to release the protein's other amino acids. In this method, tryptophan is separated from common amino acids by anion-exchange chromatography in 12 min and directly detected by integrated pulsed amperometry. The estimated lower detection limit for this method is 1 pmol [37], with negligible interference with glucose in the sample.

Of course, due to matrix effects and sample complexity, in certain applications, special care must be taken with preparation of the samples [38]. Thiele et al. developed a sample preparation method based on water/methanol extraction. Sample preparation was evaluated with four different food samples, sourdough, skim milk, lemon juice, and potato, and proved useful for quantitative analysis of amino acid and carbohydrate contents of food samples.

1.3.3 Simultaneous Separation of Amino Acids and Carbohydrates: AminoPac Columns

For many years, the standard practice for amino acid analysis was based upon cation-exchange chromatography with a step gradient of eluent concentration and pH, combined with ninhydrin postcolumn reaction [1]. Amino acids and carbohydrates are often found in the same samples and are frequently simultaneously quantified after column separation by amperometric or similar setups.

Several specialized columns have been developed for this application, most notably AminoPac series of columns, with polymeric pellicular anion-exchange resin. Capabilities of amino acid separation on an AminoPac PA10 column are shown in Fig. 1.1 and show superb resolution. Two detection modes are shown, with clear advantage of detection mode I.

Simultaneous detection is enabled by integrated pulsed electrochemical detection [28]. Samples of glycoprotein hydrolysate were analyzed, with use of several monosaccharides as standards and with superb resolution. A similar approach directed at pulsed electrochemical detection (PED) following liquid chromatographic separation has been applied to the indirect determination of amino acids and proteins. Limits of detection of amino acids were found to be 2–30 pmol using optimized potential–time waveforms at an Au electrode. Indirect PED provided much greater detection sensitivity toward amino acids than direct PED.

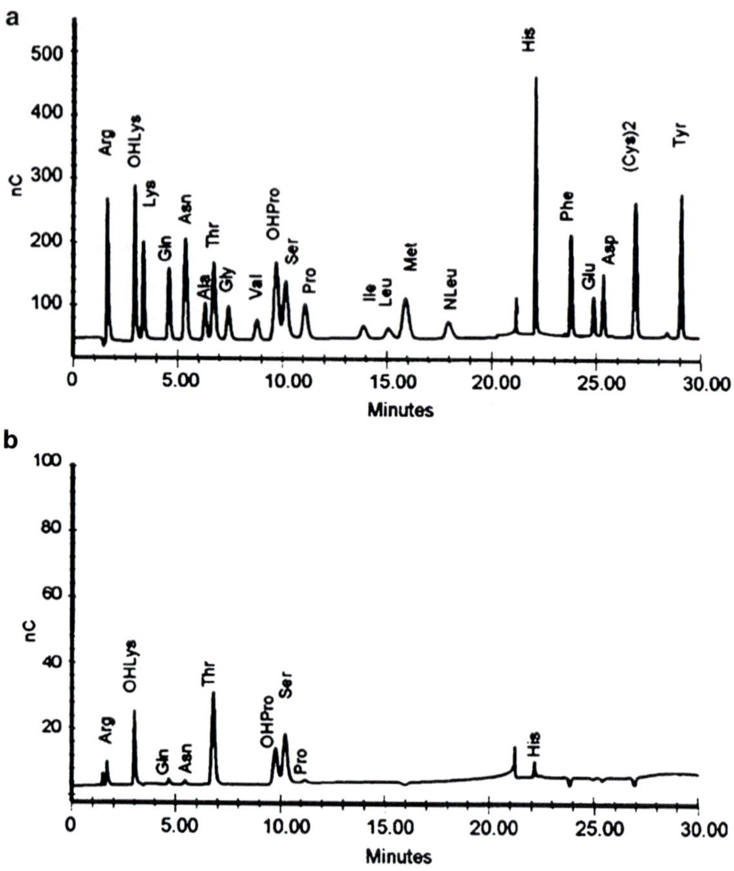

Fig. 1.1 Separation of amino acids on an Aminopac AP10 column. All components were at 200 pmol/dm³ (Nleu at 313 pmol/dm³). (**a**) Mode II detection, waveform with E2 = −50 mV. (**b**) Mode I detection, waveform with E2 = 50 mV (Reproduced from Ref. [28] with kind permission of © Elsevier Ltd. (2004))

1.3.4 Reactive IonExchange Chromatography

An interesting approach is that pioneered by Zammouri and further expanded by Harscoat et al. [39, 40]. Reactive ion exchange chromatography takes advantage of the amphoteric character of the amino acids and peptides. Reactions between acidic and basic compounds in the liquid phase allow displacement of the ion exchange equilibria to fix or elute species. Therefore, the chromatographic effect is obtained by the dissociation equilibria in solution rather than ion exchange equilibria of ionic species on the resin. In the papers in question, reactive exchange principles were employed to separate peptides and amino acids from different sources with great success. The elution is performed on classic anion-exchange stationary phases (Amberlite IRA 900B, Q Sepharose Fast Flow) with saturated aqueous solution of carbon dioxide as the eluent.

1.4 Application of Ion Exchange Chromatography in Peptide Separation

There are several conventions trying to establish an all encompassing definition of peptides. Most experts would agree that amino acid chains with less than 50 units can be called peptides. Peptides have numerous and diverse biological functions, ranging in size and function from amyloid beta peptide to glutathione.

Peptides are attributed with a number of different biological properties, as antioxidants, antimicrobials, and surfactant agents [41]. Peptides have been also investigated for their sensorial properties, which are known to determine sweet and bitter taste in foodstuff [42]. Proteins generally contain a high degree of tertiary structure, kept together by van der Waals' forces, ionic and hydrophobic interactions, and hydrogen bonding. Any conditions capable of destabilizing these forces may cause denaturation and/or precipitation. By contrast, peptides contain a low degree of tertiary structure. Their native state is dominated by secondary structures, stabilized mainly by hydrogen bonding.

Peptides tolerate a much wider range of conditions than proteins. This basic difference in native structures is also reflected in that proteins are not easily renatured, while peptides often renature spontaneously. These properties of peptides enable the use of more drastic conditions for separation, such as those used in strong cation-exchange chromatography (SCX), where the pH values can be as low as pH 2.

Sample preparation of peptides for ion exchange chromatography can be broadly divided into two areas. For all proteomics applications, samples are obtained after enzymatic (most often by trypsin) digestion of the biological (or other) samples and mixed with an appropriate buffer for the column in question. In all other cases, sample preparation is case specific and depends on the matrices used and conditions of separation.

Although reversed-phase chromatography (RPC) is often the method of choice for peptide separation, it sometimes cannot deliver the desired separation result [43].

1.4.1 Matrix Technology

All matrices used in protein separation procedures have found their place in peptide separation procedures. Some matrices have been specifically developed for the separation of peptides, such as DEAE Affi-Gel Blue gel and several composite matrices. Several mixed-mode matrix materials have been developed, with application in peptide separation.

Various properties of mixed-mode matrices are employed in peptide separation, including electrostatic and hydrophobic interactions. The possibilities of mixed matrices can be further enhanced by using them as the stationary phase in cation-exchange chromatography. Various chemistries are used, ranging from functionalized silica sorbents [44, 45] to sulfoalkylbetaine monolithic columns [46].

1.4.2 Mixed-Mode Hydrophilic Interaction Ion Exchange Separations of Peptides

Mixed-mode hydrophilic interaction/ion exchange separations of peptides have been described, achieving selectivity of reversed-phase chromatography [47].

An integrated multidimensional liquid chromatography method is demonstrated for the separation of peptide mixtures by two-dimensional HPLC coupled with mass spectrometry. The method uses strong cation-exchange and reversed-phase column sections for two-dimensional liquid chromatography. The peptide mixture was fractionated by a pH step gradient using a series of pH buffers, followed by reversed-phase chromatography. No salt was used during separation, and the integrated multidimensional liquid chromatography can be directly connected to mass spectrometry for peptide analysis. The peptides were eluted according to their pI. This method provides a technique for large-scale protein identification using existing one-dimensional HPLC systems [48].

In a recent study, a mixed-mode reversed-phase/weak anion-exchange matrix was used for the separation and purification of peptides [49]. In this paper, a novel material was developed that contains two distinct binding domains, a lipophilic alkyl chain for hydrophobic interactions with lipophilic moieties of the solute, such as in the reversed-phase chromatography, and a cationic site for anion-exchange chromatography with oppositely charged solutes, which also enables repulsive ionic interactions with positively charged functional groups. The use of this stationary phase increased yield and simplified the purification protocol.

Examples of use of ion exchange chromatography of peptides are numerous and extremely diverse – applications range from purification of antioxidative peptides from oysters [50] to proteomics applications.

1.4.3 Hybrid Applications with Capillary Electrochromatography

An interesting approach to IE based peptide separation is the use of capillary electrochromatography (CEC). In this approach, the capillary column is packed with HPLC stationary phase, and the mobile phase is driven by electroosmotic flow, as opposed to pressurized flow in HPLC. Unusually high column efficiency has been explained by the capillary electrophoretic stacking and chromatofocusing phenomena during the injection and separation of (in most cases) positively charged peptides. Again, the most common setup employs SCX as the stationary phase [51].

Comparison of pressure-supported CEC with microcolumn HPLC using the same column and instrument, CEC resulted in twofold higher column efficiency at comparable elution times and better resolution factors [52].

Application of CEC in peptide separation has been reviewed in some depth by Isaaq and coworkers [43].

1.4.4 Proteomics

With the development of high-throughput methods for proteome analysis, a growing need has emerged for better separation of the pre-MS step in shotgun proteomics, a methodological approach of combining HPLC with tandem mass spectrometry.

For very complex mixtures, the usual reversed-phase HPLC does not accomplish the desired separation necessary for MS spectra analysis. This problem is increasingly being solved by incorporating a second dimension of chromatographic separations in these methods. The combination of IE liquid chromatography in the first dimension and RPC in the second dimension are the dominant procedures for the separation of proteome digests today. The most common technique used is strong cation exchange,which is performed on a variety of matrixes. Most commonly, SCX matrices are based on silica microspheres derivatized by aromatic sulfonic acid groups. An added benefit of SCX chromatography is that it removes interfering substances such as SDS, PEG, and nonionic detergents before MS.

A schematic of the experimental strategy employed for the global analysis of the human serum proteome is shown in Fig. 1.2. Each IEF (isoelectrofocusing) fraction is further fractionated by SCX into several fractions that are analyzed by μLC–MS/MS.

In a two-dimensional separation of a proteome digest, strong cation exchange is normally the first dimension for the fractionation of peptides on-column [2] or off-column [53], while RPC is used in the second dimension. The SCX procedures, when optimized, give good separation of complex peptide mixtures.

At low pH conditions (pH bellow 3), under which SCX is performed, tryptic peptides are usually positively charged due to protonation of the N-termini, arginine, histidine, and lysine side chains, while negative charges at carboxyl groups and C-termini are neutralized. These positively charged peptides bind onto the SCX resin and are subsequently fractionated by increasing the salt concentration, stepwise or linear, or, alternatively and less common, by increasing the pH.

This topic has been investigated in some detail by Link and coworkers [54].

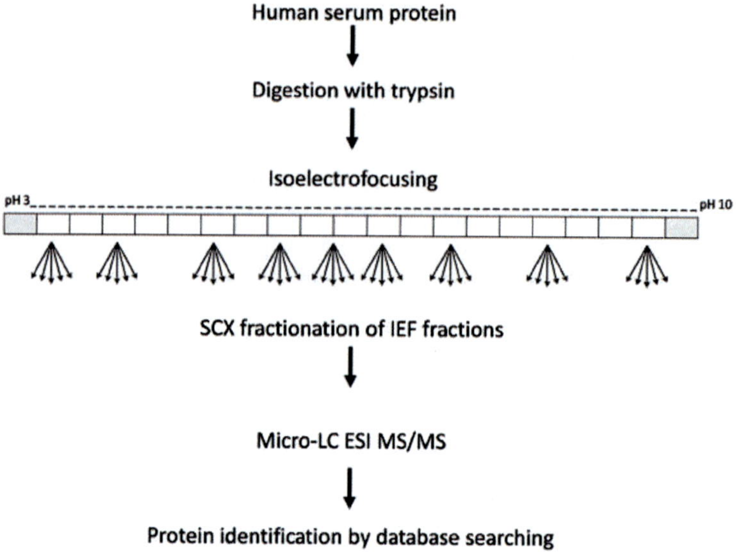

Fig. 1.2 Experimental strategy for human serum proteome analysis with use of SCX fraction of peptides (Adapted from Ref. [43] with kind permission of © Elsevier Ltd. (2009))

Similar off-line procedures were reported for the separation of peptides (protein digests) in human plasma filtrate [55], plasma peptides [56], human urine [57], and numerous other applications. Janini et al. reported a simple interface for on-line CE-MS (capillary electrophoresis-mass spectrometry) that was used as a second dimension to study peptide fractions from a SCX column [58].

Various setups for multidimensional separation of peptides prior to MS are shown in Fig. 1.3.

In cases where two-dimensional approaches do not give satisfactory results, a third orthogonal separation procedure can be used, which results in an increase in resolution and separation of large numbers of peptides [66].

An important point to consider regarding separation on SCX columns is the presence of peptide molecules modified by sialic acid or phosphate, as well as blocking of N- or C-termini. These modifications render peptides less retentive on the column, which has found application in separation of these peptides from the bulk of molecules [61, 67].

One recent application describes an integrated platform for on-line digestion, separation, and identification of proteins and peptides that contained a microcolumn with mixed weak anion and weak cation particles. That platform exhibited improved sequence coverage and significant shortening of work time [68].

All recent insights in the field of multidimensional peptide separation are described in reviews by Sandra et al. [65] and Gao et al. [69].

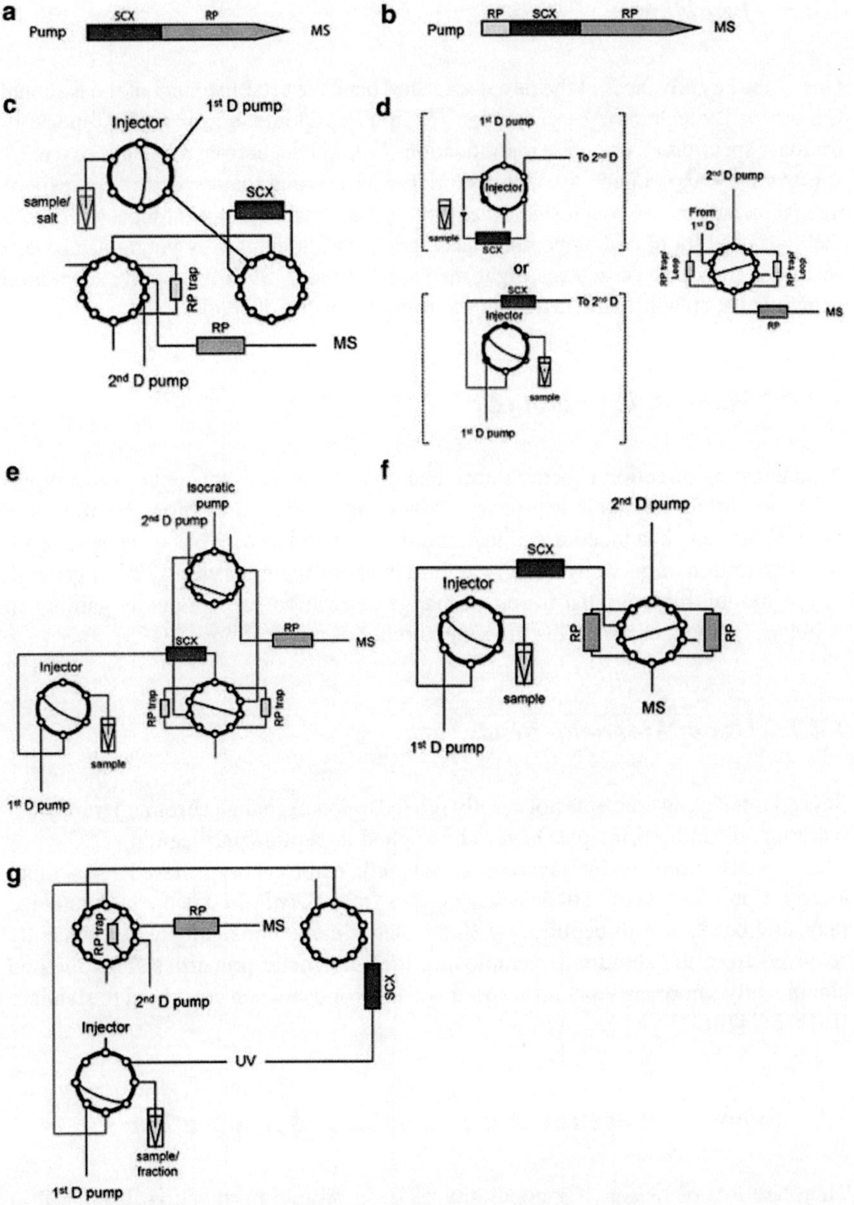

Fig. 1.3 A selection of on- and off-line two-dimensional combinations employing SCX (**a–g**) or RPLC (**h**) in the first, and RPLC in the second dimension. (**a**) Biphasic MudPIT [59, 60], (**b**) triphasic MudPIT [61], (**c**) column switching setup using salt steps [62], (**d**) and (**e**) column switching setups employing a linear gradient in the first dimension [63, 64], (**f**) column switching setup utilizing two second dimension columns, (**g**) automated off-line setup. Numerous variations on every scheme are possible (Modified from Ref. [65] with kind permission of © Elsevier .B.V. (2009))

1.4.5 Peptidome

One of the key advances of the past decade has been the establishment of the National Center for Biotechnology Information (NCBI) Peptidome, a new public repository for mass spectrometry peptide identifications [70, 71]. Data from all stages of a mass spectrometry experiment are captured, including original mass spectra files, experimental metadata, and conclusion-level results. The database allows for easy access to methods used for obtaining peptide mass spectra. Although a very young database, a search of the database returns eight instances of using SCX in peptide separation experiments, an approximate fifth of the overall number of studies.

1.4.6 Monolithic Columns

An interesting direction is being developed, introducing the concept of biocompatibility to chromatographic stationary phases for separation of biomolecules. In a paper by Yun et al., a biocompatible monolithic column system has been developed, with integration of polyethylene glycol into the poly(meth)acrylate [72]. In general, use of monolithic columns in ion exchange separation of peptides is gaining in popularity, as witnessed by the growing number of publications [73–75].

1.4.7 Use of Magnetic Beads

Several interesting concepts not strictly related to ion exchange chromatography but working on similar principles have been applied in peptide purification.

In a paper published by Bayard et al. magnetic carboxyl beads have been synthesized for purification of antibacterial peptides from *E. coli*. In this method, charged magnetic beads adsorb peptides by electrostatic interactions and are subsequently removed from the solution by employing their magnetic properties. This method significantly improves separation speed and ease of use when compared to standard IE-HPLC [76].

1.5 Separation of Proteins: Examples and Applications

There are lots of reasons for popularity of IE in protein purification. It is used in research, analysis, and large-scale purification of proteins and recombinantly produced proteins [77, 78]. Ion exchange is ideal for the initial capture of proteins because of its high capacity, relatively low cost, and its ability to survive rigorous cleaning regimes. Ion exchange is also very useful as a next step in purification of partially purified protein [18]. IEC is widely applicable because buffer conditions can be adapted to suit a broad range of proteins rather than being applicable to a single functional group of proteins [79]. IE chromatography has also found use in proteomics technology, enabling separation of complex mixtures prior the mass spectrometric analysis [80].

The protein separation is a difficult process due to the complexity of proteins itself and their biological environments: (1) Proteins are composed of a large amount of amino acids, and different amino acid sequences may lead to completely different properties and functions. (2) The same molecules may also express different activities when they are in different solution environments (e.g., different pH values, ionic strength). (3) Proteins in native environments always exist in isoforms, many of them having similar pI value and molecular weight. (4) Some target protein molecules may bind with other protein molecules, thus resulting in the difficulty of separating them by a simple size exclusion method.

Protein solutions are a type of colloidal solution, and the transport mechanisms and selective affinity of proteins are related to their molecular weight, charge characteristics, pH value of solution, and charge properties of separation media (e.g., solution, gel, membranes) and others [81]. Hence, it seems to be unfeasible to find a universal separation principle for protein separation based on a single protein property.

The development of a suitable protocol for a given protein is usually based on trial and error. Helpful hints can be obtained from structural data and existing protocols for separation of similar proteins.

It is important to consider polarity of the ion exchanger which is determined by the pI of the protein and its pH stability. The size of the protein will determine porosity of the matrix. The higher the molecular weight of the protein, the greater the porosity of the matrix should be. Highly porous matrices are commercially available from many manufacturers, such as GE Healthcare, Bio-Rad, Waters, etc. Compatibility of the additives and buffers used in IE with the matrix and biological sample should be carefully checked.

In principle, binding conditions are achieved in low ionic strength, carefully choosing the appropriate buffer, pH, and additives. The elution and separation of proteins by ion exchange chromatography is then achieved by:

1. Competition with small counterions and isocratic elution [82, 83]
2. Elution by linear or stepwise salt gradient [78, 84]
3. Elution by pH gradient [84–86]
4. Combined salt and pH-gradient elution [10]

Other principles of IE separation of proteins include selective nonadsorption preparative methods. In those methods, conditions where the protein does not bind to the stationary phase are applied [87]. The use of mixed-bed ion exchangers was also described in those separations. The conditions are set up in such a way that the majority of undesirable proteins remain bound, while the protein of interest flows through the column. It does not offer concentration of the sample or separation of proteins but can significantly enrich the concentration of protein of interest.

Affinity elution from an ion exchanger, phosphocellulose, is applicable to proteins with an affinity for phosphoryl groups and can provide a significant enrichment of the protein providing elution with phosphate or, more often, with a specific ligand [88]. The instability of this specific stationary phase does not allow wide applicability of this approach, and not many applications can be found in the literature.

Various modifications of the original separation principle, as well as its integration with other modes of protein separation are literally endless in the literature. Several applications selected here involve applications of the general separation and elution modes in IE of selected proteins of great biotechnological significance, integrated approaches, variations of IE to suite separation of unstable proteins, and applications of IE in one-step isolation and quantification of proteins.

1.5.1 Hen Egg White Fractionation by a Combination of Cation and Anion Ion Exchange Chromatography

Hen egg white represents a major raw material for the food industry because of its technological properties, especially foaming and gelling. Major hen egg white proteins have been widely studied because of their functional properties, but isolation of bioactive peptides and minor proteins is essential work for progressing in the understanding of hen egg white biological properties. An interesting approach combines anion- and cation-exchange chromatography in order to achieve preparative isolation of biotechnologically important hen egg white proteins: lysozyme, ovotransferin, and ovalbumin.

Preparative chromatographies of diluted egg white were performed with S Ceramic Hyper DF (cation exchanger from Biosepra, Cergy Saint-Christophe, France) and Q Sepharose Fast Flow (anion exchanger from Amersham Biosciences, Uppsala, Sweden) [89] (Fig. 1.4). In two successive steps, lysozyme and ovotransferrin were extracted by cation-exchange chromatography on S Hyper DF. The coproduct was then used as starting material for anion-exchange chromatography. Two other fractions, A and B, which are enriched in recently detected minor proteins in hen egg white (EW), were extracted by single-step anion-exchange chromatography on Q Sepharose FF.

1.5.2 Fractionation of Proteins from Whey Using Cation-Exchange Chromatography and a Combined Salt/pH Elution of Protein Fractions

Whey contains a mixture of proteins, each having unique attributes for nutritional, biological, and food ingredient applications [90, 91]. The major proteins present in milk include α-lactalbumin, β-lactoglobulin, immunoglobulin, bovine serum albumin, and caseins: κ-casein, β-casein, and the α-caseins. Minor but commercially

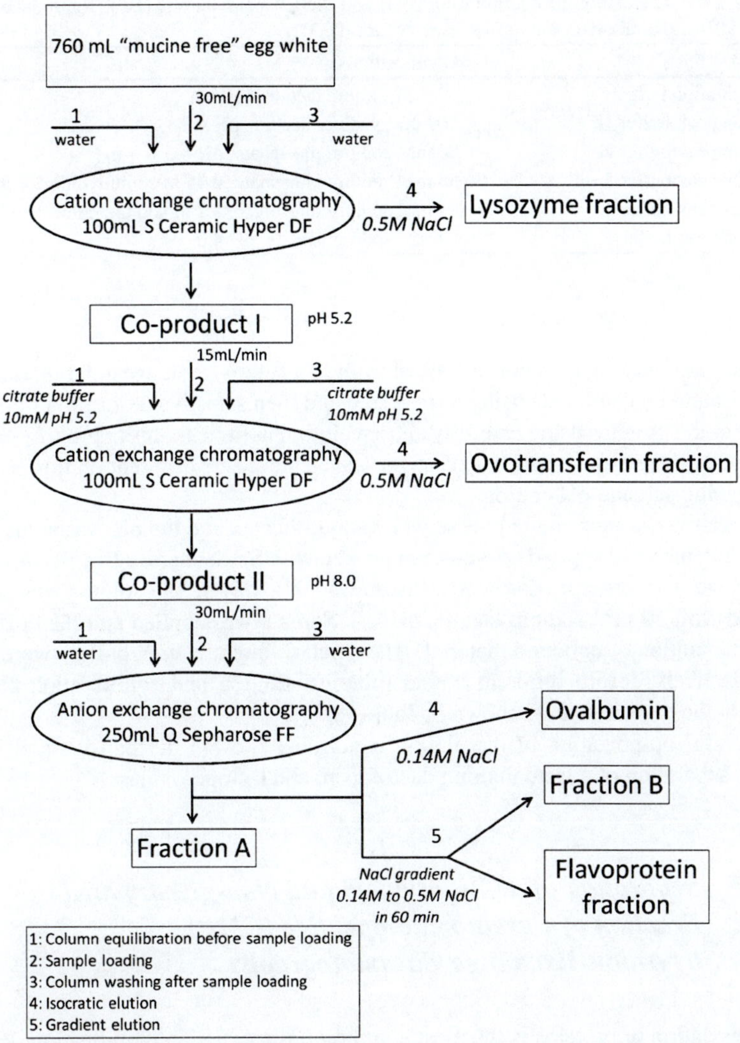

Fig. 1.4 Hen egg white fractionation process based on ion exchange chromatography. "Mucin-free" EW was "mucin-free" egg white protein; COI and COII, were coproducts I and II, respectively (Modified from Ref. [89] with kind permission of © Elsevier Ltd. (2005))

important proteins are lactoferrin and lactoperoxidase. In addition, rennet whey contains glycomacropeptide that is cleaved from κ-casein by chymosin to initiate precipitation of the caseins forming curd.

The central hypothesis of the work by Doultani et al. [90] was that a single ion exchange system could be used to switch between manufacture of different whey protein fractions simply by switching buffers. This was accomplished by capture of

Table 1.6 Composition of equilibration (EQ) and elution peak buffers (EP) (Reproduced from Ref. [90] with kind permission of © Elsevier Ltd. (2004))

Buffer	Composition
Equilibration buffer	50 mM sodium lactate, pH 4.0
Elution peak buffer 1b	100 mM sodium acetate, pH 4.9
Elution peak buffer 2b	50 mM sodium phosphate, pH 6.5
Elution peak buffer 3	50 mM sodium phosphate, 0.35 M sodium chloride, pH 6.5
Elution peak buffer 4	50 mM sodium phosphate, 1.2 M sodium chloride, pH 6.5
Elution peak buffer 1b	50 mM sodium lactate, pH 4.0

all the positively charged proteins in whey onto a cation-exchange column, removal of unbound contaminants using a rinse step and then selective desorption of one or more of the proteins using carefully chosen elution buffers. In this approach, separation of proteins from a cation exchanger was achieved by different elution buffers, combining salt and pH elution.

In that work, mozzarella cheese whey was collected and the pH was adjusted to 4.0. Proteins were applied on a column packed with SP Sepharose Big Beads cation exchanger (Amersham Bioscience, Piscataway, NJ, USA). The column was equilibrated with 50 mM sodium lactate, pH 4.0. Sample was applied and the next step was the elution of unbound material. After that, different elution buffers were used to selectively desorb different bound proteins. Elution and equilibration buffers used in the experiment are shown in Table 1.6.

Careful optimization of the elution conditions allowed inexpensive and food-grade separation of a lactoalbumin, lactoferrin, and lactoperoxidase.

1.5.3 Separation of a Glycosylated and Non-glycosylated Fraction of Caseinomacropeptide (CMP) by Anion-Exchange Chromatography

Glycosylation of proteins is the most common posttranslation modification. IE has found use in separation of glycosylated and non-glycosylated fractions of proteins.

The 64 C-terminal amino acids of κ-casein, known as caseinomacropeptide (CMP) or glycomacropeptide, are released by endopeptidase chymosin (or pepsin) cleavage of κ-casein between Phe105 and Met106 during the renneting of cheese [92]. CMP is released into the whey, whereas the remaining part of κ-casein, the para-κ-casein, precipitates into the cheese curd. CMP is a heterogeneous group of peptides regarding the degree of glycosylation and phosphorylation. CMP can effectively be fractionated into two major fractions: the glycosylated gCMP and the non-glycosylated aCMP by anion-exchange chromatography [93]. The method was applied to compare membrane adsorption chromatography (MAC) devices with classical bead-based column chromatography. The membrane adsorption chroma-

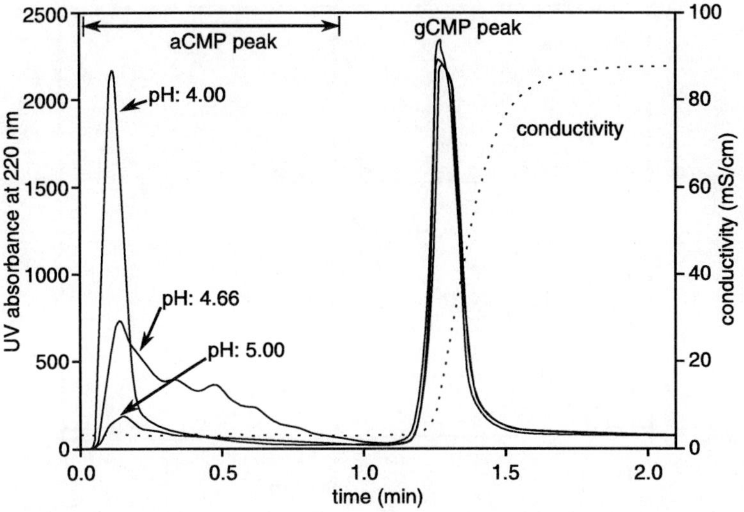

Fig. 1.5 Effect of different buffer pH on the CMP fractionation peak profile. Running conditions were Sartobind Q75 MAC device (Göttingen, Germany) using 0.02 M acetate buffer at a flow rate of 10 ml/min and a 10 column volumes NaCl gradient elution (Reproduced from Ref. [93] with kind permission of © Elsevier Ltd. (2007))

tography (MAC) module Sartobind Q 75 was obtained from Sartorius (Göttingen, Germany). The comparison of separation efficiency was performed with a strong bead-based anion-exchange resin Resource Q (GE Healthcare, Munich, Germany).

The best fractionation of total CMP into gCMP and aCMP fractions was achieved using a pH of 4.0–4.1, where the aCMP is mostly uncharged, whereas the gCMP retains a negative charge due to sialic acid content and is therefore adsorbed. When the pH is adjusted below 4.0, some of the gCMP isoforms are washed out with the aCMP fraction because these are less acidic and thus reach their isoelectric state. As can be seen in Fig. 1.5, the RP-HPLC peaks of the collected gCMP fraction are shifted to the earlier eluting, less hydrophobic gCMP peaks with decreasing pH values.

The authors also tested buffer composition on the separation of gCMO and aCMP fractions (Fig. 1.6). The differences in the quality of the peak shapes obtained for the sodium acetate and sodium citrate buffers (Fig. 1.6) might be explained by the ability of the sodium citrate to complex calcium, which might stabilize the phosphorylated CMP. The gCMP peak profile is broader with the sodium citrate buffer and the individual gCMP isoforms are not as clearly visible as with the sodium acetate buffer.

The authors showed that MAC devices can separate gCMP from aCMP, however, at a lower level of binding capacity and chromatographic resolution compared to classical chromatography. On the other hand, fractionation is achieved four times faster.

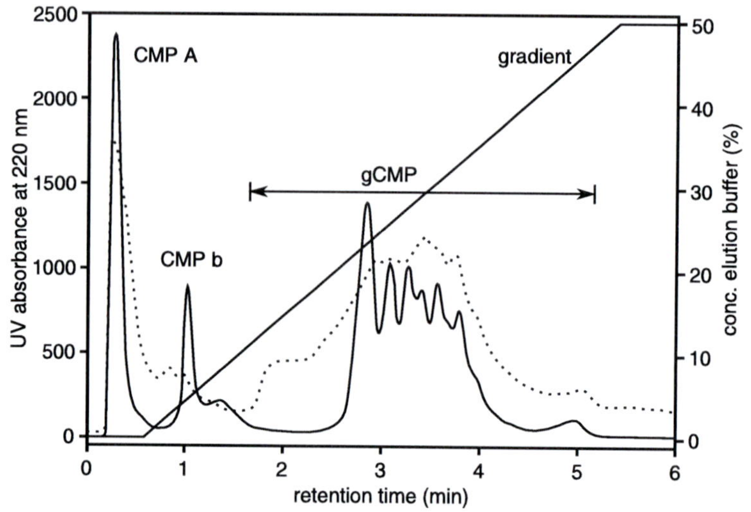

Fig. 1.6 Comparison of different buffers systems: citrate buffer (*dotted line*) and 0.03 M acetate buffer (*solid line*) at pH 4.0. Running conditions on Resource Q column were 0.03 M citrate buffer and 0.03 M acetate buffer at pH 4.0, at flow rate of 4.0 mL/min and a 24 column volumes NaCl gradient from 0 to 1 M (Reproduced from Ref. [93] with kind permission © Elsevier Ltd. (2008))

1.5.4 Purification of Unstable Proteins by Modification of Ion Exchange Chromatography

In general, binding of the proteins to the ion exchange matrix occurs in conditions of low ionic strength. However, some proteins are stable only in conditions of high ionic strength, such as halophilic proteins, usually separated by salting-out methods [94]. Separation of halophilic proteins by IE requires some modifications [95, 96]. Purification of unstable proteins from *Halobacterium salinarium* crude cell extracts has been achieved by combining cell disruption and desalting by a hollow-fiber membrane module as an access to perform IE [96]. This gentle conditioning method allows the inclusion of ion exchange chromatography in purification procedures of proteins from halophilic bacteria. A fast, one-step procedure involves both on-line cell disruption and on-line lowering of the ionic strength by cross-flow ultradialysis using a hollow-fiber membrane module prior to ion exchange chromatography. Ion exchange was performed by Q Sepharose Fast Flow (GE Healthcare). Compared to conventional sample preparation, the total recovery of protein after anion-exchange chromatography was higher when combined cell disruption-desalting was performed prior to anion-exchange chromatography.

1.5.5 Separation of Proteins by Filtration and IonExchange Chromatography: An Integrated Approach

Protein isolation and purification from a fermentation broth usually involve centrifugation, filtration, adsorption, and chromatography steps. In the study by Dai et al. [97], a simple cyclic process integrating ultrafiltration membranes and chromatographic resin beads was developed to achieve the same goal in one device. The device basically consists of a hollow-fiber filtration module whose shell side is filled with chromatographic resin beads. Systems studied in this cyclic process include the mixtures: myoglobin and beta-lactoglobulin; hemoglobin and bovine serum albumin; and myoglobin and alpha-lactalbumin. Excellent separations of the proteins were obtained. A yeast-based cellular suspension containing a mixture of myoglobin and alpha-lactalbumin was also applied to this device. The target proteins were recovered and purified successfully. The cyclic process-based device integrates clarification, concentration, and chromatographic purification of biomolecules and is suitable for both extracellular and intracellular products. Based on this approach, an integrated device was developed, achieving efficient ultrafiltration and chromatographic separation of model binary mixtures of proteins [98]. Similar approach applies adsorbent membrane devices of different functionalities. The devices made it possible to load the membrane directly with preclarified fermentation broth or cell lysate and separate the protein of interest, often in a single step, using all advantages of the membrane adsorbent technology, such as fast processing and easy scale-up [99].

1.5.6 Quantification of Proteins by IE-HPLC-UV Method

An interesting one-step purification of two allergenic plant proteins was achieved on MiniS and MiniQ, nonporous media available from GE Healthcare, suitable for analytical and micropreparative purifications of proteins [85]. The biological activity of an extract depends mainly on the content of the major allergen, the one that binds IgE from more than 50% of the allergic patients' sera. It is often the most abundant protein in the sample, quantified the most frequently by a monoclonal antibody-based immunosorbent assay (ELISA) [100–102]. Lately, by using new technologies, allergen extracts can be defined by major allergen content in mass units, and the consistency of each lot can be accurately monitored [103, 104]. High-performance liquid chromatography (HPLC), capillary electrophoresis, and mass spectrometry are methods currently used for allergen identification with potential use in standardization [105–109]. The determination of allergenic proteins by liquid chromatography and mass spectrometry has greatly advanced in recent years, and gel-free allergenomics is becoming a routinely used approach for the identification of food allergens [110]. Profiling of protein isoforms and isoallergens is also possible with proteomics tools [111]. However, new and improved standardization methods require functional testing of the contaminant food allergen in the stuff. Although enzyme-linked immunosorbent assay, polymerase chain reaction, and

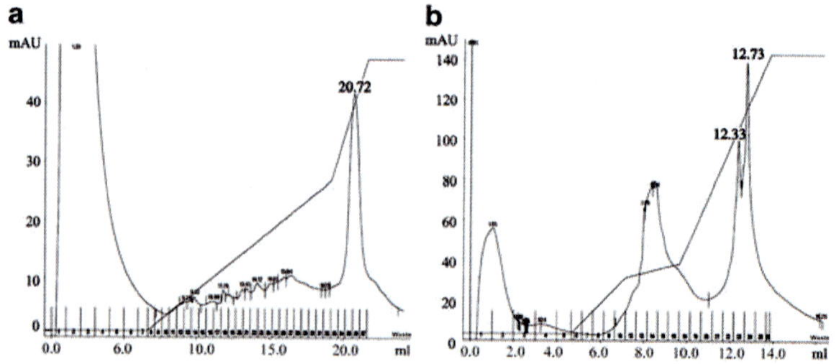

Fig. 1.7 Chromatograms of (**a**) mugwort pollen extract separated on a MiniS PE 4.6/50 column and (**b**) kiwi fruit extract separated on a MiniQ PE 4.6/50 column. One milliliter of each sample was applied on the column (Reprinted from Ref. [85] with kind permission of © Elsevier Ltd. (2007))

mass spectrometry are sensitive and specific instruments to detect trace amounts of food proteins, the ability of food constituents to trigger activation of mast cells or basophils is a tool to detect functional traces of food allergens [112, 113].

The method described an application of ion exchange HPLC in quantification of allergenic proteins. The prerequisite is one-step separation method for the protein in question and known extinction coefficient of the proteins, as the quantification is based on UV absorptivity.

Optimal conditions for binding of Art v 1, a major mugwort pollen allergen, on the MiniS PE 4.6/50 column (filling—nonporous hydrophilic polymer with methyl sulfonate ligand, column volume 0.8 ml) (Amersham Biosciences, Uppsala, Sweden) were achieved in 20 mM acetate pH 4.4, using an ÄKTA Purifier HPLC System (Amersham Biosciences, Uppsala, Sweden). A salt gradient was applied for elution of the protein. The single peak corresponding to Art v 1 was well separated from other mugwort pollen extract proteins (Fig. 1.7). Act c 1 was separated using a MiniQ PE 4.6/50 analytical column (filling—nonporous hydrophilic polymer with a quaternary amine ligand, column volume 0.8 ml) (Amersham Biosciences, Uppsala, Sweden).

For the isolation and purification of Act c 1, the major kiwi allergen, elution method was pH gradient elution. The method was optimized in order to achieve good separation of all Act c 1 isoforms from other proteins in the mixture. Act c 1 was bound to the MiniQ PE 4.6/50 column in a 20 mM Tris buffer pH 8.0. During elution with a pH gradient, Act c 1 appeared as two overlapping peaks containing different isoforms of the same protein, also well separated from other proteins.

Purity of both proteins was examined by SDS-PAGE, immunoblotting, and affinity staining for Act c 1 (Figs. 1.8 and 1.9). The quantity of the protein was estimated directly in mass units per 1 mL of extract sample by dividing the calculated *A*280 of the peak area with the theoretical extinction coefficient. The method described by Blanusa et al. could be a useful tool for standardization of allergenic extracts for clinical use.

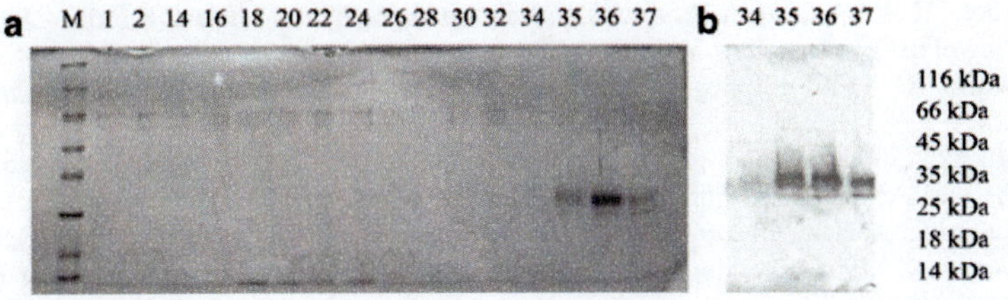

Fig. 1.8 (**a**) SDS-PAGE of mugwort pollen extract fractions from a MiniS PE 4.6/50 column, 32 μl of each fraction was applied per lane; M–protein standards. (**b**) Immunoblot of fractions developed with rabbit anti-Art v 1 antibodies (Reprinted from Ref. [85] with kind permission of © Elsevier Ltd. (2007))

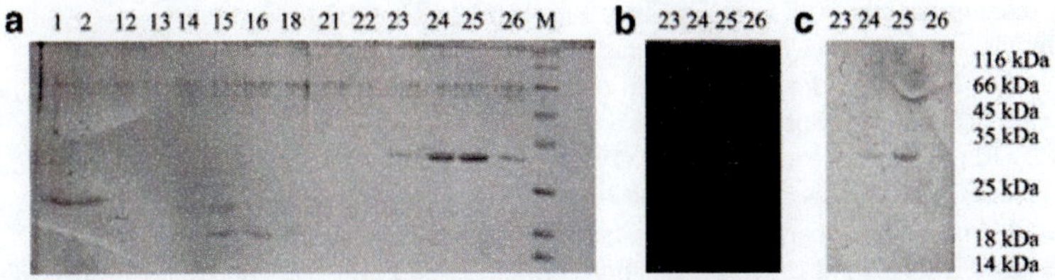

Fig. 1.9 (**a**) SDS-PAGE of kiwi fruit extract fractions separated by ion exchange chromatography on a MiniQ PE 4.6/50 column, 32 μg of the protein solution was applied per lane; M, protein standards; 10 μg of the protein was applied per lane. (**b**) Activity staining after SDS-PAGE using gelatin-co-polymerized gel. (**c**) Immunoblot of fractions developed with rabbit anti-Act c 1 antibodies (Reprinted from Ref. [85] with kind permission of © Elsevier Ltd. (2007))

Acknowledgments The authors acknowledge the support of the Ministry of Science and Technological Development of the Republic of Serbia (Grant No. 172024).

References

1. Helfferich FG (1962) Ion-exchange, Mcgraw-hill series in advanced chemistry. McGraw-Hill, New York
2. Helfferich FG (1995) Ion-exchange. Dover science books, Dover edn. Dover Publications, New York
3. LeMaster DM, Richards FM (1982) Prediction of the ph dependence of the separation of weak electrolytes by ion-exchange: amino acid chromatography with competitive buffer ion eluents. Anal Biochem 122(2):223–237. doi:Doi: 10.1016/0003-2697(82)90274-3
4. Wellner D (1985) Separation of [gamma]-glutamyl amino acids by ion-exchange chromatography. In: Alton M (ed) Methods in enzymology, vol 113. Academic, San Diego, pp 564–566
5. Yu H, Ding YS, Mou SF (2003) Some factors affecting separation and detection of amino acids by high-performance anion-exchange chromatography with integrated pulsed amperometric detection. J Chromatogr A 997(1–2):145–153
6. Stanton PG, Simpson RJ, Lambrou F, Hearn MTW (1983) High-performance liquid chromatography of amino acids, peptides and proteins: XIVII. Analytical and semi-preparative separation of several pituitary proteins by high-performance ion-exchange chromatography. J Chromatogr A 266:273–279

7. Roos PH (1999) Chapter 1 ion-exchange chromatography. In: Michael K (ed) J Chromatogr Lib vol 61. Elsevier, New York, pp 3–88
8. Scopes RK (2001) Strategies for protein purification. Curr Protoc Protein Sci Chapter 1: Unit 1 2
9. Miksík I (2000) Protein liquid chromatography. In: Kastner M (ed) J Chromatogr Lib, vol 61. Elsevier, Amsterdam, J Chromatogr B: Biomed Sci and Appl 749(1):143–144
10. Jungbauer A, Hahn R (2009) Ion-exchange chromatography. Methods Enzymol 463:349–371
11. Doonan S (1996) Protein purification protocols, vol 59, Methods in molecular biology. Humana Press, Totowa
12. Hirs CHW, Stanford M (1951) Chromatography of proteins. Ribonuclease. J Am Chem Soc 73(4):1893–1893
13. Simpson RJ (2004) Purifying proteins for proteomics: a laboratory manual. Cold Spring Harbor Laboratory Press, Cold Spring Harbor
14. Small H (1989) Ion chromatography, Modern analytical chemistry. Plenum, New York
15. Bonner PLR (2007) Protein purification, The basics. Taylor & Francis, New York/Abingdon
16. Yang HW, Viera C, Fischer J, Etzel MR (2002) Purification of a large protein using ion-exchange membranes. Ind Eng Chem Res 41(6):1597–1602
17. Jungbauer A (2005) Chromatographic media for bioseparation. J Chromatogr A 1065(1):3–12
18. Scopes RK (1994) Protein purification: principles and practice, 3rd edn, Springer advanced texts in chemistry. Springer, New York
19. Roy RN, Cramer J, Randon V et al (1998) Buffers for the physiological ph range: thermodynamics of the second dissociation of n-(2-hydroxyethyl)piperazine-n′-2-hydroxypropanesulfonic acid from 5 to 55 degrees c. J Sol Chem 27(5):425–434
20. Scopes RK (1987) Protein purification: principles and practice, 2nd edn, Springer advanced texts in chemistry. Springer, New York
21. Lage OM, Vasconcelos M, Soares H et al (1996) Suitability of the pH buffers 3- n-n-bis(hydroxethyl)amino -2-hydroxypropanesulfonic acid and n-2-hydroxyethylpiperazine-n′-2-ethanesulfonic acid for in vitro copper toxicity studies. Arch Environ Contam Toxicol 31(2):199–205
22. Roy RN, Bice J, Greer J et al (1997) Buffers for the physiological ph range: acidic dissociation constants of zwitterionic compounds (aces and ches) in water from 5 to 55 degrees c. J Chem Eng Data 42(1):41–44
23. Azab HA, Orabi AS, El-Salam ETA (1998) Apparent second-stage dissociation constants of some zwitterionic buffers for biochemical and physiological research in various hydroorganic media. J Chem Eng Data 43(5):703–707
24. Welling-Wester S, Feijlbrief M, Koedijk D, Welling GW (1998) Detergent extraction of herpes simplex virus type 1 glycoprotein d by zwitterionic and non-ionic detergents and purification by ion-exchange high-performance liquid chromatography. J Chromatogr A 816(1):29–37
25. Ertingshausen G, Adler HJ, Reichler AS (1969) Fully automated high-speed ion-exchange chromatography of amino acids. J Chromatogr A 42:355–366
26. Mondino A (1970) Automatic ion-exchange chromatography of amino acids: experimental studies for optimising resin column dimensions. J Chromatogr A 50:260–273. doi:Doi: 10.1016/s0021-9673(00)97948-2
27. Yokoyama Y, Yamasaki K, Sato H (2005) Simultaneous determination of urinary creatinine and uv-absorbing amino acids using a novel low-capacity cation-exchange chromatography for the screening of inborn errors of metabolism. J Chromatogr B 816(1–2):333–338
28. Jandik P, Cheng J, Avdalovic N (2004) Analysis of amino acid-carbohydrate mixtures by anion-exchange chromatography and integrated pulsed amperometric detection. J Biochem Biophys Methods 60(3):191–203
29. Fountoulakis M, Lahm HW (1998) Hydrolysis and amino acid composition analysis of proteins. J Chromatogr A 826(2):109–134
30. Stenberg M, Marko-Varga G, Öste R (2002) Enantioseparation of – and -amino acids by a coupled system consisting of an ion-exchange column and a chiral column and determination of -aspartic acid and -glutamic acid in soy products. Food Chem 79(4):507–512

31. Ravindran G, Bryden WL (2005) Tryptophan determination in proteins and feedstuffs by ion-exchange chromatography. Food Chem 89(2):309–314
32. Arakawa T, Tsumoto K, Kita Y et al (2007) Biotechnology applications of amino acids in protein purification and formulations. Amino Acids 33(4):587–605
33. Thiele B, Fullner K, Stein N, Oldiges M et al (2008) Analysis of amino acids without derivatization in barley extracts by lc-ms-ms. Anal Bioanal Chem 391(7):2663–2672
34. Jandik P, Clarke AP, Avdalovic N et al (1999) Analyzing mixtures of amino acids and carbohydrates using bi-modal integrated amperometric detection. J Chromatogr B Biomed Sci Appl 732(1):193–201
35. Rombouts I, Lamberts L, Celus I et al (2009) Wheat gluten amino acid composition analysis by high-performance anion-exchange chromatography with integrated pulsed amperometric detection. J Chromatogr A 1216(29):5557–5562
36. Genzel Y, Konig S, Reichl U (2004) Amino acid analysis in mammalian cell culture media containing serum and high glucose concentrations by anion-exchange chromatography and integrated pulsed amperometric detection. Anal Biochem 335(1):119–125
37. Hanko VP, Rohrer JS (2002) Direct determination of tryptophan using high-performance anion-exchange chromatography with integrated pulsed amperometric detection. Anal Biochem 308(2):204–209
38. Thiele C, Ganzle MG, Vogel RF (2002) Sample preparation for amino acid determination by integrated pulsed amperometric detection in foods. Anal Biochem 310(2):171–178
39. Harscoat C , Muhr L, Grévillot G (2003) Reactive ion-exchange chromatography: concentrations and separations of amino acids and peptides by means of an aqueous solution of carbon dioxide under pressure as displacer. Chem Eng Res Des 81(10):1333–1342
40. Zammouri A, Chanel S, Muhr L et al (1999) Displacement chromatography of amino acids by carbon dioxide dissolved in water. Ind Eng Chem Res 38(12):4860–4867
41. Gill I, Lopez-Fandino R, Jorba X et al (1996) Biologically active peptides and enzymatic approaches to their production. Enzyme Microb Technol 18(3):163–183
42. Bumberger E, Belitz HD (1993) Bitter taste of enzymic hydrolysates of casein. I. Isolation, structural and sensorial analysis of peptides from tryptic hydrolysates of beta-casein. Z Lebensm Unters Forsch 197(1):14–19
43. Issaq HJ, Chan KC, Blonder J et al (2009) Separation, detection and quantitation of peptides by liquid chromatography and capillary electrochromatography. J Chromatogr A 1216(10):1825–1837
44. Gilar M, Yu YQ, Ahn J, Fournier J et al (2008) Mixed-mode chromatography for fractionation of peptides, phosphopeptides, and sialylated glycopeptides. J Chromatogr A 1191(1–2):162–170
45. Ohyama K, Shirasawa Y, Wada M, Kishikawa N, Ohba Y et al (2004) Investigation of the novel mixed-mode stationary phase for capillary electrochromatography. II. Separation of amino acids and peptides on sulfonated naphthalimido-modified silyl silica gel. Electrophoresis 25(18–19):3224–3230
46. Wang X, Ding K, Yang C, Lin X, Lu H (2010) Sulfoalkylbetaine-based monolithic column with mixed-mode of hydrophilic interaction and strong anion-exchange stationary phase for capillary electrochromatography. Electrophoresis 31(17):2997–3005
47. Mant CT, Kondejewski LH, Hodges RS (1998) Hydrophilic interaction/cation-exchange chromatography for separation of cyclic peptides. J Chromatogr A 816(1):79–88
48. Dai J, Shieh SH, Sheng QH, Zhou H, Zeng R (2005) Proteomic analysis with integrated multiple dimensional liquid chromatography/mass spectrometry based on elution of ion-exchange column using ph steps. Anal Chem 77(18):5793–5799
49. Nogueira R, Lammerhofer M, Lindner W (2005) Alternative high-performance liquid chromatographic peptide separation and purification concept using a new mixed-mode reversed-phase/weak anion-exchange type stationary phase. J Chromatogr A 1089(1–2): 158–169
50. Qian ZJ, Jung WK, Byun HG, Kim SK (2008) Protective effect of an antioxidative peptide purified from gastrointestinal digests of oyster, crassostrea gigas against free radical induced DNA damage. Bioresour Technol 99(9):3365–3371

51. Ye M, Zou H, Liu Z, Ni J (2000) Separation of peptides by strong cation-exchange capillary electrochromatography. J Chromatogr A 869(1–2):385–394
52. Wistuba D, Banspach L, Schurig V (2005) Enantiomeric separation by capillary electrochromatography using monolithic capillaries with sol-gel-glued cyclodextrin-modified silica particles. Electrophoresis 26(10):2019–2026
53. Issaq HJ, Xiao Z, Veenstra ZD (2007) Serum and plasma proteomics. Chem Rev 107(8): 3601–3620. doi:10.1021/cr068287r
54. Link AJ, Eng J, Schieltz DM, Carmack E, Mize GJ, Morris DR et al (1999) Direct analysis of protein complexes using mass spectrometry. Nat Biotechnol 17(7):676–682
55. Raida M, Schulz-Knappe P, Heine G, Forssmann WG (1999) Liquid chromatography and electrospray mass spectrometric mapping of peptides from human plasma filtrate. J Am Soc Mass Spectrom 10(1):45–54
56. Richter R, Schulz-Knappe P, Schrader M et al (1999) Composition of the peptide fraction in human blood plasma: database of circulating human peptides. J Chromatogr B Biomed Sci Appl 726(1–2):25–35
57. Heine G, Raida M, Forssmann WG (1997) Mapping of peptides and protein fragments in human urine using liquid chromatography-mass spectrometry. J Chromatogr A 776(1):117–124
58. Janini GM, Zhou M, Yu YR, Blonder J et al (2003) On-column sample enrichment for capillary electrophoresis sheathless electrospray ionization mass spectrometry: evaluation for peptide analysis and protein identification. Anal Chem 75(21):5984–5993
59. Washburn MP, Wolters D, Yates JR (2001) Large-scale analysis of the yeast proteome by multidimensional protein identification technology. Nat Biotechnol 19(3):242–247
60. Wolters DA, Washburn MP, Yates JR (2001) An automated multidimensional protein identification technology for shotgun proteomics. Anal Chem 73(23):5683–5690
61. Dormeyer W, Mohammed S, Breukelen B, Krijgsveld J, Heck AJ (2007) Targeted analysis of protein termini. J Proteome Res 6(12):4634–4645
62. Mitulovic G, Stingl C, Smoluch M et al (2004) Automated, on-line two-dimensional nano liquid chromatography tandem mass spectrometry for rapid analysis of complex protein digests. Proteomics 4(9):2545–2557
63. Nagele E, Vollmer M, Horth P (2004) Improved 2d nano-lc/ms for proteomics applications: a comparative analysis using yeast proteome. J Biomol Tech 15(2):134–143
64. Winnik WM (2005) Continuous ph/salt gradient and peptide score for strong cation-exchange chromatography in 2d-nano-lc/ms/ms peptide identification for proteomics. Anal Chem 77(15):4991–4998
65. Sandra K, Moshir M, D'Hondt F et al (2009) Highly efficient peptide separations in proteomics. Part 2: bi- and multidimensional liquid-based separation techniques. J Chromatogr B Analyt Technol Biomed Life Sci 877(11–12):1019–1039
66. Jacobs JM, Mottaz HM, Yu LR et al (2004) Multidimensional proteome analysis of human mammary epithelial cells. J Proteome Res 3(1):68–75
67. Aivaliotis M, Gevaert K, Falb M et al (2007) Large-scale identification of n-terminal peptides in the halophilic archaea halobacterium salinarum and natronomonas pharaonis. J Proteome Res 6(6):2195–2204
68. Yuan H, Zhang L, Hou C et al (2009) Integrated platform for proteome analysis with combination of protein and peptide separation via online digestion. Anal Chem 81(21): 8708–8714
69. Gao M, Guan X, Hong G, Zhang X (2009) Advances in multidimensional high performance liquid chromatography for separation technology in proteomic study. Se Pu 27(5):551–555
70. Slotta DJ, Barrett T, Edgar R (2009) Ncbi peptidome: a new public repository for mass spectrometry peptide identifications. Nat Biotechnol 27(7):600–601
71. Ji L, Barrett T, Ayanbule O et al (2010) Ncbi peptidome: a new repository for mass spectrometry proteomics data. Nucleic Acids Res 38(Database issue):D731–735
72. Li Y, Lee ML (2009) Biocompatible polymeric monoliths for protein and peptide separations. J Sep Sci 32(20):3369–3378

73. Chen X, Tolley HD, Lee ML (2009) Polymeric strong cation-exchange monolithic column for capillary liquid chromatography of peptides and proteins. J Sep Sci 32(15–16): 2565–2573
74. Gatschelhofer C, Mautner A, Reiter F et al (2009) Ring-opening metathesis polymerization for the preparation of norbornene-based weak cation-exchange monolithic capillary columns. J Chromatogr A 1216(13):2651–2657
75. Krenkova J, Gargano A, Lacher NA et al (2009) High binding capacity surface grafted monolithic columns for cation-exchange chromatography of proteins and peptides. J Chromatogr A 1216(40):6824–6830
76. Bayard F, Raveneau A, Letourneau A et al (2009) Use of magnetic carboxyl beads to purify a cationic peptide in a batch system. Anal Biochem 384(2):350–352
77. Yamamoto S, Miyagawa E (1999) Retention behavior of very large biomolecules in ion-exchange chromatography. J Chromatogr A 852(1):25–30
78. Rege K, Pepsin M, Falcon B et al (2006) High-throughput process development for recombinant protein purification. Biotechnol Bioeng 93(4):618–630
79. Cutler P (2004) Protein purification protocols, vol 244, 2nd edn, Methods in molecular biology. Humana Press, Totowa
80. Neverova P, Van Eyk JE (2005) Role of chromatographic techniques in proteomic analysis. J Chromatogr B 815(1–2):51–63
81. Shi QS, Zhou Y, Sun Y (2005) Influence of ph and ionic strength on the steric mass-action model parameters around the isoelectric point of protein. Biotechnol Prog 21(2):516–523
82. Kelley BD, Tobler SA, Brown P, Coffman JL et al (2008) Weak partitioning chromatography for anion-exchange purification of monoclonal antibodies. Biotechnol Bioeng 101(3): 553–566
83. Bodo E, Durieux A, Saint-Hubert C, Lavallee R et al (2006) Recovery of nuclease produced by lactococcus lactis using expanded bed ion-exchange chromatography. Biotechnol Lett 28(13):1033–1039
84. Nordborg A, Zhang B, He XPZ et al (2009) Characterization of monoclonal antibodies using polymeric cation-exchange monoliths in combination with salt and ph gradients. J Sep Sci 32(15–16):2668–2673
85. Blanusa M, Perovic I, Popovic M, Polovic N et al (2007) Quantification of Art v 1 and Act c 1 being major allergens of mugwort pollen and kiwi fruit extracts in mass-units by ion-exchange hplc-uv method. J Chromatogr B-Anal Tech Biomed Life Sci 857(2):188–194
86. Tsonev LI, Hirsh AG (2008) Theory and applications of a novel ion-exchange chromatographic technology using controlled ph gradients for separating proteins on anionic and cationic stationary phases. J Chromatogr A 1200(2):166–182
87. Fonseca LP, Cabral JMS (2002) An integrated downstream processing strategy for the recovery and partial purification of penicillin acylase from crude media. J Chem Technol Biotechnol 77(10):1176–1185
88. Wang Q, Kim JS, Chung KC (2004) Purification and characterization of cyclic amp-binding protein from ganoderma lucidum. Chem Res Chinese U 20(5):588–593
89. Guerin-Dubiard MPC, Hietanen A, Quiros del Bosque A et al (2005) Hen egg white fractionation by ion-exchange chromatography. J Chromatogr A 1090:58–67
90. Doultani KNTS, Etzel MR (2004) Fractionation of proteins from whey using cation-exchange chromatography. Process Biochem 39:1737–1743
91. Smithers GW, Ballard FJ, Copeland AD et al (1996) New opportunities from the isolation and utilization of whey proteins. J Dairy Sci 79(8):1454–1459
92. Delfour A, Jolles J, Alais C, Jolles P (1965) Caseino-glycopeptides: characterization of a methionine residue and of the n-terminal sequence. Biochem Biophys Res Commun 19:452–455
93. Markus IK, Kulozika U (2008) Separation of a glycosylated and non-glycosylated fraction of caseinomacropeptide using different anion-exchange stationary phases. J Chromatogr A 1208:126–132
94. Leicht W, Pundak S (1981) Large-scale purification of halophilic enzymes by salting-out mediated chromatography. Anal Biochem 114(1):186–192

95. Kaufmann M (1997) Unstable proteins: how to subject them to chromatographic separations for purification procedures. J Chromatogr B 699(1–2):347–369
96. Heesche-Wagner K, Schwarz T, Bartholmes P, Kaufmann M (1995) Purification of unstable proteins from halobacterium salinarium crude cell extracts: combined cell disruption and desalting by a hollow-fiber membrane module as an access to perform ion-exchange chromatography. J Chromatogr A 711(1):175–179
97. Dai XP, Luo LG, Sirkar KK (1999) An integrated process for biomolecule isolation and purification. Biotechnol Prog 15(6):1095–1105
98. Xu YK, Sirkar KK, Dai XP, Luo RG (2005) A new integrated membrane filtration and chromatographic device. Biotechnol Prog 21(2):590–597
99. Kokpinar O, Harkensee D, Kasper C (2006) Innovative modular membrane adsorber system for high-throughput downstream screening for protein purification. Biotechnol Prog 22(4):1215–1219
100. Arilla MC, Gonzalez-Rioja R, Ibarrola I et al (2006) A sensitive monoclonal antibody-based enzyme-linked immunosorbent assay to quantify parietaria judaica major allergens, par j 1 and par j 2. Clin Exp Allergy 36(1):87–93
101. Jimeno L, Duffort O, Serrano C et al (2004) Monoclonal antibody-based elisa to quantify the major allergen of artemisia vulgaris pollen, art v 1. Allergy 59(9):995–1001
102. Gavrovic-Jankulovic M, Spasic M, Cirkovic Velickovic T, Stojanovic M et al (2008) Quantification of the thaumatin-like kiwi allergen by a monoclonal antibody-based elisa. Mol Nutr Food Res 52(6):701–707
103. van Ree R, Dorpema JW, Vieths S (2005) Allergy vaccines: a need for standardisation in mass units of major allergen. Pharmeuropa Bio 2005(1):27–30
104. Slater JE (2005) Characterization of allergen extracts. Dev Biol (Basel) 122:145–152
105. Cirkovic Velickovic T, Gavrovic-Jankulovic M, Jankov RM (2005) Overview of the most commonly used methods in allergen characterization. J Serb Chem Soc 70(3):347–360
106. Punzet M, Ferreira F, Briza P et al (2006) Profiling preparations of recombinant birch pollen allergen bet v 1a with capillary zone electrophoresis in pentamine modified fused-silica capillaries. J Chromatogr B Analyt Technol Biomed Life Sci 839(1–2):19–29
107. Duffort O, Palomares O, Lombardero M, Villalba M et al (2006) Variability of ole e 9 allergen in olive pollen extracts: relevance of minor allergens in immunotherapy treatments. Int Arch Allergy Immunol 140(2):131–138
108. Stephan O, Weisz N, Vieths S, Weiser T et al (2004) Protein quantification, sandwich elisa, and real-time pcr used to monitor industrial cleaning procedures for contamination with peanut and celery allergens. J AOAC Int 87(6):1448–1457
109. Helsper JP, Gilissen LJ, van Ree R, America AH et al (2002) Quadrupole time-of-flight mass spectrometry: a method to study the actual expression of allergen isoforms identified by pcr cloning. J Allergy Clin Immunol 110(1):131–138, 110
110. Faeste CK, Ronning HT, Christians U, Granum PE (2011) Liquid chromatography and mass spectrometry in food allergen detection. J Food Prot 74(2):316–345
111. Erler A, Hawranek T, Kruckemeier L, Asam C, Egger M, Ferreira F, Briza P (2011) Proteomic profiling of birch (betula verrucosa) pollen extracts from different origins. Proteomics 11(8):1486–1498
112. Tantoush Z, Stanic D, Stojadinovic M, Ognjenovic J, Mihajlovic L, Atanaskovic-Markovic M, Cirkovic Velickovic T (2011) Digestibility and allergenicity of beta-lactoglobulin following laccase-mediated cross-linking in the presence of sour cherry phenolics. Food Chem 125(1):84–91
113. Sabato V, van Hengel AJ, De Knop KJ, Verweij MM, Hagendorens MM, Bridts CH, De Clerck LS, Schiavino D, Stevens WJ, Ebo DG (2011) Human basophils: a unique biological instrument to detect the allergenicity of food. J Investig Allergol Clin Immunol 21(3):179–184

Chapter 2
Application of Ion Exchanger in the Separation of Whey Proteins and Lactin from Milk Whey

Dragana Stanic, Jelena Radosavljevic, Marija Stojadinovic, and Tanja Cirkovic Velickovic

Abstract Whey disposal represents a huge obstacle for dairy industry, being costly and problematic. On the other hand, it could be used as a starting material for isolation of some components that are valuable on the market. Whey processing is not an easy operation; it requires robust techniques of high volumetric throughput with minimal pretreatment of the starting material. For this purpose, ion exchangers can be used, due to their versatility, safety, and relative cheapness. Being a mixture of acidic and basic proteins, double-step ion exchange chromatography, involving both cation and anion exchangers, must be performed to obtain highly purified whey proteins. Development of new technologies, based on ion exchange, which can provide fast and efficient whey processing, provides maximal exploitation in environmentally safe way.

In this chapter, the literature review on the usage of ion exchangers in the separation of lactin from bovine whey proteins and fractionation of bovine whey proteins will be given.

2.1 Introduction

Dairy whey is characterized as the water-like liquid or serum that is separated from the curds during cheese manufacturing [1]. It consists of 0.7% protein, 0.05% fat, 4.9% lactose, <0.05% casein, 0.7% ash, and 6.35% total solids (Fig. 2.1). Every 100 kg of milk used in cheese manufacture results in 10–20 kg of cheese and 80–90 kg of whey [1]. For a long time, whey has been considered as a waste rather than a by-product of dairy industry because of high-cost recovery of its low concentrated compounds. However, whey contains 100% of the total lactin (e.g., lactose)

D. Stanic • J. Radosavljevic • M. Stojadinovic • T.C. Velickovic (✉)
University of Belgrade, Studentski trg 12-16, 11000 Belgrade, Serbia
e-mail: tcirkov@chem.bg.ac.rs

Inamuddin and M. Luqman (eds.), *Ion Exchange Technology II: Applications*,
DOI 10.1007/978-94-007-4026-6_2, © Springer Science+Business Media B.V. 2012

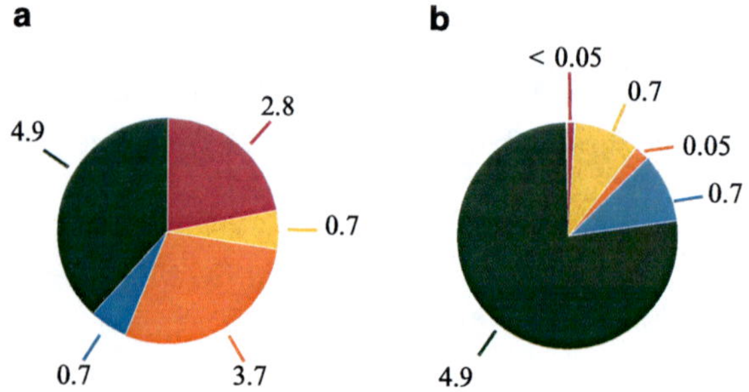

Fig. 2.1 Composition of total solids of (**a**) milk, (**b**) whey. *Green*, lactin; *purple*, casein protein; *yellow*, whey protein; *orange*, fat; *blue*, ash. Represented as percentage of weight per volume

and 20% of the total milk proteins, which if isolated could be highly valuable for the food and pharmaceutical industry. Since whey disposal is costly and problematic for cheese manufacturers, today the focus is on techniques to convert this waste product into valuable functional components.

2.2 Whey

There are two kinds of whey: sweet whey (produced when caseins are coagulated using rennet type enzymes at pH 5.9–6.3) and acid whey (resulting from cheese making after casein precipitation at pH 4.6–4.7) [2]. The latter generally has twice the amount of calcium phosphate and more lactic acid. Whey preparations commercially produced today are whey protein concentrates (WPC) and whey protein isolates (WPI), both of which contain all the whey proteins in the same proportions as in the original whey. WPCs have a protein content ranging from 34% to less than 90%. WPI is more pure and is more concentrated, accounting for approximately 90% of whey proteins and a little or none percentage of lactose and fat which makes it suitable for people suffering from lactose intolerance [1, 2]. Whey preparations are prepared usually by ultrafiltration, ion exchange processes, and spray dried as a powder. WPI and WPC have become most popular among sport enthusiasts.

2.2.1 *Whey Proteins*

Whey is a heterogeneous mixture of proteins possessing valuable technofunctional and nutritional properties. Bovine whey proteins are used in human and veterinary medicine, pharmaceutical industry, food industry, and as supplements for cell-culture media [3–5]. Major whey protein components are β-lactoglobulin and α-lactalbumin while minor whey proteins are bovine serum albumin, immunoglobulins, lactoperoxidase, lactoferrin (Table 2.1), glycomacropeptide (sweet whey only), acid phosphatase, lysozyme and ribonuclease [8], protein components of milk fat globule membrane, and proteose-peptone components [9].

Table 2.1 Data on the bovine whey proteins, complied from reference [6, 7]

Protein	Isoelectric point (pI)	Molecular weight (kDa)	Concentration (g/l)	Percentage of the whey protein fraction
Beta-lactoglobulin	5.2–5.4	18.3	3–4	56.3
Alpha-lactalbumin	4.7–5.1	14.2	1.2–1.5	25.0
Bovine serum albumin	4.9–5.1	66.0	0.3–0.6	6.3
Immunoglobulins	5.8–7.3	150.0–900.0	0.6–0.9	12.5
Lactoperoxidase	9.6	78.0	<0.06	<1
Lactoferrin	8.0	78.0	<0.05	<0.9

Beta-lactoglobulin (BLG) accounts for approximately 50% of total bovine whey proteins (Table 2.1) and is present only in trace amounts in human milk [10]. It shows excellent gel formation, emulsifying and foaming properties, and can be used as a texturizing and stabilizing agent in structuring dairy product matrices [11]. BLG has a high nutritional value due to the essential amino acid composition, especially cysteine content [3]. On the other hand, BLG has been identified as a potential source of allergic reactions to infant formulas seen in some children [10, 12, 13]; thus, selective removal of BLG may therefore be of interest in the commercial application.

Alpha-lactalbumin (ALA) is present up to 30% in both bovine and human milk. It can be used in food industry to provide whipability, high gel strength, viscosity, water binding, and high solubility [14]. ALA is used as a nutraceutical in infant formulas and meat products because of its high tryptophan content [15, 16]. It also has strong affinity for glycosylated receptors present on the surface of oocytes and spermatozoids and thus may have a potential as a contraceptive agent [17] Moreover, ALA also has a potential in reducing the risk of incidence of some cancers [18].

Immunoglobulins (IgG) are known to provide passive immunity and are effective in treatment of various infections of newborn infants [4] and can enhance the immunological properties of infant formula. They can be also used therapeutically in the treatment of animal neonates [19].

Lactoferrin (LF) has multiple biological properties including antimicrobial (due to its iron chelating capacity), anti-inflammatory, anticarcinogenic, immunomodulatory, bone-growth-stimulating [20], ion-transporting, toxin-binding, and wound-healing properties [21–23]. There is also evidence indicating a direct cell lysing effect of LF unrelated to iron binding [24, 25].

Lactoperoxidase (LP) has medical and functional benefits due to its antimicrobial properties [26]. Isolated, both LF and LP can be used as natural antibiotics and in food industries as supplement in infant formulas.

Bovine serum albumin (BSA) is of interest in a number of food and therapeutic applications. It is used as a nutrient in cell and microbial culture and also to stabilize some enzymes during digestion of DNA in vitro and to prevent adhesion of the enzymes to reaction tubes and other vessels.

Glycomacropeptide (GMP) is a peptide fragment produced by enzymatic cleavage of κ-casein by chymosin. It is highly soluble in water and remains in the whey fraction after coagulation of the milk. Glycomacropeptide has been examined for improved lipid digestion, protection against influenza, prevention of tartar adhesion to teeth, and inhibition of *E. coli* attachment to intestinal walls [17]. GMP-rich whey protein concentrate is also a good source of branched-chain amino acids, which have been shown to promote lean muscle development. Glycomacropeptide contains no aromatic amino acids and no phenylalanine, making GMP suitable for diet of patients with phenylketonuria [27].

Whey also contains proteins that serve as potent growth stimulants for many mammalian cell lines in culture. *Growth factors* present in the whey are IGF-I, IGF-II, and *des*(1–3)IGF-I [28]; transforming growth factor-P [29]; platelet-derived growth factor [30]; and heparin-binding growth factors [31]. These growth factors have been proved in a series of trials to be reliable and cost-effective supplement for fetal bovine serum which is occasionally variable in quality, and its production is becoming limited [28].

In the past, dairies were disposing whey in the environment, but since 1980s when the production of whey became enormous worldwide, whey disposal has become a huge threat to environment. In 2009, the world production of liquid whey reached over 90 million tons per year. Whey is a serious environmental pollutant because of high biochemical oxygen demand (BOD) due to the high concentration of dissolved organic compounds [6]. That is why dairy companies have to process whey or to dispose it in some environmentally acceptable manner.

On a commercial scale, for whey protein preparations, dominantly used are techniques ultrafiltration, reverse osmosis, microfiltration, and chromatographic methods such as gel filtration and ion exchange. Ultrafiltration is the most common method in dairy industry due to the various applications and possible integration with other methods for obtaining from low to high protein content whey preparations [2]. Reverse osmosis, along with ultrafiltration, has dominantly been used for concentration and partial demineralization as a step in a separation process [7], while microfiltration offers an in-place sterilization, removal of lipids, and partial protein isolation [14]. Major disadvantages of these membrane processes are high operating costs due to membrane fouling, incomplete separation, and large permeate volumes.

Various techniques have been used to isolate and purify whey proteins, e.g., selective solubility in the presence of 3% w/w trichloroacetic acid (TCA) [32, 33], solubility at low pH in the presence of salt [34], peptic hydrolysis followed by selective membrane filtration [33], salting out followed by chromatographic techniques [35, 36], ion exchange chromatography [37] or ion exchange membranes [38–40], gel filtration [41], hydrophobic chromatography, and affinity separation process [42]. To explore or exploit the potentially available functional and bioactive properties of whey proteins, it is important to isolate native proteins by a procedure that avoids all possible denaturing conditions (such as high salt conditions and low pH, protease treatment).

Lactin has been commonly separated from whey proteins by ultrafiltration and ion exchange processes [5, 30, 43].

2.3 Application of Ion Exchangers in Whey Protein and Lactin Separation

Selective separation of individual whey proteins has been proved to be difficult due to the relatively similar physicochemical properties of the different proteins (Table 2.1). Good selective separation on the basis of size requires at least a twofold difference in molecular weight and would therefore be limited to separating proteins such as ALA (14.2 kDa) and IgG (150.0–900.0 kDa). However, separating whey proteins on the basis of their isoelectric points gives two distinct groups: the major whey proteins: BLG, ALA, and BSA, which are negatively charged at the pH of sweet whey (pH 6.2–6.4); and minor whey proteins LF and LP that hold a positive net charge at the pH of whey. These distinct properties offer the possibility of selectively separating the groups using ion exchange chromatography and ion exchange membranes.

To date, ion exchange chromatography (IEC) has been used extensively for extractions of proteins from whey either in packed-bed chromatography [2, 36, 37, 44, 45] or using ionic adsorptive membranes [38, 40].

In the preparation of this chapter, old and new publications, including traditional and modern approaches, on application of ion exchangers in whey protein separation were reviewed.

2.3.1 *Advantages and Disadvantages of Ion Exchange Chromatography in Whey Protein and Lactin Separation*

Reasons for the popularity of IEC include its high resolving power, high protein-binding capacity, versatility (there are several types of ion exchangers, and the composition of the buffer and pH can be varied over a wide range), straightforward separation principle (primarily according to differences in charge), and ease of performance [46] while preserving protein's native state (depending on the pH and ionic strength of the eluting buffer) and almost absolute removal of lactin [5]. In addition to relative cheapness is its acceptance by the regulatory authorities in the production of pharmaceutical proteins. Nevertheless, IEC has several drawbacks when used on an industrial scale: production of large volumes of disposal, usage of chemicals, and diluted protein fractions [2]. New technologies such as ion exchange membranes have been developed to overcome drawbacks of classical IEC.

In order to provide industrial applicability, optimization of fractionation processes is oriented toward (a) reducing costs of isolation of proteins by choosing inexpensive buffers and operability through many cycles of purification before cleaning along with nonreduced column capacity, (b) providing economical viability of such a purification process by usage of columns of high volume and high flow rates which provide short cycles along with high recovery of proteins (over 90%), and (c) manufacturing a competitive product of high purity (at least food-grade product) [47].

Ion exchange chromatography can be operated under two conditions: selective adsorption of whey proteins and selective elution of proteins from ion exchange resin [47–49].

Selective adsorption processes are not of a great potential for development into industrial applications because they provide optimal conditions usually for binding of one protein. In this case, revenue obtained from this purification process is generated by production of one protein and should, also, cover the cost of disposal of remaining whey proteins [47, 50].This chromatographic system can be easily optimized to obtain one concentrated protein of high purity and recovery. To exploit the remaining depleted whey, this system would require more than one chromatographic column and changes of buffers.

Another approach for optimization of selective adsorption processes implies conditions that provide displacement of one protein by another protein [48, 51] or by a small molecule [52]. Under conditions where a protein is displaced by another one, it is usually impossible to produce more than one protein of high purity. The purity of obtained protein depends exclusively on the time when collecting of the fraction is stopped and switched to collecting the displacer protein. If a small molecule (usually organic polycation or polyanion) is used as displacer, then it could be possible to obtain proteins of higher purity and recovery [52].

The main disadvantage of this approach is that, for the specified volume of matrix, the strict volume of whey with no variation in content should be applied onto the column in conditions close to the maximum binding capacity of the ion exchange resin. Under these conditions, the purity of fractions depends, also, on the flow through the column. Everyday usage of the column leads to the flow decrease due to the clogging of the column and is impairing the column efficacy. Also, ability to switch to another ion exchanger of similar properties is not easily achievable because this system is operable under a very narrow range of conditions (working pH, flow, whey composition).

By displacement chromatography approach, several substances may be isolated in a row and concomitantly concentrated which may be useful for the processing of diluted starting materials, such as whey [52].

Processes based on selective elution rather than on selective adsorption are of greater interest for industry. These processes are operated in such conditions that all proteins are bound to the column and subsequently eluted providing purification of several purified proteins. The cost of production is spread among different products, and it is possible to obtain different products by changing elution conditions.

The main advantage of selective elution over selective adsorption is that by this approach, proteins from the whey are simultaneously fractionated and concentrated. Also, it is possible to concentrate all proteins simply by eluting them with one buffer. This system can be easily adapted to produce desired proteins or protein mixtures, in accordance with demands from the market, by using only one piece of equipment [47, 52].

Ion exchange chromatography can be operated as column and batch chromatography. In column chromatography, the ion exchange resin is packed in the column, and all solutions are passed through the column. Batch purifications involve stirring of the whey with ion exchange resin and, after settling, separation from the resin and subsequent elution by stirring matrix with adsorbed protein with elution buffer. Limitations of using ion exchanger in this way on large scale are necessity for big tanks for stirring that can accommodate whey and ion exchange buffer for stirring and not feasible fractionation because each elution step requires emptying of that tank. This approach does not provide high throughput and has a low recovery, both originating from establishing equilibrium between adsorbed proteins on the resin and proteins that are in solution [47]. These are the main reasons why column IEC is being favored over batch chromatography on industrial scale. On laboratory scale, supremacy of column chromatography over batch chromatography has already been proven [53]. Column chromatography separations are of particular importance to the biopharmaceutical industry because they can deliver high-purity products and can easily be scaled to the desired production level.

Almost 50 publications are listed for their technical information in order to assist reader in optimizing isolation processes to his own needs. Only a few have been discussed more thoroughly as examples of the process efficiency and utilization.

Since whey is a mixture of acidic and basic proteins, usage of both anion- and cation-exchanger chromatography for their separation is possible.

2.3.2 Anion-Exchange Chromatography in Whey Protein Separation

Acidic whey proteins are BLG, ALA, and BSA with pI values in a narrow pH range from 4.7 to 5.4 (Table 2.1). The idea for the anion-exchange separation is to develop conditions under which acidic whey proteins bind and are subsequently eluted from the column usually by changing buffer pH or increasing salt concentration. Whey basic proteins (IgG, LP, LF) would, under these conditions, pass through the column nonadsorbed and could be further fractionated by cation-exchange step.

First task in whey protein isolation is to separate two main protein fractions, ALA from BLG. Weak anion exchangers with DEAE group attached to the matrix support such as DEAE-Sephadex, DEAE-Cellulose, and DEAE-Sepharose from different manufacturers (Table 2.2) were used successfully for this purpose.

Ye et al. [37] separated ALA and BLG on DEAE Toyopearl (Toyosoda, Tokyo, Japan). Whey adjusted to pH 8.5 was passed through column equilibrated in 50 mM Tris–HCl buffer at pH 8.5. ALA was unadsorbed, and BLG was adsorbed on the column and eluted using linear gradient from 0 to 0.3 M NaCl (Fig. 2.2). This agrees with results obtained by batch chromatography using DEAE-Sepharose-CL6B (Pharmacia, Uppsala, Sweden) in a method proposed by de Jongh et al. [56], where buffer ionic strength (100 mM Tris–HCl pH 7.2) allows mainly BLG to be adsorbed to the matrix while ALA is present in the flow-thru fraction. Further purification of BLG by gel filtration was necessary because batch chromatography has less resolution than column chromatography.

Table 2.2 Ion exchange chromatography of whey proteins

Protein	Matrix	References
Beta-lactoglobulin	DEAE-Cellulose	[36, 54, 55]
–	DEAE-Sepharose CL6B	[56]
–	DEAE-Toyopearl	[37]
–	Q-Sepharose	[2]
–	Mono Q	[57]
–	Bio-Scale Q2	[52]
–	Super Q	[52]
–	CM-Cellulose	[55]
–	SP Sepharose Big Beads	[50, 58]
–	SP Sepharose FF	[48, 51]
Alpha-lactalbumin	DEAE-Cellulose	[36, 54, 55]
–	DEAE-Toyopearl	[37]
–	Mono Q	[57]
–	Bio-Scale Q2	[52]
–	Super Q	[52]
–	CM-Cellulose	[55]
–	SP Sepharose Big Beads	[50, 58]
–	SP Sepharose FF	[48, 51]
Bovine serum albumin	DEAE-Cellulose	[36]
–	Mono Q	[57]
Lactoferrin	DEAE-Cellulose	[59, 60]
–	CM-Toyopearl	[61]
–	SP Sepharose Big Beads	[44, 50]
–	Amberlite CG-50	[62]
–	Amberlite IRC	[63]
Lactoperoxidase	DEAE-Sephadex	[64]
–	DEAE-Cellulose	[65]
–	CM Sephadex C-25	[53]
–	SP-Toyopearl 650	[53]
–	SP Sepharose Big Beads	[44, 50]
–	P-Cellulose	[65]
–	Bio-Rex 70	[66]
–	Amberlite IRC-50	[67, 68]
Immunoglobulins	DEAE-Sephadex	[69, 70]
–	DEAE-Cellulose	[71, 72]
–	P-Cellulose	[71]
–	Fractogel EMD SO_3^- 650 (S)	[49]
–	Macro-Prep High S Support	[49]

It seems that ALA can be either unadsorbed, as mentioned above, or adsorbed on the DEAE-column, depending on the matrix type, sample preparation, or chromatography buffer choice. Neyestani and collaborators [36] reported a method for separation of BLG on DEAE-Cellulose (Pharmacia, Uppsala, Sweden) from BSA and ALA, where ALA was adsorbed on the column along with other proteins. ALA and BSA were coeluted with 0.1 M NaCl in 0.05 M Tris–HCl pH 6.8 and separated further by gel filtration step.

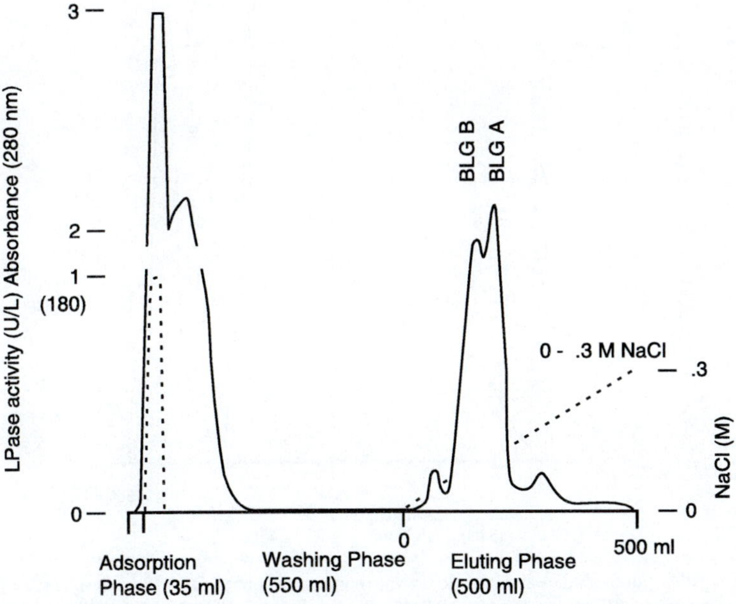

Fig. 2.2 Adsorption of BLG in bovine rennet whey onto a DEAE-TP column (1.5×18 cm). Adsorption phase, 35 ml rennet whey (pH 6.5); washing phase, 550 ml of 0.05 M Tris–HCl buffer (pH 8.5); eluting phase, 250 ml of 0.05 M Tris–HCl buffer (pH 8.5) in one chamber of the gradient mixer and 250 ml of the same buffer containing 0.3 M NaCl in the other chamber of the gradient mixer. LPase activity is shown by the *dotted line* on the *left*, while NaCl gradient is shown by the *dotted line* on the *right* (Reprinted from Ref. [37] with kind permission of © Elsevier Ltd. (2000))

Besides weak anion exchangers, in whey fractionation, strong anion exchangers were used for whey fractionation as well. Usage of QAE (quaternary aminoethyl)-Toyopearl (Toyosoda, Tokyo, Japan), Q-Sepharose (GE Healthcare, Uppsala, Sweden), and Mono Q (GE Healthcare, Uppsala, Sweden) has been reported for that purpose [2, 37, 57]. All acidic proteins are adsorbed on strong anion exchanger at buffer of pH above their pI values (Fig. 2.3). Adsorbed protein elution in separate peaks was achieved by either increasing salt concentration [37] or changing the pH of the elution buffer [2].

It should be emphasized that separation of BLG genetic variants can also be achieved by anion-exchange high-performance liquid chromatography (HPLC) [33, 36, 41] and, to some extent, by normal-pressure column anion-exchange chromatography [37].

Though anion-exchange chromatography is not usually used for whole whey fractionation, Kunz and Lonnerda [73] separated major human whey proteins (lysozyme, IgG, LF, ALA, BSA) by anion-exchange fast protein liquid chromatography (FPLC) on Mono Q column (Pharmacia, Uppsala, Sweden). Separation was obtained using 20 mM diethanolamine buffer at pH 9.5 with a NaCl linear gradient from 0 to 0.3 M.

New approaches in whey processing such as anion-exchange displacement chromatography tend to overcome classical IEC disadvantages – large amounts of

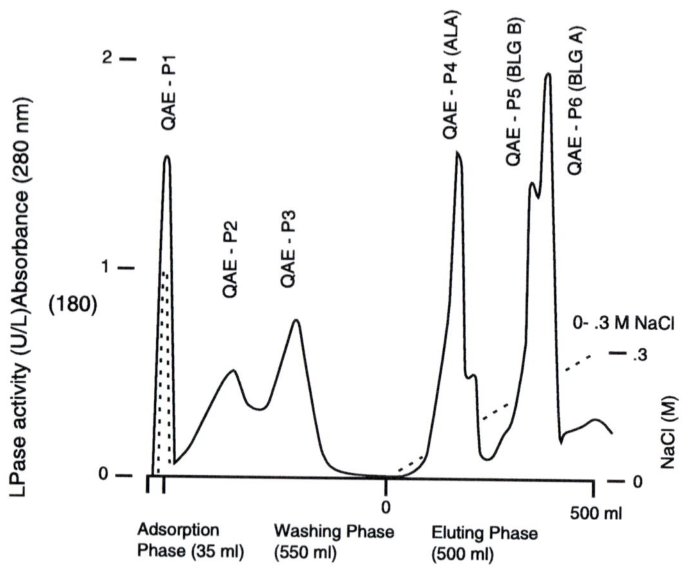

Fig. 2.3 Adsorption of ALA and BLG in bovine rennet whey onto a QAE-TP column (1.5 × 18 cm). Adsorption phase, 35 ml rennet whey (pH 6.5); washing phase, 550 ml of 0.05 M Tris–HCl buffer (pH 8.5); eluting phase, 250 ml of 0.05 M Tris–HCl buffer (pH 8.5) in one chamber of the gradient mixer and 250 ml of the same buffer containing 0.3 M NaCl in the other chamber of the gradient mixer. LPase activity is shown by the *dotted line* on the *left*, while NaCl is shown by the *dotted line* on the *right* (Reprinted from Ref. [37] with kind permission of © Elsevier Ltd. (2000))

disposable liquid and diluted protein fractions. Vogt et al. [52] optimized such a procedure by choosing column with smaller particle diameter, Bio-Scale Q2 column (10 μm particle size, Bio-Rad, Munich, Germany) instead of Super Q column (35 μm particle diameter, TosoHaas Europe, Stuttgart, Germany), appropriate polyanion displacer (polyacrylic acid), and carrier pH (pH 8.0).Under these conditions, they succeeded to obtain ALA and BLG with high recoveries of 80% and 90%, respectively. Separation of BLG variants A and B was also achieved by anion-exchange displacement chromatography [74].

2.3.3 Cation-Exchange Chromatography in Whey Protein Separation

Cation-exchange chromatography has been anticipated as a method of first choice for isolation and separation of basic whey proteins, primarily LP and LF, and also for isolating valuable IgGs. These proteins are positively charged over very broad pH range, allowing optimization of conditions in which ALA and BLG would not adsorb on the matrix [49]. This is of particular importance because IgG, LP, and LF are not present in high concentration in whey, unlike ALA and BLG (Table 2.1).

By operating under finely tuned conditions, appropriate ion exchanger could allow passing of nonadsorbed ALA and BLG. Basic proteins are concentrated by adsorbing on the matrix and subsequently fractionated by elution with carefully chosen buffers. Also, the trials to fractionate all whey proteins on cation-exchange column were performed, but the usage of one chromatographic column for separation of both basic and acidic proteins cannot be optimized easily [49, 50].

Development of matrices for separation of proteins led to more thorough investigation of possible applications of cation exchanger in separating acidic whey proteins ALA and BLG. Also, cation exchange found application in isolating growth factors which can be used as supplement media for laboratory cell cultures [26].

Turhan and Etzel [58] used cation exchange for obtaining WPI and ALA. After adjusting a pH of whey to 4.0 and application on SP Sepharose Big Beads (Amersham Biosciences, Uppsala, Sweden) packed column (equilibrated with 10 mM sodium lactate, pH 4.0), elution (by 10 mM sodium hydroxide) gave one-step preparation of WPI and recovery of 96% of all proteins from whey. By introducing one more step elution with 100 mM sodium acetate (pH 4.9), they obtained ALA of 93% purity (recovery 90% ± 5%) and WPI depleted in ALA (90% BLG, 2% ALA, 4% IgG and BSA, 97% proteins recovered). This work introduced a very convenient way for obtaining highly purified ALA, WPI, and ALA-depleted WPI of high recovery by using cheap, food-grade buffers.

To our knowledge, there is a report of separation of acidic proteins on the cation exchanger [51]. The authors used SP Sepharose FF prepacked columns (GE Healthcare, Amersham, UK) equilibrated with 0.1 M sodium acetate (pH 3.7) to fractionate whey proteins. By eluting proteins with 0.1 M Tris–HCl (pH 9.0), they obtained two fractions: first one, consisting of BLG (95% purity) and other containing 32% of ALA and 38% of BLG. Conditions for separation are carefully chosen to allow binding of all whey proteins. However, separation of BLG from other proteins present in whey is achieved by displacement of BLG by ALA, yielding BLG of satisfying purity and fraction containing all remaining proteins with some BLG. This approach does not have a good recovery of main proteins, 84% and 80% for ALA and BLG, respectively.

El-Sayed and Chase [48] made improvement of their method for isolation of whey proteins based on selective adsorption [51] by introducing second stage. The first stage process yielded two fractions eluted from SP Sepharose FF: fraction containing BLG (which is eluted first) and second containing both ALA and BLG. This second fraction from the first stage was mixed with whey concentrate solution to form the feedstock for the second stage (Fig. 2.4). In this stage, the fraction contained BLG of high purity, so further processing was not required. In both stages, the application of feedstock was terminated at the point when BLG dropped to its inlet concentration. Two fractions were collected during the loading period (the first fraction was collected up to the start of the ALA breakthrough, while the second fraction was collected up to the point where BLG dropped to its inlet concentration). During each of both stages after loading, the column was eluted by two buffers of different pH, giving two fractions of purified ALA. Almost all of the BLG in the feed was recovered, with 78% being recovered at 95% purity (first fraction) and a remaining 20% at 86%

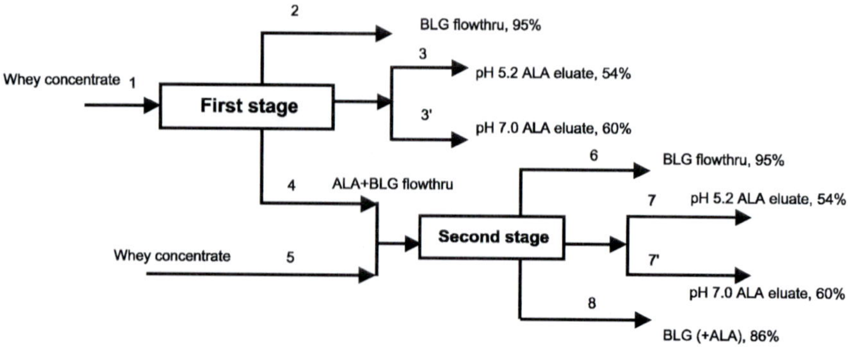

Fig. 2.4 Flow diagram of the consecutive two-stage process. El-Sayed MMH, Chase HA (2010). Purification of the two major proteins from whey concentrate using a cation-exchange selective adsorption process (Reprinted from Ref. [48] with kind permission © John Wiley & Sons, Inc. (2010))

purity (second fraction). In addition, 67% of ALA was recovered, 48% at 54% purity (first fraction) and 19% at 60% purity (second fraction).

Many trials have been performed with the goal to produce system easily operable on cation exchange which could provide more than one protein of high purity [49, 50]. However, the results never showed to be satisfactory for purification of multiple proteins, so the researchers focused more on obtaining at least one protein of satisfying purity and fractions enriched in particular proteins or one-protein depleted WPI.

Doultani and coworkers [50] tried to isolate multiple proteins from whey by optimizing elution conditions from cation exchanger. They used SP Sepharose Big Beads (Amersham Biosciences, Uppsala, Sweden) equilibrated with 50 mM sodium lactate (pH 4.0) and used stepwise elution by 100 mM sodium acetate (pH 4.9) to elute ALA. Next step, elution by 50 mM phosphate (pH 6.5) gave WPI depleted of ALA, LF, and LP. Elution with 50 mM phosphate (pH 6.5) containing 0.35 or 1.2 M sodium chloride yielded purified LP and LF, respectively. Recoveries of the proteins from the whey were satisfactory (96% ± 9%), yielding one fraction containing ALA of high purity (recovery: 96% ± 7%), and WPI consisting predominately of BLG (recovery: 107% ± 1%), with some IgG (recovery: 80% ± 30%) and BSA (recovery: 110%), while ALA was present in traces. Also, fractions containing LF and LP had present ALA and BLG in high amounts.

The main application of cation-exchange chromatography is for isolation and separation of LP and LF. Optimization of batch isolation of LP on CM Sephadex C-25 (Kabi Pharmacia, Uppsala, Sweden) (50 mM phosphate, pH 6.5), showed poor recovery of LP (from 1.2% to 47.4%), depending on whey volume, concentration of salt in buffer used for desorption, and time for adsorption and desorption of proteins [53]. However, if the column chromatography (50 mM phosphate, pH 6.5, linear gradient 0–2 M NaCl) is used to separate LP from other proteins, then recovery is higher: 48% for isolation on CM Sephadex C-25 and 92% on Toyopearl SP-650 (Toyosoda, Tokyo, Japan). The main disadvantage of such procedures is inability to separate LP from LF, which is present in LP-containing fractions [53].

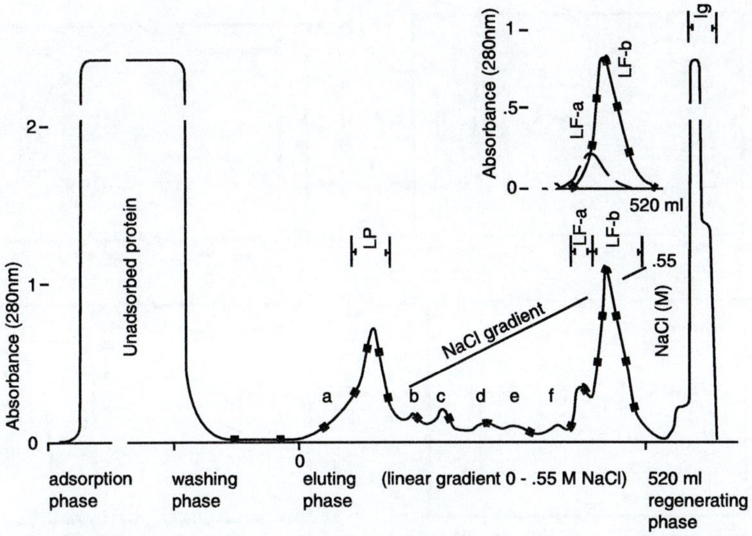

Fig. 2.5 Carboxymethyl cation-exchange chromatography of the dialyzed acid whey using CM-Toyopearl column (1.5×25 cm) and rechromatography of LF: LF-a and LF-b. Adsorption phase: 500 ml of dialyzed acid whey at pH 7.7, .0.05 M phosphate, which corresponded to 350 ml acid whey. Washing phase: 200 ml of pH 7.7, 0.05 M phosphate. Eluting phase: 260 ml of pH 7.7, 0.05 M phosphate and 260 ml of pH 7.7, 0.05 M phosphate containing 0.55 M NaCl linear gradient as indicated. Regenerating phase: 100 ml of 2 M NaOH (Reproduced from Ref. [61] with kind permission © Elsevier Ltd. (1991))

However, it is possible to separate LP from LF under very stringent conditions even from whey without prior fractionation. If the whey is applied on CM-Toyopearl (Toyosoda, Tokyo, Japan) column equilibrated with 50 mM phosphate buffer of pH 7.7, and then eluted by 0.3 M NaCl to form linear gradient over 12 column volumes, it is possible to separate clearly two peaks (Fig. 2.5). One contains LP (recovery: more than 99%, calculated to literature whey protein content) and the other contains LF (recovery: more than 99%, calculated to literature whey protein content), both being of purity more than 95% by SDS-PAGE [61]. Being of great importance due to broad applications, a lot of interest is driven toward isolation of LP and LF, not just from whey but also from milk directly. Fee et al. [44] showed that these two proteins could be isolated from raw whole milk by column chromatography on SP Sepharose Big Beads (Pharmacia, Uppsala, Sweden).

It has been shown that the milk whey could serve as a potential source of cell-growth factors. Cation-exchange chromatography found application in isolation of these valuable basic proteins that could serve as cell-culture media supplements. The main advantage of this isolation process is ability to concentrate these proteins, which are not present in high amounts in whey. It is hard to optimize the system for binding only cell-growth factors, so further purification by some other methods is necessary in order to obtain pure growth factors, free of even traces of IgG, LP, and LF [26].

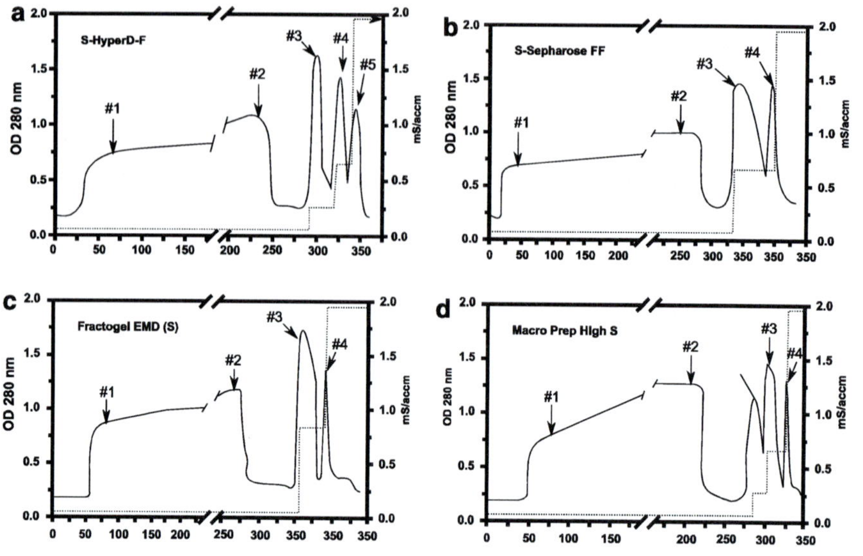

Fig. 2.6 Purification of bovine whey proteins by cation-exchange chromatography. Clarified whey was applied to four different cation-exchange columns (3.5×1.6 cm) equilibrated with 20 mM citric acid, pH 4.7. The flow rate was 3.3 ml/min (200 cm/h). Elution of bound material was carried out with sequential NaCl gradients. Unbound material and eluted fractions were characterized by analytical size-exclusion chromatography and SDS-PAGE: (**a**) S-HyperD-F, (**b**) S-Sepharose FF, (**c**) Fractogel EMD SO_3^- S, (**d**) Macro-Prep High S Support (*dashed line*, theoretical salt gradient; *full-line*, UV absorbance at 280 nm) (Reprinted from Ref. [49] with kind permission © Elsevier Ltd. (1998))

Interesting comparative study of Hahn et al. [49] focuses on separation of whey proteins using four different cation-exchange media: Macro-Prep High S Support (Bio-Rad, Richmond, CA, USA), S-Sepharose Fast Flow (Pharmacia Biotech, Uppsala, Sweden), S-HyperD-F (Biosepra, Marlborough, MA, USA), and Fractogel EMD SO_3^- 650 (S) (Merck, Darmstadt, Germany). All columns were equilibrated by 20 mM citrate buffer at pH 4.7, and diluted whey was applied. Same eluting conditions (step elution: 0.1 M, 0.3 M, 0.4 M, and 1 M NaCl) were used for protein fractionation on the all four media (Fig. 2.6).

However, the separation profile was not the same for all of used media. Flow-thru (i.e., unbound proteins) from all matrices contained predominately ALA and BLG. Separation on S-HyperD-F yields two more fractions: fraction eluted with 0.1 M NaCl (IgG, LP, and traces of BLG) and fraction eluted with 0.3 M NaCl (LP and LF). From S-Sepharose Fast Flow, proteins are eluted under higher concentration of NaCl: 0.3 M NaCl fraction (IgG, LP, significant amounts of BLG and ALA) and 1 M NaCl fraction (LP and LF). Fractogel EMD SO_3^- 650 (S) adsorbs whey proteins even stronger than S-Sepharose Fast Flow, and proteins could be eluted only at 0.4 M NaCl (IgG, LP, significant amounts of ALA) and 1 M NaCl (LF and IgG). From Macro-Prep High S Support, 0.1 M NaCl elutes some ALA and some IgG; 0.3 M NaCl – IgG, LF, LP, and some ALA; while 1 M NaCl – IgG and traces of all whey proteins. This study showed that although all matrices have the same type of ion exchange active group, they exhibit different selectivity, and a matrix can be chosen depending on the desired purpose. Of course, this approach does not produce proteins of satisfactory purity, so, in order to obtain whey protein of high purity, a second purification step must be included.

2.4 Recently Developed Concepts Based on Ion Exchange for Separation of Whey Components

Whey processing requires easily scalable techniques capable of high volumetric throughput with minimal pretreatment.

2.4.1 Simulated Moving Bed Chromatography

Simulated moving bed chromatography (SMB) concept is based on the simulation of a true countercurrent operation between the solid and the liquid phase, by valves switching over a series of columns or column movement in a carousel (Fig. 2.7). Feed entry and the analyte recovery are simultaneous and continuous, but, instead of moving the bed, the sample inlet and the analyte exit positions are moved continuously, giving the impression of a moving bed. The countercurrent operation allows a more

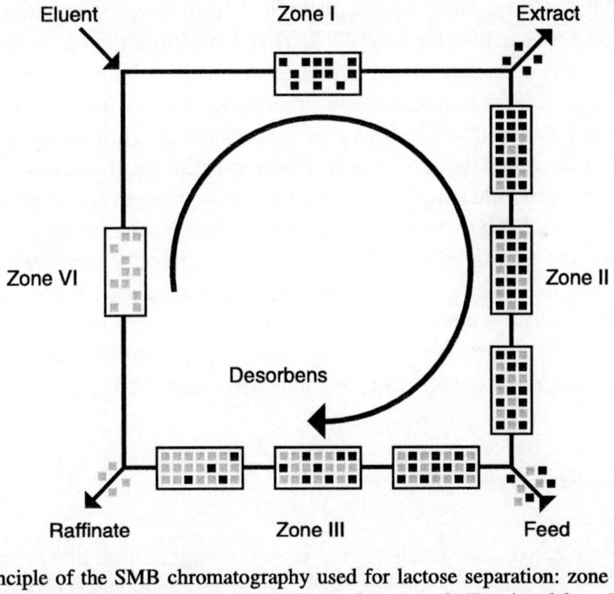

Fig. 2.7 Principle of the SMB chromatography used for lactose separation: zone 1 and zone 4 includes one column each, zone 2 and zone 3 three columns each (Reprinted from Ref. [75] with kind permission © Elsevier Ltd. (2005))

efficient use of the adsorbent and the liquid streams leading to advantages like higher productivity/smaller separation plants, higher product concentration, reduced buffer consumption, more efficient use of raw material, and higher target purity [76].

Andersson and Mattiasson [77] developed a 20-column SMB process (Streamline-SP, GE Healthcare, Uppsala, Sweden) for the separation of LP and LF from WPC. They showed, in comparison with the same 20-column setup run in nonmoving bed mode, 48% raise in productivity, a 4.3 times decrease in buffer consumption, and 6.5 times raise in target protein concentration with a better raw material utilization. The column capacity, 13 mg protein/ml of chromatographic support, and flow rates of 880 cm/h in the loading zone are high enough to be considered for a large-scale application and could be further optimized.

2.4.2 *High-Gradient Magnetic Fishing*

High-gradient magnetic fishing (HGMF) technology is based on protein adsorption to superparamagnetic (i.e., magnetically responsive particles which when the magnetic field is applied are drawn into) nonporous micron-sized supports without "magnetic memory" [78]. Because of a very small size and robust material used for manufacturing of these supports, these matrices allow very short adsorption times and are less prone to fouling than porous adsorbents when challenged with crude feedstocks [78–81]. Rapid collection of the supports in a strongly magnetized filter and subsequent easy desorption and recovery of the bound protein from the supports enable much higher processing speeds [82].

Heebøll-Nielsen et al. [79] developed superparamagnetic cation exchangers and applied HGMF for the separation of basic proteins from sweet bovine whey, obtaining LP purified 36-fold and concentrated 4.7-fold. In their further work, Heebøll-Nielsen et al. [81] described preparation of superparamagnetic anion exchangers and their use together with superparamagnetic cation exchangers in bovine whey fractionation. Crude bovine whey was treated with a superparamagnetic cation exchanger to adsorb basic protein species, and the supernatant arising from this treatment was subsequently applied on the anion exchanger. In the initial cation-exchange step, quantitative removal of LF and LP was achieved with some simultaneous binding of immunoglobulins (Ig). The immunoglobulins were separated from the other two proteins by desorbing with a low concentration of NaCl (≤0.4 M), whereas LF and LC were then coeluted in significantly purer form. The anion exchanger adsorbed BLG selectively allowing separation from the remaining proteins, but little resolution could be obtained during elution, though BLG of high purity was obtained.

2.4.3 *Selective Adsorption*

IEC based on selective adsorption has not been thoroughly used till recent dates. Ion exchangers are not selective matrices per se since they may adsorb a lot of proteins with similar pI values [83]. Improvement of selective adsorption onto an ion exchanger may be achieved by exploiting some specific properties of a target protein.

One of the properties that can be used for discrimination between proteins is their size. It is known that for protein adsorption on the most of the ion exchange resins, multipoint protein-support interactions have to be established. By using supports with low concentration of ionic groups per matrix surface (i.e., poorly activated matrices), it is possible to selectively adsorb the proteins that are large enough to establish several ionic interactions with matrix [84].

On this basis, Pessela et al. [85] developed a new protocol with only two steps for purification of IgGs, the largest proteins in dairy whey. They prepared monoamino-ethyl-N-aminoethyl (MANAE)-agarose supports, with different degrees of activation (these tailor-made supports may be currently purchased from Biotica SA, Villa Real, Spain). The ability of BSA to form complexes with other proteins, including IgG, prevented the success of the direct adsorption of IgG on mildly activated support. Because of that, chromatography on DEAE-agarose was introduced to separate BSA from IgG. Obtained IgG, contaminated only by very small proteins (ALA and BLG), was further purified by selective adsorption on low activated MANAE-agarose supports. Following this protocol, they obtained 80% recovery of purified IgGs.

However, there are no ionic-exchanger supports designed to selectively adsorb small proteins yet: large proteins would be always more strongly and more rapidly adsorbed than small ones. Bolivar et al. [86] designed a new anion-exchanger support for the selective adsorption of small proteins by activating an aminated support (prepared MANAE-Sepharose) with glutaraldehyde and further coating the support surface with BSA. In this support, "wells" are generated by two neighboring BSA molecules. On the bottom of those "wells," some of monoaminoethyl-N-aminoethyl groups are exposed out ready to bind small molecules that have a size small enough to be accommodated between two BSA molecules on the preexisting support (Fig. 2.8). Since the BSA surface was able to adsorb many proteins, thereby reducing the selectivity of the system, coating the immobilized BSA molecules with dextran reduced the adsorption of proteins on the BSA surface. This new matrix has been evaluated in the selective adsorption of small proteins from dairy whey, BLG and ALA, while other larger proteins remained in the supernatant. These two proteins are considered as major allergens of milk, and their elimination from any food by this method may be advantageous [87, 88].

2.4.4 *Displacement Chromatography*

Hydroxyapatite is a very useful ion exchange matrix having many advantages, such as low price, easy-to-make, nontoxic, with possibility to regenerate the material. Hydroxyapatite chromatography of whey is mostly based on ion displacement, where displacers, phosphate [5] or organic acids [52], form stronger complexes with calcium than carboxyl groups of the adsorbed proteins. Rossano et al. [5] presented

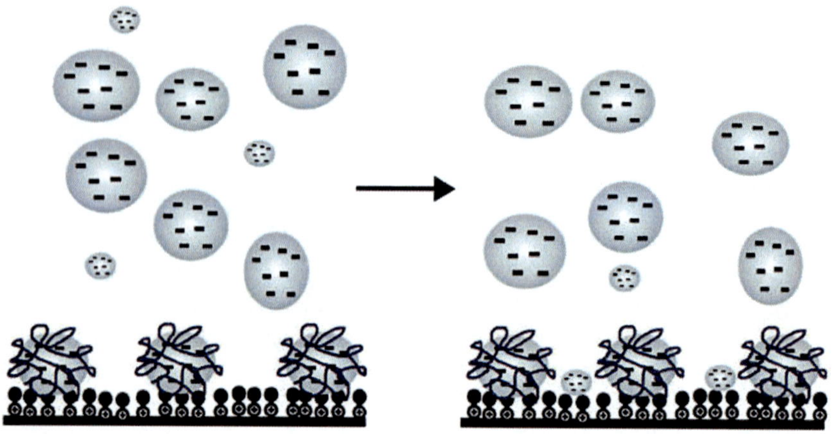

Fig. 2.8 Schematic representation of an amino-glutaraldehyde support coated with proteins modified with dextran: proteins can only interact with the support surface and only small proteins can do it (Reprinted from Ref. [86] with kind permission © Elsevier Ltd. (2010))

a method based on the use of homemade hydroxyapatite for one-step separation of lactin (unadsorbed) from bovine whey proteins (adsorbed). Total protein fraction can be eluted with 0.4 M phosphate at pH 7.0. About 56% of the proteins, primarily ALA and IgG, were eluted with 0.4 M phosphate at pH 5.0. The other major proteins, BLG and BSA, were eluted with 0.4 M phosphate at pH 6.0. Fractions were applied to a Superdex 75 column for final purification by gel filtration. The major advantages of this procedure are (a) one-step quantitative separation between lactin and proteins; (b) high flexibility in protein recovery, ranging from specific proteins to total protein fractions; and (c) purification of whey proteins in the native form.

Vogt and Freitag [52] compared the suitability of strong anion exchange (Bio-Scale Q2, Bio-Rad, Munich, Germany and Super Q, TosoHaas Europe, Stuttgart, Germany) and apatite (Bio-Rad, Munich, Germany) displacement chromatography for the processing of technical dairy whey. As a consequence of the pH sensitivity of the BLG, they found that hydroxyapatite displacement chromatography is less successful in whey separation than displacement on strong ion exchanger. At a denaturing carrier pH (>8.5), the BLG zone is broad and stretches over the entire displacement train. At a lower carrier pH, polyanionic displacers do not contribute to protein separation. However, under the same conditions, low-molecular-mass displacers could be used, though they tend to overrun and contaminate the protein zones. It appeared that organic acids separated BLG more efficiently from other whey proteins in anion-exchange displacement chromatography than by hydroxyapatite displacement chromatography [52].

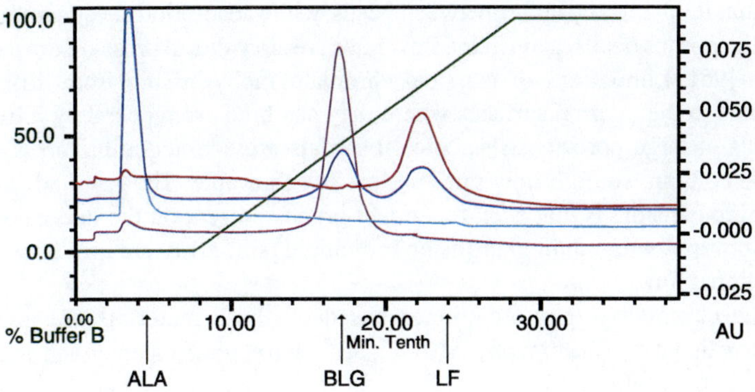

Fig. 2.9 Separation of ALA, BLG, and LF by CHT ceramic hydroxyapatite chromatography. Chromatographic profiles of purified LF (*red*), BLG (*purple*), and ALA (*turquoise*) and a mixture of the three proteins (*blue*). Elution buffer conductivity (*green*) (Reprinted from Ref. [91] with kind permission © Elsevier Ltd. (1998))

Schlatterer et al. [89] made use of the previously reported atypical elution behavior of BLG being displaced by fluoride ions from hydroxyapatite [90]. They performed the successful separation of BLG from other bovine whey proteins by hydroxyapatite chromatography (Macro-Prep Ceramic hydroxyapatite Bio-Rad, Munich, Germany) with a fluoride ion gradient in phosphate buffer as a displacement agent. BLG was completely eluted in one peak at a fluoride concentration of approximately 0.6 mol/l. The purity of BLG in this fraction was at least 96% with traces of IgG, BSA, and LF which were successfully removed by Superdex.

Ng and Yoshitake [91] described a mixed-mode ceramic hydroxyapatite chromatography method (CHT ceramic hydroxyapatite, Bio-Rad, Munich, Germany) for one-column fractionation of LF from whey. They succeeded in removing LP, which usually contaminates LF purified by other methods, due to similarities in molecular weight and isoelectric properties. In their procedure, LP was initially desorbed from the matrix under isocratic conditions. LF was obtained without LP activity and free from other major whey proteins such as ALA and BLG (Fig. 2.9). Their study showed that an 80 L hydroxyapatite column would generate 0.32 kg of LF per hour. This process has all the performance characteristics needed for effective scale-up in the biotech industry (simplicity, high throughput, and low cost).

2.4.5 High-Porosity Chromatographic Materials Providing High Flow Rate

Use of macroporous media is a new, emerging technology for extraction of proteins from feedstreams at high flow rates, which overcomes the need for multiple steps and associated high capital costs [92–94]. Classical bed-based column chromatography has two main objectives: (a) the molecules travel only by diffusion from the interstitial fluid, i.e., the buffer between beads, to the adsorption sites mainly located inside the porous beads, and (b) the flow rates are slow due to pressure drop over the column [95]. Limitations of classical chromatography arising from diffusion of molecules to the internal surfaces where they can bind are reduced by introducing materials of large porosity. Also, such materials provide negligible pressure drop over the column, so high flow rates are easily achievable. The main advantage of macroporous media is that even functional groups present on the internal surfaces adsorb proteins with minimal diffusion limitations, since they are located within the convectional flow.

Three chromatography techniques based on these materials are membrane convective liquid chromatography (MCLC), perfusion chromatography, and continuous bed chromatography.

Membrane convective liquid chromatography (MCLC) is a technique where stacked porous cellulose membranes of 1.2-um pore size are used for isolation of proteins [96]. Perfusion chromatography utilizes packed columns, in which particles have a bidisperse porous structure composed of 600–800-nm through pores transecting the particles and 50–150-nm diffusive pores that line the through pores [97]. Continuous bed technology is based on the polymerization of advanced monomers and ionomers directly in the chromatographic column [98]. The polymer chains aggregate into a dense network of nodules consisting of microparticles with an average diameter of 200 nm with the channels between the nodules large enough (7–15-um diameter) to permit a high convective flow.

Girardet et al. [99] performed rapid separation of whey proteins by three chromatography methods based on large-porosity materials: MCLC (DEAE MemSep cartridge, Millipore Co., Bedford, MA, USA), perfusion chromatography (Poros 20HQ, PerSeptive Biosystems, Framingham, MA, USA), and continuous bed chromatography (UNO Q-I, Bio-Rad Laboratories, Hercules, CA, USA). Bovine whey proteins were separated in less than 10 min with high resolution (Fig. 2.10). Due to the particular technology based on stacked membranes, MCLC seemed to be easier to handle than perfusion chromatography or continuous bed chromatography, especially for scale-up in preparative fractionations. These rapid methods are suitable in the quality control of whey and could be applied in dairy industry or in research. The behavior of membrane adsorbers may differ from conventional chromatography resins even if the same ion exchange ligand is used.

Imafidion and Ng-Kwai-Hang [100] used QAE-cellulosic anion-exchange cartridge (QAE ZetaPrep 250, LKB-Produkter AB, S-16126 Bromma, Sweden) for purification of BLG from acid whey, after precipitation with trichloroacetic acid (TCA) to remove other whey proteins [101]. The prepacked module offers a fast, simple, and efficient mass ion exchange chromatography where the liquid flows radially through the system to provide high flow rates. In one chromatographic step, at a flow rate of 30 ml/min, they obtained 4–7 g electrophoretically pure BLG from 1.8 L whey.

Cryogel monoliths with a pore diameter of 100 μm, modified to immobilize ion exchange groups, were proved to be a useful chromatographic macroporous media with low pressure drops at moderate flow velocities (300–525 cm h^{-1}) for extraction

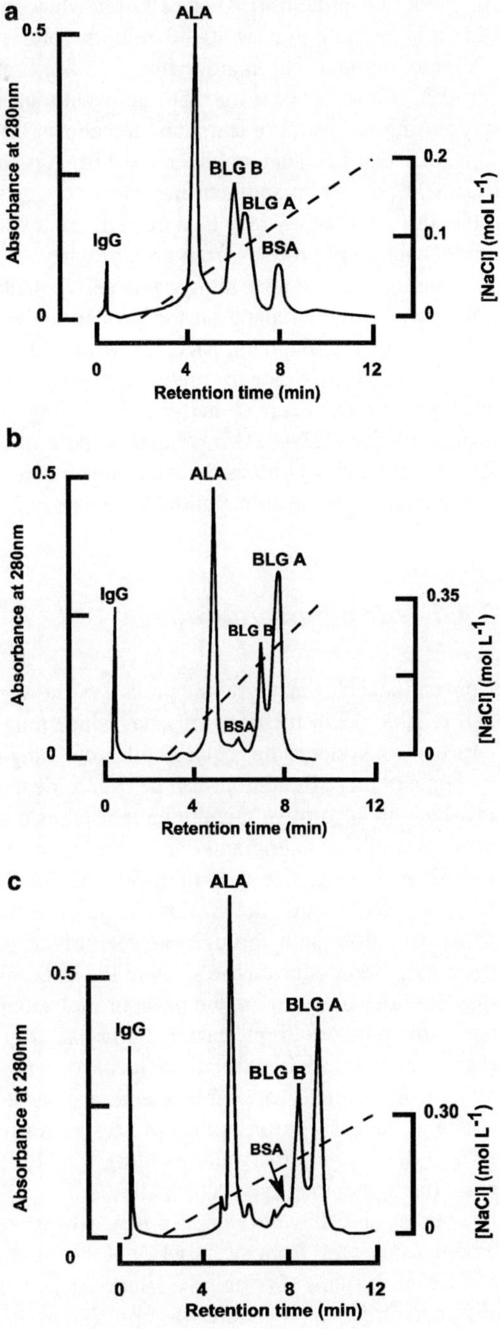

Fig. 2.10 Separation of the major whey proteins, IgGs, ALA, BLG-B, and BLG-A, respectively, and BSA by (**a**) membrane convective liquid chromatography (MCLC), (**b**) ion exchange perfusion liquid chromatography, and (**c**) continuous bed chromatography (Reprinted from Ref. [99] with kind permission © EDP Sciences (1998))

of minor milk proteins (LF and LP) from whey. Billakanti and Fee [102] demonstrated that it is possible to pass 40–50 column volumes of acid whey through a prepared cationic cryogel chromatographic column (also commercially available from Protista, Lund, Sweden) at 550 cm/h without exceeding its pressure limits if the processing temperature is maintained above 35°C. The cryogel column selectively binds LF and LP. They obtained LF with a yield of over 85% and a purity of more than 90%. Ease of manufacturing, low cost, and versatile surface chemistry of cryogels suggest that they may be a good alternative to packed-bed chromatography for direct capture of proteins from milk or whey.

Among the modern highly porous Q packings commercially available for preparative protein separation, the most attractive seems to be Poros Q/M (PerSeptive Biosystems, Framingham, MA, USA) and Q Hyper D/F (Biosepra, Marlborough, MA, USA) [103]. Since for preparative separation, the most important parameter of packing is a small decrease in dynamic binding capacity when the flow rate increases, hyperdiffusive Q Hyper D/F is superior to Poros Q/M. By using Q Hyper D/F matrix, Couriol et al. [104] obtained nearly pure fractions of the five acid whey proteins on the preparative scale at interstitial velocity of 127 cm h^{-1} (flow rate of 2 ml min^{-1}).

2.4.6 Membrane Adsorption

Conventional chromatographic processes show tremendous disadvantages for a fast and reliable downstreaming of whey, since long cycle times and complicated process control systems are required for processing of large volumes.

Some of these limitations can be overcome by using membrane chromatography, in which an adsorptive membrane is used as a stationary phase that combines the principles of chromatography and membrane separation in one device. Chromatographic membranes are modified with specific ligands or functional groups to bind target molecules and allow fast binding, since the adsorptive ligands are within the convective flow path through the membrane pores. Membrane chromatography systems generally have pore sizes in the microfiltration range (0.02–10 mm) since they are not separating on the basis of molecular size, so they can allow high flow rates, low pressure drops, easy column packing and scale-up, as well as low clogging tendency, which makes them ideal for large-scale separation of proteins [105, 106]. Commercially available membrane systems for laboratory and industrial scales on the basis of ion exchange are systems with strongly acidic (sulfonic acid), strongly basic (quaternary ammonium), weakly acidic (carboxylic acid), and weakly basic (diethyl amine) types [107, 108].

Plate et al. [109] developed a new downstream procedure for the isolation of bovine LF and LP from sweet whey at the laboratory scale, using cation-exchange membrane systems (Sartobind S, Sartorius, Goettingen, Germany). This fast, reliable, and effective procedure was upscaled to an industrially relevant scale module for the purification of sweet whey concentrate with a recovery yield for LF of more than 90% (7 g LF per cycle for 2m2 modules).

Goodall et al. [110] used weak (diethylamine) and strong anion-exchange membranes (Sartobind MA, Sartorious, Goettingen, Germany) to separate BLG, BSA, and ALA from sweet whey. When the membranes were saturated with whey, they had higher selectivity for BLG than for other major whey proteins (less than 1% of the eluate consisting of ALA or BSA). Once a certain protein loading is achieved, BLG begins to displace ALA and BSA from the membrane. These ion exchange membrane separations are viable processing method with a high selectivity producing a relatively pure BLG elution fraction and the possibility of producing a permeate fraction enriched in ALA and BSA. The next consideration would be to use these membranes in a cascade process. Two-step separation could be used to firstly remove BLG using an anion-exchange membrane, enabling recovery of ALA and BSA from the permeate by passing it through a second anion-exchange membrane [111]. Manipulation of loading volume, pH, and elution gradient can produce individual protein streams. The current limitation of the process, particularly when trying to bind all three of the major whey proteins, is the low binding capacity of commercially available membranes. Future studies are necessary to evaluate the effects of repeated cycling.

As single-stage chromatography is seldom sufficient for preparative protein separation, controllable mixed-mode phases, where several interaction types are exploited in series, may yield columns with unique separation features [112]. However, with conventional HPLC columns, mixed-mode packings are difficult to be produced in a homogeneous and reproducible manner. Membrane adsorber (MA) chromatography utilizes stacks of functionalized membranes as stationary phase and different types of membranes can be arranged in the desired pattern. Comparing to single-mode MA chromatography, a broader range of proteins, e.g., in terms of the isoelectric points, can be separated in a single chromatographic procedure on a mixed-mode system due to the protein interactions with one or both types of MA.

Freitag et al. [113] used alternating layers of cation- and anion-exchange membrane adsorber sheets in one module system for the fractionation of major whey protein components using commercial membrane adsorbers from Sartorius AG. This controlled mixed-mode interaction membrane chromatography system was able to capture major whey proteins using a single chromatographic run at pH 6.0. Prior to elution, however, the cation and anion membranes had to be disconnected and eluted separately in a salt gradient to prevent immunoglobulin from coeluting with ALA. While the flow rate shows no influence, both the pattern of arrangement and the ratio of the two types of ion exchanger influence on the separation (Fig. 2.11).

However, the production of the most chromatographic membranes involves chemical modification of the base membrane in order to attach a suitable functional group for specific types of adsorption. Sometimes harsh chemical and physical conditions must be used, which can cause undesirable and irreversible changes in the membrane structure [114]. Ion exchange mixed-matrix membranes (MMM) are prepared by incorporating an ion exchange resin into a membrane polymer solution at a specified ion exchanger resin loading prior to membrane casting [38, 39]. The polymer/ion exchanger suspension can then be cast as a flat sheet membrane or spun into a hollow fiber membrane. The concept is quite simple, and it is easy to

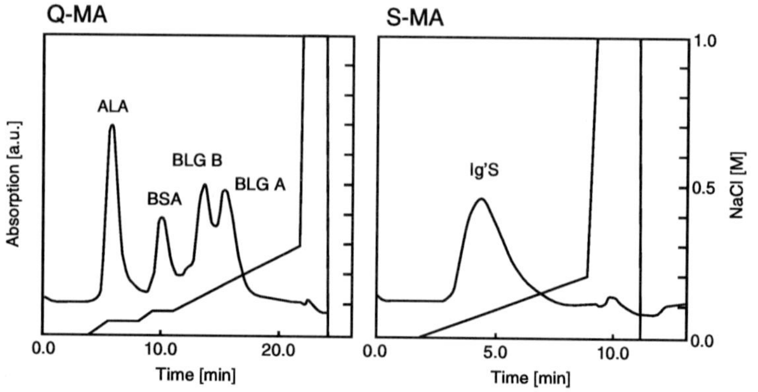

Fig. 2.11 Separation of Ig, BSA, ALA, and the genetic variants of BLG by MA chromatography on strong anion- and cation-exchange MAs. Buffer A: 20 mM piperazine, pH 6.0. Buffer B: 20 mM piperazine, pH 6.0, plus 1.0 M NaCl. Sample: 8 ml (BSA, 0.125 mg/ml; IgG, 0.05 mg/ml; ALA, 1 mg/ml; BLG, 1 mg/ml). Module: 2 Q15+1 S15 in series (Reprinted from Ref. [113] with kind permission © Elsevier Ltd. (1996))

prepare such while still giving a performance that is competitive with current chromatographic membranes. Saufi and Fee [7] described the application of MMM chromatography for fractionation of BLG from bovine whey using ethylene vinyl alcohol polymer and Lewatit anion-exchange resin (MonoPlusTM MP500) to form a flat sheet membrane. In batch fractionation of whey, the MMM showed selective binding toward BLG comparing to other proteins, and the MMM had almost the same binding capacity for whey BLG as it did for pure BLG (150 mg BLG/g membrane), which is promising for the selective capture of BLG from bovine whey.

2.5 Concluding Comments

Being a mixture of acidic and basic proteins, whey is not suitable material for fractionation by a single ion exchange chromatography step. If the main goal is to obtain all whey proteins purified, then it is recommended to use a continuous double-step ion exchange chromatography involving both cation and anion exchangers.

Acknowledgments The authors acknowledge the support of the Ministry of Science and Technological Development of the Republic of Serbia (Grant No. 172024).

References

1. Smithers GW, John Ballard F, Copeland AD et al (1996) New opportunities from the isolation and utilization of whey proteins. J Dairy Sci 79(8):1454–1459
2. Gerberding SJ, Byers CH (1998) Preparative ion-exchange chromatography of proteins from dairy whey. J Chromatogr A 808(1–2):141–151

3. de Wit JN (1998) Nutritional and functional characteristics of whey proteins in food products. J Dairy Sci 81(3):597–608
4. Hutchens TW, Magnusson JS, Yip T-T (1990) Secretory IgA, IgG, and IgM immunoglobulins isolated simultaneously from colostral whey by selective thiophilic adsorption. J Immunol Methods 128(1):89–99
5. Rossano R, D'Elia A, Riccio P (2001) One-step separation from lactose: recovery and purification of major cheese-whey proteins by hydroxyapatite – a flexible procedure suitable for small- and medium-scale preparations. Protein Expres Purif 21(1):165–169
6. Hobman PG (1984) Review of processes and products for utilization of lactose in deproteinated milk serum. J Dairy Sci 67(11):2630–2653
7. Sienkewicz T, Riedel CL (1990) Determination of the bulk density. In: Whey and whey utilization, 2nd edn. Verlag Th. Mann, Gelsenkirchen-Buer
8. Yaguchi M, Rose D (1971) Chromatographic separation of milk proteins: a review. J Dairy Sci 54(12):1725–1743
9. Strange ED, Malin EL, Van Hekken DL et al (1992) Chromatographic and electrophoretic methods used for analysis of milk proteins. J Chromatogr A 624(1–2):81–102
10. Monaci L, Tregoat V, van Hengel A et al (2006) Milk allergens, their characteristics and their detection in food: a review. Eur Food Res Technol 223(2):149–179
11. Saufi SA, Fee CJ (2009) Fractionation of beta-lactoglobulin from whey by mixed matrix membrane ion-exchange chromatography. Biotechnol Bioeng 103(1):138–147
12. El-Agamy EI (2007) The challenge of cow milk protein allergy. Small Ruminant Res 68(1):64–72
13. Suutari TJ, Valkonen KH, Karttunen TJ et al (2006) IgE cross reactivity between reindeer and bovine milk beta-lactoglobulins in cow's milk allergic patients. J Invest Allergol Clin Immunol 16(5):296–302
14. Pearce RJ, Marshall SC (1991) New ways with whey components. Aust J Dairy Technol 42(2):105–107
15. Jost R, Maire JC, Maynard F et al (1999) Aspects of whey protein usage in infant nutrition, a brief review. Int J Food Sci Technol 34(5–6):533–542
16. Muller A, Chaufer B, Merin U et al (2003) Purification of alpha-lactalbumin from a prepurified acid whey: ultrafiltration or precipitation. Lait 83(6):439–451
17. Maubois JL (1997) Current uses and future perspectives of microfiltration technology in the dairy industry. Int Dairy Fed Bull 320:37–40
18. Ganjam LS, Thornton WH Jr, Marshall RT et al (1997) Antiproliferative effects of yogurt fractions obtained by membrane dialysis on cultured mammalian intestinal cells. J Dairy Sci 80(10):2325–2329
19. Zydney AL (1998) Protein separations using membrane filtration: new opportunities for whey fractionation. Int Dairy J 8(3):243–250
20. Wong DWS, Camirand WM, Pavlath AE et al (1996) Structures and functionalities of milk proteins. Crit Rev Food Sci 36(8):807–844
21. Arnold D, Di Biase AM, Marchetti M et al (2002) Antiadenovirus activity of milk proteins: lactoferrin prevents viral infection. Antivir Res 53(2):153–158
22. Baveye S, Elass E, Mazurier J et al (2005) Lactoferrin: a multifunctional glycoprotein involved in the modulation of the inflammatory process. Clin Chem Lab Med 37(3):281–286
23. Ellison RT, Giehl TJ (1991) Killing of gram-negative bacteria by lactoferrin and lysozyme. J Clin Invest 88(4):1080–1091
24. Arnold RR, Russell JE, Champion WJ et al (1982) Bactericidal activity of human lactoferrin: differentiation from the stasis of iron deprivation. Infect Immun 35(3):792–799
25. Ellison RT, Giehl TJ, LaForce FM (1988) Damage of the outer membrane of enteric gram-negative bacteria by lactoferrin and transferrin. Infect Immun 56(11):2774–2781
26. Chiu CK, Etzel MR (1997) Fractionation of lactoperoxidase and lactoferrin from bovine whey using a cation-exchange membrane. J Food Sci 62(5):996–1000
27. van Calcar SC, MacLeod EL, Gleason ST et al (2009) Improved nutritional management of phenylketonuria by using a diet containing glycomacropeptide compared with amino acids. Am J Clin Nutr 89(4):1068–1077

28. Francis GL, Regester GO, Webb HA et al (1995) Extraction from cheese whey by cation-exchange chromatography of factors that stimulate the growth of mammalian cells. J Dairy Sci 78(6):1209–1218
29. Cox DA, Burk RR (1991) Isolation and characterisation of milk growth factor, a transforming-growth-factor-β2-related polypeptide, from bovine milk. Eur J Biochem 197(2):353–358
30. Shing Y, Klagsbrun M (1987) Purification and characterization of a bovine colostrum-derived growth factor. Mol Endocrinol 5:335–338
31. Sandowski Y, Peri I, Gertler A (1993) Partial purification and characterization of putative paracrine/autocrine bovine mammary epithelium growth factors. Livest Prod Sci 35(1–2):35–48
32. Caessens PWJR, Visser S, Gruppen H (1997) Method for the isolation of bovine beta-lactoglobulin from a cheese whey protein fraction and physicochemical characterization of the purified product. Int Dairy J 7(4):229–235
33. Konrad G, Lieske B, Faber W (2000) A large-scale isolation of native beta-lactoglobulin: characterization of physicochemical properties and comparison with other methods. Int Dairy J 10(10):713–721
34. Alomirah HF, Alli I (2004) Separation and characterization of beta-lactoglobulin and alpha-lactalbumin from whey and whey protein preparations. Int Dairy J 14(5):411–419
35. Lozano JM, Giraldo GI, Romero CM (2008) An improved method for isolation of beta-lactoglobulin. Int Dairy J 18(1):55–63
36. Neyestani TR, Djalali M, Pezeshki M (2003) Isolation of alpha-lactalbumin, beta-lactoglobulin, and bovine serum albumin from cow's milk using gel filtration and anion-exchange chromatography including evaluation of their antigenicity. Protein Expres Purif 29(2): 202–208
37. Ye XY, Yoshida S, Ng TB (2000) Isolation of lactoperoxidase, lactoferrin, alpha-lactalbumin, beta-lactoglobulin b and beta-lactoglobulin a from bovine rennet whey using ion-exchange chromatography. Int J Biochem Cell B 32(11–12):1143–1150
38. Avramescu M-E, Borneman Z, Wessling M (2003) Mixed-matrix membrane adsorbers for protein separation. J Chromatogr A 1006(1–2):171–183
39. Avramescu M-E, Gironès M, Borneman Z et al (2003) Preparation of mixed matrix adsorber membranes for protein recovery. J Membr Sci 218(1–2):219–233
40. Yang H, Viera C, Fischer J et al (2002) Purification of a large protein using ion-exchange membranes. Ind Eng Chem Res 41(6):1597–1602
41. Felipe X, Law AJR (1997) Preparative-scale fractionation of bovine, caprine and ovine whey proteins by gel permeation chromatography. J Dairy Res 64(3):459–464
42. Vyas HK, Izco JM, Jiménez-Flores R (2002) Scale-up of native [beta]-lactoglobulin affinity separation process. J Dairy Sci 85(7):1639–1645
43. Akpinar-Bayizit A, Ozcan T, Yilmaz-Ersan L (2009) Membrane processes in production of functional whey components. Mljekarstvo 59(4):282–288
44. Fee CJ, Chand A (2006) Capture of lactoferrin and lactoperoxidase from raw whole milk by cation-exchange chromatography. Sep Purif Technol 48(2):143–149
45. Savina IN, Galaev IY, Mattiasson B (2006) Ion-exchange macroporous hydrophilic gel monolith with grafted polymer brushes. J Mol Recognit 19(4):313–321
46. Janson JC, Ryden L (1998) Protein purification. Wiley, New York
47. Etzel MR (2004) Manufacture and use of dairy protein fractions. J Nutr 134(4):996–1002
48. El-Sayed MMH, Chase HA (2010) Purification of the two major proteins from whey concentrate using a cation-exchange selective adsorption process. Biotechnol Prog 26(1):192–199
49. Hahn R, Schulz PM, Schaupp C et al (1998) Bovine whey fractionation based on cation-exchange chromatography. J Chromatogr A 795(2):277–287
50. Doultani S, Turhan KN, Etzel MR (2004) Fractionation of proteins from whey using cation-exchange chromatography. Process Biochem 39(11):1737–1743
51. El-Sayed MMH, Chase HA (2008) Separation of the two major whey proteins using cation-exchange adsorption. In: The world congress on engineering and computer science, San Francisco

52. Vogt S, Freitag R (1997) Comparison of anion-exchange and hydroxyapatite displacement chromatography for the isolation of whey proteins. J Chromatogr A 760(1):125–137
53. Fweja LWT, Lewis MJ, Grandison AS (2010) Isolation of lactoperoxidase using different cation-exchange resins by batch and column procedures. J Dairy Res 77(3):357–367
54. Kalan EB, Greenberg R, Walter M et al (1964) Chemical properties of beta-lactoglobulins a, b and c. Biochem Biophys Res Commun 16(3):199–203
55. Kiddy CA, Townend RE, Thatcher WW et al (1965) Beta-lactoglobulin variation in milk from individual cows. J Dairy Res 32(02):209–217
56. de Jongh HHJ, Gröneveld T, de Groot J (2001) Mild isolation procedure discloses new protein structural properties of beta-lactoglobulin. J Dairy Sci 84(3):562–571
57. Manji B, Hill A, Kakuda Y et al (1985) Rapid separation of milk whey proteins by anion-exchange chromatography. J Dairy Sci 68(12):3176–3179
58. Turhan KN, Etzel MR (2004) Whey protein isolate and alpha-lactalbumin recovery from lactic acid whey using cation-exchange chromatography. J Food Sci 69(2):66–70
59. Gordon WG, Groves ML, Basch JJ (1963) Bovine milk "red protein": amino acid composition and comparison with blood transferrin. Biochemistry 2(4):817–820
60. Szuchet-Derechin S, Johnson P (1966) The "Albumin" fraction of bovine milk-III. The micro-heterogeneity of the red protein. Eur Polym J 2(1):29–32, 33–35
61. Yoshida S, Ye X (1991) Isolation of lactoperoxidase and lactoferrins from bovine milk acid whey by carboxymethyl cation-exchange chromatography. J Dairy Sci 74(5):1439–1444
62. Oram JD, Reiter B (1968) Inhibition of bacteria by lactoferrin and other iron-chelating agents. Biochim Biophys Acta Gen Subj 170(2):351–365
63. Gordon VG, Groves ML, Basch JS (1963) Isolation of an iron-binding protein from cow's milk. Biochim Biophys Acta 60:410
64. Carlstrom A (1969) Lactoperoxidase. Identification of multiple forms and their interrelationships. Acta Chem Scand 23:171–175
65. Groves ML (1965) Preparation of some iron-binding proteins and [alpha]-lactalbumin from bovine milk. Biochim Biophys Acta, Gen Subj 100(1):154–162
66. Rombauts WA, Schroeder WA, Morrison M (1967) Bovine lactoperoxidase. Partial characterization of the further purified protein. Biochemistry 6(10):2965–2977
67. Allen PZ, Morrison M (1963) Lactoperoxidase. IV. Immunological analysis of bovine lactoperoxidase preparations obtained by a simplified fractionation procedure. Arch Biochem Biophys 102(1):106–113
68. Morrison M, Hultquist DE (1963) Lactoperoxidase. II. Isolation. J Biol Chem 238:2847–2860
69. Kickhofen B, Hammer DK, Scheel S (1968) Isolation and characteristics of gamma-g type immunoglobulins from bovine serum and colostrum. H-S Z Physiol Chem 349:1175–1185
70. Murphy FA, Osebold JW, Aalund O (1965) Physical heterogenity of bovine gamma-globulins: characterization of gamma-M and gamma-G globulins. Arch Biochem Biophys 112:126–133
71. Groves ML, Gordon WG (1967) Isolation of a new glycoprotein-A and a gamma-globulin from individual cow milks. Biochemistry 6(8):2388–2394
72. Maki Z, Kanamori M (1969) Studies on glycopeptides from a bovine immune lactoglobulin of colostrum. Eiyo To Shokuryo 22:42
73. Kunz C, Lonnerdal B (1989) Human milk proteins: separation of whey proteins and their analysis by polyacrylamide gel electrophoresis, fast protein liquid chromatography (FPLC) gel filtration, and anion-exchange chromatography. Am J Clin Nutr 49(3):464–470
74. Liao AW, Rassi Z, LeMaster DM et al (1987) High performance displacement chromatography of proteins: separation beta-lactoglobulins A and B. Chromatographia 24:881–885
75. Geisser A, Hendrich T, Boehm G et al (2005) Separation of lactose from human milk oligosaccharides with simulated moving bed chromatography. J Chromatogr A 1092(1):17–23
76. Broughton DB, Gerhold CG (1961) Continuous sorption process employing fixed bed of sorbent and moving inlets and outlets. US Patent 2985589
77. Andersson J, Mattiasson B (2006) Simulated moving bed technology with a simplified approach for protein purification: separation of lactoperoxidase and lactoferrin from whey protein concentrate. J Chromatogr A 1107(1–2):88–95

78. Anspach FB, Johnston A, Wirth HJ et al (1989) High-performance liquid chromatography of amino acids, peptides and proteins XCII thermodynamic and kinetic investigations on rigid and soft affinity gels with varying particle and pore sizes. J Chromatogr A 476:205–225
79. Heebøll-Nielsen A, Justesen SFL, Hobley TJ et al (2004) Superparamagnetic cation-exchange adsorbents for bioproduct recovery from crude process liquors by high-gradient magnetic fishing. Sep Sci Technol 39(12):2891–2914
80. Heebøll-Nielsen A, Choewsmiddelberg APJ, Thomas ORT (2003) Efficient inclusion body processing using chemical extraction and high-gradient magnetic fishing. Biotechnol Prog 19(3):887–898
81. Heeboll-Nielsen A, Justesen SFL, Thomas ORT (2004) Fractionation of whey proteins with high-capacity superparamagnetic ion-exchangers. J Biotechnol 113(1–3):247–262
82. Hubbuch JJ, Matthiesen DB, Hobley TJ et al (2001) High-gradient magnetic separation versus expanded bed adsorption: a first principle comparison. Bioseparation 10:99–112
83. Tsonev LI, Hirsh AG (2008) Theory and applications of a novel ion-exchange chromatographic technology using controlled ph gradients for separating proteins on anionic and cationic stationary phases. J Chromatogr A 1200(2):166–182
84. Pessela BCC, Munilla R, Betancor L et al (2004) Ion-exchange using poorly activated supports, an easy way for purification of large proteins. J Chromatogr A 1034(1–2):155–159
85. Pessela BCC, Torres R, Batalla P et al (2006) Simple purification of immunoglobulins from whey proteins concentrate. Biotechnol Prog 22(2):590–594
86. Bolivar JM, Batalla P, Mateo C et al (2010) Selective adsorption of small proteins on large-pore anion-exchangers coated with medium size proteins. Colloid Surf B Biointerfaces 78(1):140–145
87. Blanc F, Bernard H, Alessandri S et al (2008) Update on optimized purification and characterization of natural milk allergens. Mol Nutr Food Res 52(S2):166–175
88. Li X, Luo ZL, Chen HB et al (2008) Isolation and antigenicity evaluation of beta-lactoglobulin from buffalo milk. Afr J Biotechnol 7(13):2258–2264
89. Schlatterer B, Baeker R, Schlatterer K (2004) Improved purification of beta-lactoglobulin from acid whey by means of ceramic hydroxyapatite chromatography with sodium fluoride as a displacer. J Chromatogr B 807(2):223–228
90. Gorbunoff MJ (1984) The interaction of proteins with hydroxyapatite. I. Role of protein charge and structure. Anal Biochem 136(2):425–432
91. Ng PK, Yoshitake T (2010) Purification of lactoferrin using hydroxyapatite. J Chromatogr B 878(13–14):976–980
92. Dainiak MB, Kumar A, Plieva FM et al (2004) Integrated isolation of antibody fragments from microbial cell culture fluids using supermacroporous cryogels. J Chromatogr A 1045(1–2):93–98
93. Jungbauer A, Hahn R (2004) Monoliths for fast bioseparation and bioconversion and their applications in biotechnology. J Sep Sci 27(10–11):767–778
94. Lozinsky V, Plieva F, Galaev I et al (2001) The potential of polymeric cryogels in bioseparation. Bioseparation 10(4):163–188
95. Andrews AT, Taylor MD, Owen AJ (1985) Rapid analysis of bovine milk proteins by fast protein liquid chromatography. J Chromatogr A 348:177–185
96. Gerstner JA, Hamilton R, Cramer SM (1992) Membrane chromatographic systems for high-throughput protein separations. J Chromatogr A 596(2):173–180
97. Afeyan NB, Gordon NF, Mazsaroff I et al (1990) Flow-through particles for the high-performance liquid chromatographic separation of biomolecules: perfusion chromatography. J Chromatogr A 519(1):1–29
98. Hjerten S, Liao JL, Zhang R (1989) High-performance liquid chromatography on continuous polymer beds. J Chromatogr A 473:273–275
99. Girardet JM, Saulnier F, Linden G et al (1998) Rapid separation of bovine whey proteins by membrane convective liquid chromatography, perfusion chromatography, continuous bed chromatography, and capillary electrophoresis. Lait 78(4):391–400

100. Imafidon GI, Ng-Kwai-Hang K-F (1992) Isolation and purification of beta-lactoglobulin by mass ion-exchange chromatography. J Dairy Res 59(01):101–104
101. Fox KK, Holsinger VH, Posati LP et al (1967) Separation of beta-lactoglobulin from other milk serum proteins by trichloroacetic acid. J Dairy Sci 50(9):1363–1367
102. Billakanti JM, Fee CJ (2009) Characterization of cryogel monoliths for extraction of minor proteins from milk by cation-exchange. Biotechnol Bioeng 103(6):1155–1163
103. Rodrigues AE, Loureiro JM, Chenou C et al (1995) Bioseparations with permeable particles. J Chromatogr B Biomed Sci Appl 664(1):233–240
104. Couriol C, Le Quellec S, Guihard L et al (2000) Separation of acid whey proteins on the preparative scale by hyperdiffusive anion-exchange chromatography. Chromatographia 52(7–8):465–472
105. Kawai T, Saito K, Lee W (2003) Protein binding to polymer brush, based on ion-exchange, hydrophobic, and affinity interactions. J Chromatogr B 790(1–2):131–142
106. Zou H, Luo Q, Zhou D (2001) Affinity membrane chromatography for the analysis and purification of proteins. J Biochem Biophys Methods 49(1–3):199–240
107. Gebauer KH, Thommes J, Kula MR (1997) Plasma protein fractionation with advanced membrane adsorbents. Biotechnol Bioeng 54(2):181–189
108. Ulber R, Plate K, Weiss T et al (2001) Downstream processing of bovine lactoferrin from sweet whey. Acta Biotechnol 21(1):27–34
109. Plate K, Beutel S, Buchholz H et al (2006) Isolation of bovine lactoferrin, lactoperoxidase and enzymatically prepared lactoferricin from proteolytic digestion of bovine lactoferrin using adsorptive membrane chromatography. J Chromatogr A 1117(1):81–86
110. Goodall S, Grandison AS, Jauregi PJ et al (2008) Selective separation of the major whey proteins using ion-exchange membranes. J Dairy Sci 91(1):1–10
111. Fuda E, Jauregi P (2006) An insight into the mechanism of protein separation by colloidal gas aphrons (CGA) generated from ionic surfactants. J Chromatogr B 843(2):317–326
112. El Rassi Z, Horvath C (1986) Tandem columns and mixed-bed columns in high-performance liquid chromatography of proteins. J Chromatogr 359:255–264
113. Freitag R, Splitt H, Reif OW (1996) Controlled mixed-mode interaction chromatography on membrane adsorbers. J Chromatogr A 728(1–2):129–137
114. Liu C, Bai R (2006) Preparing highly porous chitosan/cellulose acetate blend hollow fibers as adsorptive membranes: effect of polymer concentrations and coagulant compositions. J Membr Sci 279(1–2):336–346

Chapter 3
Application of Ion Exchangers in Speciation and Fractionation of Elements in Food and Beverages

Pawel Pohl, Helena Stecka, and Piotr Jamroz

Abstract A relationship between different species of elements in dietary products and their bioavailability from that source is very important to human safety and wholesomeness. Analytical methods outlined in this report are based on solid phase extraction with polymeric ion exchange resins and other nonionic sorbents can be applied for fractionation analysis of elements in a number of food products and beverages. Their attractiveness mostly lies in simplicity of solid phase extraction operation, versatility, and diversity of available sorbents and possibility of insight into chemical nature of element forms due to different sorption mechanisms served by these sorbents. This article surveys two main means of partitioning of elements by solid phase extraction, including procedures with single sorbents or two coupled sorbents. Attention is also drawn to operational character of fractions of species distinguished and evaluation of possible groups of species of elements. Sample pretreatment and quality assurance and control of results of fractionation analysis achieved using the mentioned solid phase extraction schemes are highlighted.

3.1 Introduction

Elements in food and beverages of natural origin can be present in various physico-chemical forms that differ from each other by chemical properties and, in consequence, expose diverse biological activity, availability, and absorbability, in addition to dissimilar positive or negative effects on the human body [1]. Apart from simple ions, metals, for example, can be bound by different groups of compounds and form

P. Pohl (✉) • H. Stecka • P. Jamroz
Division of Analytical Chemistry, Wroclaw University of Technology,
Wybrzeze Stanislawa Wyspianskiego 27, Wroclaw 50370, Poland
e-mail: pawel.pohl@pwr.wroc.pl

Inamuddin and M. Luqman (eds.), *Ion Exchange Technology II: Applications*,
DOI 10.1007/978-94-007-4026-6_3, © Springer Science+Business Media B.V. 2012

complexes of different stability [2–7]. One of more common group of metal ligands in natural dietary products is inorganic species, namely, carbonates, chlorides, sulfates, fluorides, and phosphates. Another group of species that readily complex ions of metals are low molecular weight (LMW) organic compounds, particularly amino acids and carbohydroxy acids of primary origin or being products of processes involved in food treatment and production, for instance, alcoholic fermentation. Finally, the last group of natural bioligands of metals in food and beverages are moderate and high molecular weight (MMW and HMW) organic compounds, that is, peptides, polypeptides, proteins, polysaccharides, polyphenols, and flavonoids.

It is well acknowledged that distinct physicochemical forms of elements can be available and absorbable to the human body to a different extent. This is the reason why the actual effect of elements ingested with food on human safety and well-being can be quite different from that anticipated on the basis of their total contents. Unfortunately, apparently from the literature devoted to chemical analysis of food products by means of atomic spectrometry methods, vast number of food analyses made is related to determinations of total concentrations of elements only. Respective measurements are typically carried out after dry or wet digestions of analyzed samples in order to eliminate a detrimental effect of organic matrices of these samples [8]. Rarely, food and beverage samples are directly analyzed without any special preparation or handling [8]. As a rule, information about total concentrations of elements retrieved in the course of such kind of analyses is considered for evaluating food quality and safety according to apparent health or sanitary recommendations [9, 10]. What is more, daily recommended or allowable intakes of food and beverages are usually determined on the basis of total concentrations of macro- and microelements measured in these products [11].

According to results of gastrointestinal digestion experiments, this practice mentioned above cannot be valid anymore because availability and absorbability of elements from food and beverages are crucially dependent on ultimate forms in which they exist in dietary products. Correspondingly, metals, for example, can be present in the form of hydrated free cations, hydroxyl complexes, and other positively charged complexes with different inorganic and organic ligands of different stability in addition to their labile species. All these species are acknowledged to be the most available and absorbable forms of elements for the human body due to an ease with which they can migrate and permeate through cell membranes [12–15]. Ions of metals bound to such LMW organic ligands as carboxylic acids (acetic, citric, tartaric, oxalic, malic, maleic, succinic, fumaric, ketoglutaric) are also well absorbable [4]. Ion complexes of metals with polyfunctional HMW ligands, including phenolic acids (phthalic, gallic, tannic, salicylic, vanillic, shikimic) or polyphenolic compounds (tannins) or flavonoids (flavones, isoflavones, flavanols, anthocyanidins), are very stable species, but they are established to significantly reduce their availability and absorbability from food and beverages [2–5]. Similar behavior is observed in case of complexes of metals with polysaccharides; these forms are also less absorbable by the human body [6].

3.2 Fractionation Versus Speciation Analysis

Although analysis of food and beverages products carried out in overwhelming food chemistry and technology laboratories are focused on total concentrations of different elements, this provides rather incomplete and improper information about nourishment quality and safety of analyzed samples as well as a real risk related to the presence of elements in dietary products and their effect on human health and well-being. Such kind of analysis cannot either be decisive when considering the role of elements in food and human nutrition or processes of food treatment and production. These issues require possessing the knowledge of existing physicochemical forms of elements; however, this is only possible when sample analysis is focused on identification and quantification of individual species of elements (speciation analysis) or groups (classes) of species of elements having similar chemical or physical properties (group speciation analysis, fractionation analysis).

Information about species of elements present in everyday-life human diet products should nowadays be regarded as one of the most important reasons for which trace analysis, atomic spectrometry, and sample preparation procedures are developed and improved. Such reliable and verified information about species of elements leads to compiling new analytical methods and procedures enabling speciation and fractionation analyses [16, 17].

Determination of different species of elements in food and beverages is however a challenging task that can procure many problems. Having different consistency, matrix components, and element contents, samples of food and beverages commonly require different analytical strategies of preparation and analysis, including sampling, sample treatment prior to measurements of total concentrations, and speciation forms [8, 18]. Limitations to perform speciation analysis could be detection power of an instrumental technique used for determining defined and separated species or efficiency of separation/isolation procedures applied for distinguishing species or groups of species (fractions) of elements. A great number of possible species of elements that can undergo different processes and reactions depending on sample pH, coexistence of very similar species due to structure, and finally very low contributions of separate forms of elements in relation to their summary contents usually make the speciation analysis very difficult, sometimes impossible to carry out, or unprofitable due to costs and time and work investments [8, 19]. This results in searching for new alternative approaches to speciation analysis of elements [16, 17].

One of such approaches is group speciation tending to classify and determine fractions of species of elements having corresponding physicochemical features. Another approach is operational speciation in which analytical procedures and techniques selected for analysis decide about properties of species of elements that are distinctively separated and determined with these procedures and techniques. Because of similar analytical aims of group and operational speciation analyses or means of their performance, they are very often called fractionation analysis. In both cases, distinguished and discriminated fractions of species of elements are a consequence of speciation itself, that is, existence of different species of elements of defined structure and physicochemical properties. However, fractionation analysis is not aimed at evaluating or determining the structure of these species groups. Nevertheless, information available with fractionation analysis of elements is really valuable and could be a rational base to estimating actual dietary value of food products according to various macro- and microelements as well as elements' bioavailability.

Apparently from the literature, it can be found that size-exclusion chromatography (SEC) in combination with inductively coupled plasma mass spectrometry (ICP-MS) detection is useful for fractionation of elements in milk [20–22], wine [23], or tea infusions [24, 25]. This separation/detection approach results in assessing distributions of elements between species of defined masses (LMW, MMW, HMW), different types (proteins, polysaccharides, polyphenolic substances), and functions played in the organism. Unfortunately, evident drawbacks of SEC-ICP-MS approach are relatively high costs of the necessary equipment.

A much easier and inexpensive approach to fractionate metals seems to be application of solid phase extraction (SPE) with different ion exchange and other nonionic polymeric sorbents and resins. It appears that SPE in combination with atomic spectrometric detection could be very useful for classifying and distinguishing different forms of elements due to ion exchange, formation of chelates, or adsorption, depending on sorbents and fractionation schemes used [8]. Application of sorbents of different retention mechanisms is especially convenient since it enables to differentiate species of elements according to their stability, reactivity, charge, and hydrophobicity (chemical fractionation). Additional use of other simple separation techniques like filtration and ultrafiltration enhances this possibility to distinguish species of elements due to their physical properties like size, shape, and mass (physical fractionation).

3.3 Solid Phase Extraction (SPE)-Based Fractionation Analysis

3.3.1 *Single-Sorbent Solid Phase Extraction (SPE) Fractionation Approach*

Two different SPE approaches to partitioning species of elements in food and beverages are commonly applied. There is a large group of procedures and fractionation schemes in which samples, after appropriate sample handling and preparation, are treated with one sorbent placed in a column, an extraction disk or a syringe cartridge. Procedures, in which two or three single sorbents are separately used to classify species of elements according to different sorption mechanisms, also belong to the same group of fractionation schemes [8]. Examples of such procedures are given in Table 3.1. As can be seen, sorbents commonly applied for that purpose are strongly acidic cation-exchange (CE) or basic anion-exchange (AE) resins.

Table 3.1 Fractionation analysis of food and beverages using single-sorbent SPE

Samples	Elements	Sorbents used and fractions distinguished	Partitioning procedures	Ref.
Tea infusions	Mn	PE column (70×15 mm) with strongly acidic CE LABII (2.0 g) Cationic species PE column (70×15 mm) with strongly basic AE LABV (5.0 g) Anionic species	Sample filtrates (10 mL) are passed through separate columns at 3 mL min^{-1} Mn species retained are eluted using a 2-mol L^{-1} HNO_3 solution Concentrations of Mn species in both fractions are determined by F-AAS	[18]
Tea infusions	Al	PE column (7 mm i.d.) with chelating resin Chelex-100 (0.5 g) or strongly acidic CE Dowex HCR-S (0.5 g) Stable cationic species and labile species PE column (7 mm i.d.) with strongly AE AG MP1 (0.5 g) Anionic species and neutral species (complexes with polyphenols)	Samples (2 mL) are passed through separate columns at 0.5 mL min^{-1} Al species retained are sequentially eluted using 20 mL of H_2O, 20 mL of a 0.02-mol L^{-1} HCl solution and 20 mL of a 2-mol L^{-1} HCl solution Al concentrations in distinct fractions are determined using GF-AAS	[19]
Tea infusion	Al, Ba, Ca, Co, Cu, Fe, Mg, Mn, Ni, Rb, Sr, and Zn	An Alltech Maxi-Clean IC-Na extraction disk with a strongly acidic CE (0.5 mL resin bed) Noncationic species	0.45 μm sample filtrates (1.04 mL) are injected to a H_2O carrier and passed through the SPE disk at 1 mL min^{-1} Effluents collected are analyzed on the content of elements not retained on the sorbent using ICP-MS	[24]
Beer	Al, Ba, Ca, Cd, Co, Cr, Cu, Fe, Mg, Mn, Ni, Pb, Sr, and Zn	A glass column (300×10 mm) with a strongly acidic CE Dowex 50 W×4 (1.2 g) Cationic species and labile species	Degassed samples (200 mL) are passed through the column at 2 mL min^{-1} Resin beds with retained species of elements are wet digested in 65% HNO_3 and 30% H_2O_2 in a closed-vessel microwave-assisted system Elements are determined in resulting solutions by ICP-OES to evaluate fraction contributions	[26]

(continued)

Table 3.1 (continued)

Samples	Elements	Sorbents used and fractions distinguished	Partitioning procedures	Ref.
Nongassed mineral waters	Cd, Co, Cu, Fe, Ni, Pb, and Zn	A glass column (300 × 10 mm) with a strongly acidic CE Dowex 50 W × 4 (1.2 g) Cationic species and labile species	0.45 μm sample filtrates (250 mL) are passed through the column at 2 mL min^{-1} Species of elements retained by the sorbent are recovered using 10 mL of a 4-mol L^{-1} HCl solution at 2 mL min^{-1} Eluates are collected and analyzed by ICP-OES on element concentrations to estimate fraction contributions	[27]
Water extracts of olives	Mg	A glass column (250 × 8 mm) with a strongly acidic CE Dowex 50 W × 8 (5.0 g) Cationic species A PE column with a nonionic adsorbing resin Amberlite XAD-4 (2.0 g) Polyphenolic-bound species	Sample filtrates are separately treated with both columns by passing through them at 3 mL min^{-1} Effluents are collected and Mg concentrations are determined by F-AAS to assess contributions of both fractions	[28]
Beer	Cu, Fe, and Mn	An IST Isolute SPE SCX disk with a strongly acidic CE Noncationic species An IST Isolute SPE SAX disk with a strongly basic AE Nonanionic species	0.45 μm sample filtrates (1.04 mL) are injected to a H_2O carrier and separately passed through both SPE disks at 3 mL min^{-1} Effluents (9.0 mL) are collected and concentrations of elements not retained by both sorbents are determined by GF-AAS	[31]
Honeys, fruit juices, and tea infusions	Ca and Mg	A glass column (300 × 10 mm) with a strongly acidic CE Dowex 50 W × 4 (1.2 g) or a weakly acidic CE Diaion WT01S (1.2 g) Cationic species A glass column (300 × 10 mm) with a strongly basic AE Dowex 1 × 4 (1.2 g) Anionic species	0.45 μm sample filtrates (50 or 100 mL) are separately passed through columns at 2 mL min^{-1}; effluents (20 mL) are collected and subjected to analysis by F-AAS on concentrations of elements Fraction contributions are assessed by subtracting contents found in effluents from total contents	[32]

Honeys, fruit juices, and tea infusions	Fe and Zn	A glass column (300 × 10 mm) with a strongly acidic CE Dowex 50 W × 4 (1.2 g) Cationic species	0.45 μm sample filtrates (50 or 100 mL) are passed through separate columns at 1 mL min^{-1}; species of elements retained by resins are eluted using 10 mL of a 4-mol L^{-1} HCl solution at 1 mL min^{-1}	[29]
		A glass column (300 × 10 mm) with a strongly basic AE Dowex 1 × 4 (1.2 g) Anionic species	Fraction contributions are evaluated by analyzing eluents on element contents with F-AAS	
Low-fat cow milks	Ca, Fe, Mg, and Zn	A glass column (300 × 10 mm) with a strongly acidic CE Dowex 50 W × 4 (1.2 g) Cationic species	Samples (50 mL) are passed through separate columns at 2 mL min^{-1} Species of elements retained on both columns are recovered using 10 mL of a 2-mol L^{-1} HCl solution at 2 mL min^{-1}	[30]
		A glass column (300 × 10 mm) with a strongly basic AE Dowex 1 × 4 (1.2 g) Anionic species	Fraction contributions are assessed measuring element concentrations in collected eluates by means of F-AAS	

PE polyethylene, *CE* cation exchanger, *AE* anion exchanger, *F-AAS* flame atomic absorption spectrometry, *GF-AAS* graphite furnace atomic absorption spectrometry, *ICP-MS* inductively coupled plasma mass spectrometry, *ICP-OES* inductively coupled plasma optical emission spectrometry

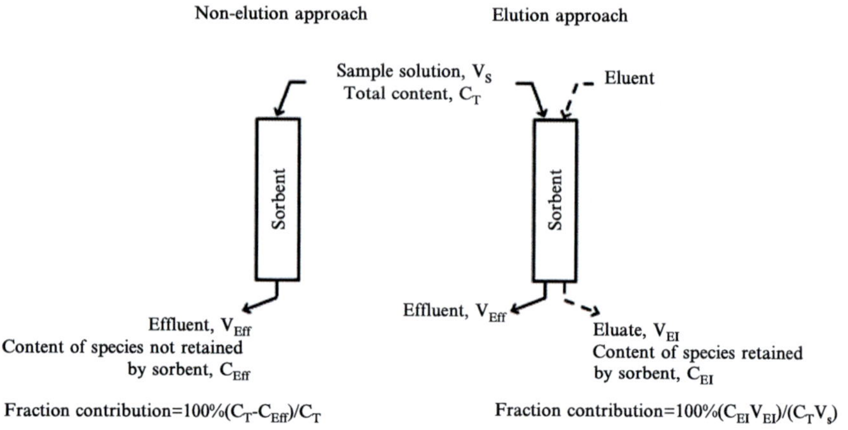

Fig. 3.1 Nonelution and elution approaches in single-sorbent SPE used for estimating fraction contributions

Accordingly, CE resins are typically used for separation and evaluation of broadly meant cationic species of elements fraction. This fraction normally includes simple ions of elements, their very stable and moderately cationic complexes with LMW inorganic and/or organic ligands and labile complexes that usually decay in contact with functional groups of resins, releasing simple ions [26, 27]. AE resins are appropriate to distinguish stable anionic complexes of elements [8] or sums of anionic and neutral species of elements in case of simultaneous retention of these forms [19]. Nonionic macroporous sorbents of high porosity are much rarely used in single-sorbent approaches, otherwise, they enable to retain hydrophobic (nonpolar) or slightly polar species of elements mostly bound to MMW and HMW compounds like polyphenols [28].

In fractionation analysis of elements with single-sorbent SPE, species of elements retained on sorbents used are usually eluted using HCl [27, 29, 30] and HNO_3 [18] solutions. Alternatively, recoveries of elements can be realized by decomposing resin beds in a closed-vessel microwave-assisted system [26]. Analysis of resulting eluates or digest solutions enables to determine concentrations of elements in different fractions and estimate fraction contributions in reference to total concentrations of elements determined in analyzed samples. A sequential elution procedure with solutions of increasing ionic strength, that is, water, a 0.02 mol L^{-1} HCl solution and a 2 mol L^{-1} HCl solution, can be used to distinguish three fractions differing by stability and/or polarity of species of elements and evaluate fraction donations [19].

Due to practical reasons, it is much easier to determine contributions of separated fractions analyzing effluents obtained from passing sample solutions through sorbents placed in columns, disks, or cartridges (Fig. 3.1). Measuring concentrations of species of elements being not retained by given sorbents, contributions of fractions of species of elements can be found by differences between contents of elements in samples and collected effluents [24, 28, 31, 32].

Fractionation analysis with single-sorbent SPE can also be a part of much complex fractionation procedures, in which other separation techniques, that is, coprecipitation, filtration, solvent extraction with water and organic solvents of different polarity, are applied apart from SPE with ionic and nonionic sorbents [18, 28]. In this case, besides chemical fractionation, physical fractionation is additionally applied for better discrimination and differentiation of possible species of elements present in analyzed food products [8]. Such approach enables to separate several fractions of species of elements differing according to chemical character (fractions of species associated with lipids, proteins, or polysaccharides) and physical properties (fractions of particle or dissolved species).

3.3.2 Dual-Sorbent Solid Phase Extraction (SPE) Fractionation Approach

A much better differentiation of existing physicochemical forms of elements in food products and beverages can certainly be obtained by fractionation procedures, in which two sorbents of different sorption mechanisms are coupled in series and applied to treat analyzed sample solutions [8]. In this kind of procedures, sample solutions are passed through a first sorbent and then resulting effluents are directly passed through a second sorbent. After passing sample solutions through both linked sorbents, retained species of elements are eluted from them using selected solvents. Resulting eluates are subsequently analyzed on contents of elements retained by different sorption mechanisms, and appropriate fraction contributions are determined. In this way, fractionation analysis performed makes possible to separate from 2 to 3 fractions of elements being dissimilar due to properties of species that these groups include (Fig. 3.2). Sometimes, because of problems related to quantitative recovery of retained species of elements from the first sorbent, a nonelution approach can be used to determine concentrations of species of elements by analyzing effluents from this sorbent.

Collecting and analyzing effluents resulting from passing sample solutions through both linked sorbents, it is also possible to determine concentrations of species of elements being not retained by both sorbents applied in dual-sorbent SPE fractionation schemes. In this case, the contribution of inert or residual fractions of species of elements can be evaluated [1, 33–36].

Usually, the following sorbent combinations are applied for partitioning species of elements with dual-sorbent SPE, that is., a strongly basic AE resin AG 1 × 8 with a chelating resin Chelex-100 [1], a weakly basic AE resin Reillex 402 with a strongly acidic CE resin Dowex 50 W × 4 [33, 35], a strongly acidic CE resin Dowex 50 W × 4 with a strongly basic AE resin Dowex 1 × 4 [30], a weakly acidic CE resin Diaion WT01S with a strongly basic AE resin Dowex 1 × 4 [36], or a nonionic adsorbing resin Amberlite XAD-7HP with a strongly acidic CE resin Dowex 50 W × 4 [33, 34]. Type and order of sorbents for fractionation analysis with dual-sorbent SPE procedures can arbitrary be selected [1], assuming, for example, that strongly basic

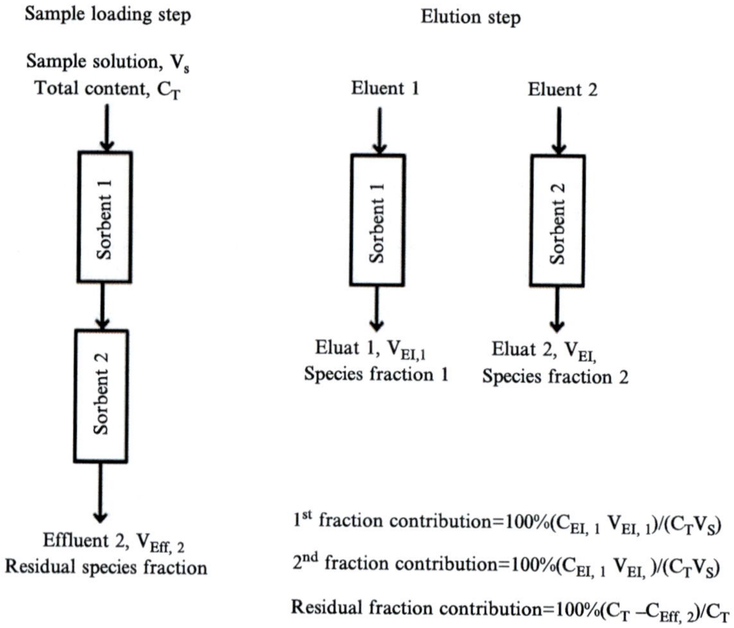

Fig. 3.2 Sample loading and elution steps in dual-sorbent SPE

AE resin placed before chelating resin would prevent from retaining labile anionic complexes of elements. More often, preliminary experiments are made using solutions of simple ions of elements or their complexes with various ligands to evaluate sorption and desorption properties of these sorbents toward different classes of species of elements [33–38].

Dual-sorbent SPE fractionation schemes with description of analyzed samples, elements determined, and partitioning strategies used are given in Table 3.2. A very advanced protocol for partitioning of Cu, Fe, and Zn in wines was proposed by Karadiova et al. [38]. In this fractionation procedure, initial sample filtration through 0.20-μm membrane filters is combined with further treatment of resulting filtrates (dissolved fraction) with SPE. At first, filtrate portions (8 mL) are passed through a SPE column containing an adsorbing resin Amberlite XAD-8 (2.0 g) at 2 mL min^{-1}. Species of elements retained by this column are eluted using 5 mL of a 2-mol L^{-1} HNO_3 solution to establish fraction contributions of elements bound to polyphenolic-polysaccharide agglomerates. Further analysis is made by passing 1-mL portions of effluents obtained after Amberlite XAD-8 treatment through columns containing a strongly acidic CE resin Dowex 50 W×8 (1.0 g), a strongly basic AE resin Dowex 1×8 (1.0 g), and an apolar adsorbing resin Amberlite XAD-2 (1.0 g) at 1 mL min^{-1}. In case of the latter column, a solution of 1,10-phenantroline is added to effluent portions in order to complex Fe(II) ions. Finally, species of elements retained on these three columns are stripped with a 2-mol L^{-1} HCl solution (ion exchangers) and methanol (adsorbent) that are passed through columns at 2 mL min^{-1}. Concentrations

Table 3.2 Fractionation analysis of food and beverages using dual-sorbent SPE

Sample	Elements	Sorbents and fractions distinguished	Partitioning procedure	Ref.
Cow milk	Cu and Mn	A microcolumn with a chelating resin Chelex-100 (0.04 g) – a microcolumn with a strongly basic AE AG 1 × 8 (0.04 g) Cationic species (column 1), anionic species (column), and residual species	Samples (1.0 mL) are injected into a H_2O carrier and passed at 3 mL min^{-1} through both columns Species of elements retained are recovered passing 1.0 mL of a 2-mol L^{-1} solution through both columns at 3 mL min^{-1} Respective eluates are collected and analyzed by ICP-OES to determine elements in fractions; additionally Effluents after passing samples through both columns are measured by GF-AAS to evaluate contributions of residual species	[1]
Beers, tea infusions	Mn and Zn	A glass column (300 × 10 mm) with a nonionic adsorbing resin Amberlite XAD-7HP (2.0 g) – a glass column (300 × 10 mm) with a strongly acidic CE Dowex 50 W × 4 (1.0 g) Hydrophobic species (column 1), cationic species (column 2), and residual species	Samples (100 mL of degassed beers, 50 mL of tea infusion filtrates) are passed through both columns at 1 mL min^{-1} Species of elements retained are eluted using 10 mL of a 2-mol L^{-1} HNO_3 solution (column 1) and 10 mL of a 4-mol L^{-1} HCl solution (column 2) at 1 mL min^{-1} Eluates collected are subjected to analysis by F-AAS to determine element concentrations and estimate contributions of fractions of species Additionally, elements are determined in effluents resulted from passing samples through both columns to assess residual species contribution	[33, 34]

(continued)

Table 3.2 (continued)

Sample	Elements	Sorbents and fractions distinguished	Partitioning procedure	Ref.
Beers	Mn and Zn	A glass column (300 × 10 mm) with a weakly basic AE Reillex 402 (2.0 g) – a glass column (300 × 10 mm) with a strongly acidic CE Dowex 50 W × 4 (1.0 g) Polyphenolic species (column 1), cationic species (column 2), and residual species	Samples (100 mL of degassed beers) were passed through both columns at 1 mL min^{-1}, additionally 5.0-mL effluent portions from both columns were collected Species of elements retained on the column 2 were recovered using 10 of a 4 mol L^{-1} HCl solution at 1 mL min^{-1} Elements were determined in effluents and eluates using F-AAS to establish fraction contributions	[35]
Honeys	Ca, Fe, Mg and Zn	A glass column (300 × 10 mm) with a weakly acidic CE Diaion WT01S (1.0 g) – a glass column (300 × 10 mm) with a strongly basic AE Dowex 1 × 4 (1.0 g) Cationic species (column 1), anionic species (column 2), and residual species	Samples solutions (100 mL) are passed through both columns at 2 mL min^{-1} Species of elements are eluted from columns using 10 mL of a 2 mol L^{-1} HCl solution at 2 mL min^{-1} Elements are determined in collected eluates by F-AAS to acquire fraction contributions Additionally, effluents obtained from passing sample solutions through both columns are analyzed to estimate residual fraction donation	[36]
Tea infusion	Al	A PE column (10 mm i.d.) with a nonionic adsorbing resin Amberlite XAD-7 (2.0 g) – a PE column (10 mm i.d.) with a chelating resin Chelex-100 Polyphenols-bound species (column 1) and cationic species (column 2)	Sample filtrates (10 mL) are passed through the column 1 at 2 mL min^{-1} and its effluents are passed through the column 2 at 1 mL min^{1} Species of elements retained are eluted using 10 mL of a 2-mol L^{-1} HNO_3 (column 1) or a 0.08-mol L^{-1} HCl solution (column 2) Elements are determined in collected eluates using F-AAS to estimate fraction contributions	[37]

CE cation exchanger, *AE* anion exchanger, *ICP-OES* inductively coupled plasma optical emission spectrometry, *GF-AAS* graphite furnace atomic absorption spectrometry, *F-AAS* flame atomic absorption spectrometry, *PE* polyethylene of Cu, Fe, and Zn in collected eluates are determined using GF-AAS, and the following fraction contributions are assessed: labile species (Dowex 50 W×8), stable anionic species (Dowex 1×8), and labile Fe(II) species. Additionally, proteins and polysaccharides are precipitated from the effluent obtained after passing samples through Amberlite XAD-8 using acetone and methanol, respectively. Precipitates can be later on dissolved in a 0.1 mol L^{-1} NaOH solution (proteins) or water (polysaccharides) and concentrations of elements associated with both classes of compounds are determined using GF-AAS.

The use of a weakly basic AE resin Reillex 402 [33, 35], being a cross-linked polyvinylpyrrolidone (PVPP) with a capability to strongly retaining HMW organic compounds, mostly flavonoids, tannins, and humic acids, is of special interest. This resin forms very strong hydrogen bonds with phenolic, hydroxylic, and carboxylic groups of the mentioned classes of organic species and does interact at all with LMW organic compounds such as amino acids and carboxylic acids. Unfortunately, it is very difficult to quantitatively elute these species. Usually, a nonelution approach is applied to evaluate donations of respective fractions of polyphenolic-bound species of elements.

The use of a nonionic adsorbing resin Amberlite XAD-7HP that is a cross-linked polyacrylic resin characterized by relatively high pore size (450 Å on average) and polarity (1.8 D) of resin beads also appears to be very promising as compared to other Amberlite resins, that is, XAD-2, XAD-4, or XAD-16, having a cross-linked styrene-divinylbenzene (SDB) copolymer framework [33, 34]. The mentioned resin has successfully been used in fractionation analysis of selected elements in various food samples, and it seems that it well enables to distinguish nonpolar and/or weakly polar species of elements associated with polyphenolics due to hydrophobic and polar interactions between sample components and the sorbent surface. In reference to other SDB-based Amberlite XAD resins, application of Amberlite XAD-7HP certainly simplifies a partitioning procedure since this resins does not require a special conditioning treatment aimed at blocking active sites of ion exchange present due to various strong polar impurities [28, 37–40]. The mentioned organic impurities have been found to retain simple ions of elements, and hence, other Amberlite XAD resins were shaken for a long time with aqueous solutions containing Bi(III) or In(III) ions before use.

3.3.3 Sorption and Desorption Behavior of Resins for Solid Phase Extraction (SPE)-Based Fractionation

Ionic and nonionic resins used for partitioning of elements by SPE-based procedures are commonly studied in terms of their sorption behavior toward certain classes of species of elements. It is important in so far as results of fractionation analysis obtained with SPE are operationally discriminated, which means that they are dependent on sorption and desorption properties of sorbents used for partitioning.

In other words, information related to a given class of species of elements that can be retained by a certain resin is very important for arranging a fractionation scheme, or evaluating and interpreting final results of this analysis [8].

Solutions of simple ions of metals are typically used to assess experimental conditions required for quantitative retention of broadly meant cationic forms (fraction of cationic species) in case of a chelating resin Chelex-100 [1, 19, 37], CE resins Dowex 50 W×4 [20, 26, 27, 29, 30, 32–35, 38], Dowex HCR-S [19], and Diaion WT01S [32, 36]. These solutions are also applied to verify an eventual retention of cationic species on other sorbents, including AE resins [19, 33, 35] and apolar adsorbing resins [33, 34, 38]. It is very useful in eliminating artifacts and errors in developed fractionation procedures and interpreting results of SPE-based element partitioning analyses.

Solutions of anionic complexes of metals with ethylenediaminetetraacetic acid (EDTA) are normally used to evaluate experimental conditions required for retaining anionic metal complexes with LMW organic ligands on AE resins, that is., AG 1×8 [1], Dowex 1×4 [29, 30, 36], or Reillex 402 [35]. Solutions of metal complexes with tartaric acid can also be applied for this purpose [38]. The mentioned solutions are also useful for verifying if other ionic and nonionic sorbents, that is, CE resins Dowex 50 W×4 [33–35] and Diaion WT01S [36] or an adsorbing resin Amberlite XAD-7HP [33, 34], retain species of metals bound to LMW organic compounds under these conditions.

Finally, solutions of metals with gallic or tannic acids are taken to assess the extent of sorption of hydrophobic and/or weakly polar complexes of metals with HMW ligands by nonionic adsorbing resins that selectively retains polyphenolics, for example, Amberlite XAD-2 [38], Amberlite XAD-7 [33, 34, 37], or a weakly AE resin Reillex 402 [35]. These solutions are also often applied to retrieve information about eventual sorption of the mentioned hydrophobic species of metals on other resins, that is, Chelex-100 [19, 37], Dowex 50 W×4 [33, 34], Dowex XCR-S [19], Diaion WT01S [36], and AG MP1 [19]. This seems to be especially important in case of dual-sorbent SPE fractionation procedures.

Because of highly ionized sulfonic ($-SO_3^-$) functional groups in a broaden pH range, Dowex 50 W×4 resin has been found to be capable of complete and reversible retaining cationic species of main (Ca, Mg) and many trace metals (Cd, Co, Cu, Fe, Mn, Ni, Pb, and Zn) from neutral and weakly (pH 2–7 [26, 29, 30, 32–35]) or strongly acidified ($pH<1$ [27], made with up to 0.5 mol L^{-1} HNO_3 or 0.2 mol L^{-1} HCl) solutions. This behavior of Dowex 50 W×4 is especially significant in case of fractionation of metals in waters. Water samples are commonly acidified to pH ~2 so as to minimize losses of species of metals due to their adsorption on the surface of containers used for transporting and handling [41–45]. Unfortunately, such a treatment usually requires changing pH of water samples before subsequent preconcentrations with chelating resins, for instance, Chelex-100, Metalfix Chelamine [46–50], applied for quantitative retention of traces of metals from analyzed water samples. It is usually made by adding relatively high amounts of adequate buffers and can result in a subsequent water contamination or an alteration of equilibrium existing between different forms of metals. In this case, application of a strongly acidic CE resin such as Dowex 50 W×4 can eliminate this inconvenience and enable to determine metals present at 10^{-1}–10^{+1} μg L^{-1} levels in waters of high mineralization [27].

A weakly acidic CE resin Diaion WT01S with carboxylic functional groups can also be very useful for fractionation analysis of metals [29, 32, 36]. This resin has higher exchange capacity as compared to commonly applied Dowex 50 W resins and can retain ions of metals, for example, of Ca, Mg, Fe, and Zn, at pH 4.0–6.5 due to plain ion exchange ($2R\text{–}COOH + M^{2+} = (R\text{–}COO)_2M + 2\ H^{+}$) and some donor-acceptor interactions between donor oxygen atoms of carboxylic groups and cations of hard (Ca and Mg) and transition (Fe and Zn) metals [51, 52].

3.4 Samples and Their Treatment for Solid Phase Extraction (SPE)-Based Fractionation

Fractionation of elements by means of SPE with ionic and nonionic sorbents has been applied so far in analysis of the following food products and beverages: fresh milk [1], nonfat ultrahigh-temperature (UHT)-processed milks [30], infusions of green and black teas [18, 19, 24, 29, 32, 33, 37], nongassed mineral waters [27], fruit juices [29, 32], bottled and canned beers of lager and pilsner types [26, 31, 34, 35], bottled and barreled white and red wines [38], bee honeys (monofloral, multifloral, honeydew, honeydew-floral) [29, 36], and olives [28].

Apparently, from the literature, an initial treatment of liquid samples is aimed at separating dispersed particles that can clog pores of resin beads. Accordingly, samples are filtered through membrane filters while resulting filtrates (dissolved fraction) are further subjected to fractionation analysis [8]. Filters with retained particles (particle fraction) can be mineralized while resultant solutions can be analyzed to evaluate contributions of this fraction.

Casein is removed from milk by precipitating it; usually other macromolecules, mostly lipids, are coprecipitating under these conditions [1]. It can be done by adding to milk a buffer of pH corresponding to the isoelectric point of casein (pH ~4.8). Afterward, precipitated casein can be removed by filtrating through a 0.3-μm membrane filter. This leads to receiving a milk solution free of colloidal components that can be further diluted (by 10) and analyzed [1]. Milk samples can also be only diluted (by 25) with water. This treatment does not result in changing pH but is responsible for dissolving colloidal calcium phosphate (CCP) associated with a micellar structure of casein and releasing Ca(II) and Mg(II) cations that are adsorbed on CCP clusters and electrostatically stabilize casein structure [26, 53, 54]. It is required since casein micelles (500–2,500 Å) can affect column process by clogging pores of resin beads.

Fruit juices and mineral waters are filtered through 0.45-μm membrane filters just after opening. In this way, any particles are removed, and resulting filtrates containing dissolved species of elements can freely be treated with SPE fractionation procedures [27, 29, 32].

Beer is degassed passing Ar through a portion of beer [31] or exposing it to action of ultrasounds in an ultrasonic bath [26]. A similar effect can be reached by passing beer samples through 0.45-μm membrane filters [34, 35].

3.4.1 *Total Content Analysis*

Preparation procedures for total content sample analysis accompanying fractionation analysis of food and beverages by means of SPE-based separation schemes are similar to those that are commonly used to determine total concentrations of different metals in tea infusions, beers, wines, and honeys by atomic spectrometry methods [54–66].

Unquestionably, milks, tea infusions, beers, wines, and honeys are complex matrices that contain relatively high concentrations of various LMW and HMW organic compounds. These compounds can potentially be a source of vast physical and chemical interferences (known as matrix effects) in determinations of element concentrations. As a result, sample decompositions aimed at eliminating matrix presence and releasing mineral components are carried out at a sample preparation stage using wet digestion with appropriate oxidizing agents and under high temperature conditions [50, 56].

Apparently, samples of milk, mineral water, beer, and wine are typically decomposed in open-vessel systems using mixtures of concentrated HNO_3 and 30% H_2O_2 [30, 31, 34, 35, 38], concentrated H_2SO_4 and HNO_3 [26], or concentrated HNO_3 only [27]. In two last cases, a CE resin is additionally applied to preconcentrate traces of elements by passing resulting digest solutions through columns with a strongly acidic CE resin Dowex 50 W×4. Elements retained are then stripped with an HCl solution or by decomposing resin beds in a closed-vessel microwave-assisted system using concentrated HNO_3 and 30% H_2O_2 [26, 27].

Beverages and extracts of solid samples can also be directly analyzed with their minimal handling and modification. Such a practice can certainly limit a contamination risk and loss of elements during sample decomposition and additionally reduce time of analysis [8]. Usually, possible matrix effects can be minimized by appropriate dilutions of sample solutions prepared before measurements as it is the case of tea infusions [19, 33, 37] and water extracts of olives [22] subjected to direct analysis. Samples can also be acidified to release elements from their complexes with different dissolved organic components. The following examples can be found in related literature: dilution of beer samples [31], dilution and acidification of milk samples [1], acidification and dilution of samples of wines [38], tea infusions [18, 29, 32], fruit juices [29, 32], and honeys [29, 32, 36].

In order to assess extraction efficiencies made with water or organic solvents, total concentrations of elements have to be determined in analyzed materials, that is, tea leaves, olives, using open- or closed-vessel digestion systems, and mixtures of concentrated HNO_3 and $HClO_4$ [18, 37], concentrated HNO_3 and HCl [19] or concentrated HNO_3 [29, 32, 33], or concentrated HNO_3 only [28].

Finally, membrane filters with retained particles can be decomposed in concentrated HNO_3 with admixture of 30% H_2O_2, while resulting digest solutions can be subjected to analysis on the content of elements associated with particle fractions present in analyzed samples [27, 29, 30, 32]. Additionally, contributions of dissolved and particle fractions can be evaluated on the base of analysis of filtrates obtained by passing sample solutions through 0.20- and 0.45-μm membrane filters [28, 38].

Mineral components of solid and semisolid samples have to be transferred from their matrices to solutions. This can be done by simply dissolving them in water or by extracting with different solvents [8]. Resulting sample solutions or extracts are subsequently analyzed by certain SPE fractionation procedures. Accordingly, tea infusions are prepared by pouring leaves or bags with boiling water (deionized, doubly distilled) and brewing them for 5 [24] to 15 min [18, 29, 32]. Leaves can also be poured with water and heated to boil and then, resulting infusions are left to cool [37]. Ratios of tea masses (in g) to water volumes (in mL) are usually 1–100 [18, 24, 29]. However, other ratios, for example, 1–250 [19], 1–80 [32, 33], or 1–20 [37], are also used. Before fractionation with SPE, infusions are filtrated through paper filters [18, 29, 32, 33, 37] or 0.45-μm syringe filters [24].

Honeys are dissolved and 20-fold diluted in water (deionized, doubly distilled), and this treatment does not lead to altering their pH [29, 32, 36]. Afterward, the resulting honey solutions are filtered through 0.45-μm membrane filters.

Olives are washed in distilled water, seeded and cut into small pieces, and kept at −28°C [28]. Water extracts of such samples (1/25 g/mL) prepared for SPE fractionation analyses are filtered through filter papers and then 0.45-μm membranes.

3.5 Quality Assurance (QA) and Control (QC) of Fractionation Analysis Results

Quality assurance and control (QA and QC) of analytical results are obvious and significant requirements of every analysis undertaken in a laboratory. Considering fractionation analysis made with SPE-based procedures, it is rather difficult to verify dependability and reliability of their results since no certified reference materials (CRMs) with well-defined and quantified fractionation forms of elements are available in commerce [8]. As a result, QA and accuracy of total concentrations are mostly assessed in case of direct analysis methods carried out with no sample handling and preparation or procedures applied for sample matrix decomposition. Very often recovery tests of elements added to sample solutions are undertaken for that purpose. Samples are spiked with elements in the form of simple ions and subjected to fractionation procedures. Recoveries of added elements are used to evaluate correctness and consistency of applied sample treatments like decomposition and enrichment and analysis conditions in reference to total concentrations of elements [27–32].

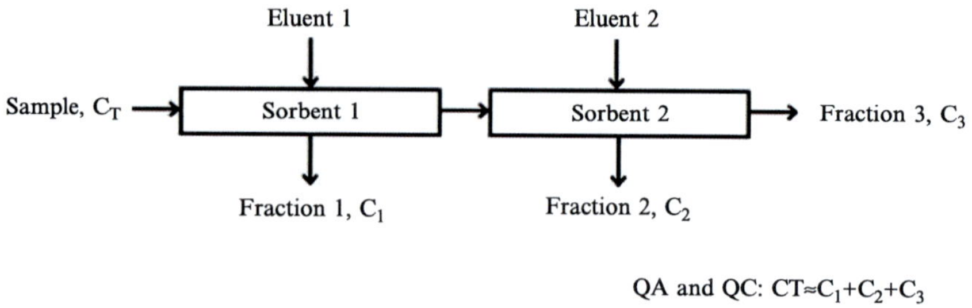

Fig. 3.3 QA and QC in fractionation analysis with dual-sorbent SPE

Regardless lack of appropriate CRMs, some alternative means of QA and QC of results achieved with SPE-based element partitioning procedures have been proposed. Studying of recovery is one of such methods. Analyzed samples can be spiked with investigated elements in the form of simple ions and treated with SPE fractionation procedures. Then, concentrations of added elements are determined in distinct operationally defined fractions of species of elements distinguished with a given procedure. Sums of these concentrations are related to initial contents added to samples in order to verify any loss of elements or secondary contamination that may appear during partitioning of elements by SPE [27, 31–34, 36].

Another way of QA and QC of fractionation analysis is based on comparing contributions of certain fractions of species of elements separated with the aid of two different sorbents, that is, cationic species of elements distinguished with a strongly acidic CE resin Dowex 50 W×4 and a weakly acidic CE resin Diaion WT01S [32], polyphenolic-bound species of elements discriminated with a weakly basic AE resin Reillex 402 and a nonionic adsorbing resin Amberlite XAD-7HP [34].

Finally, in case of dual-sorbent SPE fractionation procedures, an additional residual species of elements fraction can be differentiated and quantified. Usually, this class of species of elements is evaluated analyzing column effluents resulting from passing samples through linked sorbents. Accordingly, contents of species of elements not retained by both sorbents are attained. By relating sums of these concentrations found in all three fractions distinguished (Fig. 3.3), information about reliability of SPE partitioning procedures can be reached [1, 30, 33–36].

3.6 Element Fractionation Patterns in Food and Beverages

Speciation of elements using different fractionation approaches, in which certain analytical procedures and operations enable to discriminate, separate, and determine groups of species of elements having similar chemical and/or physical properties, is rather difficult since this kind of speciation analysis does not identify these species in reference to their structures or defined chemical forms. Unique fractionation patterns obtained for studied elements are indeed consequences of their speciation, but fractionation analysis is not aimed at an unequivocal distinguishing of chemical forms of species of elements in separated operationally defined fractions. Fractionation and partitioning of elements enable to classify and separate classes of species of elements of distinct properties, important in relation to their living functions, nutritional value, and effects on human body. In this way, it indirectly can serve for assessing bioavailability of elements in analyzed food products and beverages. Usually, results of fractionation analysis are individual fractionation patterns of elements obtained according to procedures selected for separation of species of elements (e.g., ion exchange, adsorption, filtration, ultrafiltration, size exclusion, solvent extraction) and sample treatments used (dilution, filtration, pH adjustment).

A rational and acute examination of results of fractionation analysis is very significant for their interpretation and understanding. A sample preparation stage and possible alteration of original speciation of elements have to be considered. Knowledge about sorption and desorption features of sorbents toward different groups of species of elements, order of sorbents used in dual-sorbent SPE procedures, or general information about sample components and possible interactions between bioligands and ions of elements also matter. Unquestionably, this information of fractionation patterns of elements in food and beverages is more appropriate for estimating dietary quality and safety of samples and considering availability and absorbability of elements than their usually measured total concentrations. Certain fractionation patterns of elements established could also be useful in tracing authenticity and provenience of natural food products. It could be expected that any alterations in food processing, for example, by adding forbidden substances or components, would result in changing these fractionation patterns [8].

Recapitulating results of fractionation analysis of elements in different food products and beverages, it can be seen that the first and second group of metals, including Rb, Ca, and Mg, are exclusively present in the form of cationic species, most likely as simple ions, their complexes with LMW inorganic and/or organic ligands. Behavior of transition metals like Cu, Fe, and Zn is quite different. These elements seem to readily form complexes with different LMW and HMW organic compounds and can be characterized by fairly complex distributions of discriminated fractions of species. In this case, except for cationic forms, other species of elements can be found, including hydrophilic anionic forms, inert species, and hydrophobic and/or weakly polar species. Table 3.3 shows information of various fractions of species of elements and their contributions to total concentrations of elements found in different food products and beverages using single- and dual-sorbent SPE partitioning procedures.

3.6.1 Tea Infusions

Almost 95% of total Mn is presumed to be simple ions [18]. Remaining 5% of this metal seems to be related to Mn forms associated with nonpolar and/or weakly polar

Table 3.3 Fractionation distributions of elements in food products and beverages obtained with single- and dual-sorbent SPE

Sample	Detection	Distribution of species of elements fractions, %				Ref.
		Cationic	Anionic	Residual	Hydrophobic	
Aluminum						
Tea infusions	F-AAS, GF-AAS, ICP-OES	10–66	34–75	–	28–70	[19, 24, 26, 37]
Beer	ICP-OES	10	–	–	–	[26]
Barium						
Tea infusion	ICP-MS	–	46	–	–	[24]
Beer	ICP-OES	68	–	–	–	[26]
Calcium						
Tea infusion	F-AAS, ICP-MS	96–99	4	–	–	[24, 32]
Beer	ICP-OES	100	–	–	–	[26]
Fruit juices	F-AAS	90–100	1–14	–	–	[32]
Honeys	F-AAS	97–100	2–6	–	–	[32, 36]
Milks	F-AAS	95–100	0.1–1	<0.1–2	–	[30]
Cadmium						
Beer	ICP-OES	–[a]	–	–	–	[26]
Mineral waters	ICP-OES	96–98	–	–	–	[27]
Cobalt						
Tea infusion	ICP-MS	–	30	–	–	[24]
Beer	ICP-OES	11	–	–	–	[26]
Mineral waters	ICP-OES	<59	–	–	–	[27]
Chromium						
Beer	ICP-OES	8	–	–	–	[26]
Copper						
Milk	ICP-OES	41	–	4	–	[1]
Tea infusion	ICP-MS	–	23	–	–	[24]
Beers	GF-AAS, ICP-OES	6–34	72	–	–	[26, 31]

Mineral waters	ICP-OES	<100	–	–	–	[27]
Wines	GF-AAS, F-AAS	38–48	5–11	–	26–42	[38]
Iron						
Tea infusions	F-AAS, ICP-MS	41–52	27–39	9–31	–	[24, 29]
Beers	GF-AAS, ICP-OES	–[a]	81	–	–	[26, 31]
Mineral waters	ICP-OES	74–98	–	–	–	[27]
Fruit juices	F-AAS	23–59	40–55	1–22	–	[29]
Honeys	F-AAS	15–40	12–40	29–73	–	[29, 36]
Milks	F-AAS	46–66	22–48	<5–13	–	[30]
Wines	F-AAS, GF-AAS	22–57	2–18	–	16–26	[38]
Magnesium						
Tea infusions	F-AAS, ICP-MS	98–100	<0.3	–	–	[24, 32]
Beer	ICP-OES	40	–	–	–	[26]
Olives	F-AAS	15–20	–	–	<4	[28]
Fruit juices	F-AAS	96–100	1–2	–	–	[32]
Honeys	F-AAS	98–100	1–5	–	–	[32, 36]
Milks	F-AAS	92–99	0.1–5	<0.1–7	–	[29]
Manganese						
Milk	ICP-OES	48	11	26	–	[1]
Tea infusions	F-AAS, ICP-MS	53–95	<0.2	0.4–3	3–46	[18, 24, 33]
Beers	F-AAS, GF-AAS, ICP-OES	28–100	–[a]	3–42	<0.3–37	[26, 31, 34, 35]
Nickel						
Tea infusion	ICP-OES	–	47	–	–	[24]
Beer	ICP-OES	23	–	–	–	[26]
Mineral waters	ICP-OES	72–99	–	–	–	[27]
Lead						
Beer	ICP-OES	–[a]	–	–	–	[26]
Mineral waters	ICP-OES	100	–	–	–	[27]

(continued)

Table 3.3 (continued)

Sample	Detection	Distribution of species of elements fractions, %				Ref.
		Cationic	Anionic	Residual	Hydrophobic	
Rubidium						
Tea infusion	ICP-MS	–	<0.1	–	–	[24]
Strontium						
Tea infusion	ICP-MS	–	13	–	–	[24]
Beer	ICP-OES	87	–	–	–	[26]
Zinc						
Tea infusions	F-AAS, ICP-MS	59–86	4–9	<2–29	10–39	[24, 29, 33]
Beers	F-AAS, ICP-OES	25–86	–	13–54	<2–53	[26, 34, 35]
Mineral waters	ICP-OES	56–72	–	–	–	[27]
Fruit juices	F-AAS	89–95	<0.3–3	<4–8	–	[29]
Honeys	F-AAS	41–100	2–10	5–50	–	[29, 36]
Milks	F-AAS	33–50	<0.5–2	42–59	–	[30]

F-AAS flame atomic absorption spectrometry, *GF-AAS* graphite furnace atomic absorption spectrometry, *ICP-OES* inductively coupled plasma optical emission spectrometry, *ICP-MS* inductively coupled plasma mass spectrometry

[a]Not detected

organic molecules. Similar fractionation pattern is reported elsewhere [24], where over 99% of the total Mn content is likely to be present in the form of cationic species, being unbound Mn(II) ions or their complexes with LMW organic ligands (<1 kDa). The fraction of cationic species of Mn could contain free Mn(II) cations and its stable cationic complexes with carboxylic acids (oxalic, citric, tartaric, malic, succinic). However, the contribution of this fraction could be overestimated due to retention of Mn species bound to polyphenolics on CE resins [18, 24]. Hence, the actual contribution of this fraction can be lower (53–82%), while the mentioned fraction of polyphenolic species, containing nonpolar and/or weakly polar compounds of Mn with polyphenolic substances, accounts for 13–46% of total Mn in tea infusions [33].

Calcium and Mg are entirely present in the form of cationic species (96–100% of total contents) [24, 32]. In a corresponding way, cationic species are predominant class of species for Rb, Sr, and Zn; contributions of this fraction are from 87% in case of Sr to 100% in case of Rb [24]. What is more, SEC-ICP-MS experiments indicate that Rb is associated with organic compounds of molecular weight (MW) <1 kDa, Sr with molecules of MW 22–27 kDa and >75 kDa, while Zn with molecules of MW <1 kDa and 2–4 kDa. A more complicated fractionation pattern for Zn provides other works [29, 33]. Except for fraction of cationic species contributing 59–86% to total Zn, other groups of species are distinguished, including fraction of polyphenolic species (10–39%), fraction of anionic species (6–9%), and the less abundant fraction of residual species.

Fraction distributions achieved for Ba and several transition metals (Co, Cu, Fe, and Ni) appears that cationic species of these metals have the highest contribution to their total concentrations, that is, from 54% (Ba) to 77% (Cu) [24]. Noncationic forms found for these metals can refer to complexes with organic ligands of MW 4–6 kDa (Cu, Fe, Ni), 23–28 kDa (Ba), and >75 kDa (Ba). In case of Fe, the fraction of cationic species can reach 41–52% in reference to its total content, while noncationic species can include anionic forms (28–39%) and residual species (9–31) [29].

The most noncationic character has Al. It seems that more than 75% of its total content can be found in the form of different noncationic species, likely associated with the presence of Al bound to organic species of MW 4–8 kDa [24]. Indeed, the contribution of cationic fraction for Al is very low (10–18% of total Al), while the contribution of fraction of HMW species, possibly corresponded to Al bound to polyphenols, can reach 28–32% in relation to its total content [24]. Analogous donations of noncationic forms of Al (40–60% of total Al) are also reported; they are presumed to be complexes of this metal with organic molecules and/or different cationic forms [19].

3.6.2 *Mineral Waters*

Cadmium, Cu, and Pb are entirely present in the form of cationic species that are likely to be simple ions of metals, stable cationic complexes with inorganic and/organic ligands, and labile species [27]. Cationic species are also predominant forms of Co, Fe, Ni, and Zn that contribute from 56% (Zn) to 74% (Fe) to total metal concentrations. However, these metals can be present as well in the form of anionic and/or neutral species, presumably being complexes with MMW and HMW organic ligands such humic acids [27].

3.6.3 Milks

Calcium and Mg are found to primarily exist as cationic species (92–199%), likely to be free cations and cations related to the micellar structure of casein or cationic complexes with different anions, for example, HCO_3^{2-}, Cl^-, HSO_4^-, $H_2PO_4^-$ [30]. Experiments with SEC-ICP-MS indicate that Ca and Mg can be associated with noncasein fraction of MW <2 kDa [21, 67, 68]. Fractions of anionic and residual species of Ca and Mg, related to their complexes with anions of orthophosphoric and citric acids (H_3Cyt), that is, HPO_4^{2-}, PO_4^{3-}, $HCyt^{2-}$, Cyt^{3-}, rather negligibly contribute to total contents of these metals (<7%) [30].

Copper is mainly distributed between two fractions, that is, cationic and casein-associated species; the latter obtained after precipitating casein [1]. Fraction of cationic Cu species can be simple ions and labile Cu(II) species. Remaining species of Cu are likely to be neutral or nonpolar; anionic Cu species have not been found [1].

The predominant fraction of Mn species is also cationic in nature (~50% of total Mn) [1]. Other fractions distinguished, that is, casein-associated, anionic, and neutral, are also present but their contributions are lower than this of cationic species.

Iron can be present in three different classes of species [30]. Cationic species are predominant (46–66%) and can include unbound simple Fe(III) ions, and those attributed to the micellar structure of casein. Additionally, cationic complexes of Fe(III) with anions of orthophosphoric and citric acids, that is, $FeHPO_4^+$, $FeH_2PO_4^{2+}$, $FeHCyt^+$, can be included [30]. The fraction of anionic Fe species, contributing 22–48% to total Fe, is likely to be associated with stable complexes of this metal with lactose and/or other oligosaccharides that strongly bind ions of metals to form anionic complexes [6]. It is quite possible that this fraction can contain Fe species bound to β-lactoglobulin or α-lactalbumin [21]. Residual species are the less abundant fraction of Fe species, possibly linked to the presence of simple Fe complexes like $FePO_4$ and FeCyt [30].

Nonpolar and/or residual, inert Zn species contribute 42–59% to its total and are the most abundant class of Zn species. It can contain inert complexes of Zn(II) ions with LMW and HMW ligands, that is, $HCyt^{2-}$, HPO_4^{2-}, immunoglobulins [21]. The contribution of cationic Zn species is also high (22–50%). These species can be unbound simple Zn(II) ions, cations released from caseins micelles, cationic complexes with LMW inorganic and organic ligands, and labile complexes of Zn [30].

3.6.4 Fruit Juices

Calcium and Mg can almost completely be found in the form of cationic species (93–100%), including simple ions and cationic complexes with organic acids and labile species [32]. A corresponding distribution of this fraction can be found for Zn. In this case, cationic species contribute 90–95% to total Zn. Remaining forms are possibly anionic species accounting for <1–8% [29]. In case of Fe, three equally abundant fractions of cationic, anionic, and residual species are established [29].

3.6.5 Beers

Forms of Mn in beer are mostly attributed to the presence of free Mn(II) ions [31]. A more complex distribution of fractions of Mn is reported elsewhere [26, 34, 35]. Contributions found for cationic species range from 60% to 90%; however, it could even be 38% [26, 34, 45]. Manganese can also be present in the form of complexes with polyphenolic substances [69, 70]. These species contribute 1–8% to total Mn [34, 35]. Remaining Mn species are anionic and residual species, possibly being forms of Mn bound to amino acids and organic acids, that is, citric, tartaric, malic, lactic, ketoglutaric, fumaric acid, of primary origin or secondary products of alcoholic fermentation [34, 35, 69, 70]. These results prove that unbound cationic forms of Mn can take part in O_2 activation processes in a similar way as it is acknowledged for Cu and Fe species. As a result, reactive oxygen species (ROS), including O_2^-, OOH$^{\cdot}$, and OH$^{\cdot}$, can be formed and oxidize different organic constituents of beer, that is, polyphenols, iso-humulons, alcohols, amino acids, fatty acids, iso-α-acids, α-acids, β-acids, and other substances [71–76]. In consequence of these reactions, alterations of beer taste, color, smell, and colloidal stability can happen. In other words, it can be stated that equilibrium existing between free Mn(II) ions and its complexing forms with organic bioligands affects stalling reactions in beer and its quality. Strongly binding Mn(II) ions, HMW polyphenolic substances of beer, seems to be a natural process decreasing ROS yield formation and enhancing colloidal stability of beer during maturation and storage.

Iron is entirely present as anionic species, and it points out that this metal can be bound to different LMW and HMW compounds [26, 31]. A very similar behavior can be observed for Cd and Pb [26]. More than 70% of total Cu is present as anionic forms; remaining species of Cu are cationic [31]. An even lower donation of cationic species for Cu (~6%) is also reported [26]. SEC-ICP-MS experiments confirm that Fe is strongly associated with organic compounds of MW 4–9 kDa. In case of Cu, complexes found are species of MW 9–12 kDa which could be forms of Cu bound to polyphenolic substances, that is, tannins, flavonoids, and phenolic acids [31].

Cationic species of Al, Co, Cr, Ni, and Zn seem to be the less abundant forms of these metals. In reference to total concentrations of metals, they account for 8%

(Cr) to 25% (Zn). This shows that these metals are likely to be mostly associated with different HMW organic ligands [26]. In case of Zn, it can also be observed that donations of cationic and polyphenolic species are related to quality and brand of beer [34, 35].

Finally, Ca, Ba, and Sr are present to a high degree (70–100%) in forms of cationic species, mostly as simple ions and/or LMW complexes with inorganic and organic compounds [26].

3.6.6 Wines

Copper is present in the form of labile species (38–48%). Additionally, this metal is bound to polyphenols and proteins (26–42%) [38]. Similarly, labile species of Fe account for 22–57%, while the fraction containing forms of Fe bound to polyphenols and proteins contributes 16–26% [38]. In case of both metals above, anionic species can be found as well, that is, 5–11% for Cu and 2–8% for Fe.

Cationic species of Zn have the highest contribution among other fractions of species of this metal, that is, 47–84%. Forms of Zn bound to polyphenols and proteins change from 8% to 33% in reference to total Zn. There is no anionic species for this metal.

3.6.7 Honeys

Calcium and Mg entirely exist in the form of cationic species that contribute 96–100% to their total concentrations [32]. These forms could be simple Ca(II) (11–18%) and Mg(II) (36–62%) ions and positively charged stable complexing ions of these metals with LMW organic acids. Donations of stable anionic complexes of Ca and Mg are lower than 6% [32].

High contributions of cationic species are also established for Zn, that is, from 40% to 100% referring to total Zn [29]. This fraction could be related to simple Zn(II) ions, stable cationic complexes of Zn with inorganic ligands, and/or its labile species. Zinc is also found to be present in the form of residual (inert) species (up to 50%), possibly being forms of Zn bound to HMW phenolic acids (gallic, tannic, phytic, salicylic, vanillic, shikimic) and some flavonoids (rutin, quercetin, myricetin, catechin), anthocyanins, and tannins. The latter compounds are well known to form with ions of metal complexes that are hydrophobic and inert in nature [2–5]. The mentioned nonpolar HMW Zn species could also be associated with complexes of Zn with polysaccharides [6].

Fractions of species determined for Fe includes cationic, anionic, and residual groupings that are rather equally abundant [29].

3.6.8 Olives

Magnesium in olives is distributed among species dissolved in water (20–33%), lipids (<1%), and chlorophyll (2–21%). Among dissolved species, cationic and Mg bound to polyphenols and polysaccharide forms can be separated [28].

3.7 Conclusions

Considering different rates of complexation of elements by endogenous bioligands of natural food products [2–6], it can be stated that cationic species of elements are the most bioavailable and bioabsorbable to humans during food digestion mostly due to the presence of simple ions, stable cationic complexes with LMW ligands, and labile species of elements [77, 78]. Information about contributions of this fraction is more important than total concentrations of elements when estimating nourishment quality and safety of food products. Evaluation of cationic species of elements could also be helpful for accurate estimating of daily element uptakes and covering of daily recommended values for elements by different food and beverages. It should be however noted that contribution of fractions of cationic species assessed with SPE procedures can be overestimated, especially when using single-sorbent SPE with strongly acidic CE resins that are capable of retaining polyphenolic species of elements. Dual-sorbent SPE procedures support obtaining more proper results of fractionations because nonpolar and weakly polar forms of elements bound to polyphenolic macromolecules are separated at first in these schemes. The latter species along with other macromolecular substances of food products, that is, proteins and polysaccharides, which strongly attach ions of elements through donor oxygen atoms of hydroxylic, phenolic, carboxylic, or sulfonic moieties, seem to be the less bioavailable to human body. These neutral or negatively charged species of nonpolar and/or weakly polar character commonly reduce availability of elements from dietary products, making them not absorbable in gastrointestinal digestion [6]. For that reason, evaluation of these hydrophobic and residual fractions of species of elements is also crucial for better estimation of elements' bioavailability from food and beverages.

It should also be concluded that fractionation procedures based on use of ion exchange and other nonionic sorbents can provide a useful knowledge about quality and nutritional value of foodstuff and beverages. In terms of identification of botanical or geographical origins, SPE procedures may be helpful in evaluating authenticity of natural origin dietary products and tracing all fake or substitute products. Any modifications of sample composition would be reflected by changes in integrity of samples and characteristic fractionation patterns of elements. It seems as well that new type of sorbents and sorptive materials appearing in commerce can be applied in SPE-based fractionation schemes and bring quite new information about existing species of elements due to additional interactions between species and solid substrates used.

References

1. Abollino O, Aceto M, Bruzzoniti MC, Sarzanini C (1998) Speciation of copper and manganese in milk by solid-phase extraction/inductively coupled plasma-atomic emission spectrometry. Anal Chim Acta 375:299–306
2. Khokhar S, Apenten RKO (2003) Iron binding characteristics of phenolic compounds: some tentative structure – activity relations. Food Chem 81:133–140
3. O'Coinceanainn M, Bonnely S, Baderschneider B, Hynes MJ (2004) Reaction of iron(III) with theaflavin: complexation and oxidative products. J Inorg Biochem 98:657–663
4. South PK, Miller DD (1998) Iron binding by tannic acid: effects of selected ligands. Food Chem 63:167–172
5. Umeta M, West CE, Futa H (2005) Content of zinc, iron, calcium and their absorption inhibitors in foods commonly consumed in Ethiopia. J Food Comp Anal 18:803–817
6. Debon SJJ, Tester RF (2001) In vitro binding of calcium, iron and zinc by non-starch polysaccharides. Food Chem 73:401–410
7. Singhal RS, Kulkarni PR, Rege DV (1997) Handbook of indices of food quality and authenticity. Woodhead, Cambridge
8. Pohl P (2007) Fractionation analysis of metals in dietary samples using ion-exchange and adsorbing resins. Trends Anal Chem 26:713–726
9. Karavoltsos S, Sakellari A, Dassenakis M, Scoullos M (2008) Cadmium and lead in organically produced foodstuffs from the Greek market. Food Chem 106:843–851
10. Lopez FF, Cabrera C, Lorenzo ML, Lopez MC (2002) Aluminium content of drinking waters, fruit juices and soft drinks: contribution to dietary intake. Sci Total Environ 292:205–213
11. Onianwa PC, Adeyemo AO, Idowa OE, Ogabiela EE (2001) Copper and zinc contents of Nigerian foods and estimates of the adult dietary intakes. Food Chem 72:89–95
12. Berton G (1995) Handbook of metal-ligand interaction in biological fluids. Marcel Dekker, New York
13. Fishbein L (2004) Multiple sources of dietary calcium-some aspects of its essentiality. Reg Toxicol Pharm 39:67–80
14. Maret W, Sandstead HH (2006) Zinc requirements and the risks and benefits of zinc supplementation. J Trace Elem Med Biol 20:3–18
15. Uchida T, Oda T, Sato K, Kawakami H (2006) Availability of lactoferrin as a natural solubilizer of iron for food products. Int Dairy J 16:95–101
16. Hulanicki A (2001) Modern analytical chemistry. Polish Scientific Publishers PWN, Warsaw
17. Hulanicki A (2006) Speciation and speciation analysis. In: Jarosz M (ed) Modern analytical techniques. Publishing Office of Warsaw University of Technology, Warsaw
18. Ozdemir Y, Gucer S (1998) Speciation of manganese in tea leaves and tea infusions. Food Chem 61:313–317
19. Ruszczynska A, Pyrzynska K, Bulska E (2004) On the use of solid-phase extraction for the fractionation of aluminium species. Chem Anal (Warsaw) 49:19–28
20. Sanz Medel A, Montes Bayon M, Fernandez Sanchez ML (2003) Trace element speciation by ICP-MS in large biomolecules and its potential for proteomics. Anal Bioanal Chem 377:236–247
21. Rivero Martino FA, Fernandez Sanchez ML, Sanz Medel A (2002) Multi-elemental fractionation in milk whey by size exclusion chromatography coupled on line to ICP-MS. J Anal At Spectrom 17:1271–1277
22. Coni E, Bocca B, Galoppi B, Almonti A, Caroli S (2000) Identification of chemical species of some trace and minor elements in mature breast milk. Microchem J 67:187–194
23. Szpunar J, Pellerin P, Makarov A, Doco T, Williams P, Medina B, Lobinski R (1998) Speciation analysis for biomolecular complexes of lead in wine by size-exclusion high-performance liquid chromatography-inductively coupled plasma mass spectrometry. J Anal At Spectrom 13:749–754

24. Odegard KE, Lund W (1997) Multi-element speciation of tea infusion using cation-exchange separation and size-exclusion chromatography in combination with inductively coupled plasma mass spectrometry. J Anal At Spectrom 12:403–408
25. Matsuura H, Hokura A, Katsuki F, Itoh A, Haraguchi H (2001) Multielement determination and speciation of major-to-trace elements in black tea leaves by ICP-AES and ICP-MS with the aid of size exclusion chromatography. Anal Sci 17:391–398
26. Pohl P, Prusisz B (2004) Pre-concentration of Cd, Co, Cu, Ni and Zn using different off-line ion exchange procedures followed by the inductively coupled plasma atomic emission spectrometric detection. Anal Chim Acta 502:83–90
27. Pohl P, Bogdal Z, Prusisz B (2005) Preconcentration and fractionation of Cd, Co, Cu, Ni, Pb and Zn in natural water samples prior to analysis by inductively coupled plasma atomic emission spectrometry. Microchim Acta 150:253–259
28. Yasar SB, Gucer S (2004) Fractionation analysis of magnesium in olive products by atomic absorption spectrometry. Anal Chim Acta 505:43–49
29. Pohl P, Prusisz B (2007) Simple and versatile operational fractionation of Fe and Zn in dietary products by solid phase extraction on ion exchange resins. Talanta 71:411–418
30. Pohl P, Prusisz B (2007) Determination of Ca, Mg, Fe and Zn partitioning in UHT cow milks by two-column ion exchange and flame atomic absorption spectrometry detection. Talanta 71:715–721
31. Svendsen R, Lund W (2000) Speciation of Cu, Fe and Mn in beer using ion exchange separation and size-exclusion chromatography in combination with electrothermal atomic absorption spectrometry. Analyst 125:1933–1937
32. Pohl P, Prusisz B (2006) Fractionation of calcium and magnesium in honeys, juices and tea infusions by ion exchange and flame atomic absorption spectrometry. Talanta 69:1227–1233
33. Pohl P, Prusisz B (2007) Fractionation analysis of manganese and zinc in tea infusions by two-column solid phase extraction and flame atomic absorption spectrometry. Food Chem 102:1415–1424
34. Pohl P, Prusisz B (2007) Fractionation analysis of manganese and zinc in beers by means of two sorbent column system and flame atomic absorption spectrometry. Talanta 71:1616–1623
35. Pohl P (2007) Manganese and zinc operational fractionation in beer by means of tandem ion exchange column assemblage and flame atomic absorption spectrometry. Microchim Acta 159:325–332
36. Pohl P, Prusisz B (2007) Fractionation of calcium, iron, magnesium and zinc in bee honeys by means of tandem column ion exchange and flame atomic absorption spectrometry. Can J Anal Sci Spectrosc 52:207–214
37. Erdemoglu SB, Pyrzynska K, Gucer S (2000) Speciation of aluminum in tea infusion by ion-exchange resins and flame AAS detection. Anal Chim Acta 411:81–89
38. Karadjova I, Izgi B, Gucer S (2002) Fractionation and speciation of Cu, Zn and Fe in wine samples by atomic absorption spectrometry. Spectrochim Acta B 57:581–590
39. Terada K (1992) Preconcentration by sorption. In: Alfassi ZB, Wai CM (eds) Preconcentration techniques for trace elements. CRC Press, Boca Raton
40. Burba P (1994) Labile/inert metal species in aquatic humic substances: an ion-exchange study. Fresenius J Anal Chem 348:301–311
41. Tessier A, Turner DR (1996) Metal speciation and bioavailability in aquatic systems. Wiley, New York
42. Jiann KT, Presley BJ (2002) Preservation and determination of trace metal partitioning in river water by a two-column ion exchange method. Anal Chem 74:4716–4724
43. Gueguen C, Belin C, Thomas BA, Monna F, Favarger PY, Dominik J (1999) The effect of freshwater UV-irradiation prior to resin preconcentration of trace metals. Anal Chim Acta 386:155–159
44. Blain S, Appriou P, Handel H (1993) Preconcentration of trace metals from sea water with the chelating resin Chelamine. Anal Chim Acta 272:91–97

45. Pohl P (2006) Application of ion-exchange resins to the fractionation of metals in water. Trends Anal Chem 25:31–43
46. Alvarez MB, Malla ME, Batistoni DA (2004) Performance evaluation of two chelating ion-exchange sorbents for the fractionation of labile and inert metal species from aquatic media. Anal Bioanal Chem 378:438–446
47. Grotti M, Abelmoschi ML, Soggia F, Frache R (2003) Determination of ultratrace elements in natural waters by solid-phase extraction and atomic spectrometry methods. Anal Bioanal Chem 375:242–247
48. Gueguen C, Dominik J, Perret D (2001) Use of chelating resins and inductively coupled plasma mass spectrometry for simultaneous determination of trace and major elements in small volumes of saline water samples. Fresenius J Anal Chem 370:909–912
49. Vasconcelos TMSD, Leal MFC (1997) Speciation of Cu, Pb, Cd and Hg in waters of the Oporto coast in Portugal, using pre-concentration in a Chelamine resin column. Anal Chim Acta 353:189–198
50. Groschner M, Appriou P (1994) Three-column system for preconcentration and speciation determination of trace metals in natural waters. Anal Chim Acta 297:369–376
51. Pesavento M, Biesuz R, Cortina JL (1994) Sorption of metal ions on a weak acid cation-exchange resin containing carboxylic groups. Anal Chim Acta 298:225–232
52. Camel V (2003) Solid phase extraction of trace elements. Spectrochim Acta B 58:1177–1233
53. Philippe M, Le Graet Y, Gaucheron F (2005) The effects of different cations on the physico-chemical characteristics of casein micelles. Food Chem 90:673–683
54. Bellido-Milla D, Onate-Jaen A, Palacios-Santander J (2004) Beer digestions for metal determination by atomic spectrometry and residual organic matter. Microchim Acta 144:183–190
55. Alcazar A, Pablos F, Martin MJ, Gonzalez AG (2002) Multivariate characterisation of beers according to their mineral content. Talanta 57:45–52
56. Asfaw A, Wibetoe G (2005) Direct analysis of beer by ICP-AES: a very simple method for the determination of Cu, Mn and Fe. Microchim Acta 152:61–68
57. Llobat-Estelles M, Mauri-Aucejo AR, Marin-Saez R (2006) Detection of bias errors in ETAAS: determination of copper in beer and wine samples. Talanta 68:1640–1647
58. Onate-Jaen A, Bellido-Milla D, Hernandez-Artiga MP (2006) Spectrophotometric methods to differentiate beers and evaluate beer ageing. Food Chem 97:361–369
59. Moreno IM, Gonzalez-Weller D, Gutierrez V, Marino M, Camean AM, Gonzalez AG, Hardisson A (2008) Determination of Al, Ba, Ca, Cu, Fe, K, Mg, Mn, Na, Sr and Zn in red wine samples by inductively coupled plasma optical emission spectroscopy: evaluation of preliminary sample treatments. Microchem J 88:56–61
60. Moreno IM, Gonzalez-Weller D, Guterrez V, Marino M, Camean AM, Gonzalez AG, Hardisson A (2007) Differentiation of two Canary DO red wines according to their metal content from inductively coupled plasma optical emission spectrometry and graphite furnace atomic absorption spectrometry by using probabilistic neural networks. Talanta 72:263–268
61. Fernandez PL, Pablos F, Martin MJ, Gonzalez AG (2002) Multi-element analysis of tea beverages by inductively coupled plasma atomic emission spectrometry. Food Chem 76:483–489
62. Gallaher RN, Gallaher K, Marshall AJ, Marshall AC (2006) Mineral analysis of ten types of commercially available tea. J Food Comp Anal 19(Suppl):S53–S57
63. Terrab A, Recamales AF, Hernanz D, Heredia FJ (2004) Characterisation of Spanish thyme honeys by their physicochemical characteristics and mineral contents. Food Chem 88:537–542
64. Rashed MN, Soltan ME (2004) Major and trace elements in different types of Egyptian mono-floral and non-floral bee honeys. J Food Comp Anal 17:725–735
65. Jodral-Segado AM, Navarro-Alarcon M, Lopez-G de la Serrana H, Lopez-Martinez MC (2003) Magnesium and calcium contents in foods from SE Spain: influencing factors and estimation of daily dietary intakes. Sci Total Environ 312:47–58
66. Simpkins WA, Louie H, Wu M, Harrison M, Goldberg D (2000) Trace elements in Australian orange juice and other products. Food Chem 71:423–433

67. Silva FV, Lopes GS, Nobrega JA, Souza GB, Nogueira ARA (2001) Study of the protein-bound fraction of calcium, iron, magnesium and zinc in bovine milk. Spectrochim Acta B 56:1909–1916
68. Bocca B, Alimonti A, Coni E, Di Pasquale M, Giglio L, Piccioli Bocca A, Caroli S (2000) Determination of the total content and binding pattern of elements in human milk by high performance liquid chromatography-inductively coupled plasma atomic emission spectrometry. Talanta 53:295–303
69. Montanari L, Perretti G, Natella F, Guidi A, Fantozzi P (1999) Organic and phenolic acids in beer. LWT Food Sci Technol 32:535–539
70. Gorinstein S, Caspi A, Zemster M, Trakhtenberg S (2000) Comparative contents of some phenolics in beer, red and white wines. Nutr Res 20:131–139
71. Mochaba F, O'Connor-Cox ESC, Axcell BC (1996) Effects of yeast quality on the accumulation and release of metals causing beer instability. J Am Soc Brew Chem 54:164–171
72. Pohl P (2008) Metals in beer. In: Preedy VR (ed) Beer in health and disease prevention. Academic, Amsterdam
73. Bamforth CW (1999) Enzymic and non-enzymic oxidation in the brewhouse: a theoretical consideration. J Inst Brew 105:237–242
74. Vanderhaegen B, Neven H, Verachtert H, Derdelinckx G (2006) The chemistry of beer aging – a critical review. Food Chem 95:357–381
75. Cortacero-Ramirez S, Hernainz-Bermudez de Castro M, Segura-Carretero A, Cruces-Blanco C, Fernandez-Gutierrez A (2003) Analysis of beer components by capillary electrophoretic methods. Trends Anal Chem 22:440–445
76. Nardini M, Ghiselli A (2004) Determination of free and bound phenolic acids in beer. Food Chem 84:137–143
77. French P, Gardner MJ, Gunn AM (1989) Dietary aluminium and Alzheimer's disease. Food Chem Toxicol 27:495–498
78. Fox PF, McSweeney PLH (1998) Diary chemistry and biochemistry. Chapman and Hall, New York

Chapter 4
Applications of Ion Exchangers in Alcohol Beverage Industry

Cristina Lasanta, Juan Gómez, and Ildefonso Caro

Abstract There are numerous applications of ion exchange process in the alcohol beverage industry. Among the most widely used techniques are, on the one hand, the use of cation exchange in acid cycle to remove potassium and stabilize wine against the sediments of potassium bitartrate crystals and, on the other hand, a mixed anion and cation-exchange treatment of wine for cation and anion removal from grape must in grape sugar production. Today, only the two aforementioned ion exchange techniques are authorized by the European Union legislation. Unlike laws in the EU, the legislation in the member countries of the World Wine Trade Group (WWTG) is less restrictive and allows the use of ion exchange in other applications. In this way, cation exchange in acid cycle is used for must/wine acidification and chelating resins for the metal removal from wine. Last but not the least, anion-exchange resins are also employed for vinegar decolourization, protein content and volatile acidity reduction, and extraction of wine compounds of commercial interest. Ion exchange treatment is a cost-effective selective process in the alcohol beverage industry, thus becoming a very attractive option nowadays.

4.1 Introduction

Ion exchange resins have several applications in oenology. First, cation-exchange resins in hydrogen and/or sodium form, in alternate regeneration cycles, are used to achieve tartaric stabilization of wine without excessive content of sodium or a very low wine pH. Second, anion-exchange resins are used to remove the excess of wine acidity or reduce volatile acidity. A combination of both resins is used in the elaboration of rectified concentrated must or 'grape sugar' to remove cations and

C. Lasanta (✉) • J. Gómez • I. Caro
Department of Chemical Engineering and Food Technology, University of Cádiz,
Campus Rio San Pedro, 40, Puerto Real 11510, Spain
e-mail: cristina.lasanta@uca.es

Inamuddin and M. Luqman (eds.), *Ion Exchange Technology II: Applications*,
DOI 10.1007/978-94-007-4026-6_4, © Springer Science+Business Media B.V. 2012

anions from the must and retain the sugar content. In addition, both chelating and cation-exchange resins are used to remove metal ions. Last but not the least, the use of adsorbent resins is aimed at removing polyphenols and products derived from their oxidation and unstable proteins.

The use of ion exchange in oenology dates back to 1945, Italy [1, 2]. The first resins employed in the wine industry were cationic resins in Na^+ form for K^+ removal in tartrate stabilization. Although such a method was temporarily permitted in several European countries, soon, two arguments in favour of its prohibition were put forth, namely, the sodium enrichment and the possible solubility of styrene in wine [3]. Nowadays, technical issues were definitely solved. Ion exchange practices have always been subject of controversy and led to confrontation between traditional and new wine-producing countries. While, the former, France, Italy and Spain, have always taken a restrictive stance on oenological practices, the latter, the United States of America, Australia, South Africa and Chile, favour a less strict approach.

Today, ion exchange treatment is permitted only for tartrate stabilization of wine and production of rectified concentrated must [4–6] by both the European Union (EU) and the international organisation of vine and wine (OIV). WWTG member governments have a more permissive approach towards ion exchange treatment, which is reflected in the agreement of oenological practices signed in Toronto in 2001, by which they agree to allow import of wine that has been produced following the regulations regarding oenological practices set by the export country [7]. In the United States, legislation authorizes the use of cationic, anionic and nonionic resins, though it imposes some limitations on the use of these resins in relation to the intensity of their effects (Electronic Code of Federal Regulations, e-CFR) [8]. The following section provides a detailed description of the aforementioned ion exchange applications.

4.2 Tartrate Stabilization, Grape Sugar Production and Acidification

4.2.1 Application to Tartrate Stabilization

Formation of potassium bitartrate sediments, formed through the reaction between the bitartrate ion (HT^-) and the potassium ion (K^+), in bottled wine is one of the most frequent causes of wine instability. Thus, wine stabilization is usually performed before bottling. The most common stabilization method is cold stabilization by chilling in which the wine is chilled to a temperature just above its freezing point and cold storage in insulated tanks at that temperature for approximately 1 week. Tartrate stabilization of wine by ion exchange can be performed in three different ways [9, 10]:

- Use of a cation-exchange resin in hydrogen, sodium and magnesium cycle; the potassium cation of the resin is exchanged with hydrogen, sodium or magnesium. This exchange results in tartaric acid, sodium bitartrate or magnesium bitartrate, respectively. A mixed resin, in sodium and hydrogen form for example, could also be used to reduce the final effect on final pH and sodium concentration.

- Use of an anion-exchange resin in hydroxyl (OH^-) form; the (OH^-) ion is exchanged for the tartrate anion or other anions, such as the citrate anion.
- Mixed treatment exchanging potassium and tartrate for (H^+) and (OH^-), respectively, with two resins, one cationic and another anionic, with the aim of exchanging some potassium bitartrate for water.

The first method, with cation-exchange resins in acid cycle, in particular, is the only procedure for treating wine permitted in the EU [4]. This is because the other two methods may significantly affect the quality attributes and organoleptic properties of the wines under treatment. In the case of wine treatment with cation-exchange resins in hydrogen form, pH decreases rapidly (below 2.0), remaining stable until the loss in total exchange capacity of the resin. Likewise, potassium content decreases to approximately 0. If treatment continues after the resin exchange capacity loss, then pH returns to its initial value, and retained potassium starts to elute, displaced for other strongly retained cations [11].

Because of the pronounced effect on the concentrations of the cations that intervene in stability, the ion exchanged wine is often blended back with a fraction of untreated wine in order to obtain a stable wine blend. To determine the adequate fraction of the wine blend, either the empirical method of cooling test [12] to the blends or the conductimetric method suggested by Gómez Benítez et al. [13] may be employed. The treated wine fractions to be mixed ranged from 10% to 20% for both white or red table wines, as well as other wines such as Sherry wines [10, 14]. In these tests, the maximum reduction in pH was 0.2 units.

The results obtained from cation-exchange treatment of wine do not depend upon the physicochemical properties of wine [15], which is an advantage when compared to the traditional cold stabilization method since it is affected by the colloidal structure of wine [16]. Also, cold treatment is not effective with wines with certain physicochemical properties, such as red wines with high levels of pH and potassium. Other authors [3] have supported the good results yielded by ion exchange applied under these conditions for red, rosé and white wines, in which a blend of 40% of untreated wine, with 3.0–3.43 pH values in the blends. Other studies [11] have reported that stabilization by ion exchange does not affect the organoleptic properties of wine, though they have observed a slight colour reduction which they attributed to pH reduction.

4.2.2 *Production of Rectified Concentrated Must (Grape Sugar)*

Concentrated rectified must is mainly used to enrich wines. One to two grammes per litre of sulphur dioxide (SO_2) is added to the first grape must to achieve a better keeping of the must, while achieving a spontaneous clarifying effect. The clear liquid fraction is passed through diatomaceous earth or a crossflow filter. The resulting clear must free from colloidal substances is then passed on cation and anion-exchange resins so that organic and mineral cations and anions are retained. The remaining phenolic and colouring substances are removed by activated charcoal. Last, the must is passed through the evaporator system so that sulphur dioxide is removed and sugars concentration increased [17, 18] up to 61.7% (minimum set by EU regulations).

4.2.3 *Applications to Must and Wine Acidification*

From the oenological point of view, pH plays an important role in winemaking, since pH levels lower than 3.5 do not only have a protective effect against bacterial infection but also contribute significantly to the sensory appreciation of wine [19]. High levels of pH are obtained when harvest is delayed to attain better phenolic maturation [16] and in warm climate wine regions [20]. Consequently, reducing the pH of the must or wine by acidification under the aforementioned circumstances is a very common oenological practice. In the European Union, only the use of L(+) tartaric acid, DL-malic acid and lactic acid is allowed, limited to 1.5 g/L for must and 2.5 g/L for wine (expressed in tartaric acid) [4, 5]. The OIV sets similar limits with no further distinction between must and wine [21, 22].

Acidification of must and wine by ion exchange has good prospects. Some studies [23, 24] show that ion exchange treatment can lower pH to 3.5 without having detrimental effects either on wine colour or sensory properties of wine. Palacios et al. [25] obtained excellent pH adjustment outcomes through the application of the aforementioned technique in the wine region of Jerez. In this study, the authors achieved a pH reduction of 3.2 by ion exchange, thus showing the feasibility and economic advantage of the technique. It is worth pointing out that acidification by ion exchange has the added benefit of allowing for partial tartrate stabilization of wine and reducing formation of potassium bitartrate sediments on the winery equipment walls (e.g. hoses, tanks and wooden barrels) that are in contact with wine.

4.3 Metal Removal

One of the problems that can affect the quality of finished wine is the presence of precipitates due to metal content, which is known as 'mettalic casse'. Such a phenomenon may take place several months after bottling and noticeably alter the organoleptic properties of wine [26, 27]. Metal content present in wine may come both from grapes and machinery used. Metal content may even increase due to the effects of environmental pollution, treatments applied in the vineyard or during winemaking, in which varying quantities of mineral, namely, iron, copper, lead, cadmium, zinc, manganese, arsenic or mercury can be found.

Iron is the most abundant metal in must and wine (2–10 ppm). Normally, Fe^{3+} is the responsible for ferric casse in wines, which is the formation of insoluble ferric complexes with tannins and phosphates. On broad lines, wines with iron concentrations below 5–7 ppm are thought to remain stable against haze formation [28]. Another metal that affects wine quality is copper. Although copper is usually found in smaller amounts, it causes the development of the so-called copper casse, which is the formation of an unstable colloid of copper sulphide and its later interaction with the proteins present in wine. The copper threshold concentration to inhibit this phenomenon is between 0.3 and 0.5 ppm [29].

It has also been proved that certain metals are catalysts for wine oxidation. Again, the most important metals involved in this process are iron and copper, and manganese and zinc, to a lesser degree [30, 31]. Usually, the amount of manganese present in wine is very small, depending on the grape region and type of winemaking [32]. Zinc is present in grapes, and even though small amounts of zinc (0.1–0.4 ppm) can be found in grapes, greater zinc concentrations can be the result of contamination during winemaking. Therefore, metal removal has always remained an important practice in the process of winemaking and performed in wine cellars by addition of potassium hexacyanoferrate (II) or blue fining [33]. However, fining with potassium hexacyanoferrate (II) entails a series of problems such as possible formation of cyanhydric acid and waste with cyanide compounds [34], which require appropriate removal and disposal, taking into account the related complications and expenditure underlying this operation.

Several studies on the application of ion exchange to metal removal from must and wine have been carried out. The first results show a reduction of 82% and 98% in iron and copper, respectively [35], for white wines, and for red wines, a reduction of 61% in iron and 88% in copper by means of a chelating ion exchange resin containing 8-hydroxyquinoline acid functional group. This type of resin does not influence potassium, calcium, sodium and manganese concentrations, and slightly varies wine pH. However, this ion exchange resin does affect both colour intensity and polyphenol content of wine. The study described below also reported a reverse relationship between process efficiency and the flow employed. When the technique is applied to treat non-fermented must, a reduction of 73% in iron and 98% in copper was achieved, and in the case of high alcohol content beverages, a 69% and 93% reduction was obtained, respectively. In this respect, it has been proved that the effectiveness of the process can be affected by several factors, namely, the alcohol degree of the wine and the ionic strength of the medium [36].

When a chelating resin is used to remove metal from non-fermented must [37], depending on the initial amount of iron and process conditions, almost full iron and copper removal occurs without affecting must pH. Nevertheless, the results from the study show that partial inhibition of yeast growth takes place in the demetallized must, which may affect later fermentation. Further studies [38], in which chelating resin containing iminodiacetic acid groups were used, suggest that costs can be reduced up to approximately 60% when compared to blue fining. In fact, a reduction of 99% and 97% of iron and copper, respectively, was achieved in white wines. In addition, the technique does not have significant effects either on wine pH or on sodium, potassium, calcium and magnesium content. The technique does not affect the colour or protein index of wine.

When a chelating resin containing iminodiacetic acid groups and a cationic one are compared in white wines, it appears that both types of resin are quite effective in reducing the iron (between 81% and 99%, depending on the wine treated and resin used) and copper content (between 51% and 96%). However, it appears that chelating resins reduce the amount of polyphenols to a greater extent, which translates into a decrease in the tendency towards browning. But cationic resins cannot be used for removing metals because they eliminate not only metals but also other cations of the wine [39].

Among other resins that can be used for metal removal in wine are PVI/PVP resins (polyvinylimidazole/polyvinylpyrrolidone copolymers). However, these resins seem to be less effective for iron (72% in whites and 47% in red wines) and copper (58% in whites and 74% in red wines) removal under the study conditions [40]. The abovementioned resins retain noticeable amounts of aluminium (40% in whites and 46% in red wines) and lead (14% and 24% in white and red wines, respectively). As regards the characteristics of the wines treated, a slight effect upon the phenolic composition and colour, a decrease in total acidity and an increase in pH level were observed. On the other hand, no changes were observed regarding the sensory characteristics of the wines treated.

Furthermore, studies in which different resins were employed to carry out a selective removal of lead concentrations coming from any contaminating source were conducted [41]. In these studies, it was shown that better or worse results in the removal of a certain metal are associated to the type of resin used and operation conditions. In this case, anion-exchange and adsorption resins do retain lead effectively. However, while the former raise pH in wines, the latter increase the amount of chlorine concentration. Chelating-type resins become effective when they are 'regenerated' in acid cycle and 'weak acid', and cation-exchange resins are not useful if selective removal of lead is intended.

4.4 Other Applications

In addition to the aforementioned main ion exchange practices in the oenology industry, there exist other applications of this basic operation that should be pointed out due to its relative importance and use: the application of ion exchange technology to decolourization of wine, vinegar and by-products; protein and volatile acidity removal; and extraction of compounds of commercial interest.

4.4.1 Decolourization

Ion exchange technology for decolouring wine is used due to the great adsorption capacity of certain anion-exchange resins over anthocyanins and other wine polyphenols which are responsible for wine colour (especially in red wines) [42]. However, the aforementioned resins are also responsible for the colour in brandy, vinegar or other liquors aged for a long time in wood barrels [43]. Certain oxidative polymerization reactions of such polyphenols result in excessive colouring, thus causing the aforementioned products to be brown or brownish. Such chemical reactions are usually the result of processes in which there is excessive exposure to heating and air. Since excessive colour may cause beverages to be unsuitable for consumers, colour reduction is necessary. What is more, the alteration of the original flavour of wine is a frequent occurrence as a result of the aforementioned browning processes.

In this respect, for instance, data from industrial continuous vinegar decolourization with Lewatit S-5328-A exchange resins [44, 45] have been made available. Lewatit S-5328-A is an anion-exchange resin that contains quaternary ammonium functional groups and has great polyphenol adsorption capacity. In its normal state, it is a resin in the Cl^- form. The characteristics of the equipment used in the industrial process operation were as follows: column length, 1.5 cm; column diameter, 25 cm; microsphere diameter, 0.5 mm; and porosity, 0.34. The total vinegar volume treated was 150 bed volume (BV); the decolourization efficiency (DE) and the total polyphenol index removed (TPIR) were of 75.5% and 60%, respectively. DE is the difference in colour intensity obtained after treatment divided by the initial colour intensity, expressed in percentage. Colour intensity is the sum of absorbances measured at different wavelengths. TPIR is the difference in the total polyphenol index obtained after treatment, divided by the initial polyphenol index, expressed in percentage. On the other hand, as expected, the different types of polyphenols were retained at different rates. In this way, the hydroxycinnamic acid retention was of 99%, while catechins (+) and gallic acid retention was of 85%. The retention percentages corresponding to other polyphenols were lower.

4.4.2 Protein Removal

The excessive concentration of proteins in wine may cause precipitation during storage, which, in turn, may lead to visual and taste wine faults, a common cause for wine rejection. Depending on several factors, the natural protein concentration present in wine may vary from 15 to 275 mg/L, with molecular weight between 10 and 160 kDa, and has an isoelectric point (p*I*) of 3.1–9.2 [46–48]. Although most proteinaceous fractions in wine are usually sensitive to proteolytic enzyme treatment [49], not all fractions are fully digestible. Besides, wines usually have other concentrations, such as polyssacharides and mannoproteins that usually interfere with proteolytic activity to a lesser or great extent [50]. Consequently, treatment with sodium bentonite is the most widely used fining technique to remove proteins from wine. Even though sodium bentonite is a convenient and inexpensive technique, it has its drawbacks: wine loss and production of solid waste which becomes difficult to deal because, after addition of bentonite, wine has to be passed through diatomaceous earth to remove precipitates.

An alternative technique to bentonite fining employed to achieve protein removal from wine is the use of anion-exchange resins (generally with sulphonic functional groups). Among these resins are Macro-Prep 50S (Bio-Rad, USA), Streamline LS (Pharmacia Bioprocess Tech., Sweden) or S-Spherosil (Institut Merieux, France). There is no comprehensive understanding of the phenomenon underlying the removal of proteins by the matrix of these resins since there are numerous factors influencing the fractioning process, namely, the temperature, wine pH and isoelectric point of the resins to be removed, concentration of tannins, polysaccharides, mannoproteins, metallic cations and other wine constituents [51]. Nevertheless, some kinematic models which achieve considerable precision for such industrial operations have been published [52].

Good outcomes were achieved in a series of tests in which protein retention in wine was performed with ion exchange resins under different conditions. In most cases, the total protein retention (TPR) rate was 80%, and in some other cases, 96% [53]. TPR is the difference in the protein index obtained after treatment, divided by the initial protein rate, expressed in percentage. Thus, S-Spherosil and Streamline LS have a mean TPR of 90% versus the standard TPR value obtained with bentonite treatment, which is usually 82%.

4.4.3 Volatile Acidity Reduction

When the amount of wine volatile acidity reaches 1 g/L approximately, wine starts to develop negative organoleptic properties, which are easily detected by the majority of consumers and lead to wine rejection. Such an excessive amount of acetic acid is often the result of undesirable bacterial activity (acetic acid bacteria or lactic bacteria), which requires its removal. A relatively new method that has been suggested as an alternative for reduction of acetic acid in wine is the use of anion-exchange resins.

Such a procedure mainly involves two stages: the first stage is a reverse osmosis stage and, the second one, an ion exchange stage. In the first stage, the smallest molecules in wine such as H_2O, ethanol and acetate anions with acidic protons are separated from wine by passing the permeate stream through the filter membrane with the appropriate molecular cut-off weight. The remaining constituents of wine with greater molecular size are retained shortly after, in the second stage, this permeate is passed through an anion (OH^- from) exchange resin. This process causes the exchange of the acetate anions for hydroxyl anions, thus, allowing for definite removal of excess acetate acid from the permeate which can be then recombined with the retentate [54].

4.4.4 Compound Extraction

With respect to the extraction of compounds of commercial interest with ion exchange resins, there is relevant literature in relation to the adsorption of procyanidins (a type of tannin) using different types of ion exchange resins (D101, LSA800B and Amberlite XAD-16). Procyanidins come from the same polyphenol group as those mentioned in Sect. 4.4.1, and given their varied properties, they present a number of health benefits. Therefore, procyanidins concentration and extraction from wine ferment is of commercial interest [55]. In general, LSA800B is the resin that presents higher procyanidin retention rates under a wide range of working conditions.

4.5 Conclusion

The main advantage of ion exchange when it is used in the manufacture of alcoholic beverages and products is its selectivity, which usually allows very specific treatments. Furthermore, ion exchange also shows other advantages over the traditional techniques. In the tartaric stabilization of wine, it gives results in much less time than cold treatment and it is not affected by the composition of wine. In acidification of musts and wines, it allows simultaneous partial tartaric stabilization. In metal removal, it prevents the formation of dangerous residues of ferric hexacyanoferrate generated during the traditional blue fining. In other cases, as in the production of grape sugar or in the reduction of volatile acidity, it is simply necessary. In addition, it has usually lower costs. However, the prediction of the adsorptive behaviour over other components of alcoholic beverages is not easy and must be considered in each case.

References

1. Garino-Canina E (1953) Equilibri físico-chimici del vino e scambiatori ionici. Anuario della Stazione Enologica Sperimentale di Asti 2(4):3–13
2. Garino-Canina E (1953) L'emploi des échangeurs dions dans la technique enologique moderne. Anuario della Stazione Enologica Sperimentale di Asti 2(4):17–27
3. Díaz-Yubero F (1993) Empleo de las resinas de intercambio iónico en Enología. La Semana Vitivinícola 2441–2442(22/29):1921–1947
4. European Union (2009) Commission Regulation (EC) N 606/2009 of 10 July 2009 laying down certain detailed rules for implementing council regulation (EC) No 479/2008 as regards the categories of grapevine products, oenological practices and the applicable restrictions. L 193:1–59
5. European Union (2008) Council Regulation (EC) No 479/2008 of 29 April 2008 on the common organisation of the market in wine. L 148:1–61
6. OIV (2009) International code of oenological practices. OIV, Paris, 2009 Issue
7. WWTG (2001) Agreement on mutual acceptance of oenological practices. http://trade.gov/td/ocg/eng_agreement.htm Accessed on 1 Feb 2011
8. ECFR (2010) Title 27: alcohol, tobacco and firearms PART 24-WINE. §24.248 Processes authorized for the treatment of wine, juice, and distilling material. http://federal.eregulations.us/cfr/rulehome/1/13/2012/Title27/ChapterI/Part24/Section24.248.html#%c2%a7+24.248 Accessed on 26 Mar 2012
9. Mira H, Leite P, Ricardo-Da-Silva J, Curvelo-Garcia AS (2006) Use of ion exchange resins for tartrate wine stabilization. J Int des Sci de la Vigne et du Vin 40(4):223–246

10. Mourgues J (1993) Utilisitation des résines échangeuses d'ions. Revue des Oenologues, n° 69S:51–54
11. Hernández P, Mínguez S (1997) Utilisation des resines d'échange ionique en oenologie. Stabilisation tartrique. Rev Franc d'Oenol., n° 162:32–35
12. Brugirard A, Rochard J (1992) Prévention des précipitations tartriques. In: Aspects pratiques des traitements thermiques des vins. Bourgogne-publications, Chaintré, pp 74–105
13. Gómez Benítez J, Palacios Macías VM, Veas López R, Pérez Rodríguez L (2003) Prediction of tartrate stability of sherry wines by a conductimetric system with rapid response. Food Chem 81:457–462
14. Gomez Benítez J, Palacios Macías VM, Sánchez Pazo JA, Pérez Rodríguez L (2002) Industrial development of proton exchange for tartrate stabilization of sherry wines. Eur Food Res Technol 214:418–422
15. Rankine BC, Bond RD (1955) Prevention of potassium bitartrate deposition in wine by cation-exchange resins. Aust J Appl Sci 6:541–549
16. Ribereau-Gayon P, Dubourdieu D, Donèche B, Lonvaud A (2006) Handbook of enology, 2nd edn. Wiley, Chichester
17. Pompei C (1982) Grape sugar: technological aspects (production and utilization). Bulletin de l'OIV 55(611):25–52
18. Razzari F, Brambilla C, Peri C, Pompei C (1978) Grape sugar production for wines and foods. VII. Use of ionic exchange resins and synthetic decolorizing polymers in the production of grape sugar. Rivista di Viticoltura e di Enologia 31(3):105–120
19. Drysdale G, Fleet G (1988) Acetic acid bacteria in winemaking: a review. Am J Enol Viticult 39:143–154
20. Valcarcel MJ, González P, Pérez L, Asencio A, Domecq B (1990) Control de calidad del estado sanitario de la uva en la zona del jerez. Vitivinicultura, n° 5:42–49
21. O.I.V (1999) Resolution OENO 3/99 concerning chemical acidification of must
22. O.I.V (2001) Resolution OENO 13/2001 concerning maximum limits for acidification of musts
23. Walker T, Morris J, Threlfall R, Main G (2004) Quality, sensory and cost comparison for pH reduction of Syrah wine using ion exchange or tartaric acid. J Food Quality 27(6):483–496
24. Walker T, Morris J, Threlfall R, Main G (2002) pH Modification of cynthiana wine using cationic exchange. J Agric Food Chem 50(22):6346–6352
25. Palacios VM, Caro I, Pérez L (2001) Application of ion exchange techniques to industrial process of must acidification. Separ Purif Method 30(1):143–156
26. Boulton R, Singleton V, Bisson L, Kunkee R (1995) Principles and practices of winemaking. Chapman and Hall, New York, pp 320–351
27. Navrotsky VI, Avakiants SP (1986) La Theorie du Viellissement du Vin. In: IV Congress of Wine, Tradition, Economy and Health. O.I.V. Varna, Bulgaria
28. Wurdig G, Woller R (1989) Chemie des wines. Ed Ulmer, Stuttgart
29. Eder R, Screiner A, Schlager G, Wendelin S (2003) Réduction de la teneur en métaux dans les vins à l'aide de résins sélectives. Bull OIV 76:243–260
30. Cacho J, Castells JE, Esteban A, Laguna B, Sagristá N (1995) Iron, Koper, and manganese influence on wine oxidation. Am J Enol Viticult 46:380–380
31. Li H, Guo A, Wang H (2008) Mechanisms of oxidative browning of wine. Food Chem 108:1–13
32. Cabrera-Vique C, Teissedre PL, Cabanis MT, Cabanis JC (2000) Manganese determination in grapes and wines from different regions of France. Am J Enol Viticult 51(2):103–107
33. Tarantola C (1963) Traitement des vins par le ferrocyanure de potassium. Annales de Technologie Agricole 12:36–41
34. Sánchez-Pineda MT, Martín López E (1997) Metales pesados en el Vino: alternativas al tratamiento con Hexacianoferrato (II) de Potasio. Alim Equip Technol 3:111–115
35. Loubser GJ, Sanderson RD (1986) The removal of copper and iron using a chelating resin. S Afr J Enol Vitic 7(1):47–51
36. Lasanta C, Caro I, Pérez L (2005) Theoretical model for ion exchange of iron (III) in chelating resins: application to metal ion removal from wine. Chem Eng Sci 60:3477–3486

37. Feng M, Mei J, Hu S, Janney S, Carruthers J, Holbein B, Huber A, Kidby D (1997) Selective removal of iron from grape juice using an iron (III) chelating resin. Sep Purif Technol 11:127–135
38. Palacios V, Caro I, Pérez L (2001) Application of ion exchange techniques to industrial process of metal ion removal from wine. Adsorption 7:131–138
39. Benítez P, Castro R, Barroso C (2002) Removal of iron, copper and manganese from white wines trough ion exchange techniques: effects on their organoleptic characteristics and susceptibility to browning. Analitica Chimica Acta 458:197–202
40. Mira H, Leite P, Catarino S, Ricardo-Da-Silva JM, Curvelo-Garcia AS (2007) Metal reduction in wine using PVI-PVP copolymer and its effects on chemical and sensory characters. Vitis 46(3):138–147
41. Minguez S, Hernández P, Gonzalo M (1996) Selective extraction of lead from wine with ionic exchange resins. In: Proceedings of the 5th symposium international Oenolologie. Tec & Doc Lavoisier, Paris
42. Shahidi F, Naczk M (1995) Food phenolics: sources, chemistry, effects and applications. Technomic Publishing Co, Lancaster, pp 136–149
43. Lamuela RM, Buxaderas S, Singleton VL, de la Torre MC, Lacueva CA, Ibern M (2000) Browning of Cava (sparkling wine) during aging in contact with lees due to the phenolic composition. Am J Enol Viticult 51(1):29–36
44. Achaerandio I, Güell C, López F (2002) Continuous vinegar decolourisation with exchange resins. J Food Eng 51:311–317
45. Achaerandio I, Güell C, López F (2007) New approach to continuous vinegar decolourisation with exchange resins. J Food Eng 78:991–994
46. Bayley FC, Berg HW (1967) Grape and wine proteins of white wine varietals. Am J Enol Viticult 17:18–32
47. Murphey JM, Spayd SE, Bewers JR (1989) Effect of grape maturation on soluble protein characteristics of Gewurztraminer and white Riesling juice and wine. Am J Enol Viticult 40:99–207
48. Moio I, Addeo F (1989) Focalizzazione isoelettrica delle protein dei mosti i dei vini. Vignevini 4:53–57
49. Dizy M, Besson L (1999) White wine protein analysis by capillary zone electrophoresis. Am J Enol Viticult 50:120–127
50. Maine-Ledoux V, Dubourdieu D (1999) An invertasa fragment responsible for improving the protein stability of dry white wine. J Sci Food Agric 79:537–543
51. Santoro M (1995) Fractionation and characterization of must and wine proteins. Am J Enol Viticult 46:250–254
52. Sarmiento MR, Oliveira JC, Slatner M, Boulton RB (1999) Kinetic of the adsorption of bovine serum albumin contained in a model wine solution by non-swelling ion-exchange resins. J Food Eng 39:65–71
53. Sarmiento MR, Oliveira JC, Slatner M, Boulton RB (1999) Effect of ion-exchange adsorption on the protein profiles of white wine. Food Sci Technol Int 7(3):217–224
54. Smith CR (1996) Apparatus and method for removing compounds from a solution. US Patent 5:480,665
55. Wang W, Yue J, Li H, Gao H (2008) Adsorption of procyanidins from grape wine's sedimentation by three macroporous absorbent resins. Food Sci Technol 6:146–149

Chapter 5
Use of Ion Exchange Resins in Continuous Chromatography for Sugar Processing

Viviana M.T.M. Silva, Pedro Sá Gomes, and Alírio E. Rodrigues

Abstract In this chapter, the use of ion exchange resins in continuous sugar process industry is reviewed. A particular focus is given to chromatographic methods and, in particular, to continuous annular chromatography and simulated moving-bed (SMB) apparatus. The use of ion exchange resins for the separation of a fructose and glucose mixture (by means of an SMB unit) and the separation into pure components of a ternary mixture of sugars (by means of a Pseudo-SMB technique) is detailed. The use of ion exchange resins in integrated reactors (separation in situ with reaction) is also addressed in this chapter, presenting several advantages of these promising intensified processes for the sugar industry.

5.1 Introduction

Sugar, and its derivatives, is considered one of the more relevant commodities in the world's food supply chain. Its market growth average rate was 2.5% per year, in the last decade (see Fig. 5.1).

V.M.T.M. Silva (✉) • A.E. Rodrigues
Department of Chemical Engineering, Laboratory of Separation and Reaction Engineering, Associate Laboratory LSRE/LCM, University of Porto, Rua Dr. Roberto Frias, 4200-465 Porto, Portugal
e-mail: viviana.silva@fe.up.pt; arodrig@fe.up.pt

P. Sá. Gomes
Department of Chemical Engineering, Laboratory of Separation and Reaction Engineering, Associate Laboratory LSRE/LCM, University of Porto, Rua Dr. Roberto Frias, 4200-465 Porto, Portugal

CAT Catalytic Center – Institute for Technical and Macromolecular Chemistry, Rheinisch-Westfälische Technische Hochschule – RWTH Aachen University, Worringerweg 1, D-52074 Aachen, Germany
e-mail: pedro.sa.gomes@fe.up.pt

Inamuddin and M. Luqman (eds.), *Ion Exchange Technology II: Applications*,
DOI 10.1007/978-94-007-4026-6_5,

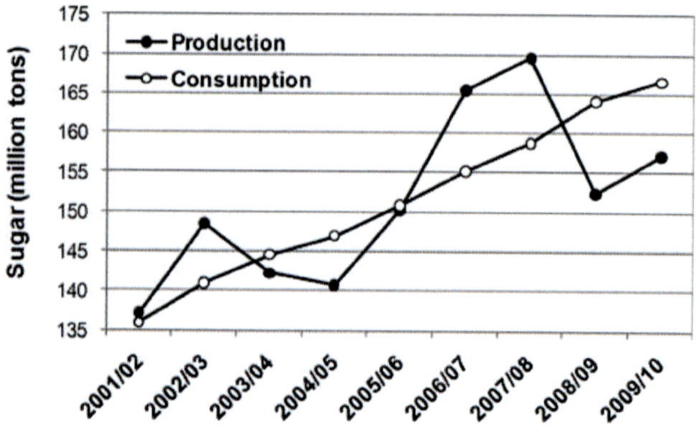

Fig. 5.1 World sugar production and consumption, raw value (Data source: ISO, World sugar balances)

Among other sugars, sucrose is the most common one obtained from natural sources, being mainly produced from sugarcane (*Saccharum officinarum*) and sugar beet (*Beta vulgaris*). Sucrose production uses about 28 million hectares of agricultural ground, 88% planted with sugarcane and the rest with sugar beet. The major beet producers are the European Union (with France, Germany, United Kingdom, and Poland representing 64% of the EU-27 production), United States, Russian Federation, and Turkey, representing half of the beet sugar production. Brazil and India cane sugar production represented half of the total world production in 2009, as shown in Fig. 5.2a. According to the Food and Agriculture Organization of the United Nations (FAOSTAT), the sugar production has more than doubled since 1960 due to an increase in the sugarcane culture. The beet sugar covered area continues to decrease since the mid-1970s, as shown in Fig. 5.2b. In fact, since the 1920s, sugar beet industry is facing the competition of cane sugar production, and from that time, it has been largely supported by tariff protection.

Sucrose is mainly used as sweetener but also as a preservative or a flavor enhancer. Sucrose is a disaccharide constituted by simple sugars (monosaccharides): glucose (dextrose) and fructose (levulose). Therefore, sucrose hydrolysis, also known as sucrose inversion, produces an equimolar mixture of glucose and fructose, called as invert sugars (Fig. 5.3). Other oligosaccharides are also used as sweeteners, as maltose (a disaccharide constituted by two units of glucose; it appears usually in corn syrups), lactose (a disaccharide found in milk and is formed from galactose and glucose), and raffinose (a trisaccharide composed by galactose, fructose, and glucose; commonly find in beet sugars).

Sugar's sweetness depends on its physical properties, concentration, temperature, pH and the presence of other substances [1]. A sugars comparison in terms of sweetness is shown in Table 5.1, considering sucrose as the reference. The need for sugar substitutes arose from dietary proposes aiming to enhance the sweet taste and, simultaneously, decrease the caloric content. Most of the approved sugar substitutes are artificial sweeteners (saccharin, aspartame, acesulfame potassium,

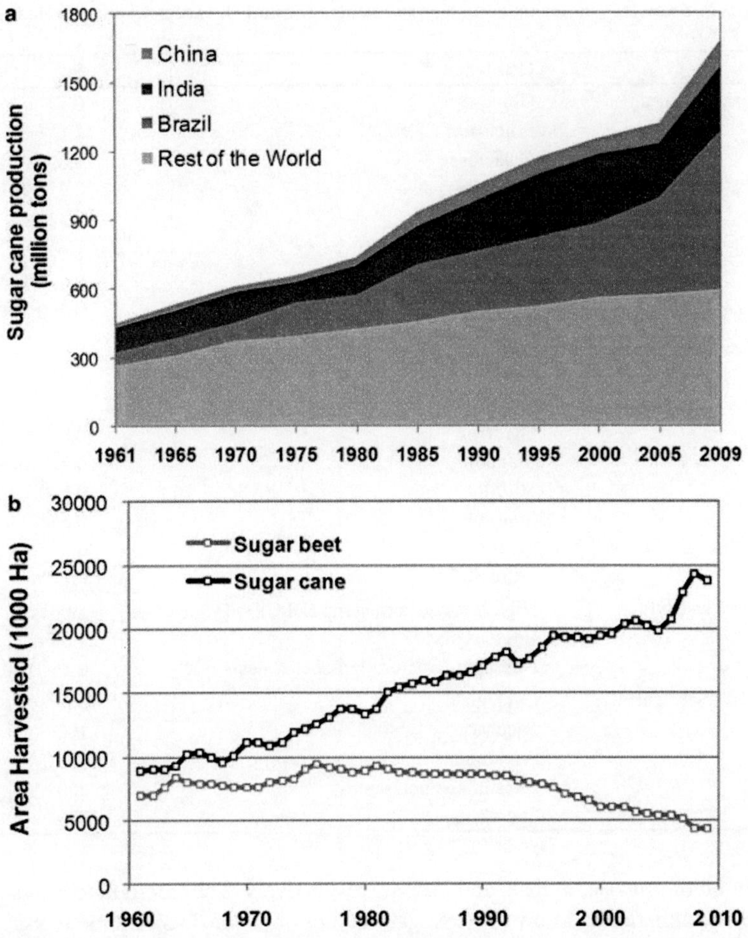

Fig. 5.2 World sugar evolution, 1961–2010: (**a**) sugar cane production (*on the left*), (**b**) harvested areas of sugar beet and cane (*on the right*) (Data source: FAOSTAT, Agricultural Production Crops Primary)

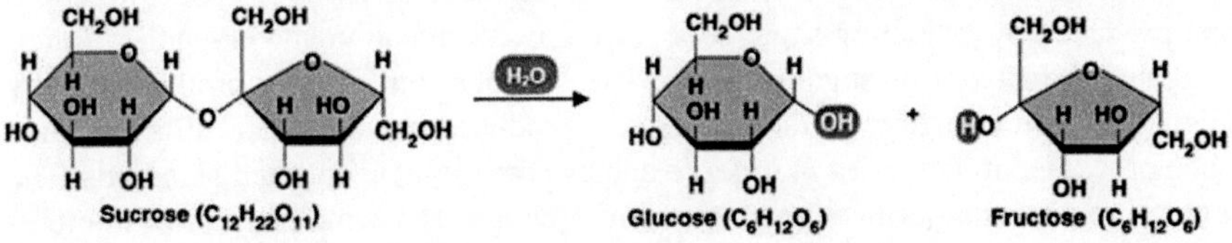

Fig. 5.3 Sucrose inversion to form glucose and fructose

and sucralose), which lately have been associated with some health risks. Therefore, natural alternatives are preferable, such as sorbitol, xylitol, and lactitol, which, in spite of being found in some fruits and vegetables, are commercially produced by

Table 5.1 Sweetness of sugars and other sweeteners (From [1], except when indicated: & [3]; $ [4]; # [5])

Type	Compound	Sweetness
Monosaccharides	Glucose	0.7
	Fructose	1.1–1.8
	Galactose	0.6; 0.21–0.32[&]
	Xylose	0.7
	Mannose	0.32[&]
Disaccharides	Sucrose	1
	Lactose	0.4; 0.16[&]
	Maltose	0.4; 0.32[&]
Trisaccharides	Raffinose	0.01
	Maltotriose	0.18[$]
Polyols	Xylitol	1
	Glycerol	0.8
	Mannitol	0.7
	Sorbitol	0.5–0.7[#]
	Maltitol	0.63[&]
	Lactitol	0.34[&]
	Cellobitol	0.11[&]
Mixtures and syrups	High fructose corn syrup (HFCS)	1.0–1.6[#]
	Invert syrup	1.05[#]
	Neosugar (fructo-oligosaccharides – FOS)	0.46–0.60[#]
Artificial sweeteners	Cyclamate	30
	Saccharin	100
	Aspartame	180
	Acesulfame potassium	200
	Sucralose	600

reduction of glucose, xylose, and lactose, respectively. The fructooligosaccharides (FOS), commercially know as neosugar, are not only considered to be low-calorie sweeteners but also classified as prebiotics due to their beneficial effects for intestinal flora [2].

Still, sucrose is the most abundant and economic viable sweetener, usually supplied as raw, white, or brown sugars. Raw sugars, with a yellowish/brown color, are produced by extracting sugar beet pulp or by purification and crystallization of the juice extracted from sugarcane. Nowadays, beet sugar factories produce directly high-grade white sugars, while raw sugar is produced in cane sugar mills, which is then processed in refineries in order to achieve the high purity sugar standards. The conventional sugar purification process involves successive melting of raw intermediate products, followed then by recrystallization. This process is robust and reliable for large-scale production of very high-quality products, but still presents some inefficiency. Therefore, "new" technologies have been introduced to meet the increasing global demand from sugar industry, aiming to reduce production costs and increase sugar recovery, without compromising the product quality. Therefore, this work will review the use of cost-effective ion exchange processes for sugar industry, focusing on continuous chromatographic apparatus. Two examples, using simulated moving-bed-based technologies for sugar separations, are detailed by simulations obtained from a well-stated mathematical model.

5.2 Ion Exchange Resins for Sugar Processing

Sugar industry competes in a global economy, and therefore, new processes are needed to decrease production costs and increase its final product quality. In this context, ion exchange processes are regarded as the most promising technique to achieve such goals, allowing higher sugar yields, reducing wastes. and meeting high-quality requirements.

5.2.1 Sugar Juices Softening

Before being evaporated, sugar beet juices are softened in order to prevent scale formation during evaporation stages, produce soft molasses; or prevent thick juices during storage [6]. The softening process consists on exchanging alkaline earth ions (as Ca^{2+}) for alkali ions (as Na^{+}), using weak or strong cation exchange resins (WAC or SAC resins, respectively). In the past years, there has been a considerable interest in such technique, as one can observe by the number of patent applications presented in Table 5.2.

In the conventional process, calcium is removed prior to evaporation using a SAC resin (in Na^{+} form) of macroporous (as Lewatit S 2528 and Amberlite™ FPC22) or gel (as Lewatit S 1468, Amberlite™ FPC14, and Dowex™ HCR-S) type, which are then regenerated with NaCl. The major drawbacks of this regeneration procedure are the dilution of treated juice, hot water consumption, salt consumption, losses of sugar in molasses, and wastewater generation. Alternatively, WAC resins (in H^{+} form) might be used and then regenerated with strong inorganic acids (as sulfuric acid). The WAC process has higher throughput then the SAC process and leads to significant water savings, reducing not only the production cost but also the wastewater treatment cost. However, SAC resins, since are sodium substituted, do not incur the risk of sucrose inversion and enable the removal of all cations, contrarily to WAC resins that are in acid form. Therefore, alternative regeneration steps have been developed. The so-called Gryllus process, which uses thick juice (green syrup II) as regenerant [7], reduces reactant consumption and wastewater generation but implies third-strike precipitation of calcium salts as well as the Standard Liquor one recycling and turbidity in the first-strike sugar. The "New Regeneration System" (NRS) process (also called Akzo) uses a mixture of the concentrated softened juice and a little amount of caustic soda, which avoids salt addition with zero pollution, resulting in one of the most efficient decalcification processes, still presenting some disadvantages, such as soda consumption and

Table 5.2 Sugar juices softening: patented processes using ion exchangers

Patent	Title	Assignee	Inventors
1975 US3887391	Process for the decalcification of sugar beet juice	The Amalgamated Sugar Company	Schoenrock, K. W.R. Richey, P. Rounds, H. G.
1976 US3982956	Process for the purification of impure sugar juice	The Amalgamated Sugar Company	Schoenrock, K. W.R. Gupta, A. Rounds, H. G.
1982 US4329183	Decationization of aqueous sugar solutions	Rohm and Haas Company	Rousseau, G. Lamotte, C.
1992 US5096500	Process for decolorization and decalcification of sugar solutions	Rar-Refinarias de Açúcar Reunidas, S.A.	Bento, L. R.
1994 EP0629707	Process for softening a sugar containing juice such as molasses and the application thereof to a process for the recovery of the sugars contained in this juice	Applexion	Lancrenon, X. Saska, M.
1995 US5443650	Process for softening a sugar-containing aqueous solution, such as sugar juice or molasses	Board of Supervisors of Louisiana State University and Agricultural; Societe Nouvelle de Recherches et D'Applications Industrielles	Saska, M. Lancrenon, X.
1995 US5454875	Softening and purification of molasses or syrup	Board of Supervisors of Louisiana State University and Agricultural	Clarke, S. J.
1998 EP0832986	Process for regenerating ion exchange resins in the process of decalcifying sugar juices	Generale Sucriere SA	Rousseau, G.
1999 US5893947	Process for purifying sugar solutions	Advanced Separation Technologies Incorporated	Pease, S. F. Weiss, S. M.
2004 US6709527	Treatment of sugar juice	Ufion (Pty) Limited	Fechter, W. L. Brewer, P. A. Van De, P. G. Smith, I. A.

increased coloration of clarified juices. A more recent process, which uses dilute molasses for regeneration, claim s to have solved the drawbacks of the NRS and Gryllus processes [8].

5.2.2 *Sugar Yield Enhancement*

5.2.2.1 Demineralization of Sugar Solutions

The beet sugar yield can be increased by replacing alkali metal cations (as K^+ and Na^+) by alkaline earth cations (as Ca^{2+} and Mg^{2+}) [9] and consequently promoting the sucrose crystallization. This can be explained by the fact that sucrose solubility is largely decreased by replacing alkali cations by alkaline earth cations. The first patent concerning this phenomena reports the use of barytes to enhance the process of extracting sugars [10]. Subsequent processes replace alkalis (K, Na) by calcium or other alkaline earths (Mg, Ba, and Sr) using zeolites or silicates [11, 12]. Later, Quentin process was established by choosing magnesium as the best ion to replace alkalis, since the sucrose solubility is lower in the presence of Mg than in the presence of other alkaline earth metals [13]. In spite of the viscosity of sugar solutions containing magnesium being higher, the overall crystallization process is improved and the amount of sugar in the molasses is reduced to half. Examples of resins used for the Quentin process are Amberlite IR 122S, Amberlite FPC22, Amberlite FPC23 H, Lewatit S 2528, and Dowex Monosphere™ C600B. Further developments in the demineralization process are being pursued by several companies, such as The Dow Chemical Company, USA [14]; Suiker Unie, NL [15]; and Organo Corporation, USA [16]; among others.

5.2.2.2 Sugar Recovery from Molasses

The sugar yield of sugar beet industry was significantly enhanced by molasses desugarization (containing about 50% sucrose on solids and only 1–2% invert sugars). In this process, sucrose is separated from final molasses (nonsugars, as glutamic acid, betaine, and raffinose) by ion exclusion chromatography. However, the process of sugarcane molasses desugarization (about 20–30% sucrose on solids and 15–25% invert sugars) is not so well established, since the separation of sucrose present in the sugar fraction is regarded as difficult and expensive, when compared with the state-of-the-art processes applied to the sugar beet one [17]. Nevertheless, more recent developments proved that the sugarcane molasses desugarization can be processed by means of the ion exchange technology in a continuous way, using simulated moving-bed (SMB) processes with ion exchange resin in Na^+ or Ca^{2+} form [18, 19]. This process requires low concentrations of calcium and magnesium in molasses, in order to avoid competition for the resin active sites.

5.2.3 Color Removal from Sugar Syrups and Juices

Sugarcane color depends essentially on its nonsugar content, since sucrose, glucose, and fructose are white crystalline materials. The color is due to the presence of some extractables from sugarcane (flavonoids and polyphenolics) or sucrose degradation compounds. Typically, the decolorization process was performed by means of traditional bone char or granular activated carbon systems. However, anionic resins proved to be more economical, simple, and flexible decolorizers, being now widely used in the sugar industry [20, 21]. Porous quaternary ammonium resins (strong base resins) are particularly good for sugarcane juices and syrups decolorization, being then regenerated with brine solutions; while weak base resins of low cross-linking degree are more suitable for beet sugars [22]. The resins properties are quite important for the color removal performance and for the resin regeneration efficiency. For instance, anionic styrenic resins are more efficient decolorizing materials than acrylics, but are also more difficult to regenerate [23]. Therefore, the sugar decolorization has been improved not only by developments in terms of resins production [24, 25] but also by improving the resin regeneration process [26, 27].

5.2.4 Sucrose Inversion

Liquid sugars can be essentially supplied in two types: (1) essentially pure sucrose (produced by melting granulated sugar or from decolorized, high-grade refinery process liquors) and (2) mixture of sucrose and invert sugars (produced by partial hydrolysis of sucrose). Sugar inversion occurs under acidic conditions (acid catalysis) or in the presence of certain enzymes, called invertases, as the β-d-fructofuranidase [1]. Concerning acid catalysis, heterogeneous catalysts are usually preferable than homogenous, in order to avoid the neutralization step of the mineral acid and subsequent removal of the formed. However, diffusional resistances inside porous catalysts may limit the reaction kinetics, particularly for viscous solutions. Considering cation exchange resins (in acid form), the kinetics of sucrose inversion will depend on resin activity and mass transfer limitations. Apart operating conditions that affect both phenomena (as solution composition, temperature, hydrodynamic conditions, and residence time), the resin properties (such as the nature of functional groups, active sites concentration, polymer matrix, and cross-linking density) have major impact either in the reaction kinetics as in the internal mass transfer. Regarding resin functionalization, sulfonic cations are more active than carboxylic cations [22]. The influence of other resin properties onto the sucrose inversion kinetics was performed by considering two different resins: Amberlyst 15 and Amberlyst XN-1010 [28]. Both are of macroporous sulfonic acid ion exchange resins of polystyrene-divinylbenzene copolymer, but with different properties (see Table 5.3). The effect of particle size (from 200 to 650 μm) and reaction temperature (from 50°C to 70°C) on the initial reaction rates allowed to conclude that

Table 5.3 Properties of macroporous sulfonic acid ion exchange resins [29]

Properties	Amberlyst 15	Amberlyst XN-1010
Active sites concentration (meq/g)	4.4–4.8	3.3–3.6
Surface area (m^2/g)	45–55	540–600
Cross-linking (wt% divinylbenzene)	20	80
Macroporosity	0.32–0.36	0.47–0.50
Macropore average diameter (nm)	20–40	4–5
Microparticle average diameter (nm)	70–90	7–8

Amberlyst XN-1010 is a more effective catalyst than Amberlyst 15. Additionally, the existence of transport limitations inside the particle was observed, being more significant to Amberlyst 15.

5.3 Preparative Chromatographic Techniques for Sugar Separation and Purification

Chromatographic processes are gaining more and more importance in sugar processing industry, in particular, for fine sugars purification. Either for preparative or industrial scale, continuous processes are usually preferable to batch ones, since the adsorbent is used more efficiently. Over the years, the continuous chromatographic processes passed through a large number of improvements, and it is now possible to process a desired separation in many different ways. In this section, a particular insight is given to two of these continuous processes: the continuous annular chromatography (CAC) [30] and the simulated moving bed (SMB) [31]. Both techniques are based on a countercurrent mode of operation and thus maximizing the mass transfer driving force which results in a better utilization of the adsorbent, reducing the solvent consumption (less diluted products) and increasing productivity and purity, when compared to conventional batch chromatography (Fig. 5.4).

5.3.1 Continuous Annular Chromatography

The continuous annular chromatography (CAC) is a technology available at lab and preparative scales, which allows the separation of simple binary as complex multicomponent mixtures in its pure components [32]. The CAC consists in a bed packed positioned within a narrow annular space between two concentric cylinders with, Fig. 5.5.

The feed is continuously introduced at a stationary point at the top of the bed while eluent flows over the remainder of the annulus. The rotation of the sorbent bed, together with the simultaneous solvent elution, causes the separation of the components present in the feed stream, in the form of helical bands, which can be driven toward stationary exit points [33]. The components are separated based on

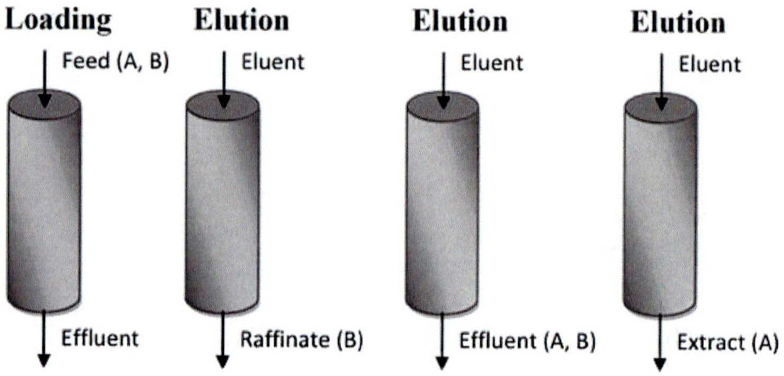

Fig. 5.4 Schematic representation of batch chromatography

Fig. 5.5 Schematic representation of annular chromatography

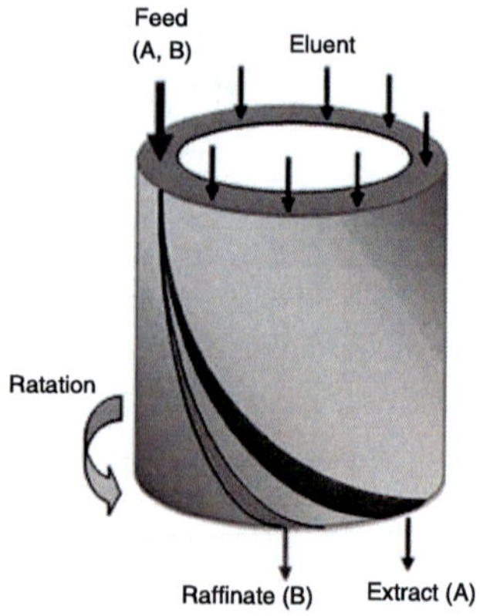

the difference between their affinities to the stationary phase: the most adsorbed species describes the longest path, while nonadsorbed species have a linear descending trajectory, going with the mobile phase. In the bottom of the apparatus are several outlet ports which collect the separated products as well as the eluent which may be recycled to the top. This technology has the main advantage of allowing the separation of multicomponent mixtures in several purified fractions. However, the scale-up of this technique, usually achieved by increasing annulus thickness, is frequently difficult to perform, which represents one major drawback for this kind of processes [34]. Nevertheless, the CAC process has been successfully applied for sugar separations at lab scale, using the calcium form Dowex 50W-X8 ion exchange resin to separate fructose, glucose, and sucrose mixtures [33, 35, 36] and mixtures of fructose, mannitol, and sorbitol [33, 37, 38]. Considering the separations based on the Na^+ coordination, a cation exchange resins in sodium form was used for the desugarization of beet molasses [39], and the resin Dowex 50W-X4 (sodium form) was used to separate fructooligosaccharides (1-kestose, nystose, 1-p-fructonystose) from monosaccharides (glucose and fructose) and disaccharide (sucrose) [40].

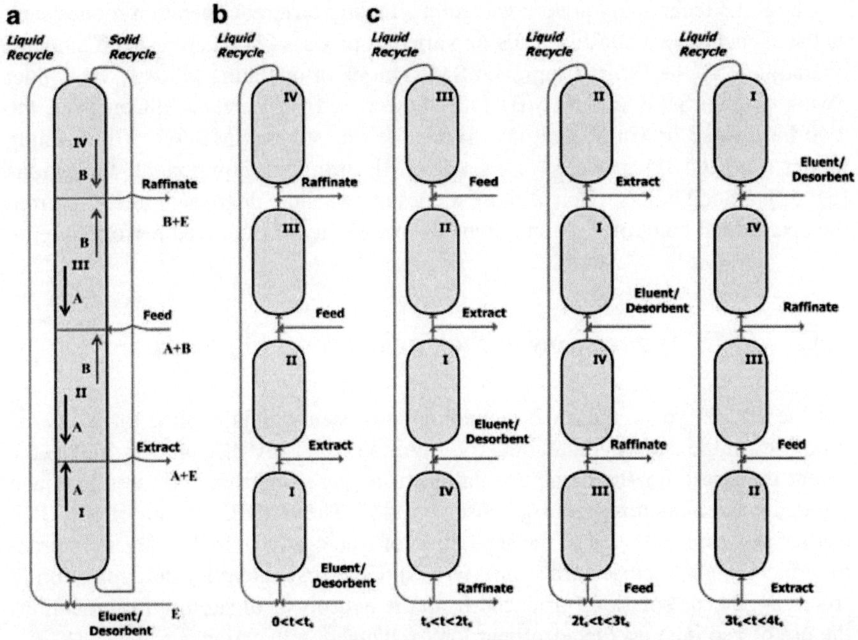

Fig. 5.6 Schematic representation of (**a**) a TMB unit and a 4-column SMB unit operating over a complete cycle, from 0 to 4 t_s (with t_s representing the ports switching time), (**b**) period of the SMB first switch, (**c**) period of the second switch

5.3.2 *Simulated Moving Bed*

The simulated moving bed (SMB) is one of the most interesting chromatographic technologies applied for large-scale sugar separation and purification. It was mentioned before that, frequently, the use of continuous countercurrent adsorption processes allows a more "efficient" use of the adsorbent bed. However, to operate an adsorption process in countercurrent, it is necessary to circulate not only the fluid phase but also the packing material. The solid motion inside of the column presents some technical problems, namely, equipment abrasion, mechanical erosion of adsorbent, and difficulties in maintaining plug flow for the solid (especially in beds with large diameter). In order to avoid these issues, a sequence of fixed-bed columns was conceived in which the solid phase is at rest in relation to a fixed referential [31], but where a relative movement between both phases is experienced by switching all the inlet and outlet streams to and from the columns from time to time (in the direction of the fluid flow). In the simplest operating mode, the period that a certain operating configuration prevails is called the switching time, t_s. Since the solid flow is avoided, although a kind of countercurrent movement is created relatively to the fluid, this technology is called simulated moving bed. The ideal case where the solid actually moves is called true moving bed (TMB), Fig. 5.6.

The SMB technology is very versatile, allowing different operation modes such as the asynchronous shifting SMB or Varicol® process [41], Partial-Feed mode of operation [42], the ISMB (Improved SMB) mode of operating [43, 44], the Outlet Swing Stream-SMB (OSS-SMB) [45], PowerFeed [46] and ModiCon [47], the Two-Feed SMB or MultiFeed [45, 48] and the side stream SMB [49, 50] operating modes, and the JO process (or Pseudo-SMB) for multicomponent separations [51, 52], among others. Each of these modes allows more degrees of freedom, from the operational point of view and, thus, the possibility of improved performance.

5.3.3 SMB Applications in the Sugar Processing Industry

For the past 35 years, the SMB technology has been widely applied for industrial processes in the carbohydrate industry, from the "old" SAREX process [53] to the recent processes for the betaine isolation from sugar industry molasses [54] and separation of mannitose from glucose [55, 56]. The SAREX is one of the UOP Sorbex processes, applied to the separation of fructose from mixed sugars in order to produce high-fructose corn syrup (HFCS). UOP first patent for the separation of fructose–glucose solutions produced from the hydrolysis of sucrose [57] mentions the use of zeolite Y as the adsorbent ion exchanged with cations of metals (K, Cs, Mg, Co, Sr, Ba+K, and Ba+Sr), varying the adsorbent selectivity for fructose from 1.4 to 6.2. Later, the process was extended to use ion exchange resins as adsorbent [58–60]. In sugar processing, another important application of the SMB technology is the desugarization of molasses [61] and the separation of fine sugars from biomass hydrolysates [62]. Several are the companies using SMB processes or related technologies (multicolumn apparatus with ports shift) for sugar applications:

- The Amalgamated Sugar Company improved the SMB technology [63, 64] and applied for the separation of sugars from nonsugars [65].
- The Improved Simulated Moving-Bed (ISMB) process patented by Mitsubishi Chemical Industries Ltd. [66], originally developed for the continuous separation between fructose and glucose in the high-fructose corn syrup (HFCS) production process, is now being developed by Nippon Rensui for new processes in the field of separation and purification of sugar and oligosaccharides, as well as being applied by Eurodia/Ameridia for the fructose/dextrose separation.
- The Organo Corporation JO process [67, 68] for the fractionation of beet molasses into raffinose, sucrose, glucose, and betaine, and separations of high purity saccharides and high purity maltitol.
- The "Finnsugar Applexion Separation Technology" (FAST) now commercialized by Novasep (France) was presented for the first time for the beet molasses separation [69, 70].
- The Sequential Simulated Moving Bed – SSMB technology, commercialyzed by Novasep (France) was applied for several sugar applications as the fractionation of sucrose solutions [71].

Table 5.4 SMB unit characteristic parameters for the fructose/glucose separation [81]

Model parameters	SMB operating conditions
Pe_j=2000 (Peclet number)	$C_i^F = 30\text{g.l}^{-1}$ (feed inlet concentrations)
$\varepsilon_b = 0.4$ (bulk porosity)	$C_i^E = 0\text{g.l}^{-1}$ (eluent inlet concentrations)
k_{LDF}=0.1 s^{-1} (mass transfer coefficient, LDF)	t_s=105 s (switching time period)
	$Q_E = 8.23\ \text{ml.min}^{-1}$
SMB columns	$Q_F = 1.28\ \text{ml.min}^{-1}$
$n_j = [3\ 3\ 3\ 3]$ (columns distribution)	$Q_X = 5.62\ \text{ml.min}^{-1}$
$L_c = 11.5\,\text{cm}$ (columns length)	$Q_R = 3.89\ \text{ml.min}^{-1}$
$d_c = 2.6\,\text{cm}$ (columns diameter)	$Q_{IV}^* = 19.89\ \text{ml.min}^{-1}$ (Recycle flowrate)

- The ISEP/CSEP technology developed by Advanced Separation Technologies Inc., now owned by Calgon Carbon Corporation, USA [72], and, similarly, the SepTor technology from Outotec Oyj, Finland [73, 74], are applied for sugar softening, glucose/fructose fractionation, and glucose demineralization.

The principles of SMB technology continues to be used in a wide-spread range of applications in the field of carbohydrate separations, existing at least 5 SAREX units and 100 ISMB plants in the world. The isolation and purification of fine sugars by SMB are being largely researched; examples are the production of fructooligosaccharides [2] and sorbitols [75], and the separations of Psicose [76], water-soluble polydextrose [77], L-arabinose [78], and betaine [79, 80].

5.3.3.1 Binary Separation of Fructose/Glucose by SMB

The simulation of the separation of fructose (A)/glucose (B) by means of an SMB unit using cation ion exchange resin Dowex monosphere 99/Ca (d_p=320 μm) of gel type as adsorbent and water as eluent [81, 82] is hereby detailed as an example of a classic SMB sugar separation. The adsorption phenomenon for this example was characterized by the following linear isotherms [81]:

$$q_A^{eq} = 0.5634C_{b_A} \tag{5.1}$$

$$q_B^{eq} = 0.3401C_{b_B} \tag{5.2}$$

Considering, C_{b_i} the bulk concentration in $g{\cdot}l^{-1}$ and q_i^{eq} the quantity of i species adsorbed onto the resin in $g{\cdot}l_{adsorbent}^{-1}$.

Considering a lab-scale SMB unit having 12 columns distributed equally for all four sections (see Table 5.4), the fructose/glucose separation was performed at optimized operating conditions considering the model parameters presented in Table 5.4.

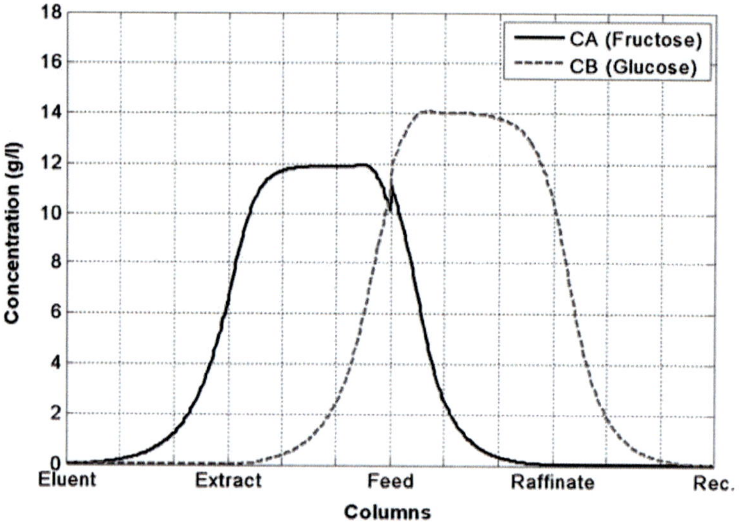

Fig. 5.7 SMB bulk concentration profiles of fructose (A) and glucose (B), at half of the switching time in the CSS

Table 5.5 SMB performance parameters values at the cyclic steady-state: purity (PU) and recovery (RE) in extract (X) and raffinate (R)

PU_X (%)	PU_R (%)	RE_X (%)	RE_R (%)
99.5	99.6	99.6	99.5

The concentration profiles at half switching time in the cyclic steady state (CSS), shown in Fig. 5.7, and the SMB performances, presented in Table 5.5, were obtained by simulation with the parameters mentioned before and using the mathematical model referred in Annex (1). It is possible to conclude that both fructose as glucose can be recovered at high purity (above 99.0%), in the extract and raffinate, respectively.

5.3.3.2 Ternary Separation of Sugars Mixture by Pseudo-SMB

Among other so-called nonconventional SMB operating modes, which have been applied to sugar processing industry, is the Pseudo-SMB, also known as JO process (Japan Organo, www.organo.co.jp). This modus operandi is based in two major stages: step 1 and step 2. During the first step, a multicomponent mixture, in this case ternary, is fed to the unit, while the intermediary species is being collected in the end of section II as consequence of the introduction of new eluent 1 in section I (Fig. 5.8a). The second step operates as a classical SMB, except the feed that is now closed and the only inlet is the eluent 2 stream, also in the beginning of section I (Fig. 5.8b) [51].

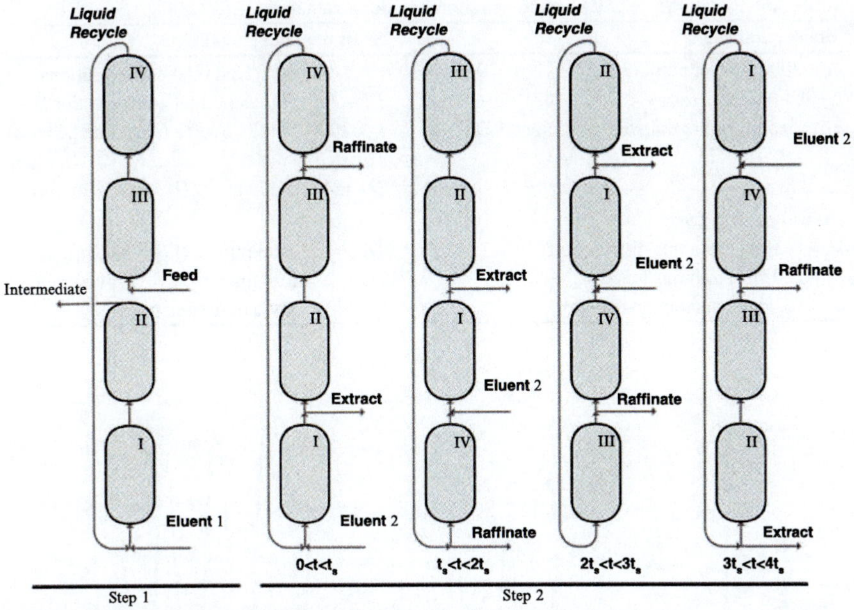

Fig. 5.8 Pseudo-SMB operation scheme during step 1 and during step 2, operating as a 4-column SMB unit for a complete cycle

While step 1 can be considered as a simple fixed bed, where there is any solid motion, the second step operates as an SMB unit. The simulation of such operating mode must take into account such discontinuities (see Annex 2 for model details on this operating mode). The example detailed here considers the separation of a ternary sugar mixture as presented before by Mata and Rodrigues [51], considering linear adsorption isotherms:

$$q_A^{eq} = 0.65C_{b_A} \tag{5.3}$$

$$q_B^{eq} = 0.39C_{b_B} \tag{5.4}$$

$$q_C^{eq} = 0.19C_{b_C} \tag{5.5}$$

Therefore, the most/least adsorbed sugars (A/C) will be collected in extract/raffinate; while the moderated adsorbed sugar (B) is collected in the intermediary purge. The unit characteristics, the model parameters, and the operating conditions are presented in Table 5.6. The outlet concentration history obtained in the Pseudo-SMB presents an oscillating behavior, as can be seen in Fig. 5.9. The averaged values for the concentration in the outlet streams over a complete cycle are presented in Table 5.7. One can observe, from purity and recovery values at the outlet streams reported in Table 5.8, that it is possible in a single unit to obtain three different components with

Table 5.6 Pseudo-SMB unit characteristics and model parameters

Model parameters	SMB operating conditions
Pe_j =2000 (Peclet number)	$C_i^F = 100\,\text{g.l}^{-1}$ (feed inlet concentrations)
ε_b=0.4 (bulk porosity)	$C_i^E = 0\,\text{g.l}^{-1}$ (eluent inlet concentrations)
k_{LDF}=0.5 s^{-1} (mass transfer coefficient, LDF)	t_{s1}=1090.2 s; t_{s2}=3960 s (step time periods)
	For step 1
	$Q_{E_1} = 422.9\,\text{ml.min}^{-1}$; $Q_F = 350.0\,\text{ml.min}^{-1}$
Pseudo-SMB columns	*For step 2*
$n_j = [3333]$ (columns distribution)	$Q_{E_2} = 451.4\,\text{ml.min}^{-1}$; $Q_{IV}^* = 900.0\,\text{ml.min}^{-1}$
$L_c = 120$ cm (columns length)	$Q_X = 274.0\,\text{ml.min}^{-1}$; $Q_R = 177.4\,\text{ml.min}^{-1}$
$d_c = 10.84$ cm (columns diameter)	t_s=340 s; (switching time periods)

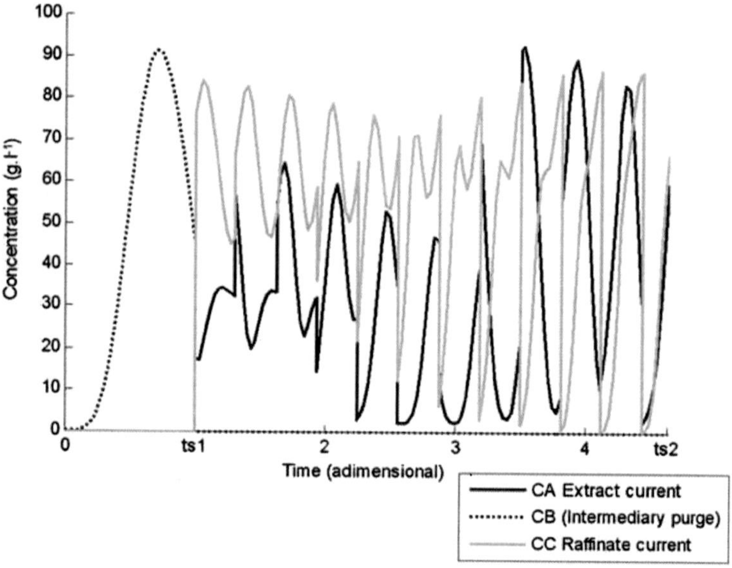

Fig. 5.9 Pseudo-SMB concentration history at the outlet streams, over a complete cycle at the CSS

Table 5.7 Pseudo-SMB average concentrations at the outlet streams

	Extract	Raffinate	Intermediary purge
A (g/l)	33.54	1.28	1.03
B (g/l)	0.16	0.00	45.06
C (g/l)	0.00	54.16	0.12

Table 5.8 Pseudo-SMB performance parameters

	Purity (%)	Recovery (%)
A Extract	99.53	95.38
B Intermediary purge	97.51	99.51
C Raffinate	97.70	99.71

purities above 97%. This proves the versatility of SMB-based technologies, for ternary separations, competing with the CAC process that is not so well established for large-scale production.

5.3.4 *Multifunctional Reactors for Sugar Processing*

The integration of reaction and separation steps in one single unit has the obvious economical advantage of reducing the cost of unit operations for downstream purification steps. Regarding bioreactions, these are usually inhibited by high product concentration; its integration with products separation usually leads to higher substrate conversion, product yield, and productivity since higher substrate concentration might be processed without product inhibition [83]. Besides reactive gas stripping, reactive extraction, or membrane reactors, the combination of biochemical reaction with a chromatographic separator has been subject of considerable attention in the last years [84], mainly based on the annular chromatography (CACR) and simulated moving-bed (SMBR) technologies.

5.3.4.1 Continuous Annular Chromatography Reactor

The process intensification using the CACR technology was applied to the sucrose inversion [85] and to the saccharification of modified starch to maltose [86]. In the case of sucrose inversion into glucose and fructose using an invertase enzyme, the products are separated by the Dowex 50W-X4 ion exchange resin (calcium form). It is possible to obtain complete sucrose conversion using feed concentrations up to 50% w/v sucrose. The starch hydrolysis maltose, using a maltogenase enzyme, produces maltose as main product, together with small amounts of glucose and unreacted dextrin. By separating the unreacted substrate and enzyme (low adsorption affinity), from the maltose formed product (high adsorption affinity) it is possible to increase the starch conversion and the yield of pure maltose.

5.3.4.2 The Simulated Moving-Bed Reactor

Since many biological reactions are characterized by inhibition caused by products or other intermediates, it is useful to combine bioreactions with product separations in order to enhance the rate and the yield of reaction. The reaction occurs either in the mobile or stationary phase, by supporting the catalyst or immobilizing the enzyme in the solid adsorbent, which promotes the separation of the reaction products. Therefore, several applications have been published considering the SMBR: enzymatic inversion of sucrose and separation of the obtained invert sugars (glucose + fructose) [83, 87–90], production of higher-fructose syrup by combining the adsorption of fructose and glucose reaction using immobilized isomerase [91, 92], inversion of sucrose with simultaneous biosynthesis of dextran from sucrose [93], and separation of mixtures containing sucrose–KCl and sucrose–betaine for higher utilization of molasses [94].

5.4 Conclusions

Among other technologies, the use of ion exchange resins for the sugar processing industry has been remarkable in the past years. Following the ever constant demand on these commodities, the number of novel and more efficient applications using ion exchange resins has increased over the last decades.

In this chapter, the more relevant, and the latest, technologies, based on ion exchange resins for sugar processing, were reviewed. Two continuous chromatographic processes, the continuous annular chromatography and simulated moving bed, well known for its actual, as future, range of application, were further described. Two sugar separations examples, a classical SMB and a nonconventional Pseudo-SMB apparatus, were modeled and simulated to better describe the operation mode of such techniques, showing its full potential for the high performance separation of elevated purity sugars by means of ion exchange resins.

Acknowledgments P. Sá Gomes gratefully acknowledges the support from Fundação para a Ciência e Tecnologia, Ministry of Science, Technology and Higher Education of Portugal (postdoc grant ref.: SFRH/BPD/63764/2009).

Annexes

Annex (1)

The mathematical model approach, hereby used to simulate an SMB unit, represents it as a sequence of columns described by the usual system equations for an adsorptive fixed bed (each column k). All these columns are then linked by the so-called node equations, stated to each section j , making use of the equivalence between the interstitial velocity in the TMB and SMB, and thus one has for

Eluent (E) node : $u_I^* = u_{IV}^* + u_E$

Extract (X) node : $u_{II}^* = u_I^* - u_X$

Feed (F) node : $u_{III}^* = u_{II}^* + u_F$

Raffinate (R) node : $u_{IV}^* = u_{III}^* - u_R$

Due to the switch of the inlet and outlet lines, the boundary conditions to a certain column are not constant during a whole cycle, but change after a period equal to the switching time. Since the model equations are set to each column k, the concentration of i species at the beginning of each column k, $C^0_{b_{i,k}}$, must be recalculated for each switching time period following the next node mass balances: $t = 0$ to t_s:

$$k=1:\ C_{b_{i,\Sigma_{j=I}^{IV} n_j}}\Big|_{z=L\Sigma_{j=I}^{IV} n_j} = \frac{u_I^*}{u_{IV}^*}C^0_{b_{i,1}} - \frac{u_E}{u_{IV}^*}C_i^E$$

$$k=2 \text{ to } (n_I+n_{II}):\ C_{b_{i,(k-1)}}\Big|_{z=L_{(k-1)}} = C^0_{b_{i,k}}$$

$$k=(n_I+n_{II}+1):\ C_{b_{i,(n_I+n_{II})}}\Big|_{z=L_{(n_I+n_{II})}} = \frac{u_{III}^*}{u_{II}^*}C^0_{b_{i,(n_I+n_{II}+1)}} - \frac{u_F}{u_{II}^*}C_i^F$$

$$k=(n_I+n_{II}+2) \text{ to } \sum_{j=I}^{IV} n_j:\ C_{b_{i,(k-1)}}\Big|_{z=L_{(k-1)}} = C^0_{b_{i,k}}$$

$t = t_s$ to $2t_s$:

$$k=1:\ C_{b_{i,\Sigma_{j=I}^{IV} n_j}}\Big|_{z=L\Sigma_{j=I}^{IV} n_j} = C^0_{b_{i,1}}$$

$$k=2:\ C_{b_{i,1}}\Big|_{z=L1} = \frac{u_I^*}{u_{IV}^*}C^0_{b_{i,2}} - \frac{u_E}{u_{IV}^*}C_i^E$$

$$k=3 \text{ to } (n_I+n_{II}+1):\ C_{b_{i,(k-1)}}\Big|_{z=L_{(k-1)}} = C^0_{b_{i,k}}$$

$$k=(n_I+n_{II}+2):\ C_{b_{i,(n_I+n_{II}+1)}}\Big|_{z=L_{(n_I+n_{II}+1)}} = \frac{u_{III}^*}{u_{II}^*}C^0_{b_{i,(n_I+n_{II}+2)}} - \frac{u_F}{u_{II}^*}C_i^F$$

$$k=(n_I+n_{II}+3) \text{ to } \sum_{j=I}^{IV} n_j:\ C_{b_{i,(k-1)}}\Big|_{z=L_{(k-1)}} = C^0_{b_{i,k}}$$

This set of equations continues to progress in a similar way (shifting one column per t_s), until $\sum_{j=I}^{IV} n_j t_s$, repeating then from the first switch.

Considering the dimensionless variables $x = \frac{z}{L_c}$ and $\theta = \frac{t}{t_s}$, and a linear driving force (LDF) approximation [95] to the mass transfer (set in terms of the adsorbed phase in equilibrium with the bulk concentration and the average adsorbed phase concentration: $\alpha_{i,k}\left(q^{eq}_{i,k} - \langle q_{i,k}\rangle\right)$, where $\alpha_k = k_{LDF}t_k$), one can establish a set of mass balances equations for each species i in each column k, in each section j, as follows for the bulk fluid phase:

$$\frac{\partial C_{b_{i,k}}}{\partial\theta} = \gamma_j^*\left\{\frac{1}{Pe_j^*}\frac{\partial^2 C_{b_{i,k}}}{\partial x^2} - \frac{\partial C_{b_{i,k}}}{\partial x} - \frac{(1-\varepsilon_b)}{\varepsilon_b}\alpha_{i,k}\left(q^{eq}_{i,k} - \langle q_{i,k}\rangle\right)\right\}$$

and for the mass balance in the particle,

$$\frac{\partial \langle q_{i,k} \rangle}{\partial \theta} = \gamma_j^* \alpha_{i,k} \left(q_{i,k}^{eq} - \langle q_{i,k} \rangle \right)$$

with the respective initial:

$$\theta = 0 : C_{b_{i,k}}(x,0) = \left\langle q_{i,k}(x,0) \right\rangle = 0$$

and boundary conditions:

$$x = 0 : C_{b\ i,k}^{\ 0} = C_{b_{i,k}}\Big|_{x=0} - \frac{1}{Pe_j^*} \frac{\partial C_{b_{i,k}}}{\partial x}\Bigg|_{x=0}$$

$$x = 1 : \frac{\partial C_{b_{i,k}}}{\partial x}\Bigg|_{x=1} = 0$$

where $\gamma_j^* = \frac{u_j^*}{u_s}$ is the ratio between SMB fluid and the solid interstitial velocities and $Pe_j^* = \frac{u_j^* L_c}{D_b}$ the column Peclet number. As consequence, one obtains discontinuous solutions, reaching not a continuous steady state (SS) but a cyclic steady state (CSS).

Performance Parameters

The definitions of purity and recovery of these performance parameters, for the case of a binary mixture, are given below:

Purity (%) of the more retained (A) species in extract and the less retained one (B) in the raffinate streams, over a complete cycle (from θ to $\theta + \sum_{j=I}^{IV} n_j$):

$$PU_X = \frac{\int_{\theta}^{\theta + \sum_{j=I}^{IV} n_j} C_A^X \, d\theta}{\int_{\theta}^{\theta + \sum_{j=I}^{IV} n_j} C_A^X \, d\theta + \int_{\theta}^{\theta + \sum_{j=I}^{IV} n_j} C_B^X \, d\theta}$$

$$PU_R = \frac{\int_{\theta}^{\theta + \sum_{j=I}^{IV} n_j} C_B^R \, d\theta}{\int_{\theta}^{\theta + \sum_{j=I}^{IV} n_j} C_A^R \, d\theta + \int_{\theta}^{\theta + \sum_{j=I}^{IV} n_j} C_B^R \, d\theta}$$

Recovery (%) of more retained (A) species in extract and the less retained one (B) in raffinate streams, again over a complete cycle:

$$RE_X = \frac{\int_{\theta}^{\theta+\sum_{j=I}^{IV} n_j} C_A^X d\theta \cdot Q_X}{\sum_{j=I}^{IV} n_j Q_F C_A^F}$$

$$RE_R = \frac{\int_{\theta}^{\theta+\sum_{j=I}^{IV} n_j} C_B^R d\theta \cdot Q_R}{\sum_{j=I}^{IV} n_j Q_F C_B^F}$$

Annex (2)

As mentioned before, the JO process (or Pseudo-SMB) is characterized by a discontinuous operating mode, mainly divided into two major steps. The mathematical model to simulate such process also replicates such discontinuity.

Step 1 (from 0 to t_{Step1})

During step 1, the feeding is performed and the intermediary product is purged, as mentioned, before leading to the node balances:

Eluent (E) node : $u_I^{1*} = u_{IV}^{1*} + u_{E_1}$

Extract (X) node : $u_{II}^{1*} = u_I^{1*}$

Intermediary (It) : $u_{It} = u_{II}^{1*}$

Feed (F) node : $u_{III}^{1*} = u_F$

Raffinate (R) node : $u_{IV}^{1*} = u_{III}^{1*}$

and for each species i :

$$j = I : C_{b_i,IV}\Big|_{x=1} = \frac{u_I^{1*}}{u_{IV}^{1*}} C_{b_i,I}^0 - \frac{u_{E1}}{u_{IV}^{1*}} C_i^E$$

$$j = III : C_{b_i,III}^0 = C_i^F$$

$$j = II, IV : C_{b_{i,(j-1)}}\Big|_{x=1} = C_{i,j}^0$$

with u_j^{1*} representing interstitial velocity in section j during step 1.

Within each column, the concentration profiles are simulated by means of the well-known fixed-bed equations, initial and boundary conditions, similar to those expressed for the LDF approach for each column in the SMB modeling strategy (see Annex 1).

Step 2 (from t_{Step1} to t_{Step2})

In the second operation stage (step 2), one has the following nodes balances:

$$\text{Eluent (E) node}: u_I^{2*} = u_{IV}^{2*} + u_{E_2}$$

$$\text{Extract (X) node}: u_{II}^{2*} = u_I^{2*} - u_X$$

$$\text{Feed / Intermediary (F / It) node}: u_{III}^{2*} = u_{II}^{2*}$$

$$\text{Raffinate (R) node}: u_{IV}^{2*} = u_{III}^{2*} - u_R$$

and for each species i:

$$j = I : C_{b_i,IV}\Big|_{x=1} = \frac{u_I^{2*}}{u_{IV}^{2*}} C_{b_i,I}^0 - \frac{u_{E2}}{u_{IV}^{2*}} C_i^E$$

$$j = II, III, IV : C_{b_{i,(j-1)}}\Big|_{x=1} = C_{i,j}^0$$

During step 2, similar mass balances and boundary conditions stated for the classical SMB unit (Annex 1) are used to simulate the operation of the Pseudo-SMB unit.

Performance Parameters (Pseudo-SMB)

The definitions of extract, intermediary, and raffinate purities (PU_X, PU_{It}, PU_R, %) and recovery at the extract, intermediary, and raffinate ports (RE_X, RE_{It}, RE_R, %) of the more intermediary and less retained components, respectively, are stated as follows:

Component A (during step 2):

$$PU_X = \frac{\int_{t_{Step1}}^{t_{Step2}} C_{b_A}^X d\theta}{\int_{t_{Step1}}^{t_{Step2}} C_{b_A}^X d\theta + \int_{t_{Step1}}^{t_{Step2}} C_{b_B}^X d\theta + \int_{t_{Step1}}^{t_{Step2}} C_{b_C}^X d\theta}$$

$$RE_X = \frac{\int_{t_{Step1}}^{t_{Step2}} C_{b_A}^X d\theta \cdot u_X}{\int_{\theta}^{\theta + t_{Step1}} C_A^F d\theta \cdot u_F}, \text{ during Step 2}$$

Component B (during step 1):

$$PU_{It} = \frac{\int_{\theta}^{\theta + t_{Step1}} C_{b_A}^{It} d\theta}{\int_{\theta}^{\theta + t_{Step1}} C_{b_A}^{It} d\theta + \int_{\theta}^{\theta + t_{Step1}} C_{b_B}^{It} d\theta + \int_{\theta}^{\theta + t_{Step1}} C_{b_C}^{It} d\theta}$$

$$RE_{IT} = \frac{\int_{\theta}^{\theta+t_{Step1}} C_{b_B}^{It} d\theta \cdot u_{It}}{\int_{\theta}^{\theta+t_{Step1}} C_{B}^{F} d\theta \cdot u_{F}}$$

Component C (during step 2):

$$PU_R = \frac{\int_{t_{Step1}}^{t_{Step2}} C_{b_C}^{R} d\theta}{\int_{t_{Step1}}^{t_{Step2}} C_{b_A}^{R} d\theta + \int_{t_{Step1}}^{t_{Step2}} C_{b_B}^{R} d\theta + \int_{t_{Step1}}^{t_{Step2}} C_{b_C}^{R} d\theta}$$

$$RE_R = \frac{\int_{t_{Step1}}^{t_{Step2}} C_{b_C}^{R} d\theta \cdot u_R}{\int_{\theta}^{\theta+t_{Step1}} C_{C}^{F} d\theta \cdot u_F}$$

References

1. Clarke MA (1993) Sugars and nonsugars in sugarcane. In: Chen JCP, Chou C-C (eds) Sugar cane handbook. Wiley, New York, pp 21–39
2. Vaňková K, Onderková Z, Antošová M, Polakovič M (2008) Design and economics of industrial production of fructooligosaccharides. Chem Pap 62(4):375–381
3. Robyt JF (1998) Essentials of carbohydrate chemistry, Springer advanced texts in chemistry. Springer, New York
4. Mitsuiki S, Utsunomiya H, Nakama Y, Sakai M, Mukae K, Moriyama Y, Goto M, Furukawa K (2005) Purification and characterization of maltotriose-producing amylases from an alkaliphilic nocardiopsis sp. Toa-1. J Appl Glycosci 52(2):95–99
5. Clarke MA (2000) Sugar, cane sugar. Kirk-othmer encyclopedia of chemical technology. John Wiley & Sons, Inc
6. Coca M, Mato S, González-Benito G, Ángel Urueña M, García-Cubero MT (2010) Use of weak cation exchange resin lewatit s 8528 as alternative to strong ion exchange resins for calcium salt removal. J Food Eng 97(4):569–573
7. Gryllus V (1967) Verfahren zur regenerierung des ionenaustauschers und zur verringerung des alkali-ionengehaltes von zuckerfabriksabläufen bei der enthärtung von dünnsaft mittels ionenaustausches. AT Patent 258230
8. Rousseau G (1999) Method of regenerating ion exchange resins in the process of decalcification of sugar factory juices. US Patent 5958142
9. Mindler AB (1948) Demineralization of sugar cane juice. Ind Eng Chem 40(7):1211–1215.
10. Massay R (1850) Improvements in defecating sugar. US Patent 7342
11. Urban K (1925) Improvements in the purification of sugar solutions. GB Patent 240253
12. Elbogen S (1925) Procédé de purification des solutions sucrées au moyen de silicates ou autres produits artificiels, qui possèdent la propriété d'échanger les alcalino-terreux contre des alcalins des solutions sucrées. FR Patent 584967
13. Quentin G (1961) Verfahren zur erhoehung der kristallisationsfaehigkeit des zuckers in loesungen der zuckerfabrikation. DE Patent 974408
14. LaBrie RL, Bharwada UJ (1992) Process for demineralizing a sugar-containing solution. US Patent 5094694

15. Bakker A, Schepers GJJM, Koerts K (1989) Method for demineralizing beet sugar thin juice. US Patent 4799965
16. Kouji T, Fumihiko M, Kikuzo K, Makoto T, Takayuki M (2001) Process for demineralizing a sugar solution. US Patent 6224683
17. Colonna WJ, Samaraweera U, Clarke MA, Cleary M, Godshall MA, White JS (2000) Sugar. Kirk-othmer encyclopedia of chemical technology. John Wiley & Sons, Inc
18. Broughton DB (1983) Sucrose extraction from aqueous solutions featuring simulated moving bed. US Patent 4404037
19. Broughton DB (1984) Extraction of sucrose. EP Patent 0103406
20. Bento LSM (1998) Ion exchange resins for decolorization. Int Sugar J 100(1191):111–117
21. De Lataillade J, Rousset F (2002) Ion-exchange decolorization: a flexible way to modernization and capacity extension. Int Sugar J 104(1237):5–9
22. Kunin R (1972) Ion exchange resins. Robert E. Krieger, New York
23. Coca M, García MT, Mato S, Cartón A, González G (2008) Evolution of colorants in sugarbeet juices during decolorization using styrenic resins. J Food Eng 89(4):429–434
24. L'Hermine GJA, Lundquist EG (1999) Decolorization of sugar syrups using functionalized adsorbents. US Patent 5972121
25. Soest H-K, Klipper DR, Schnegg DU, Gladysch M (2005) Sugar juice colour removal using monodispersed anion exchangers. EP Patent 1205560
26. Bento LR (1991) Processing for regenerating sugar decolorizing ion exchange resins, with regenerant recovery. US Patent 5019542
27. Bento LR (1998) Process for regeneration of ion-exchange resins used for sugar decolorization, using chloride salts in a sucrose solution alkalinized with calcium hydroxide. US Patent 5932106
28. Ihm S-K, Oh I-H (1984) Correlation of a two-phase model for macroreticular resin catalyst in sucrose inversion. J Chem Eng Jpn 17(1):58–64
29. Dooley KM, Gates BC (1984) Superacid polymers from sulfonated poly(styrene-divinylbenzene): preparation and characterization. J Polymer Sci Polymer Chem Ed 22(11):2859–2870
30. Martin AJP (1949) Summarizing paper. Discuss Faraday Soc 7:332–336
31. Broughton DB, Gerhold CG (1961) Continuous sorption process employing fixed bed of sorbent and moving inlets and outlets. US Patent 2985589
32. Uretschläger A, Jungbauer A (2002) Preparative continuous annular chromatography (p-cac), a review. Bioprocess Biosystems Eng 25(2):129–140
33. Byers CH, Sisson WG, Decarli JP, Carta G (1989) Pilot-scale studies of sugar separations by continuous chromatography. Appl Biochem Biotechnol 20–21(1):635–654
34. Wolfgang J, Prior A (2002) Continuous annular chromatography. In: Freitag R (ed) Modern advances in chromatography, vol 76, Advances in biochemical engineering/biotechnology. Springer, Berlin/Heidelberg, pp 233–255
35. Howard AJ, Carta G, Byers CH (1988) Separation of sugars by continuous annular chromatography. Ind Eng Chem Res 27(10):1873–1882
36. Byers CH, Sisson WG, DeCarli JP, Carta G (1990) Sugar separations on a pilot scale by continuous annular chromatography. Biotechnol Prog 6(1):13–20
37. Bart HJ, Messenböck RC, Byers CH, Prior A, Wolfgang J (1996) Continuous chromatographic separation of fructose, mannitol and sorbitol. Chem Eng Process: Process Intensificat 35(6):459–471
38. Wolfgang J, Prior A, Bart HJ, Messenböck RC, Byers CH (1997) Continuous separation of carbohydrates by ion-exchange chromatography. Sep Sci Technol 32(1–4):71–82
39. Barker PE, Bridges S (1991) Continuous annular chromatography for the separation of beet molasses. J Chem Technol Biotechnol 51(3):347–359
40. Takahashi Y, Goto S (1994) Continuous separation of fructooligosaccharides using an annular chromatograph. Sep Sci Technol 29(10):1311–1318
41. Adam P, Nicoud RM, Bailly M, Ludemann-Hombourger O (2004) Process and device for separation with variable-length chromatographic zones. US Patent 6712973

42. Zang Y, Wankat PC (2002) Smb operation strategy – partial feed. Ind Eng Chem Res 41(10):2504–2511
43. Lutin F, Bailly M, Bar D (2002) Process improvements with innovative technologies in the starch and sugar industries. Desalination 148(1–3):121–124
44. Rajendran A, Paredes G, Mazzotti M (2009) Simulated moving bed chromatography for the separation of enantiomers. J Chromatogr A 1216(4):709–738
45. Sá Gomes P, Rodrigues AE (2007) Outlet streams swing (oss) and multifeed operation of simulated moving beds. Sep Sci Technol 42(2):223–252
46. Zhang Z, Mazzotti M, Morbidelli M (2003) Powerfeed operation of simulated moving bed units: changing flow-rates during the switching interval. J Chromatogr A 1006(1–2):87–99
47. Schramm H, Kaspereit M, Kienle A, Seidel-Morgenstern A (2002) Improving simulated moving bed processes by cyclic modulation of the feed concentration. Chem Eng Technol 25(12):1151–1155
48. Kim JK, Abunasser N, Wankat PC (2005) Use of two feeds in simulated moving beds for binary. Korean J Chem Eng 22(4):619–627
49. Mun S (2006) Enhanced separation performance of the simulated moving bed process with two raffinate and two extract products. J Chem Eng Jpn 39(10):1054–1056
50. Beste YA, Arlt W (2002) Side-stream simulated moving-bed chromatography for multicomponent separation. Chem Eng Technol 25(10):956–962
51. Mata VG, Rodrigues AE (2001) Separation of ternary mixtures by pseudo-simulated moving bed chromatography. J Chromatogr A 939(1–2):23–40
52. Borges da Silva EA, Rodrigues AE (2006) Design of chromatographic multicomponent separation by a pseudo-simulated moving bed. AICHE J 52(11):3794–3812
53. Bieser HJ, deRosset AJ (1977) Continuous counter-current separation of saccharides with inorganic adsorptions. In: Paper presented at the 28th starch convention, Detmold, 27–29 Apr 1977
54. Giacobello S, Storti G, Tola G (2000) Design of a simulated moving bed unit for sucrose-betaine separations. J Chromatogr A 872(1–2):23–35
55. Zhang Z, Wang J, Feng Q (2004) Process for separating mannitose and glucose by analog moving bed. CN Patent 1528769
56. Zhang Z, Wang J, Liu J (2004) High-yield manna sugar preparation process. CN Patent 1528728
57. Neuzil RW, Priegnitz JW (1980) Process for separating a ketose from an aldose by selective adsorption. US Patent 4226977
58. Landis AM, Broughton DB, Fickel RG (1981) Simulated countercurrent sorption process employing ion exchange resins with backflushing. US Patent 4293346
59. Fickel RG (1982) Simulated countercurrent sorption process employing ion exchange resins with periodic backflushing. US Patent 4319929
60. Landis AM, Broughton DB, Fickel RG (1983) Simulated countercurrent sorption process employing ion exchange resins with backflushing. EP Patent 0075611
61. Rearick DE, Kearney M, Costesso DD (1997) Simulated moving-bed technology in the sweetener industry. ChemTech 27(9):36–40
62. Saari P, Häkkä K, Jumppanen J, Heikkilä H, Hurme M (2010) Study on industrial scale chromatographic separation methods of galactose from biomass hydrolysates. Chem Eng Technol 33(1):137–144
63. Kearney MM, Mumm MW (1991) Chromatographic separator sorbent bed preparation. US Patent 4990259
64. Kearney MM, Hieb KL (1992) Time variable simulated moving bed process. US Patent 5102553
65. Kearney MM, Kochergin V, Peterson KR, Velasquez L (1995) Sugar beet juice purification process. US Patent 5466294
66. Ando M, Hirota T, Shioda K (1980) Adsorption separation method and apparatus therefore. EP Patent 10769
67. Masuda T, Kawano K, Miyawaki I (1995) Process for production of starch sugars. US Patent 5391299

68. Tanimura M, Tamura M (1996) Method of separation into three components using a simulated moving bed. US Patent 5556546
69. Hyöky G, Paananen H, Cotillon M, Cornelius G (1999) Presentation of the fast separation technology. In: Paper presented at the 30th general meeting ASSBT-American Society of Sugar Beet Technologysts, Orlando, 13 Feb 1999
70. Heikkila H, Hyoky G, Kuisma J (2000) Method for the fractionation of molasses. US Patent 6093326
71. Hyoky G, Paananen H, Monten K-E, Heikkila H, Kuisma J (1998) Fractionation method of sucrose-containing solutions. US Patent 5795398
72. Ahlgren BK, Snyder CB, Fawaz I (2004) Fluid-directing multiport rotary valve. US Patent 6719001
73. Wijnberg BP (2005) Device for carrying out a chemical or physical treatment. WO Patent 2005025738
74. Evers JA (2006) Rotary distributor valve for chemical/physical treatment of a fluid. WO Patent 2006036062
75. Theoleyre M-A, Konetzke GDrnDC, Schmidt KDrnDC, Weidemann RDI (2002) Method of preparation of sorbitols from standard-glucose. EP Patent 1176131
76. Chang C-H (1989) Process for separating psicose from another ketose. EP Patent 0302970
77. An SC, Jee HS, Doh MH (1998) Process for obtaining, by separation, highly pure water-soluble polydextrose. US Patent 5831082
78. Antila J, Ravanko V, Walliander P (1998) Method of preparing l-arabinose from sugar beet pulp. WO Patent 1999/010542
79. Heikkila H, Hyoky G, Kuisma J (1989) Method for the recovery of betaine from molasses. EP Patent 345511
80. Kikuzo K, Takayuki M, Kohei S, Kouji T, Fumihiko M (2000) Process for recovering betaine. US Patent 6099654
81. Leão CP, Rodrigues AE (2004) Transient and steady-state models for simulated moving bed processes: numerical solutions. Comput Chem Eng 28(9):1725–1741
82. Azevedo DCS (2001) Separation/reaction in simulated moving bed. Ph.D. Thesis, University of Porto, Porto
83. Azevedo DCS, Rodrigues AE (2001) Design methodology and operation of a simulated moving bed reactor for the inversion of sucrose and glucose-fructose separation. Chem Eng J 82(1–3):95–107
84. Cen P, Tsao GT (1993) Recent advances in the simultaneous bioreaction and product separation processes. Sep Technol 3(2):58–75
85. Sarmidi MR, Barker PE (1993) Simultaneous biochemical reaction and separation in a rotating annular chromatograph. Chem Eng Sci 48(14):2615–2623
86. Sarmidi MR, Barker PE (1993) Saccharification of modified starch to maltose in a continuous rotating annular chromatograph (crac). J Chem Technol Biotechnol 57(3):229–235
87. Meurer M, Altenhöner U, Strube J, Untiedt A, Schmidt-Traub H (1996) Dynamic simulation of a simulated-moving-bed chromatographic reactor for the inversion of sucrose. Starch-Starke 48(11–12):452–457
88. Minceva M, Rodrigues AE (2005) Simulated moving-bed reactor: reactive-separation regions. AICHE J 51(10):2737–2751
89. Minceva M, Silva VMT, Rodrigues AE (2005) Analytical solution for reactive simulated moving bed in the presence of mass transfer resistance. Ind Eng Chem Res 44(14):5246–5255
90. Sá Gomes P, Leão CP, Rodrigues AE (2007) Simulation of true moving bed adsorptive reactor: detailed particle model and linear driving force approximations. Chem Eng Sci 62(4):1026–1041
91. Hashimoto K, Adachi S, Noujima H, Ueda Y (1983) A new process combining adsorption and enzyme reaction for producing higher-fructose syrup. Biotechnol Bioeng 25(10):2371–2393
92. Borges da Silva EA, Ulson de Souza AA, de Souza SGU, Rodrigues AE (2006) Analysis of the high-fructose syrup production using reactive smb technology. Chem Eng J 118(3):167–181

93. Barker PE, Ganetsos G, Ajongwen J, Akintoye A (1992) Bioreaction-separation on continuous chromatographic systems. Chem Eng J 50(2):B23–B28
94. Bubnik Z, Pour V, Gruberova A, Starhova H, Hinkova A, Kadlec P (2004) Application of continuous chromatographic separation in sugar processing. J Food Eng 61(4):509–513
95. Glueckauf E (1955) Theory of chromatography. Part 10. Formulae for diffusion into spheres and their application to chromatography. Trans Faraday Soc 51:1540–1551

Chapter 6
Application of Ion Exchange Resins in the Synthesis of Isobutyl Acetate

Alime Çıtak

Abstract Esterification reactions are one of the industrially important reactions. Esterification of carboxylic acids with alcohols in the presence of acid catalysts has been the subject of investigation by many scientists. The esters, products of these reactions, are commonly used in chemical industries such as solvents of paints, adhesives, plasticizers, flavors and fragrances, pesticides, and emulsifiers. The aim of this study is to investigate the application of the ion exchange resins as catalysts in the synthesis of isobutyl acetate. Heterogeneous catalysts such as Dowex 50W, Amberlite IR-120, and Amberlite IR-122 as catalysts were used in the experiments. Eley–Rideal model for both Amberlite IR-120 and Amberlite IR-122 was developed from the experimental data. The kinetic model for Dowex 50W was correlated by pseudo-homogeneous catalysis. The activation energy and the equilibrium constant of this reaction were calculated. As a result, Dowex 50W×2 showed the best performance among the studied catalysts.

6.1 Introduction

Both homogeneous and heterogeneous catalysts have been used to catalyze the esterification reactions since these reactions are extremely slow in the absence of catalysts. Catalysts such as mineral acid, p-toluenesulfonic acid, and ion exchange resins are always used in liquid-phase esterification reactions to raise the product yields [1–3].

Although homogeneous catalysts have high catalytic activity, using the heterogeneous solid catalysts can eliminate many of the disadvantages involved with use

A. Çıtak (✉)
Department of Chemical Engineering, Eskişehir Osmangazi University, Eskişehir 26480, Turkey
e-mail: acitak@ogu.edu.tr

Inamuddin and M. Luqman (eds.), *Ion Exchange Technology II: Applications*,
DOI 10.1007/978-94-007-4026-6_6, © Springer Science+Business Media B.V. 2012

of homogeneous catalysts. The use of a heterogeneous catalyst has the following inherent advantages over catalysis affected by dissolved electrolytes: (a) they eliminate the corrosive environment; (b) the catalyst can be easily removed from the reaction mixture by filtration; and (c) the purity of the products is higher since the side reactions can be completely eliminated or are less significant [4–8].

In industrial processes, heterogeneous catalysts such as zeolites and other metal oxides, ion exchange resins, etc., are being taken into consideration as ecological materials for hydration, esterification, etherification, etc. They can be used repeatedly over a long duration without any deactivation.

A variety of reactor configurations and modes of operation can be realized for solid–fluid catalytic reactions, thereby leading to better process control and economics. Among solid catalysts, ion exchange resins have been used for relatively low temperature reactions and are available in a variety of forms and strengths. Beside their convenience in separation and purification of products, ion exchange resins have been used as catalysts in the manufacture of some important chemicals such as MTBE, ETBE, and TAME [6, 9–11].

In order to start the esterification reaction, a proton transfers from acid catalyst to the alcohol and forms an alkyloxonium ion. Acid is protonated on its carbonyl oxygen. Nucleophilic addition of a molecule of the alcohol to carbonyl carbon of acid forms an intermediate product. This intermediate product is protonated on one of its hydroxyl oxygens. Intermediate product loses a water molecule and proton to give an ester [12–18]. There are a number of studies in which a solid resin is used as catalyst in the esterification reactions [8, 9, 19]. In the research on the esterification of acrylic and lactic acids with butanol in the presence of Amberlyst 15, it was found that the total change was controlled by the surface reaction and that alcohol and water had an inhibiting effect on the reaction rate [20, 21]. The similar results were also observed for the esterification of lactic and salicylic acids with methanol in the presence of Dowex 50W resin [22, 23]. The only contrary result, which the reaction carries out between the adsorbed alcohol molecules and acid molecules in the bulk solution, was reported by Bart et al. [24]. They showed that the adsorption of acid on the catalyst is also significant. Levesque and Craig [25] claimed that the esterification reactions catalyzed by a solid resin is basically second-order reversible after an initial slow period. However, many researchers explained that the kinetic expressions from the experimental results are based on the reactant and product adsorption on the catalyst [26, 27]. It was also showed that in the kinetic model, the use of activities gives a better result than that of concentrations [26].

Yu et al. (2004) studied esterification of acetic acid with methanol using Amberlyst 15 as a catalyst. Authors found the adsorption equilibrium constants, dispersion coefficients, and kinetic parameters. They also suggested a mathematical model for a quasi-homogeneous kinetics [28]. Huang and Sundmacher (2007) investigated the reaction between acetic acid and n-propanol catalyzed by Amberlyst 15. They tested pseudo-homogeneous, Rideal–Eley, and Langmuir–Hinshelwood–Hougen–Watson kinetic models to evaluate the obtained kinetic data [29].

Gangadwala et al. (2003) studied the esterification of acetic acid with n-butanol in the presence of Amberlyst 15. They investigated the influence of various parameters and proposed the rate equation model. Authors also investigated the kinetics for the side-reaction etherification [30]. Synthesis of n-butyl acetate has been also studied by Blagow et al. (2006). They compared three different ion exchange resin catalysts (Purolite CT 269, Amberlyst 46, and Amberlyst 48) [31]. Lee et al. (2000) chose the Dowex 50Wx8 as a catalyst in their esterification reaction between acetic acid and amyl alcohol. They showed that the kinetic data were correlated with the quasi-homogeneous, Langmuir–Hinshelwood, Eley–Rideal, and modified Langmuir–Hinshelwood models [32]. Ali et al. (2007) also used Dowex 50Wx8 as a catalyst for the reaction of propionic acid with 1-propanol. Several kinetic models were tested by these authors to correlate the kinetic data [33]. Liu and Tan (2001) studied the esterification of propionic acid with n-butanol catalyzed by Amberlyst 15, Amberlyst 35, Amberlyst 39, and HZSM-5 pellets. Rideal–Eley theory was found as the possible mechanism according to the kinetic data obtained from the reaction catalyzed by Amberlyst 35 [34].

Ion exchange resins have frequently been seen in esterification reactions and the hydrolysis of esters such as the esterification of propionic acid with n-butanol [34, 35] and esterification of acetic acid with butanol [30, 31]. Since this work is concerned with the application of the ion exchange resins as catalysts in the esterification reactions, it will be pertinent to give a few examples from the previous studies of the area concerned.

6.2 The Synthesis of Isobutyl Acetate in the Presence of Amberlite IR-120 Catalyst

Altiokka and Citak (2003) found the kinetic model representing the esterification of acetic acid with isobutanol in the presence of Amberlite IR-120. In this study, the esterification of acetic acid with isobutanol was carried out first, without adding a catalyst and then as heterogeneously catalyzed, and this reaction can be shown by:

$$CH_3COOH + C_4H_9OH \overset{H+}{\leftrightarrow} CH_3COOC_4H_9 + H_2O \tag{6.1}$$

Isobutyl acetate, as the reaction product, resembles butyl acetate and methyl isobutyl ketone in solvent performance. Isobutyl acetate, also known as 2-methylpropylethanoate (IUPAC name) or β-methylpropylacetate, is a common solvent. It is used as a solvent for lacquer and nitrocellulose. Like many esters, it has a fruity or floral smell at low concentrations and occurs naturally in raspberries, pears, and other plants. At higher concentrations, it can have an unpleasant odor and may cause symptoms of central nervous system depression such as nausea, dizziness, and headache [36].

6.2.1 *Experimental Section*

All kinetic experiments were taken place in the batch reactor consisted of a three-necked Pyrex flask of 400 ml capacity fitted with a reflux condenser, a sampling device, and a thermometer. The temperature was controlled within ± 0.1 K by circulating water from a thermostat into a cylindrical water jacket of the reactor. The reaction mixture was magnetically stirred at about 600 rpm. The equilibrium studies were carried out using sealed tubes of 5 ml capacity.

6.2.1.1 Procedure

Isobutanol of >98.5% purity (Merck) and acetic acid of 99.8% purity (Merck) were used as the reactants. The cation exchange resin in the H^+ form, Amberlite IR-120, was purchased from Rohm and Haas Co. In a typical run, the reaction was carried out in the solution of dioxane of 99.0% purity (Merck). Dioxane and one reactant were placed in the reactor. A known amount by weight of the catalyst was added and the reactor contents were being well mixed. After a steady value of the desired temperature was attained, the second reactant was added at the same temperature, and this was taken as zero time for a run. Two milliliters of liquid sample was withdrawn from the reactor at regular intervals for analysis. Unconverted acetic acid was determined by standardized 0.1 N NaOH. Water content of the reaction mixture was measured by Karl Fischer titration (784 model Titrino).

6.2.2 *Results*

The effects of the variables such as reactant molar ratio, water concentration, and temperature on the reaction rate were studied with and without catalyst. The influence of external or pore diffusion can be neglected as reported in previous studies [20, 21]. It was also stated that the reaction rate is linearly dependent on resin concentration [22, 37]. The preliminary experiments indicated that the catalyst deactivation was insignificant in the period of 35 days. The equilibrium constant, being independent of temperature ranging from 318 to 368 K, was determined to be 4, in the sealed tubes.

6.2.2.1 Reaction Rate Without Catalyst

The experiments were carried out without catalyst by using the equimolar reactant concentration (1.32 mol/L) in dioxane. The integrated form of the rate expression for the simple second-order reversible reaction, with no product initially, is given by:

$$Ln\left[X_{A,e} - (2X_{A,e} - 1)X_A\right] / (X_{A,e} - X_A) = 2k_1\left[1/X_{A,e} - 1\right]C_{A,0}t \qquad (6.2)$$

where C_{A0}, X_A, and X_{Ae} are the initial concentration, conversion, and the equilibrium conversion of acetic acid, respectively.

Applying Eq. 6.2 to the experimental data collected at temperatures of 348 and 368 K, straight lines were obtained. From the slopes of the lines, the respective forward reaction rate constants (k_1) were found to be 4.5×10^{-4} and 1.37×10^{-3} L/(mol h). Substituting these values in Arrhenius equation, the temperature dependency of the rate constant was calculated to be:

$$k_1\left(\mathrm{L/(mol.h)}\right)=\exp\left(12.78-7130/\mathrm{T}\right) \tag{6.3}$$

6.2.2.2 Reaction Rate with Catalyst

The experiments were carried out in a differential reactor by changing one of the reaction parameters while keeping others constant. Thus, the effect of the parameter on the initial reaction rate was observed. From analysis, it was concluded that the reaction mechanism can be represented by Eley–Rideal model, that is, the reaction takes place between adsorbed molecules of alcohol and the molecules of acid in the bulk solution. However, water molecules adsorbed by resin has an inhibiting effect on the reaction rate. The adsorption of solvent (dioxane) and ester was reported to be negligible [20–22]. Depending on this approach, the general reaction rate expression can be given by:

$$-r_A=\frac{\mathrm{k_f(m/v)}\left(\mathrm{C_AC_B}-\frac{\mathrm{C_EC_W}}{\mathrm{K_E}}\right)}{\mathrm{(1+K_BC_B+K_WC_W)}} \tag{6.4}$$

where subscripts A, B, E, and W refer to acid, alcohol, ester, and water, respectively; k_f is forward reaction rate constant; K_e the equilibrium constant of the reaction; *K*s the adsorption equilibrium constants, *m* the quantity of dry resin; and *V* the volume of the reaction mixture.

Some equations were obtained according to the initial reaction rate. Then, solving these equations by the "Method of Averages" [38], k_f, K_B, and K_W were determined at different temperatures. The resulting values are shown in Table 6.1.

6.2.2.3 Effect of Temperature

Applying the Arrhenius equation to the values given in Table 6.1, the temperature dependencies of the constants were found as follows:

$$k_f=\exp\left(8.22-4390/\mathrm{T}\right)\mathrm{L^2(g-dry\ resin)^{-1}.mol^{-1}.hr^{-1}} \tag{6.5}$$

$$K_B=\exp\left(2700/\mathrm{T}-8.77\right)\mathrm{L.mol^{-1}} \tag{6.6}$$

$$K_W=\exp\left(3550/\mathrm{T}-9.53\right)\mathrm{L.mol^{-1}} \tag{6.7}$$

Table 6.1 The average values of k_f, K_B, and K_W at different temperatures (Reproduced from Ref. [4] with kind permission of © Elsevier (2003))

T (K)	k_f (l^2/(g dry resin mol h))	K_B (l/mol)	K_W (l/mol)
318	0.00384	0.688	5.06
333	0.00815	0.460	3.20
348	0.01250	0.334	1.93

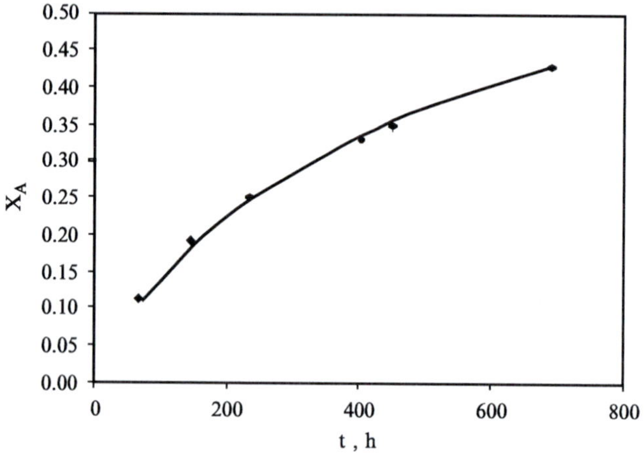

Fig. 6.1 Experimental points (♦) and calculated curve using Eq. 6.8 for initial stoichiometric amounts of alcohol and acid (1.32 mol. L^{-1}) at 318 K temperatures (Reproduced from Ref. [4] with kind permission of © Elsevier (2003))

6.2.2.4 Integrated Rate Expression

If the rate of the reaction given in Eq. 6.4 is expressed in terms of conversion of acetic acid, the following equation can be obtained:

$$\frac{dX_A}{dt}=\frac{\frac{k_f C_{A0} m}{V}\left[(1-X_A)(M-X_A)-\frac{X_A(N+X_A)}{K_e}\right]}{\left[1+K_B C_{Ao}(M-X_A)+K_W C_{Ao}(N+X_A)\right]} \tag{6.8}$$

A numerical integration by the Runge–Kutta method gave several points (X_i, t_i) which allowed us to plot the calculated curve shown in Fig. 6.1 for initial stoichiometric amounts of alcohol and acid (1.32 mol. L^{-1}) at 318 K, under the conditions of $M=C_{Ao}/C_{Bo}=1$; $N=C_{wo}/C_{Ao}=0$; $K_e=4$; m/V = 0.775 g dry resin/L.

6.2.2.5 Comparison of Reaction Rate with and Without Catalyst

The forward reaction rate constants with and without catalyst are given by Eqs. 6.5 and 6.3, respectively. If the frequency factors in both equations are equalized to each other, it can be calculated that the activation energy reduced from 59.300 to 49.000 kJ. mol^{-1} by using catalyst. In addition to this reduction in activation energy, the reaction rate will also increase linearly with the ratio of m/V.

The equilibrium constant (K_e), which is independent of temperature ranging from 318 to 368 K, was determined to be 4. As a result, it was concluded that the acidic ion exchange resin, Amberlite IR-120, is a suitable catalyst for this reaction since the activation energy reduced from 59.300 to 49.000 kJ/mol in the presence of this catalyst [4].

6.3 A Comparison of Acidic Ion Exchange Resins for Synthesis of Isobutyl Acetate

In this study, the esterification reaction of acetic acid with isobutanol was studied in a heterogeneous reaction system using a variety of solid acid catalysts. Three ion exchange resin catalysts were tested in a stirred batch reactor in the presence of 1,4-dioxane as a solvent. The tested ion exchange resins were Amberlite IR-120, Dowex 50W×2, and Amberlite IR-122 in this esterification reaction.

The effects of catalyst weight and temperature on the initial reaction rate were investigated in the activity of the different catalysts. Furthermore, the conversion of isobutanol as a function of process time was studied using various ion exchange resins as catalyst. The apparent equilibrium constant, being independent of temperature ranging from 318 to 348 K, was experimentally determined to be 4 for mole ratio M = 1/1.

The weight-based activity of the heterogeneous catalysts was in the following order: Dowex 50W×2 > Amberlite IR-120 > Amberlite IR-122. Furthermore, heterogeneous solid acids have been shown to be able to effectively catalyze the esterification reaction.

6.3.1 Experimental Section

6.3.1.1 Catalysts

Ion exchange resins which were used in this study were purchased from Rohm and Haas Co. and Dow Chemical Co. These are cross-linked three-dimensional structures of polymeric material obtained by sulfonation of a copolymer of polystrene and divinylbenzene. These resins are heat sensitive and loose activity above temperatures given in Table 6.2. The specific surface area (BET), average pore diameter, and total pore volume were found using a Quantchrome Autosab 1-C instrument. Surface area and pore size distribution analysis for all samples was carried out by the N_2 adsorption/desorption method at 353 K. For the determination of BET surface area, the value of p/p_0 in the range $0.01 < p/p_0 < 0.15$ was used. The catalysts were first pretreated. Dowex 50W×2 was dried under vacuum at 348 K overnight after

Table 6.2 Physical properties of the ion exchange resins used in this study (Reproduced from Ref. [4] with kind permission of © Elsevier (2003); from Ref. [39] with kind permission of © Elsevier Science (2007), and from Ref. [40] with kind permission of © Taylor & Francis (2009))

Physical property	Dowex 50W×2	Amberlite IR-120	Amberlite IR 122
Matrix type	Styrene-DVB	Styrene-DVB	Styrene-DVB
Operating pH	0–14	0–14	0–14
Ionic form	H^+	H^+	H^+
Total exchange capacity			
(a) meq/g(dry)	4.8	4.4	4.3
(b) meq/ml(wet)	0.6	1.9	2.15
Moisture (%)	~78	44–48	44–49
Crosslinkage (% DVB)	2	8	10
Max. operating temperature (K)	423	393	393
Particle size range (μm)	80–150	300–840	300–1,200
Surface area ($m^2.g^{-1}$)	6.940	1.532	2.127
Total pore volume (cm^3/g)	1.48×10^{-2}	1.10×10^{-2}	8.778×10^{-3}
Average pore diameter (Å)	85.34	288.5	355.2

washing with deionized water to eliminate poisoning or fouling. On the other hand, Amberlite IR-120 and IR-122 were dried at 363 K over 48 h to remove moisture completely after being washed with methanol and then deionized water. The properties of the catalysts used at the present work are listed in Table 6.2. Then, these resins were stored under airtight conditions before using.

6.3.2 Analysis

All samples were analyzed by gas chromatography (Agilent 6890 instrument with FID, N_2 as the carrier gas, DB-WAX 30 m×0.53 mm column, split 10:1). The injector and detector were kept at 523 K. The oven temperature was controlled from 363.15 to 473.15 K. Initial column temperature was 313 K for 5 min. Temperature was increased from 313 to 463 K by 10 K/min and maintained at 463 K for 1 min. M-xylene is used as an internal standard for quantitative analysis of samples. Water contents of samples were analyzed by Karl Fischer method (Schott TA05 Plus).

6.3.3 Results and Discussion

The esterification reaction of acetic acid with isobutanol was studied in a batch reactor without adding a catalyst and in the presence of acidic ion exchange resin

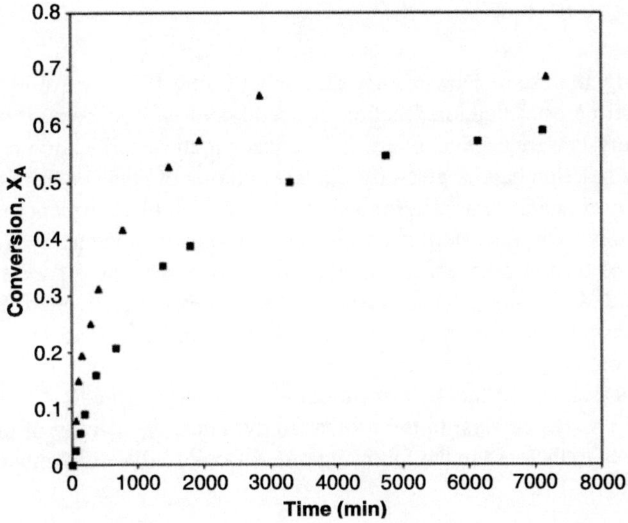

Fig. 6.2 Effect of different resins on the conversion of acetic acid. Resin loading = 5 g dry resin/L, mole ratio (acetic acid/isobutanol) = 1:1, T = 333 K, speed = 500 rpm, ▲, Dowex 50W×2; ■, Amberlite IR-120 (Reproduced from Ref. [4] with kind permission of © Elsevier (2003))

catalysts such as Dowex 50W×2, Amberlite IR-120, and Amberlite IR-122. A pseudo-homogeneous model has been used to describe the esterification reaction catalyzed by Dowex 50W×2 ion exchange resin because Dowex 50W×2 resin has higher activity than Amberlite IR-120 and IR-122 resins. The effects of temperature, catalyst loading, catalyst type, and the speed of agitation were studied.

6.3.3.1 Catalyst Performance

In Fig. 6.2, a comparison of the behavior of the Dowex 50W×2 and Amberlite IR-120 catalysts is shown. Catalysts were used to assess their efficacy in the esterification reaction of acetic acid with isobutanol. These were shown to be effective catalysts without adding resins in order to synthesize isobutyl acetate. The catalytic activity of Dowex 50W×2 is higher than that of Amberlite IR-120. It is an expected result as Dowex 50W×2 has higher total exchange capacity, surface area, and pore volume than those of Amberlite IR-120. Therefore, the kinetic study for the synthesis of isobutyl acetate was carried out in the presence of Dowex 50W×2 ion exchange resin, at reactants mole ratio of 1:1 and in temperatures ranging 318–348 K. In addition to these results, Amberlite IR-122 was also tested for this esterification reaction.

6.3.4 Conclusions

In this study, the use of Dowex 50W×2, Amberlite IR-120, and Amberlite IR-122 catalysts in the esterification reaction of acetic acid with isobutanol was investigated. Catalysts were used to assess their efficacy in the esterification reaction. The rate of this reaction can be given by Eq. 6.8 in terms of conversion of acetic acid. The activation energy was determined as 59.300 kJ/mol from experimental data without catalyst for this esterification reaction. The activation energy was found to be 1.745, 49.0, and 50.5 kJ/mol in the presence of Dowex 50W×2, Amberlite IR-120, and Amberlite IR-122, respectively. It means that Dowex 50W×2 ion exchange resin has very high activity in the synthesis of isobutyl acetate. As a result, the activation energy value decreased from 59.3 kJ/mol, without using catalyst, to 49.0 kJ/mol and 50.5 kJ/mol by using the acidic resin Amberlite IR-120 and IR-122, respectively, as the catalyst in the present study. Thus, the activity of the catalysts was observed to increase in the following order: Dowex 50W×2 > Amberlite IR-120 > Amberlite IR-122.

Acknowledgment The experimental studies in this chapter were financially supported as a Project (199915009) by Research Fund of Osmangazi University.

References

1. Lilja J, Aumoa J, Salmi T et al (2002) Kinetics of esterification of propanoic acid with methanol over a fibrous polymer-supported sulphonic acid catalyst. Appl Catal A-Gen 228:253–267
2. MaEki-Arvela P, Salmi T, Sundell M et al (1999) Comparison of polyvinylbenzene and polyolefin supported sulphonic acid catalysts in the esterification of acetic acid. Appl Catal A-Gen 184:25–32
3. Kolah AK, Asthana NS, Vu DT et al (2007) Reaction kinetics of the catalytic esterification of citric acid with ethanol. Ind Eng Chem Res 46:3180–3187
4. Altiokka MR, Citak A (2003) Kinetics study of esterification of acetic acid with isobutanol in the presence of amberlite catalyst. Appl Catal A-Gen 239:141–148
5. Yadav GD, Rahuman MSMM (2002) Cation-exchange resin-catalysed acylations and esterifications in fine chemical and perfumery industries. Org Process Res Dev 6:706–713
6. Yadav GD, Thathagar MB (2002) Esterification of maleic acid with ethanol over cation-exchange resin catalysts. React Funct Polym 52:99–110
7. Teo HTR, Saha B (2004) Heterogeneous catalysed esterification of acetic acid with isoamyl alcohol: kinetic studies. J Catal 228:174–182
8. Chen X, Xu Z, Okuhara T (1999) Liquid phase esterification of acrylic acid with 1-butanol catalyzed by solid acid catalysts. Appl Catal A-Gen 180:261–269
9. Chakrabarti A, Sharma MM (1993) Cationic ion exchange resins as catalyst. React Polym 20:1–45
10. Yadav GD, Joshi A (2002) A green route for the acylation of resorcinol with acetic acid. Clean Technol Environ 4:157–164
11. Yadav GD, Mujeebur Rahuman MSM (2003) Synthesis of fragrance and flavour grade esters: activities of different ion exchange resins and kinetic studies. Clean Technol Environ 5: 128–135

12. Carey FA (1990) Advanced organic chemistry. Plenum, New York
13. Morrison RT, Boyd RN (1992) Organic chemistry, 6th edn. Prentice Hall, New Delhi
14. Liu Y, Lotero E, Goodwin JG Jr (2006) A comparison of the esterification of acetic acid with methanol using heterogeneous versus homogeneous acid catalysis. J Catal 242:278–286
15. Liu Y, Lotero E, Goodwin JG Jr (2006) Effect of water on sulfuric acid catalyzed esterification. J Mol Catal A-Chem 245:132–140
16. Lilja J, Warna J, Salmi T et al (2005) Esterification of propanoic acid with ethanol, 1-propanol and butanol over a heterogeneous fiber catalyst. Chem Eng J 115:1–12
17. Gubicza L, Kabiri-Badr A, Keoves E et al (2000) Large-scale enzymatic production of natural flavour esters in organic solvent with continuous water removal. J Biotechnol 84:193–196
18. Zhao Z (2000) Synthesis of butyl propionate using novel aluminophosphate molecular sieve as catalyst. J Mol Catal A-Chem 154:131–135
19. Dupont P, Védrine JC, Paumard E et al (1995) Heteropolyacids supported on activated carbon as catalysts for the esterification of acrylic acid by butanol. Appl Catal A-Gen 129:217–227
20. Darge O, Thyrion FC (1993) Kinetics of the liquid phase esterification of acrylic acid with butanol catalysed by cation exchange resin. J Chem Technol Biot 58:351–355
21. Dassy S, Wiame H, Thyrion FC (1994) Kinetics of the liquid phase synthesis and hydrolysis of butyl lactate catalysed by cation exchange resin. J Chem Technol Biot 59:149
22. Choi JI, Hong WH, Chang HN (1996) Reaction kinetics of lactic acid with methanol catalyzed by acid resins. Int J Chem Kinet 28:37–41
23. Bochner MB, Gerber SM, Vieth W, Rodger AJ (1965) Ion exchange resin-catalyzed esterification of salicylic acid with methanol. Ind Eng Chem Fund 4:314–317
24. Bart HJ, Kaltenbrunner W, Landschützer H (1996) Kinetics of esterification of acetic acid with propyl alcohol by heterogeneous catalysis. Int J Chem Kinet 28:649–656
25. Levesque CL, Craig AM (1948) Kinetics of an esterification with cation-exchange resin catalyst. Ind Eng Chem 40:96–99
26. Pöpken T, Götze L, Gmehling J (2000) Reaction kinetics and chemical equilibrium of homogeneously and heterogeneously catalyzed acetic acid esterification with methanol and methyl acetate hydrolysis. Ind Eng Chem Res 39:2601–2611
27. Chakrabarti A, Sharma MM (1991) Esterification of acetic acid with styrene: ion exchange resins as catalysts. React Polym 16:5159
28. Yu W, Hidajat K, Ray AK (2004) Determination of adsorption and kinetic parameters for methyl acetate esterification and hydrolysis reaction catalyzed by Amberlyst 15. Appl Catal A-Gen 260:191–205
29. Huang YS, Sundmacher K (2007) Kinetics study of propyl acetate synthesis reaction catalyzed by Amberlyst 15. Int J Chem Kinet 39:245–253
30. Gangadwala J, Mankar S, Mahajani S et al (2003) Esterification of acetic acid with butanol in the presence of ion-exchange resins as catalysts. Ind Eng Chem Res 42:2146–2155
31. Blagov S, Parada S, Bailer O et al (2006) Influence of ion-exchange resin catalysts on side reactions of the esterification of n-butanol with acetic acid. Chem Eng Sci 61:753–765
32. Lee MJ, Wu HT, Lin H (2000) Kinetics of catalytic esterification of acetic acid and amyl alcohol over Dowex. Ind Eng Chem Res 39:4094–4099
33. Ali SH, Tarakmah A, Merchant SQ et al (2007) Synthesis of esters: development of the rate expression for the Dowex 50 Wx8-400 catalyzed esterification of propionic acid with 1-propanol. Chem Eng Sci 62:3197–3217
34. Liu WT, Tan CS (2001) Liquid-phase esterification of propionic acid with *n*-butanol. Ind Eng Chem Res 40:3281–3286
35. Lee MJ, Chiu JY, Lin HM (2002) Kinetics of catalytic esterification of propionic acid and *n*-butanol over Amberlyst 35. Ind Eng Chem Res 41:2882–2887
36. http://en.wikipedia.org/wiki/Isobutyl_acetate
37. Seo Y, Hong WH (2000) Kinetics of esterification of lactic acid with methanol in the presence of cation exchange resin using a pseudo-homogeneous model. J Chem Eng Jpn 33:128–133
38. Jenson VG, Jeffreys GV (1977) Mathematical methods in chemical engineering. Academic, New York

39. Izci A, Bodur F (2007) Liquid-phase esterification of acetic acid with isobutanol catalyzed by ion-exchange resins. React Funct Polym 67:1458–1464
40. Izci A, Uyar E, Izci E (2009) Determination of adsorption and kinetic parameters for synthesis of isobutyl acetate catalyzed by Amberlite IR-122. Chem Eng Commun 196:56–67

Chapter 7
Therapeutic Applications of Ion Exchange Resins

Fazal-Ur-Rehman and Sheeba Nuzhat Khan

Abstract Ion exchange resins (IER) are water insoluble, cross-linked, chemically inert and free from local and systemic side effects. Copolymers based on vinyl, divinyl benzene and polystyrene having extensive charged functional sites have extensive binding sites leading to very high drug-loading ability. The ability of IER to bind to drug helps in masking the bitter taste of drug, drug release, drug stabilization, etc. and has made it a popular choice with pharmaceutical companies. Some IER are used as disintegrants due to their rapid water uptake. It also helps in stabilization of drug which otherwise will deteriorate on storage. Thus, IER can be used to overcome various pharmaceutical problems. IER have also found applications in parenteral drug delivery, site-specific drug delivery, as gastric retentive systems. Site-specific drug delivery is important for cancer treatment. Chewing gums also act as mobile drug delivery system specifically for treating dental caries, smoking cessation, etc. These versatile applications of IER are being used for various therapeutic applications such as neutralizing the gastric juices (hyperacidity) in treatment of gastric ulcer, Na and K supplement/depletion in oedema of cardiac, hepatic, pancreatic and nephritic origin, i.e. in a condition of Na and water retention of cardiac failure, hepatic diseases (cirrhosis of liver), pancreatic, toxaemia of pregnancy and nephritic origin. Cation-exchange resins (CER) have been used to reduce phosphate in renal end-stage disease, also as adjuvant in the treatment of hyperkalaemia associated with oliguria/anuria secondary to ARF. Anion-exchange resins (AER) have also been used in treatment of hyperglycaemia, treatment of high cholesterol level, drug-binding defects in uraemia, to reduce penicillin G binding.

Fazal-Ur-Rehman (✉)
Department of Anatomy, Jawaharlal Nehru Medical College,
Aligarh Muslim University, Aligarh 202002, India
e-mail: fazal.rehman72@yahoo.com

S.N. Khan
Department of Biochemistry, Aligarh Muslim University,
Aligarh 202002, India

Inamuddin and M. Luqman (eds.), *Ion Exchange Technology II: Applications*,
DOI 10.1007/978-94-007-4026-6_7, © Springer Science+Business Media B.V. 2012

7.1 Introduction

Ion exchange resins are water-insoluble, cross-linked polymers containing salt-forming groups in repeating position on the polymer chain and exchange their counter-ions with same charged ions in the surrounding ionic medium [1, 2]. Ion exchange resins are vinyl, divinyl benzene and polystyrene copolymers available as high molecular weight polyelectrolytes having extensive charged functional sites [1]. An ion exchange resin is normally in the form of small (1–2 mm diameter) beads, usually white or yellowish, fabricated from an organic polymer substrate backbone [3]. The material has a highly developed structure of pores on the surfaces from where the ions are trapped or released. The trapping of ions takes place only with simultaneous release of other ions; thus, the process is called ion exchange [4]. Mainly, they can be classified as cationic or anionic exchange resins depending upon the nature of the exchangeable ions of the resin as a cation or anion, respectively. Duolite AP143 is an example of anionic resin, whereas Amberlite IRP64, Amberlite IRP69 and Amberlite IRP88 are examples of cationic resins. The prerequisite for applicability of ion exchange resin to a drug is the presence of charged groups on the drug. Examples of drugs that have used ion exchange resins in formulation include dextromethorphan, diclofenac, ibuprofen and paroxetine. Apart from taste, masking resins have been used in several different functions including modified drug release [5], drug stabilization [6], tablet disintegration [5, 6] and even in therapeutics too [6]. The degrees of cross-linking and particle size of the resin substantially modify its properties and applications. Long-term physico-chemical stability [5, 6] and safety [6] of ion exchange resins have provided an additional benefit to consider them as drug carriers for a wide array of applications. Thus, ion exchange resins can be used to overcome various pharmaceutical formulation problems including bitter taste, poor stability, deliquescence and poor dissolution of the drugs. Resins have also been used as super-disintegrants in tablet formulations because of their swelling properties. Modified release of drugs from resinate (drug–resin complexes) is another potential application of ion exchange resins. Because of the versatile utility of ion exchange resins, they are being used for various drug delivery and therapeutic applications. Electrostatic and hydrophobic drug–resin interaction has been seen in drug–resin complex. Many natural substances such as proteins, cellulose, living cells and soil particles exhibit ion exchange properties, which play an important role in the way they function in nature.

7.2 History

Ion exchange resins were first developed in the middle of the twentieth century for applications in wastewater purification. Ion exchange materials based on coal and phenolic resins were first introduced for industrial use during the 1930. Until 1934, natural and synthetic siliceous materials, known as zeolites, were available for use as IE adsorbents for the purification of water [3]. In 1934, Adams and Holmes synthesized phenol-formaldehyde resin and showed that the resin can be used as a substitute for zeolites. In 1939, chemical companies began investigations into the synthesis and production of ion exchange resins under the original Adams and Holmes patent. In fact, it paved the way for the application of IER to several industrial processes and biomedical applications. Medical applications of ion exchange resins were proposed in by Segal et al. [7] and by Dock in 1946. The former group proposed the use of a polyamine-formaldehyde resin (specifically, Amberlite IR-4) for the purpose of neutralizing gastric hyperacidity, while Dock proposed the use of cation-exchange resins as a means of withdrawing or withholding sodium from the body. In 1946, Dock fed rats on a cation exchanger to remove sodium from the gut. Since then, many clinical studies of the use of cation-exchange resins have appeared. Clinically, ion exchangers have been used for neutralizing the gastric juice and for removing sodium and potassium from the intestinal contents. It was not until 1950, however, that IER were studied for pharmaceutical and biomedical applications. At this time, Saunders and Srivastava studied the uptake and release of alkaloids from IER and suggested that these resins might act as a suitable chemical carrier for the development of sustained-release formulations [8]. IER have since been extensively explored in the field of drug delivery, leading to some important patents [1, 9, 10].

7.3 Types of Ion Exchange Resins

7.3.1 Cation Exchangers (Anionic Resin)

Many cation-exchange resins are prepared by the copolymerization of styrene and divinyl benzene and have sulphonic acid groups (-SO_3H) introduced into most of benzene rings [11]. The functional group of these resins undergoes reaction (exchange) with the cations in the surrounding medium.

The mechanism of cation exchange can be depicted by the following reaction:

$$\text{Resin}^- - \text{ex}^+ + \text{C}^+ \rightarrow \text{Resin}^- - \text{C}^+ + \text{ex}^+$$

where, Resin^- indicates polymer with SO_3^- sites available for bonding with exchangeable cation (ex^+), and C^+ indicates cation in the surrounding solution being exchanged.

7.3.2 Anion Exchangers (Cationic Resin)

These are the polyelectrolytes undergoing reaction with the anions of the surrounding solutions. They are prepared by first chlor-meythylating the benzene rings of styrene–divinylbenzene copolymer to attach CH_2Cl groups and then causing these to react with tertiary amine such as triethylamine [11].

The mechanism of anion exchange can be depicted by the following reaction:

$$\text{Resin}^+ - \text{ex}^- + \text{A}^- \rightarrow \text{Resin}^+ - \text{A}^- + \text{ex}^-$$

where, $Resin^+$ indicates polymer with N^+ sites available for bonding with exchangeable anion (ex^-), and A^- indicates cation in the surrounding solution being exchanged.

7.3.3 Advantages of Resins

1. Resins being polyelectrolytes have extensive binding sites leading to very high drug-loading ability.
2. They are chemically inert and free from local and systemic side effects.
3. Because of ion exchange ability, they have been used in taste masking, modified-release and therapeutic applications.
4. All conventional solid, semisolid and liquid dosage forms can be prepared using resins.
5. They have been used in selective separation/recovery of pharmaceuticals from mixtures.
6. Being stable to all sterilization means can be formulated in to all sterile dosage forms.

7.3.4 Factors Affecting Resin Performance

Degree of Cross-Linking [2, 12]
Extent of cross-linkage has tremendous effect on drug loading efficiency. If resin has a less degree of cross-linking, then it is more porous, and the extent of swelling due to hydration is more, whereas if it is has higher degree of cross-linking, then it has less swelling. Because of this, the drug-loading ability of less cross-linked resin is higher than the more cross-linked one. However, the drug release from the former is rapid and sustained than the latter.

Particle Size [11]
The smaller resin beads offering more surface area have shown rapid exchange of ions but shorter diffusion path length. Whereas, the larger beads have more diffusional path length leading to sustained release.

pH [11]
pH is an important factor affecting drug complexation efficiency and drug release from drug–resin complex. If the pH of surrounding medium is acidic, then it promotes dissociation of basic drugs leading to formation of more ionic species available for drug loading, and towards alkaline pH, the dissociation of acidic

drugs will be promoted. With increase in dissociation, the complexation efficiency is improved. With increase in pH, protonated fraction of cationic drug decreases and, hence, interaction with resin or loading with beads decreases.

Size of Exchanging Ion [11]
It has been observed that with increase in size of exchanging ion, slower was drug loading, and less diffusion rate followed by slow release.

Selectivity of Counter-Ion [2, 11]
The ions with low selectivity for resin such as H^+ get replaced faster by cationic drugs resulting in higher loading.

Mixing Time [2, 13]
With increase in the mixing time, the swelling of resin goes on increasing and ultimately drug loading. It has been reported that complete drug loading can be achieved in 1 h only by batch process.

Effect of Temperature
For certain resins, the effect of temperature on drug loading has been reported. High temperature may also cause swelling of resin [6]. Cation-exchange resin does not get significantly affected by temperature changes unlike anion exchangers [11].

pKa [11]
The pKa value of the resin is having significant influence on the rate at which the drug is released from the resinate in the gastric fluids. The anionic resins having sulphonic, phosphonic or carboxylic acid exchange groups have approximate pKa values of <1.6. Moreover, cationic resins containing quaternary, tertiary or secondary ammonium groups have pKa values of 5–13 and greater than 13. The pKa of drug also decides the extent of dissociation and complexation with the resin.

Stability
At ordinary conditions of temperature, oxygen, light and humidity, resins are found stable and inert in nature [5, 6].

Toxicity
Chemically, resins are polymerized styrene, vinyl and divinyl benzene molecules. Hence, although polymerized form of resin is inert in nature, purification of resin needs to be involved in their usage in drug formulations.

Thus, chemical deterioration of the resins is the result of several different processes. Breaking of covalent bonds between the resin matrix and functional groups results in losses of the groups and reduction (or even total loss) of the ion exchange capacity. Organic substances sorbed can be recycled, i.e. regenerated and washed, can physically block the functional groups and/or alter the resin non-exchange sorption capacity.

7.4 Applications

7.4.1 *Water Purification Using Ion Exchange Resins*

The aim is usually either to soften the water or to remove the mineral content altogether. The water is softened by using a resin containing Na^+ cations which binds Ca^{2+} and Mg^{2+} more strongly than Na^+. As the water passes through the resin, the resin takes up Ca^{2+} and Mg^{2+} and releases Na^+ making for a 'softer' water. If the water needs to have the mineral content entirely removed, it is passed through a resin containing H^+ (which replaces all the cations) and then through a second resin containing OH^- (which replaces all the anions). The H^+ and OH^- then react together to give more water.

7.4.2 *Pharmaceutical Applications of Ion Exchange Resins*

Ion exchange resins are used in the manufacturing of pharmaceuticals not only for catalyzing certain reactions but also for isolating and purifying pharmaceutical active ingredients. Three ion exchange resins, sodium polystyrene sulphonate, colestipol and cholestyramine, are used as active ingredients. Sodium polystyrene sulphonate is a strongly acidic ion exchange resin and is used to treat hyperkalaemia. Colestipol is a weakly basic ion exchange resin and is used to treat hypercholesterolaemia. Cholestyramine is a strongly basic ion exchange resin and is also used to treat hypercholesterolaemia. Colestipol and cholestyramine are known as bile acid sequestrants.

7.4.3 *Taste Abatement by Ion Exchange Resins*

Most of the bitter drugs have amine as a functional group, which is the cause of their obnoxious taste. If the functional groups are blocked by complex formation, the bitterness of the drug reduces drastically [14–16]. This type of complex formation is being done by converting the drug to stearates and estolate. However, nowadays, the use of ion exchange is done to block the functional group responsible for causing the bitter taste by forming complex between ion exchange resin and the drug. Thus, ion exchange resin complexes mask the taste of bitter drug (e.g. azithromycin, clarithromycin). Drug can be bound to the ion exchange resin by either repeated exposure of the resin to the drug in a chromatographic column or by prolonged contact of resin with the drug solution. The resins form insoluble adsorbates or resonates through weak ionic bonding with oppositely charged drugs. The exchange from counter-ions from resin is competitive. Taste masking in the present-day pharmaceutical industry has become a potential tool to improve patient compliance, provide brand recognition to combat private-label competition and increase commercial success of the product [8]. The technique of masking the unacceptable taste of an orally administered pharmaceutical includes filling in capsules, coating with water in soluble polymers or pH-dependent water-soluble polymer, adsorption on ion exchange resin, microencapsulation with various polymers complexing with cyclodextrin, and chemical modification such as use of insoluble prodrugs, effervescent systems, salt formation and use of excipient like flavours, sweeteners, gelatin, gelatinized starch and surfactants. Agents like sodium chloride, phosphatidic acid and peppermint flavour are known to inhibit bitterness by selected active pharmaceutical ingredient (API) molecules via a mechanism that takes place at the bitterness receptors in the taste buds.

Drug release from ion exchange resin depends upon two factors: (1) the ionic environment (i.e. pH and electrolyte concentration) within the gastrointestinal tract and (2) the properties of resin. Strong acid cation resins can be used to mask the taste of basic drugs having bitter taste, as they function throughout the entire pH range. Weak acid cation-exchange resins function at pH values above 6. Similarly, strong-base anion-exchange resins function throughout the entire pH range, while the weak-base anion-exchange resins function well below pH 7.0 [17]. Polystyrene matrix cation-exchange resins (e.g. Indion CRP-244 and Indion CRP-254) have been used to mask the bitter taste of chlorpheniramine maleate, ephedrine HCl and diphenhydramine HCl [18]. The bitter taste of ranitidine and buflomedil were masked successfully by using Amberlite IPR-69. Ion exchange (India Ltd.) has successfully masked the taste of drugs like azithromycin by using various grades of Indion ion exchange resins.

7.4.3.1 Polymers Used for Taste Masking by Coating

The polymers used for the purpose of taste masking are selected based on the fact that they should permit the rapid release of drug in the saliva, but allows it in the gastric cavity or in the duodenal region where the drug is expected to be absorbed. Polymers may be water soluble or dispersible (e.g. hydroxy ethyl methyl cellulose, methyl cellulose) for *dissolution* in stomach. It may be enteric polymer that does so in small intestine (e.g. cellulose acetate phthalate). The polymer may be insoluble at pH of the gastrointestinal tract (GIT), and such coating can be viewed as diffusion coating [19].

7.4.4 Disintegrant/Super-Disintegrant

Majority of tablet disintegrants have rapid water uptake capacity due to swelling. Because of their unusually large swelling capacities, polymethacrylic carboxylic acid ion exchange resins have found usage as tablet disintegrants.

7.4.5 *Improved Dissolution*

Drug–resin complexation converts drug to amorphous form [20]. Hence, drugs with poor solubility, during the process of desorption, are immediately released, leading to improved drug dissolution.

7.4.6 *Drug Stabilization*

Vitamin B12 deteriorates on storage. The stability of vitamin B12 can be prolonged by complexing it with a weak acid cation-exchange resin (Indion-264). This complex is as effective as the free form of the vitamins [21]. Thus, the introduction of Indion-264 in the formulation significantly reduces the deterioration.

7.4.7 *Powder Processing Aid*

Handling quality of hygroscopic drugs and drugs having caking tendency have been improved due to complexation with the resin. The fixed rigid structure of resins was thought to be responsible for the improved quality.

7.4.8 *High-Purity Water*

There will always be a need for purified water in the production of pharmaceuticals. Water softening uses a cation-exchange resin to exchange principally calcium and magnesium ions for sodium ions and so prevent the formation of calcium carbonate precipitates on reverse osmosis membranes. Feed water is passed down through the resin bed, which is in the sodium form. The polyvalent cations are absorbed onto the resin, and the sodium ions are released. After a preset volume, or time, the system is taken off line and automatically regenerated with a brine solution.

7.5 Applications of Ion Exchange Resins in Drug Delivery

7.5.1 *Oral Modified-Release Formulations*

The use of ion exchange resins into drug delivery systems have been encouraged because of their physico-chemical stability, inert nature, uniform size, spherical shape assisting coating and equilibrium-driven reproducible drug release in ionic environment [5]. Several preparations involving strong resinates of sulphuric acid (cation-exchange resins) provided more moderate release than the weak resinates of carboxylic acid [22]. Hence, resinates of strong cationic drugs are formulated as sustained-release suspension, tablets, capsules and microparticles [22]. Mainly, amphetamine [23], propranolol HCl [24], dextroamphetamine [25], codeine [25] and chlorpheniramine (Penn kinetic system by Pennwalt Corporation USA) [25] are complexed with strong cation exchanger for slow release. Polymeric coatings using ethyl cellulose and polyethylene glycol-4000 have been reported for exercising control over drug release [12]. Microencapsulated tramadol–resin complex with 10, 25 and 45 cps viscosity grade ethyl cellulose has shown slow release. The prolonged release of the active drug is accomplished by providing a semipermeable coating around discrete, minute, ion exchange resin particles with which the drug component has been complexed to form an insoluble drug–resin complex. The semipermeable coating creates a diffusion barrier, and the thickness of which can be adjusted to provide the desired level of retardation of drug availability in the gastrointestinal tract over a period of time [24]. The drug release from *hydroxypropyl methylcellulose* (HPMC) tablets containing drug–resin complexes was significantly slower than from *hydroxypropyl methylcellulose* tablets containing drug without resin [26]. Nicorette chewing gum is an example in which the nicotine–cation-exchange resin complex is encapsulated into a sorbitol-based matrix which releases nicotine slowly during chewing, providing the minimal supply of nicotine to facilitate smoking cessation [26, 27]. The drug-resinate complex of bromhexine in suspension form has been formulated to control drug release and reduce the frequency of dosage administration. Simultaneously, bioadhesive properties of ion exchange resins may be useful in mucoadhesive systems for local treatment of stomach infections like H. pylori for prolonging the gastric residence of amoxicillin and cimetidine.

7.5.1.1 Simple Resinates

Resinates alone are the simplest forms of controlled- or sustained-release delivery systems. Resinates can be filled directly in a capsule, suspended in liquids, suspended in matrices or compressed into tablets. Drug from the resinate will be slowly released and absorbed as compared to the drug particles but will also be significantly faster than the modified resinates (coated or microencapsulated). The release of diclofenac at the desired rate to avoid gastric irritation was achieved for arthritic patients [28].

7.5.1.2 Microencapsulated or Coated Resinates

Microencapsulation of resinates provides better control over the drug release because of the presence of a rate-controlling membrane [29, 30]. The absorption of the drug from coated resinates is a consequence of the entry of the counter-ions into the coated resinate, release of drug ions from the drug–resin complex by the IE process and diffusion of drug ions through the membrane into the surrounding absorption environment. Microencapsulation of resinates can be achieved by air suspension coating (Wurster process), interfacial polymerization, solvent evaporation or pan coating.

7.5.1.3 Hollow Fibre System

Hollow fibre systems have advantages of high surface area to volume ratio, loading flexibility, membrane permeability and potentially slower GIT transit time. These characteristics could provide a method to obtain controlled release for drugs in the small intestine and/or in the colon. Hollow fibres made from suitable polymeric materials are filled with resinate to obtain a controlled- or sustained-release profile. In vitro and in vivo release of phenylpropanolamine (PPA) from polyurethane fibres filled with PPA-Dowex 50 W complex (resinate) has shown sustained effect [16].

7.5.2 Application in Parenteral Drug Delivery

Microspheres of resin complexed with cytotoxic agents like adriamycin and doxorubicin upon intra-arterial administration are entrapped in capillary beds [31]. This works on the principle of chemo-embolization of a drug loaded microsphere via the tumour arterial supply. As a result, microspheres containing cytotoxic drug can be delivered to well vascularize tumour tissues in a sustained way specifically into tumour locale. This suppresses toxicity of chemotherapeutic agents due to their unavailability to entire systemic circulation. Recently, microspheres of resin complexed with radioisotopes have been used via same route for internal irradiation of tumour cells.

7.5.3 Site-Specific Drug Delivery Applications

Delivering drugs at the desired biological location or site could have several advantages in therapeutics, such as:

- Localizing the required drug concentration to maintain a minimum effective concentration throughout the treatment
- Reducing the systemic toxicity, especially with cytotoxic anticancer drugs
- Bifurcating the hostile environment of the drugs to prevent the drug degradation. Several studies have reported the use of IER for drug delivery at the desired site of action [32, 33]

7.5.3.1 Gastric Retentive Systems

Prolonged gastric retention of the drug formulations could improve the bioavailability and reduce drug wastage, especially for those predominantly absorbed from the stomach. Examples of such drugs are furosemide, cyclosporin, allopurinol and ciprofloxacin. Floating dosage forms are one of the alternatives designed to prolong gastric residence of drugs. A novel floating extended-release system consisting of a bicarbonate-charged resin coated with a semipermeable membrane was studied for improving gastric-residence time [32]. Some IER, especially anion-exchange resins (AER), such as cholestyramine, possess bio-/mucoadhesive properties, which might be caused by their electrostatic interaction with the mucin and epithelial cell surface. The use of such bioadhesive IER is another attractive approach in the development of targeted formulations for the GIT. This approach would enhance the localized delivery of antibiotics, such as tetracycline, to the sites of helicobacter pylori colonization (fundus), which conventional dosage forms fail to reach. Similar types of applications were found with bioadhesive IER [34, 35]. These systems are also useful in the delivery of diagnostics.

7.5.3.2 Sigmoidal-Release Systems

The drug release should be controlled in accordance with the therapeutic purpose and the pharmacological properties of the active substance [36]. A sigmoidal-release system rapidly releases the drug from a multiple-unit device after a predetermined lag time and can achieve both time-controlled and rhythmic release. IER were studied in the development of sigmoidal-release systems. Eudragit RS (Röhm, Darmstadt, Germany), an AER with limited quaternary ammonium groups, is coated over beads with a sugar core surrounded by organic acid and drug mixture. The ionic environment, induced by the addition of an organic acid to the system, was found be responsible for pulsatile release [37, 38].

7.5.3.3 Site-Specific Delivery of Drugs for Cancer Treatment

Entrapment of anticancer drugs within the particulate carriers (microspheres, microcapsules) is a popular approach for the development of delivery systems for cancer treatment. Several anticancer drugs (e.g. doxorubicin) are ionic in nature and can be complexed with IER. Recently, various attempts have been made to deliver some of these drugs in a controlled-release fashion to the cancer cells with the help of IER [39, 40]. These studies revealed that the drug loading is at its maximum level with the IER complex approach. Sawaya et al. [36] studied the mechanism of complexation of doxorubicin with ion exchange albumin microcapsules.

7.5.4 Application in Ophthalmic Drug Delivery

Ophthalmic drug delivery of betaxolol from resin involves release of drug from drug–resin complex. The flocculated suspension prepared in *Carbopol* 934P as 0.25% betaxolol in 0.25% BETOPTICS ophthalmic suspension. Microparticulates of ion exchange resin and betaxolol (antiglaucoma drug) are reported for sustained ophthalmic drug delivery [19]. Another example of ion exchange in ocular drug delivery was reported by Moreau et al. for delivery of ciprofloxacin complexed with polystyrene sulphonate for the treatment of eye infections [17].

7.5.5 *Application in Nasal Drug Delivery*

Amberlite IRP69, cation-exchange resin having particle size 10–150 μm, has been reported for nasal delivery of nicotine. Biphasic, extended drug delivery has been obtained by Hinchcliffe et al. [41] for nicotine delivery.

7.5.6 *Chewing Gums as Mobile Drug Delivery Systems*

Chewing gum is a mixture of natural or synthetic gums (polyvinyl acetate and similar polymers) and resins, sweetened with sugar, corn syrup, artificial sweeteners and may also contain colouring agents and flavour. It is a potentially useful means of administering drugs either locally or systemically via the oral cavity. The medicated chewing gum has gained increasing acceptance as a drug delivery system. Several ingredients are now incorporated in medicated chewing gum, e.g. fluoride for prophylaxis of dental caries, chlorhexidine as local disinfectant, nicotine for smoking cessation, aspirin as an analgesic and caffeine as a 'stay alert' preparation. Fast onset is due to rapid release of active ingredients in buccal cavity and subsequent absorption in systemic circulation. Aspirin, dimenhydrinate and caffeine showed faster absorption through medicated chewing gums than tablets. The use of sugar-free gum to counteract dental caries by stimulation of saliva secretion has led to a more widespread use and acceptance of gums. It is well known that using non-medicated chewing gums increases plaque pH, stimulates saliva flow and decreases decay. Synthetic medicated chewing gums (SMCGs) containing chlorhexidine for treatment of gingivitis and plaque have been available. Medicated chewing gums for systemic effect in conditions like vitamin C deficiency, pain and fever, alertness, motion sickness, smoking cessation, as well as for local effect in the conditions like plaque acid neutralization, fresh breath, disinfection, anticaries, antiplaque, antifungal, and antibacterial are available.

7.5.7 *Chewing Gum for Glycol Absorption*

Nicorette gum is a widely used patented product for smoking cessation programme [42, 43]. It contains nicotine adsorbed on an ion exchange resin with carboxylic acid functionality and formulated in a flavoured chewing gum base, which provides gradual drug release through glycol mucosa as the gum is chewed offering fresh saliva as solvent for elution.

7.6 Use of Ion Exchange Resins as Therapeutics

Ion exchange resins have notably found use in the treatment of various pathological states such as hyperacidity; ulcer; Na and K supplement depletion; nephrotic, pancreatic and cardiac oedema, etc. Cationic resins have also been used as oral therapeutics to reduce phosphate levels for end-stage renal disease patients. Cation-exchange resin (sodium polystyrene sulphonate) has been reported as an adjuvant in the treatment of hyperkalaemia associated with oliguria or anuria secondary to acute renal failure [44]. Anion-exchange resins have been used in the treatment of hyperglycaemias. At present, cholestyramine and colestipol (AER) are used in the treatment of type II hyperlipoproteinaemia and familial hyperlipoproteinaemia in children and young adults [15]. Colestipol hydrochloride acts by increasing fractional carbolic rate of low-density lipoproteins (LDL), thereby decreasing LDL [45, 46].

Toothpastes, lacquers and prostheses coming in contact with tooth surface are rendered anticarious by incorporating ion exchange resins containing fluorine, phosphate or calcium ions [47].

7.6.1 Application in Treatment of High Cholesterol

Cholestyramine was the first polymeric resin-based drug that was approved for the treatment of high cholesterol. The resin in this case serves as a sequestrant to bind bile acids in the gastrointestinal tract. The binding and effective removal of bile acids force the liver to consume cholesterol to synthesize more bile acids. This leads to the indirect reduction in cholesterol levels. The advantage of this therapy was that it did not use conventional drugs and hence had lower side effects as compared to conventional therapies. The drawback of this therapy was the high-dose requirements for the first-generation resins (four tablespoons administered via fruit juice).

Colestimide, a 2-methylimidazole-epichlorohydrin polymer, is a new bile acid sequestering resin that is fourfold as powerful at lowering low-density lipoprotein cholesterol (LDL-C) as the conventional resin (cholestyramine). Colestimide has been reported to lower blood glucose levels in patients with type 2 diabetes complicated by hypercholesterolaemia [48].

7.6.2 Application of Non-absorbable Ion Exchange Resin in Pruritus

These molecules are not absorbable from GIT and release chloride and bind bile acid in the lumen of the intestine, diminishing their enterohepatic recirculation. Non-absorbable ion exchange resin augments cholesterol excretion via enhanced conversion to bile acids. These drugs can decrease the bile acid pool by ca. 40%. Trihydroxy bile acid dissociates rapidly from these molecules and can be absorbed in the ileum.

Cholestyramine (orange-flavoured granules) has been used for some decades and has been proven to be effective and safe. However, palatability is poor, limiting the tolerability to this drug. Through nasogastric tube, administration of cholestyramine through this pathway has been shown to be well tolerated and successful in decreasing serum cholesterol levels and improving pruritus. Colestipol in granules or tablets has not been used frequently in children. Colesevelam hydrochloride (in the form of non-absorbable hydrogel or tablets) is probably the best tolerated of these drugs. This drug has enhanced the specificity, greater affinity and higher capacity for binding bile acids compared with the above two non-absorbable IER. Non-absorbable IERs can interfere with intestinal absorption of other drugs, like ursodeoxycholic acid (UDCA). So other drug should be given 1 or 4–6 h after cholestyramine administration to avoid interaction with other drug including fat-soluble vitamins and to allow a maximal capture of bile acids. Non-absorbable IER dose is 8–16 g orally 2–3 daily, increased progressively by 2 g daily. Side effects of non-absorbable IER include constipation and hyperchloraemic acidosis due to large quantity of chloride released in GIT and absorbed instead of bicarbonate.

7.6.3 Adsorption and Recovery of Rifamycin B and Rifamycin S

The antibiotic rifamycin S is a physiologically active derivative of the rifamycin B [49]. Strongly or weakly basic anion-exchange resins are reported for recovery of antibiotic rifamycin S from rifamycin B, by curtailing complex, multi-step processes involved in separation.

7.6.4 Other Clinical Applications of Ion Exchange Resins

Sulphonated and carboxylic resins with a polystyrene backbone are most widely used in clinical medicine. The pharmacological activity of these resins is attributed to their ability to adsorb ions, which are more selective to the resin than the counter-ion of the resin. Resins are mostly used in conditions of sodium and water retention, such as in cardiac failure, renal disease (nephrotic syndrome), toxaemia of pregnancy and cirrhosis of the liver [14]. In hypertension and oedema, dietary restriction of sodium to less than 0.5 g/day is difficult. IER have been used as reinforcement of a low sodium diet or to enable high salt intake in the diet. Certain cation-exchange resins in the H^+ or NH^+ form may be used for the relief and control of oedema. Doses of about 50 g/day are usually required, in conjunction with a limited salt intake. IER have also been used for haemoperfusion and management of drug overdoses (poisoning).

7.6.5 Cation-Exchange Resins in the Treatment of Oedema

The resins have greater affinity for calcium and magnesium than for potassium and for potassium than for sodium. Under suitable conditions, the resin may replace one cation in a solution by another, though the extent to which this occurs depends on several factors such as the relative concentrations, the pH of the medium and the time allowed for exchange. Treating oedema, either carboxylic or sulphonic resins can be given by mouth, in either the hydrogen or ammonium form. The greater part of the exchange of cations appears to take place in the upper part of the alimentary canal and involves mainly the cations ingested in the food. For each milliequivalent of cation taken up by the resin, 1 mEq of ammonium of hydrogen is exchanged and absorbed together with 1 mEq of anion. The ammonia is converted to urea, while the hydrogen ions are buffered by bicarbonate to form H_2CO_3, which is removed by increased pulmonary ventilation. The anions absorbed are accommodated in the extracellular fluid at the expense of bicarbonate. The latter is lowered still further by loss of fixed base in the urine, but the kidney reduces this loss by secreting a highly acidic urine (containing H_2PO_4 in place of HPO_4) and, after a few days' delay, by forming a large quantity of ammonia. In this way, disposable cations are absorbed in place of sodium, which is removed in the faeces. The corresponding anions are excreted by the kidney with little fixed base and with a minimum change in plasma pH (compensated metabolic acidosis such compensation cannot be effected if renal function is severely impaired: in this case, sodium and other cations will be lost in the urine and severe acidosis will develop. The amount of sodium removed in the faeces by a given dose of resin decreases with decreasing dietary sodium. As its bulk limits the quantity that can be given, there is a definite limit to the amount of sodium that can be removed and it is rare for the faeces to contain more sodium than the diet.

7.6.6 Application of Cation-Exchange Resin (Resodec) in the Treatment of Cardiac Oedema

'Resodec' (a balanced ammonium-potassium carboxylic resin) in congestive heart failure produces clinical results similar to those of dietetic sodium restriction but sometimes with undesirable side effects which are due to hyperchloraemia or to development of abnormal serum potassium levels. Biochemical control is essential in the early stages of treatment. This is especially important in patients with renal impairment and in those who have been on sodium-restricted diet or on mercurial diuretics; such patients may have abnormal blood electrolyte levels resulting from previous treatment. These abnormalities are sometimes corrected but sometimes aggravated by resin therapy. Subject to biochemical control, the resin can be used as a substitute for low sodium diet or in conjunction with the diet to reinforce the latter. When given in addition to low sodium diet, the best result was obtained with repeated short courses. Ammonium chloride should be discontinued during periods of resin administration.

7.6.7 Complications of Cation-Exchange Resin Therapy Used in the Treatment of Oedema

1. Acidosis: A compensated metabolic acidosis develops in every case treated with cation-exchange resins. In a great majority, this produces no undesirable effects, though it is possible that it might play a part in producing skeletal decalcification over a long period, especially if hypocalcaemia were also produced.
2. Sodium deficiency: The low salt syndrome may be produced if sodium removal is too drastic or if very much sodium is removed in the urine. This sodium depletion may be associated with signs of cellular over-hydration, vascular collapse and renal failure and finally death.
3. Potassium deficiency: This is particularly liable to occur with low sodium diets, for resins will then remove more potassium than sodium in the faeces. The simplest safeguard is to give extra potassium with the resin, i.e. to give it partly in the potassium cycle. The potassium is exchanged for hydrogen in the stomach and is available for absorption in the intestine.
4. Calcium deficiency: In few cases, this appears to be a real hazard after very long periods of treatment. It must be remembered that if the kidney cannot form ammonia, extra losses of sodium, potassium and calcium in the urine may cause deficiencies of these ions. On the other hand, if renal failure or oliguria is present, hyperkalaemia may occur if a potassium resin is given.
5. Anorexia and nausea, abdominal pain, diarrhoea, constipation and even faecal impaction have been noted in a few of the cases treated.
6. Urinary casts nearly always appear when highly acidic urine is being formed [41], but there is no evidence that they impair renal function. A high fluid intake has been recommended.
7. Deficiencies of iron, magnesium, thiamine and riboflavin might theoretically occur but have not been reported. Martz et al. [50] found a normal response to iron in one patient with hypochromic anaemia while on resin therapy.

7.6.8 Anion-Exchange Resins as Antacids (Amberlite IR-4)

Anion-exchange resin is effective in neutralizing acid and inactivating pepsin of both ulcer and non-ulcer patients (divided doses of about 25 g/day are usually recommended). Segal and co-workers [7] demonstrated that the weakly basic resins are capable of raising the pH of gastric juice from pH 1–5. The advantages claimed for this type of treatment include (a) great speed of action; (b) greater neutralizing power when used in practical application; (c) complete inhibition of peptic activity; (d) no removal of PO_4^-; (e) lack of acid rebound; (f) no constipating effect; and (g) no chloride removal.

7.6.9 Anion-Exchange Resin Treatment of Drug-Binding Defects in Uraemia

Serum protein binding of weakly acidic drugs is impaired in uraemia, but that of basic drugs tends to be normal. Treatment of uraemic serum with anion-exchange resin (e.g. Amberlite CG-400, acetate form) corrected binding defects for three acidic drugs (e.g. nafcillin, salicylate and sulfamethoxazole) but did not affect the binding of two basic drugs (trimethoprim and quinidine). Resin treatment of normal human serum did not alter the binding of these five drugs. Extraction of the acetate buffer eluate from resin exposed to uraemic serum with n-butyl chloride at acidic pH (3.0) resulted in a fraction that could induce similar binding defects in normal human serum. The factor(s) responsible for binding defects in uraemia appears to be lipid soluble, weakly acidic and dialyzable. It is believed to be tightly bound to albumin at physiologic pH but dissociates from it at acidic pH.

The effect of resin treatment on penicillin G binding was quite unpredictable, and the results were essentially identical to those obtained from either activated charcoal treatment [51] or n-butyl chloride extraction at acidic pH [52]. Recently, the effect of free fatty acids on protein binding of antimicrobials was investigated, and a strong positive correlation between the levels of free fatty acids and the degree of penicillin G binding was demonstrated [18]. Anion-exchange resin treatment of serum specimens was shown to remove free fatty acids from serum and therefore to reduce penicillin G binding.

Removal of bilirubin by haemoperfusion with ion exchange resin partides is suggested to replace exchange transfusion of blood of jaundiced infants in some cases of unconjugated hyperbilirubinaemia. The haemoperfusion system developed here consists of a packed bed of a macroreticular resin, which is made biocompatible by a coating of a monomolecular layer of albumin. The choice of the appropriate ionic form of the resin and the proper albumin coating and cross-linking procedure assures a high bilirubin adsorption capacity and excellent blood compatibility of the resin. The albumin-coated resin removes in vitro 80–90% of the bilirubin initially present in the plasma.

7.7 Regulations of Ion Exchange Resins for the Food, Water and Beverage Industries

The FDA has the responsibility to define conditions under which safe food additives may be used in the production and preparation of foods and beverages.

The three major conditions laid out by this law are as follows:

1. The resins must be one of a pre-approved generic list of resin compositions (listed in 21 CFR 173.25), of which the ingredients used to produce the resins comply with FDA food additive regulations.

2. The resin must be 'subjected to pre-use treatment by the manufacturer and/or user in accordance with the manufacturer's directions'.
3. The resin must be 'found to result in no more than 1 ppm of organic extractives' [53].

Aside from the commercially motivating factors to develop acceptable tasting products, the pharmaceutical industry is also monitored by the regulatory agencies for paediatric drug products. Currently, a drug must comply with the Food and Drug Administration's (*FDA's*) campaign to improve labelling for paediatric products. The 23 *FDA* regulations, known as the paediatric rule, may mandate a drug firm to conduct paediatric clinical studies as part of the new drug applications. In the pharmaceutical industry, taste-making science broadly covers physiological and physiochemical approaches to prevent active pharmaceutical ingredient (API) or drugs from interacting with taste buds, thereby eliminating or reducing negative sensory response. Physiological approaches consist of inhibiting or modifying an API-mediated bitterness response by incorporating agents into a pharmaceutical formulation. Agents like sodium chloride, phosphatidic acid and peppermint flavour are known to inhibit bitterness by selected API molecules via a mechanism that takes place at the bitterness receptors in the taste buds.

Acknowledgements The authors are thankful to the Department of Anatomy, JNMC, Aligarh, for generous support.

References

1. Kasture AV, Wadodkar SG, Mahadik KK, More HN (2002) A textbook of pharmaceutical analysis and instrumental methods, 8th edn. Pierce Scientific, Rockford, pp 39–47
2. Bhalekar M, Avari JG, Jaiswal SB (2004) Cation-exchanger in pharmaceutical formulation. Indian J Pharm Sci 38:184–187
3. Mantell CL (1951) Ion-exchangers. In: Adsorption. McGraw-Hill, New York, pp 185–216
4. Adams BA, Holmes EL (1935) Synthetic ion-exchange resins. J Soc Chem Ind 54:1T
5. Hughes L, Ion-exchange resin unique solution to formulation problem; pharmaceutical technology Excipients and solid dosage form 2004. http://www.pharmtech.com/
6. Chaubal MV (2006) Synthetic polymer based ion-exchange resin. Drug Del Technol 2006:6–8
7. (a) Segal HL, Hodge H, Watson JS, Scott WJ (1945) The influence of anion exchange resins on experimentally produced oral drugs Gostroenlerology 4:484. (b) Dock W (1946) Trans Assoc Am Phys 59:282
8. Chaudhary NC, Saunders L (1956) Sustained release of drugs from ion-exchange resins. J Pharm Pharmacol 8:975–986
9. Keating JW (1961) Pharmaceutical preparations comprising cation exchange resin adsorption compounds and treatment therewith. US Patent 2,990,332
10. Hays EE (1961) Pharmaceutical compositions containing a resin narcotic compound. US Patent 3,035,979
11. Jain NK (2005) Advanced drug delivery system, 1st edn. Antares, CBS Publication. pp 290–302
12. Pisal S, Zainnuddin R, Nalawade P, Mahadik K, Kadam S (2004) Molecular property of Ciprofloxacin–Indion 234 complexes. AAPS Pharm Sci Tech 5:62–63
13. Pisal S, Zainnuddin R, Nalawade P, Mahadik K, Kadam S (2004) Drug released property of polyethylene–glycol treated ciprofloxacin–Indion234 complexes. AAPS Pharm Sci Tech 5(4):64

14. Payne WW (1956) Ion-exchange resins in clinical medicine. J Pharm Pharmacol 8:397–402
15. Witzman JL (1996) Drug used in the treatment of hyperlipoproteinemias. In: Hardman JG, Limbird LE (eds) Goodman & Gilman's the pharmacological basis of therapeutics, pp 875–897
16. Hussain MA, Diluccio RC et al (1989) Hollow fibres as an oral sustained release delivery system. Pharm Res 6:49–52
17. Moreau JM, Green LC, Engel LS, Hill JM, O'Callaghan RJ (1998) Effectiveness of ciprofloxacin-polystyrene sulfonate (PSS), ciprofloxacin and ofloxacin in a *Staphylococcus keratitis* model. Curr Eye Res 17(8):808–812
18. Suh B, Craig WA, England AC, Elliot RL (1981) Effect of free fatty acids on protein binding of antimicrobial agents. J Infect Dis 143:609–616
19. Jani R, Gan O, Ali Y, Rodstrom R, Hancock S (1994) Ion-exchange resins for ophthalmic delivery. J Ocul Pharmacol 10(1):57–67
20. Hughes L, Gehris A, A new method of characterizing the buccal dissolution of drug. Rohm and Haas Research Laboratory Spring House, Philadelphia http://www.rohmhaas.com/ionexchange/pharmaceuticals/Formulations_doc/buccal_dissolution.pdf
21. Van Abbe NJ, Rees JT (1998) An approach for development of oral sustained release suspension. J Am Pharm Assoc Sci 47:487
22. Douglas SJ (1963) Drug adsorbates. US Patent 5,219,563
23. Sheumaker JL (1998) Liquid prolonged released pharmaceutical formulation containing ionic constituent. US Patent 4,762,709
24. Akkaramonkolporn P (2001) Molecular property of propranolol HCl prepared in drug resin complex. Drug Dev Ind Pharm 27:359–364
25. Raghunathan Y (1980) Prolong released pharmaceutical preparations. US Patent 4,221,778
26. Sriwangjanya M, Bodmeier R (1998) Effect of ion–exchange resin on drug released from matrix tablet. Pubmed 46(3):321–327. Bellamy SA, Hughes L (2003) Method for the anhydrous loading of nicotine onto ion-exchange resins. US Patent 6,607,752
27. Litchenckert S, Landgren C, Ferno O (1975) Chewable smoking substitute composition. US Patent 3,901
28. Kurowski M et al (1994) The efficacy and relative bioavailability of diclofenac resinate in rheumatoid arthritis patients. Int J Clin Pharmacol Ther 32:433–440
29. Amsel LP, Hinsvark ON, Rotenberg K et al (1984) Recent advances in sustained release technology using ion-exchange polymers. Pharm Technol 8:28–48
30. Sayed UG, Bajaj AN (2000) Oral controlled release bromhexine ion-exchange resinate suspension formulation. Indian Drugs 37:185–189
31. Yan Chen, Gray BN (1999) Controlled released preparation for cytotoxic or cytostatic drug. US Patent 5,307,162
32. Fell TJ, Conitehead L, Collett JH (2000) Prolonged gastric retention using floating dosage forms. Pharm Technol 24:82–88
33. Atyabi F, Sharma HL, Mohammad HAH (1996) Controlled drug release from coated floating ion-exchange resin beads. J Control Release 42:25–28
34. Irwin WJ et al (1974) Drug delivery by ion-exchange Part VII: release of acidic drugs from AniKo, H. and Royer, M.E., In vitro binding of drugs to Colestipol hydrochloride. J Pharm Sci 63:1914–1920
35. Burton S et al (1995) Intragastric distribution of ion exchange resins: a drug delivery system for the topical treatment of the gastric mucosa. J Pharm Pharmacol 47:901–906
36. Sawaya A et al (1987) Binding mechanism of doxorubicin in ion exchange albumin microcapsules. J Pharm Sci 76:475–480
37. Narisawa S et al (1994) An organic acid induced sigmoidal release system for oral controlled release preparations. Pharm Res 11:111–116
38. Narisawa S et al (1994) An organic acid-induced sigmoidal release for oral controlled release preparations. 2. Permeability enhancement of Eudragit RS coating led by the physicochemical interactions with organic acid. J Pharm Sci 85:184–188
39. Chen Y et al (1991) Evaluation of ion-exchange microspheres as carriers for the anticancer drug Doxorubicin: in-vitro studies. J Pharm Pharmacol 44:211–215

40. Li VHK et al (1987) Influence of drug properties and routes of drug administration on the design of sustained and controlled release systems. In: Robinson JR, Lee VHL (eds) Controlled drug delivery, Marcel Dekker, New York. pp 3–94
41. Cheng YH, Watts P, Hinchcliffe M, Hotchkiss R, Nankervis R, Faraj NF, Smith A, Davis SS, Illum L (2002) Development of a novel nasal nicotine formulation comprising an optimal pulsatile and sustained plasma nicotine profile for smoking cessation. J Control Release 79:243–254
42. Bellamy, Simon A, Hughes L (2003) Method for the anhydrous loading of nicotine onto ion-exchange resins. US Patent 6,607,752
43. Sriwangjanya M, Bodmeier R (1998) Effect of ion-exchange resin on drug released from matrix tablet. Pubmed 46(3):321–327
44. Pandit JK, Singh S, Muthu MS (2006) Controlled released formulation in neurology practice. Ann Indian Acad Neurol 9:207–216
45. Zemp HN (1983) Anionic ion-exchange resins with cholesterol decreasing properties. US Patent 4,393,145
46. Kunin R (1988) Blood cholesterol reducing pharmaceutical composition. US Patent 5,840,339
47. Naumann G, Pieper G, Rehberg H-J (1976) Dental composition and appliances containing anti-carious ion-exchange resin. US Patent 3,978,206
48. Tatsuya S, Kenzo O, Yoshimasa I, Noriaki M, Kentaro W, Shoko F, Hiroko Y, Motoshi O, Kazunari S, Yoshiaki K, Hiroshi N (2007) Colestimide lowers plasma glucose levels and increases plasma glucagon-like peptide-1 (7–36) levels in patients with type 2 diabetes mellitus complicated by hypercholesterolemia. J Nippon Med Sch 74(5):338–343
49. Gilpin JA, Patric J, Roger A (1983) Adsorption and recovery of rifamycine B and rifamycine S using basic ion-exchange resins. US Patent 4,402,877
50. Martz BL, Kohlstaedt KG, Helmer OM (1952) Use of anion and cation exchange resins in odemea. Circulation 5:524–533
51. Craig WA, Evenson MA, Sarver KP, Wagnild JP (1976) J Lab Clin Med 87:637
52. Lichtenwalner DM, Suh B, Lorber B, Rudnick MR, Craig WA (1981) J Lab Clin Med 97:72
53. Jay A, Miers JR (1995) Regulation of ion exchange resins for the food, water and beverage industries Reactive Polymers 24:99–107

Chapter 8
Application of Ion Exchange Resins in Kidney Dialysis

Fazal-Ur-Rehman and Sheeba Nuzhat Khan

Abstract Besides the proved applications of ion exchange resins (IER) in various industries, biochemists have found their uses in medicines also. The development and use of synthetic ion exchange resins for kidney dialysis is a relatively recent achievement. The artificial kidney uses cellulose membranes in place of the phospholipid-bilayer membranes used by real kidneys to separate the components of blood. Polymeric ion exchange resins are insoluble, so when taken orally, pass through gastrointestinal tract (GIT) without being absorbed. In malfunctioned kidney, sodium and calcium polystyrene sulfonate resins are designed to exchange sodium for potassium in the colon, for use in the treatment of hyperkalemia. It is predicted that additional therapeutic applications may be found for ion exchange resins in the coming years.

8.1 Introduction

Ion exchange membranes are one of the advanced separation membranes, used in electrodialysis and in artificial dialysis (such as artificial kidney) in medical field. The function of the ion exchange membrane process is based on the Donnan membrane equilibrium principle, i.e., recovery of valuable ions and removal of undesirable ions from wastewater. Artificial kidney is a machine that performs treatment known as hemodialysis. The artificial kidney, or hemodialyzer, provides a means

Fazal-Ur-Rehman (✉)
Department of Anatomy, Jawaharlal Nehru Medical College,
Aligarh Muslim University, Aligarh 202002, India
e-mail: fazal.rehman72@yahoo.com

S.N. Khan
Department of Biochemistry, Aligarh Muslim University, Aligarh 202002, India

Inamuddin and M. Luqman (eds.), *Ion Exchange Technology II: Applications*,
DOI 10.1007/978-94-007-4026-6_8, © Springer Science+Business Media B.V. 2012

for removing certain undesirable substances from the blood or of adding needed components to it, i.e., hemodialysis (HD) membranes are used to remove accumulated uremic toxins, excess ions, and water from the patient via the dialysate and to supply (deficit) insufficient ions from the dialysate. In a dialysis machine, a person's blood flows between partially permeable membranes. Waste urea passes out from the blood into the dialysis fluid. Treatment by dialysis restores the concentrations of dissolved substances in the blood to normal levels and has to be carried out at regular intervals. During dialysis, it is important that useful substances in the blood, such as glucose and mineral ions, are not lost. To prevent this, dialysis fluid contains the same concentrations of these substances as blood. This ensures that only waste substances and excess ions and water diffuse into the dialysis fluid. By these processes, it can control the acid–base balance of the blood and its content of water and dissolved materials. Modern dialyzers rely on two physicochemical principles, dialysis and ultrafiltration.

The membranes first used in dialysis were obtained from animals or prepared from collodion; cellophane has been found to be more suitable, and tubes or sheets of it are used in many dialyzers. In the late 1960s, hollow filaments of cellulosic or synthetic materials (polymers) were introduced for dialysis; bundles of such filaments provide a large membrane surface in a small volume, a combination advantageous in devising compact dialyzers. Almost all dialyzers now in use are of the hollow-fiber type. A hollow-fiber dialyzer contains a bundle of approximately 10,000 hollow fibers, each with an inner diameter of about 200 μm when wet. The membrane thickness is about 20–45 μm, and the length is160–250 mm. The walls of the hollow fibers function as the dialysis membrane.

Dialysis was first described by Thomas Graham in 1854, which was first used to treat human patients in 1945 – replaces or supplements thc action of the kidneys in a person suffering from acute or chronic *renal failure* or from poisoning by diffusible substances, such as aspirin, bromides, or barbiturates.

8.2 Terminology

Biocompatible membrane – defined as one that elicits the least degree of inflammatory response in patients exposed to it.

Semipermeable membrane – is a thin layer of material that contains various size holes, or pores; it allows the smaller solute and fluid pass through membrane, but the membrane block the passage of large substances (e.g., RBC, large proteins).

Ion exchange polymer – polymer that is able to exchange ions (cations or anions) with ionic components in solution; ion exchange (IE) polymer in ionized form may also be referred as a polyanion or a polycation. Synthetic ion exchange organic polymers are often network polyelectrolytes. A membrane having ion exchange groups is called an ion exchange membrane.

Diffusion – diffusion describes a property of substances in water. Substances in water tend to move from an area of high concentration to an area of low concentration.

Dialysis – is used to provide an artificial replacement for lost kidney.
Dialyzer – is used in hemodialysis and is an external filter that contains a semipermeable membrane.

8.3 Electrodialysis

Electrodialysis (ED) is a very versatile technology for the separation of difficult mixtures. Electrodialysis is defined as an electromembrane process in which ions are transported through ion-permeable membranes from one solution to another under the influence of a potential gradient. The electrical charges on the ions allow them to be driven through the membranes fabricated from ion exchange polymers.

8.4 Kidney

Healthy kidneys maintain the body's internal equilibrium of water and minerals (sodium, potassium, chloride, calcium, phosphorus, magnesium, and sulfate). The acidic metabolism end products which the body cannot get rid of via respiration are also excreted through the kidneys. The kidneys also function as a part of the endocrine system producing erythropoietin and calcitriol. Erythropoietin is involved in the production of red blood cells, and calcitriol plays a role in bone formation. If we compare the dialysis with kidney, then dialysis is an imperfect treatment to replace kidney function because it does not correct the endocrine functions of the kidney. Dialysis treatments replace some of these functions through diffusion (waste removal) and ultrafiltration (fluid removal).

8.4.1 Essential Functions of the Major Segment of the Kidney

Nephron segment	Function
Bowman's capsule	Filtration

Glomerulus filters proteins and cells from the blood. All other blood components pass into Bowman's capsule, then into the tubule.

U-shaped tubule	Reabsorption and secretion

Semipermeable membranes surrounding the tubule allow selective passage of particles back into the blood (reabsorption) or from the blood into the tubule (secretion). The fluid entering the tubule is identical to the blood, except that it contains no proteins or cells.

Collecting duct collects all material that has not returned to the blood through the tubular membranes. This material exits the kidney as urine.

The tubule functions as a dialysis unit, in which the fluid inside the tubule is the internal solution and the blood (in capillaries surrounding the tubule) acts as the external solution. Particles may pass through the membrane and return to the bloodstream in the process known as reabsorption, which is analogous to the movement of particles from the internal to the external solution in the dialysis experiment you performed in laboratory. The reabsorption of many blood components is regulated physiologically, as discussed below. Alternatively, particles may pass through the membrane from the blood into this tubule in the process known as secretion, which is analogous to the movement of particles from the external solution into the dialysis bag in the experiment you performed in lab. The most important particles, which are secreted from the blood back into the tubules, are H^+ and K^+ ions, as well as organic ions from foreign chemicals or the natural by-products of the body's metabolism.

8.5 Diffusion and Concentration Gradients

The direction of the passage of particles through the channel is also dependent on concentration gradients. Diffusion is defined as the mixing of two different substances that are placed in contact, which continues until the concentrations of the two solutions are equal. This state is known as dynamic equilibrium. When the two solutions are in dynamic equilibrium, particles continue to move between the two solutions, but there is no net flow in any one direction, i.e., the concentrations do not change.

In biological systems such as the kidney, the two solutions are often separated by a membrane; protein channels in the membrane allow particles to cross the membrane, flowing "down the concentration gradient" until equilibrium is reached. Once the channels are closed, particles will not travel across the membrane, even if there is a strong concentration gradient. (In effect, the two solutions are no longer in contact.) There are some proteins in the membranes that act like "pumps" using energy to move particles "against the concentration gradient" (i.e., the concentrated solution becomes even more concentrated).

8.6 Dialysis

Dialysis is primarily used to provide an artificial replacement for lost kidney function in people with renal failure. It may be used for acute renal failure or chronic kidney disease stage (previously chronic renal failure or end-stage kidney disease). In dialysis, blood flows by one side of a semipermeable membrane, and a dialysate, or special dialysis fluid, flows by the opposite side.

Dialysis membranes clear toxins such as urea from the blood and preserve water balance and serum protein levels in blood, all while leaving red and white blood cells intact.

Dialysis is the movement of molecules by diffusion from high concentration to low concentration through a semipermeable membrane. Only those molecules that are small enough to fit through the membrane pores are able move through the membrane and reach equilibrium with the entire volume of solution in the system. Once equilibrium is reached, there is no further net movement of the substance because molecules will be moving through the pores into and out of the dialysis unit at the same rate. By contrast, large molecules that cannot pass through the membrane pores will remain on the same side of the membrane as they were when dialysis was initiated. To remove additional unwanted substance, it is necessary to replace the dialysis buffer so that a new concentration gradient can be established. Once the buffer is changed, movement of particles from high (inside the membrane) to low (outside the membrane) concentration will resume until equilibrium is once again reached. With each change of dialysis buffer, substances inside the membrane are further purified by a factor equal to the volume difference of the two compartments [1–4].

8.6.1 Goals of a Dialysis Procedure

1. To get rid of the water that was ingested and produced (during metabolism); this is done by convection
2. To get rid of salts (e.g., sodium chloride, potassium chloride)
3. To maintain acid–base balance
4. To get rid of nitrogenous waste products (e.g., urea)

8.6.2 Factors Affecting Dialysis Rate

The factors include (1) dialysis buffer volume, (2) buffer composition, (3) the number of buffer changes, (4) time, (5) temperature, and (6) particle size vs. pore size. Substances that are very much smaller than the pore size will reach equilibrium faster than substances that are only slightly smaller than the pores.

Dialysis works on the principles of the diffusion of solutes and ultrafiltration of fluid across a semipermeable membrane.

8.6.3 Types of Dialysis

8.6.3.1 Hemodialysis

The principles that underlie the HD procedure are simple in practice. Blood and dialysate are circulated on opposite sides of a semipermeable membrane, thereby permitting the passage of solutes elevated as a consequence of renal failure but restricting the transfer of blood proteins and cellular elements. The device containing the semipermeable membrane is the hemodialyzer. Removal of water occurs by controlling the hydrostatic pressure gradient across the semipermeable membrane and is augmented by increasing the osmolality of the dialysate fluid.

HD removes wastes and water by circulating blood outside the body through an external filter, called a dialyzer, that contains a semipermeable membrane. The dialyzer is composed of thousands of tiny synthetic hollow fibers. The fiber wall acts as the semipermeable membrane. Blood flows through the fibers, dialysis solution flows around the outside of the fibers, and water and wastes move between these two solutions. The cleansed blood is then returned via the circuit back to the body. Ultrafiltration occurs by increasing the hydrostatic pressure across the dialyzer membrane. This is usually done by applying a negative pressure to the dialysate compartment of the dialyzer. This pressure gradient causes water and dissolved solutes to move from blood to dialysate and allows the removal of several liters of excess fluid during a typical 3- to 5-h treatment. The blood flows in one direction, and the dialysate flows in the opposite. The countercurrent flow of the blood and dialysate maximizes the concentration gradient of solutes between the blood and dialysate, which helps to remove more urea and creatinine from the blood. The concentrations of solutes (e.g., potassium, phosphorus, and urea) are undesirably high in the blood, but low or absent in the dialysis solution and constant replacement of the dialysate ensures that the concentration of undesired solutes is kept low on this side of the membrane. The dialysis solution has levels of minerals like sodium and calcium that are similar to their natural concentration in healthy blood. For another solute, bicarbonate, dialysis solution level is set at a slightly higher level than in normal blood, to encourage diffusion of bicarbonate into the blood, to act as a pH buffer to neutralize the metabolic acidosis that is often present in these patients. The levels of the components of dialysate are prescribed and monitored by a nephrologist according to the requirement of the individual patient. Thus, artificial kidney dialysis uses the same chemical principles that are used naturally in the kidneys to maintain the chemical composition of the blood. The main aim of HD is the restoration of normal ion concentrations. Normally, the levels of individual ions in the dialysate can be set to their desired plasma levels; however, in some instances, dialysate levels are set for the diffusible fraction of the ion found in plasma.

The goal of HD in patients with end-stage renal disease (ESRD) is to restore the body's intracellular and extracellular fluid environment toward the body composition of healthy individuals with functioning kidneys to the extent possible. On a biophysical level, the use of HD as renal replacement therapy is accomplished via solute removal from the blood into the dialysate, as exemplified by intradialytic removal of potassium, urea, and phosphorous, as well as the addition of solute from the dialysate into the blood, as is exemplified by bicarbonate and calcium. An additional goal of the dialysis procedure is the elimination of excess water volume from the patient via ultrafiltration (UF.)

Thus, diffusion across semipermeable membranes, polarity, and concentration gradients are central to the dialysis process for both natural and artificial kidneys [5, 6].

Dialyzer Choice

In making a decision about the choice of dialyzer, the three most critical determinants are its capacity for solute clearance, capacity for UF or fluid removal, and the nature of dialyzer membrane interactions with components of the blood and their potential clinical sequelae, i.e., biocompatibility. The ideal HD membrane would have high clearance of LMW and middle-molecular-weight uremic toxins, negligible loss of vital solutes, and adequate UF in an effort to maximize efficiency and reduce adverse metabolic effects from the HD procedure. Additional characteristics of the ideal dialyzer would be a low blood volume compartment, beneficial biocompatibility effects, high reliability, and low cost. In evaluating dialyzer solute clearance characteristics, urea is the solute most often used owing to its relevance to kinetic models of dialysis adequacy.

Anticoagulation for Hemodialysis

Interaction of plasma with the dialysis membrane produces activation of the clotting cascade, characterized by the development of thrombosis in the extracorporeal circuit, thrombin deposition in dialyzer hollow fibers, and resulting dialyzer dysfunction. Dialyzer thrombogenicity is determined by dialysis membrane composition, surface charge, surface area, and configuration. In addition, the rate of blood flow through the dialyzer, the UF rate prescribed (owing to hemoconcentration), and the length, diameter, and composition of blood lines all affect thrombogenicity. In addition, a number of patient-specific variables influence thrombogenicity and determine anticoagulation requirements during HD. These include acquired and inherited coagulopathies, neoplasia, malnutrition, hemoglobin concentration, and the presence or absence of CHF.

By far the most widely used anticoagulant for dialysis is heparin. Heparin is easy to administer, has a low cost, and has a relatively short biologic half-life.

Summary Hemodialysis Procedure

Kidneys filter the blood, clearing it of waste products. Twenty percent of the heart's output is directed to the kidneys, which filter ~180 L plasma per day. Kidneys also regulate water and salt balance. Loss of kidney function leads to death within days due to excessive buildup of nitrogenous waste, acid, potassium, sodium, and water. Kidney function can be partially replaced by hemodialysis, first performed by Willem Kolff in the early 1940s. Procedure is typically performed for 4 h thrice weekly (e.g., Monday, Wednesday, Friday), in which blood flows into tubing that divides into thousands of parallel hollow fibers. Each fiber is a semipermeable membrane. Outside of the fibers runs the "dialysate" solution.

Clearance during dialysis is done by convection (negative pressure applied, water and dissolved small solutes (<40 kDa) pass across membrane into the dialysate fluid which is then discarded) and by diffusion (solutes travel across membrane down concentration gradient). Sodium moves down its concentration gradient (in either direction, depending on plasma concentration; dialysate sodium usually fixed). Blood and dialysate flow in opposite directions, maximizing concentration gradient. The parameters for the hemodialysis are given below:

Blood flow ~400 mL/min
40% red blood cells, 60% plasma
Plasma: 93% water, 7% protein and lipids
Dialysate flow ~800 mL/min
Typically 2–4 L of "ultrafiltration" (volume removed during a 4-h procedure)

8.6.3.2 Peritoneal Dialysis (PD)

Waste fluids and water are removed from the blood inside the body using the peritoneal membrane of the peritoneum as a natural semipermeable membrane. Wastes and excess water move from the blood, across the peritoneal membrane, and into a special dialysis solution, called dialysate, in the abdominal cavity which has a composition similar to the fluid portion of blood.

Peritoneal Dialysis Principle

- Peritoneum (capillary endothelium, matrix, mesothelium) = semipermeable dialysis membrane through which fluid and solute move from blood to dialysis solution via diffusion and convection
- Effective peritoneal surface area = perfused capillaries closed to peritoneum (decreased in peritonitis)
- Ultrafiltration (movement of water) enabled by osmotic gradient generated by glucose or glucose polymers (isodextrin)

Composition of Standard Peritoneal Dialysis Solution

Na	132 mmol/L
Ca	1.25 mmol/L
Mg	0.5 mmol/L
Cl	100 mmol/L
Lactate	35 mmol/L
Glucose	1.36–4.25 g/dL
Osmolarity	347–486
pH	5.2
GDP (degradation products of glucose)	

Advantage of PD to HD

Better maintenance of residual renal function
Saves vascular access
Preferred for children (APD)

Absolute Contraindications of PD

Peritoneal fibrosis and adhesions following intra-abdominal operations
Inflammatory gut diseases

8.6.4 *Primary and Secondary Types of Dialysis*

Primary: Hemodialysis, peritoneal dialysis, and hemofiltration
Secondary: Hemodiafiltration and intestinal dialysis

8.6.5 *Indications for Dialysis*

8.6.5.1 In the Patient with Acute Kidney Injury

1. Metabolic acidosis in situations where correction with sodium bicarbonate is impractical or may result in fluid overload.
2. Electrolyte abnormality, such as severe hyperkalemia, especially when combined with acute kidney injury.
3. Fluid overload not expected to respond to treatment with diuretics.
4. Complications of uremia, such as pericarditis, encephalopathy, or gastrointestinal bleeding.
5. Intoxication, i.e., acute poisoning with a dialyzable substance. These substances can be represented by the mnemonic SLIME: salicylic acid, lithium, isopropanol, magnesium-containing laxatives, and ethylene glycol.

8.6.5.2 Chronic Indications for Dialysis

1. Symptomatic renal failure
2. Low glomerular filtration rate (GFR) (renal replacement therapy often recommended to commence at a GFR of less than 10–15 mL/min/1.73 m^2). In diabetics, dialysis is started earlier.
3. Difficulty in medically controlling fluid overload, serum potassium, and/or serum phosphorus when the GFR is very low.

When the kidneys do not function properly, dialysis must be performed artificially. Without this artificial kidney dialysis, toxic wastes build up in the blood and tissues and cannot be filtered out by the ailing kidneys. This condition is known as uremia, which means literally "urine in the blood"; this waste buildup leads to death.

8.7 Ion-Permeable Membranes

The ion-permeable membranes used in electrodialysis are essentially sheets of ion exchange resins. They usually also contain other polymers to improve mechanical strength and flexibility. The resin component of a cation-exchange membrane would have negatively charged groups (e.g., $-SO_3^-$) chemically attached to the polymer chains (e.g., styrene/divinylbenzene copolymers). Ions with a charge opposite to the fixed charge (counterions) are freely exchanged at these sites. The concentration of counterions (e.g., Na^+) is relatively high; therefore, counterions carry most of the electric current through the membrane. The fixed charges attached to the polymer chains repel ions of the same charge (co-ions), in this case the anions. Since their concentration in the membrane is relatively low, anions carry only a small fraction of the electric current through a cation-permeable membrane. Attachment of positive fixed charges (e.g., $-NR_3^+$ or $C_5H_5N^+R$ where commonly $R=CH_3$) to the polymer chains forms anion-permeable membranes, which are selective to transport of negative ions, because the fixed $-NR_3^+$ groups repel positive ions. This exclusion, as a result of electrostatic repulsion, is called Donnan exclusion.

Ion exchange membranes have been used in various industrial processes, e.g., in the electrodialytic concentration of seawater to produce edible salt, as a separator for electrolysis, in the desalination of saline water by electrodialysis, in the separation of ionic materials from nonionic materials by electrodialysis, in the recovery of acid and alkali from waste acid and alkali solution by diffusion dialysis, in the dehydration of water-miscible organic solvent by pervaporation, etc. For these wide applications, successful ion exchange membranes must have the following properties: (1) high permeable selectivity – an ion exchange membrane should be highly permeable to counterion and impermeable to co-ions; (2) low electrical resistance – permeability for the counterions under the driving force of an electrical potential gradient should be as high as possible; (3) good mechanical stability and should have a low degree of swelling or shrinkage in transition from dilute to concentrated ionic solutions; and (4) high chemical stability – membrane should be stable over a wide pH range from 0 to 14 and in the presence of oxidizing agents.

Thus, for industries and for day-to-day life, innumerable membranes have been developed for the use in reverse osmosis, nanofiltration, ultrafiltration, microfiltration, pervaporation separation, electrodialysis, and in medical use such as artificial kidney. Among these membranes, ion exchange membranes are one of the advanced separation membranes. The basic applications of the ion exchange membrane process are based on the Donnan membrane equilibrium principle and have been paid attention to solve two environmental problems: (a) recovery and enrichment of valuable ions, and (b) removal of undesirable ions from wastewater, especially to extract toxic metal ions.

Ion exchange polymers such as poly(styrene sulfonic acid) are water soluble, so cross-linking is needed to prevent dissolution of ion-permeable membranes. Divinylbenzene is used to cross-link polystyrene chains. The degree of cross-linking and the fixed-charge density affect the membrane's properties in opposite ways. Higher cross-linking improves selectivity and membrane stability by reducing swelling, but it increases electrical resistance. High charge density reduces resistance and increases selectivity, but it promotes swelling and thus necessitates higher cross-linking. A compromise between selectivity, electrical resistance, and dimensional stability is achieved by proper adjustment of cross-linking and fixed-charge densities [7].

8.8 Bipolar Membrane

Bipolar membranes consist of an anion-permeable membrane and a cation-permeable membrane laminated together. There are substantial advantages to water splitting with bipolar membranes. Since there are no gases evolved at the surface or within the bipolar membranes, the energy associated with conversion of water to O_2 and H_2 is saved, and the power consumption is about half that of electrolytic cells. Compared to the electrodes used in conventional electrolytic cells, the bipolar membranes are inexpensive. Where dilute (e.g., <1 M) acids or bases are needed, bipolar membranes offer the prospect of low cost and minimum unwanted by-products.

8.9 Artificial Membranes

There are three types of membranes currently used to manufacture dialyzers: cellulose, substituted cellulose, and synthetic noncellulose.

Cellulose – Cellulose, primarily manufactured as cuprophan (or cuprophane), is a polysaccharide-based membrane obtained from pressed cotton. Cuprammonium is primarily used in the manufacturing process of this membrane (hence the name), but other methods of manufacturing exist.

The artificial kidney uses cellulose membranes in place of the phospholipid-bilayer membranes used by real kidneys to separate the components of blood. Cellulose is a polymer of glucose molecules that form long, straight chains, i.e., chains of glucosan rings with abundant free hydroxyl groups. Parallel chains form linkages with one another by hydrogen bonding to make strong fibers. These fibers in turn interact to form the strong, sheet-like structure of the membrane.

The arrangement of the cellulose fibers may contain gaps, creating tunnels through the membrane. These form the pores through which particles may pass from one side of the membrane to the other. The size of the gaps determines the size of the particles that will be able to pass through the membrane.

In present time, more than 30 different polymers or polymer blends are used as materials for dialysis membranes. Alternatively, they can be categorized as scheme of a family tree of hemodialysis membranes. The trunk represents membranes from regenerated cellulose; major branches show either synthetically modified cellulose membranes or membranes manufactured from synthetic polymers. As the latter are standardly hydrophobic, small branches elucidate the technique on how these materials have been rendered partially or completely hydrophilic.

Complications may arise, when comparing membranes only following their polymer names, such as polysulfone, polyacrylonitrile, or polyamide. Due to varying polymer compositions, membranes with the same polymer names may differ in their hemocompatibility, flux properties, and adsorption characteristics. Proteins, e.g., beta-2-microglobulin, fibrinogen, and coagulation factors, complement proteins, or hormones like parathyroid hormone and erythropoietin are differently adsorbed by dialysis membranes, and thus, adsorption contributes to the removal characteristics.

8.9.1 Dialysis Membrane Summary

- Hollow fibers ~12,000 in parallel, 20–24 cm length, diameter 180–220 um, 6–15 um thickness, pores: average diameter 30 Å, 10^9 in number
- Old: Cellobiose (saccharide)
- New: Synthetic membranes (e.g., polysulfone, polyamide, polymethylmethacrylate)

Manufactured polymers classified as thermoplastics

Both the artificial kidney and ion exchange resins afford means for radioelement removal. The advantages of the ion exchange column over the artificial kidney are its simplicity, its potentialities for more widespread use, and its greater selectivity for polyvalent cations over divalent and monovalent cations.

Ion exchange resins have been used in many instances, where it has been necessary to remove ions from solutions of uncharged molecules, and it was thought that the same methods might be used for charged molecules provided these were too large to penetrate the resin particles. The use of ion exchange resins would then be an alternative to normal dialysis.

Experiments were carried out to see if proteins could be "dialyzed" in this way, and it was found that egg albumin could be freed from ions by passing it through columns of cation- and anion-exchange resins without significant loss. The nonabsorption of proteins by resins has been applied to the fractionation of blood serum by Reid and Jones (1949). It was also found that proteins can be damaged when they came in contact with the resins. Resins can be used for dialysis if special precautions are taken; membrane must be used to separate the resins from the solution to be dialyzed, and the tendency of the solution to become acid must be prevented by modification of the membrane or by other means. It is then possible to remove ions from enzyme solutions with resins without denaturation.

The healthy kidneys continuously remove potassium. In malfunctioned kidneys, it may be necessary to remove potassium from the intestinal tract by artificial means. This can be achieved by using polystyrene sulfonates, in either the sodium or calcium form (Purolite C100NaMR and Purolite C100CaMR are examples for sodium and calcium polystyrene sulfonates resins, respectively). Kayexalate (sodium polystyrene sulfonate [SPS]), an ion exchange resin designed to exchange sodium for potassium in the colon, is approved for use in the treatment of hyperkalemia. As the resins pass through the intestinal tract, they exchange the sodium or calcium on the resin for potassium. The adsorbed potassium cannot pass into the blood and continues through the body without being released. Such resins are now widely used in the treatment of acute and chronic hyperkalemia and in controlling serum potassium levels in patients undergoing renal dialysis. The powder resin is flavored and prepared in doses to be taken orally.

Hyperkalemia is a common clinical problem that is most often a result of impaired urinary potassium excretion due to acute or chronic kidney disease and/or disorders or drugs that inhibit the renin–angiotensin–aldosterone axis. The most serious manifestations of hyperkalemia are muscle weakness or paralysis, cardiac conduction abnormalities, and cardiac arrhythmias including sinus bradycardia, sinus arrest, slow idioventricular rhythms, ventricular tachycardia, ventricular fibrillation, and asystole. These manifestations usually occur when the serum potassium concentration is $\geq$7.0 mEq/L with chronic hyperkalemia or possibly at lower levels with an acute rise in serum potassium.

8.10 Phosphorus Control in Chronic Kidney Disease

Hyperphosphatemia was identified as a risk predictor of mortality in patients with chronic kidney disease (CKD) on hemodialysis. In addition to its potential to trigger parathyroid hormone secretion in the development of secondary hyperparathyroidism, it became clear that hyperphosphatemia may indirectly and directly promote vascular calcification. Hyperphosphatemia appears to be one of the key threats to the event-free survival of dialysis. The mainstay of therapy is the use of phosphorus binders, while management of hyperphosphatemia by balanced diets should be used in conjunction. Combinations of the different P binders may improve P control and consequently limit side effects (e.g., Ca load, gastrointestinal complaints). Nonaluminum-, noncalcium-based binders such as sevelamer hydrochloride and lanthanum carbonate seem to offer advantages in the context that no bone and brain damage and no cardiovascular calcification progression occur as does in aluminum-based binding and calcium (Ca)-containing binders, respectively [8]. They are also being closely observed with regard to their gastrointestinal tolerability and potential hepatic accumulation, respectively [9].

References

1. McPhie P (1971) Dialysis Method Enzymol 22:23–33
2. McCready RM, Hassid WZ (1943) J Am Chem Soc 65:1154
3. Reid AF, Jones F (1949) Am J Clin Path 19:10
4. Richardson RW (1949) Nature (Lond) 164:916
5. Slinin Y, Foley RN, Collins AJ (2005) Calcium, phosphorus, parathyroid hormone, and cardiovascular disease in hemodialysis patients: the USRDS waves 1, 3, and 4 study. J Am Soc Nephrol 16:1788–1793
6. Block GA, Hulbert-Shearon TE, Levin NW, Port FK (1998) Association of serum phosphorus and calcium x phosphate product with mortality risk in chronic hemodialysis patients: a national study. Am J Kidney Dis 31:607–617
7. Haagen K, Helfferich F (1961) Ion-exchange membrane. Germany Patent 971,729
8. Block GA, Spiegel DM, Ehrlich J et al (2005) Effects of sevelamer and calcium on coronary artery calcification in patient of hemodialysis. Kidney Int 68:1815–1824
9. Markus K (2006) European Renal & Genito - Urinary Disease, pp 18–21

Chapter 9
Zeolites as Inorganic Ion Exchangers for Environmental Applications: An Overview

Sadaf Zaidi

Abstract Zeolites are rendering important service in areas crucial for the sustenance of modern civilization. Their attractive attributes of ion exchange, adsorption, dehydration-rehydration, and catalysis make their use invaluable in a number of applications of great significance. In this chapter, starting with the current understanding of the general aspects of zeolites, an attempt has been made to provide an overview to the reader about the role of zeolites as inorganic ion exchangers for environmental applications. In this regard, the developments in the use of zeolites for water and wastewater treatment, detergent building, and radionuclide separation have been reviewed.

9.1 Introduction

The spectacular story of zeolites, from the status of a mineral collector's prized possession to the world's very important industrial commodity, began to unfold in 1756 when the Swedish mineralogist Cronstedt discovered the mineral stilbite. He christened it "zeolite" (derived from two Greek words, *zeo* meaning "to boil" and *lithos* meaning "a stone") on account of the intumescence shown by its crystals when heated in a blowpipe flame. St. Claire Deville is credited with synthesizing the zeolite levynite by the hydrothermal route for the first time in 1862 [1]. Initially, the zeolites were discovered in small quantities with a polyminerallic nature that mitigated against their commercial application. Their outwardly attractive appearance attracted them to mineral collectors. The late 1950s marked the beginning of major geologic discoveries of huge deposits of sediments containing a number of natural zeolites in the United States and many other parts of the world. Side-by-side efforts were going on to develop synthetic analogs of the natural zeolites and new zeolites. Hence, the

S. Zaidi (✉)
Department of Chemical Engineering, Aligarh Muslim University,
Aligarh, UP 202002, India
e-mail: sadaf63in@yahoo.com

Inamuddin and M. Luqman (eds.), *Ion Exchange Technology II: Applications*,
DOI 10.1007/978-94-007-4026-6_9,

renewed interest in the development of applications for them based on their characteristic properties has been continuing. Zeolites, both natural and synthetic, are endowed with attractive properties for cation exchange, adsorption, dehydration-rehydration, catalysis, and for acting as hosts in nanocomposite materials. It is these properties that dictate the use of zeolites in a number of important applications [2]. Some of the distinct advantages of zeolites are versatility in their application, low cost of extraction (natural zeolites are easily obtained), easy availability in large quantities, and excellent chemical, thermal, and radiation stability, and finally they can be easily bestowed with desired physical and chemical properties by well-established processes[3–9]. Zeolites are being used in the manufacture of cement, as lightweight aggregates; in drying of acid gases; in the removal of ammonia from drinking and municipal water; in the manufacture of highly selective and durable catalysts for use in the process industries, notably petroleum refining and petrochemical industries; in the separation of oxygen from air; in the removal of heavy metal cations from industrial effluents; in the removal of radionuclides from nuclear waste; and in the mitigation of nuclear fallout. They have ably replaced the phosphates as builder from detergents. As dietary supplements, they tend to improve animal production and prevent flatulence in both animals and humans. When added to soil, zeolites improve the cation-exchange capacities and water sorption capacities. They can totally replace soil in soilless greenhouses or on space missions. Zeolites provide great relief as deodorizers of athletic footwear, animal litters, hatcheries, barns, refrigerators, and other such equipment. They also act as bactericides and insecticides [2]. There is a significant research effort to introduce molecular sieves in novel areas such as sensors, membranes, optoelectronics, electrorheological fluids, and functional nanomaterial fabrication [10].

Among these applications of zeolites, the ones that are related to environment and are based on ion exchange principles include water and wastewater treatment for the removal of NH_3 and heavy metal cations, nuclear waste treatment, and detergent building. Given the continuous and rapid stream of literature pouring in the field of zeolites, it is not possible to discuss it in totality. However, in this chapter, it has been my endeavor to first introduce the reader to the current understanding of the general aspects of zeolites. This is followed by an up-to-date overview of zeolites as ion exchangers for the aforementioned environmental applications.

9.2 Zeolites: General Aspects

9.2.1 Definition

Zeolites are crystalline aluminosilicates of group IA and IIA elements such as sodium, potassium, magnesium, and calcium [1]. They are a subset of the molecular sieve family. Structurally, zeolites are complex, crystalline inorganic polymers based on an infinitely extending three-dimensional, four-connected framework of

AlO_4 and SiO_4 tetrahedra linked to each other by the sharing of oxygen ions. IUPAC has provided detailed guidelines [11] for specifying the chemical formula of zeolites. In its general form, the formula for most zeolites is given as

$$|M_x(H_2O)_\gamma|[Al_xSi_{(t-x)}O_{2t}]-IZA,$$

where the guest species are listed between the braces and the host framework is contained in the brackets. *M* represents a univalent charge-balancing cation, *x* is the number of framework Al atoms in the unit cell, γ is the number of adsorbed water molecules, *t* is the total number of framework tetrahedral atoms in the unit cell (Al + Si), and IZA is the three-letter code of the framework type assigned by the Structure Commission of the International Zeolite Association [12]. IUPAC [11] defines a material as a zeolite if it is microporous or mesoporous containing voids arranged in an ordered manner and with a free volume larger than a 0.25-nm diameter sphere. The Structure Commission of the International Zeolite Association defines zeolites as materials with a framework density (T-atoms per 1000Å^3) ranging from 19 to 21 [12], where T-atoms are the Al or Si atoms. Over the years, the term "zeolite" has acquired a new connotation to include nonaluminosilicate compositions and structures with T-atoms other than Al and Si but which display the characteristic zeolite properties.

The ability of zeolites to exchange cations is due to the isomorphous replacement of tetravalent framework cation (silica) by a cation of lower charge (normally aluminum). This substitution gives rise to a negative charge on the framework of the zeolite. Loosely held cations that are situated in the intercrystalline channels or interconnected voids of the zeolite with ample water maintain the electroneutrality of the zeolite. Some of those cations are amenable to cation exchange, and this property is exploited to carry out ion exchange for various commercial applications.

9.2.2 *Zeolite Types and Structures*

The framework type of a zeolite describes just the connectivity (topology) of the framework tetrahedral atoms in the highest possible symmetry without regard to its chemical composition. All zeolites possessing the same topology belong to one particular type of zeolite framework. As of March 2011, there were 197 known framework types recognized by the Structure Commission of the International Zeolite Association as per the information available on their website http://www.iza.online.org. New framework types are in the process of being discovered continually. Some of the more exploited types of natural zeolite minerals include clinoptilolite, mordenite, chabazite, erionite, and phillipsite. Among the synthetic zeolites, the prominent types are A, X, Y, L, "Zeolon" mordenite, and ZSM-5, to name a few. A zeolite framework type defines the shape and size of the pore apertures, the dimensionality of the channel system (whether one-, two-, or three-dimensional), the volume and arrangements of the cages (polyhedra whose faces are no larger than six rings), and the types of cation sites available. Realizing the fundamental importance of zeolite framework structures in understanding zeolite chemistry, Meier and Oslon [13] first proposed a classification of zeolites based on framework type and not on framework structure, which has been widely accepted. The composition of the framework, the observed symmetry, and the actual unit cell dimensions are not considered, but only the topology of the tetrahedrally coordinated atoms (T-atoms) in the highest possible symmetry is described. This facilitates the grouping of many different materials under one type for the purpose of classification. The confirmed framework types are assigned a three-letter code by the Structure Commission of the International Zeolite Association according to the rules set by the IUPAC Commission on Zeolite Nomenclature [14, 15]. The pore openings in a framework are characterized by the size of the ring that defines the pore, designated as an n-ring, where n is the number of T-atoms in the ring. The opening is calculated by subtracting the diameter of an oxygen ion in the silicate structures from the interatomic distance between two opposing oxygen atoms across the ring. The basic building unit in a zeolite framework consists of a TO_4 tetrahedron where the central T-atom is typically Al or Si and the peripheral atoms are oxygen atoms. Combining several basic building units gives rise to composite building units (CBUs). Polyhedral CBUs and cages can be described in different ways: by using three letter codes (such as *d6r* or *hpr*); the descriptors [$a1^{b1}$, $a2^{b2}$…] where b1 is the number of a1 rings, b2 is the number of a2 rings, etc., for example, the *d6r* CBU has the descriptor [$4^6 6^2$] as it is defined by six equivalent four rings and two equivalent six rings. The basic building units can also be connected in infinite chains forming chain CBUs. Besides, some other structural features like channels and sheets are also associated with many different zeolite framework types. In some frameworks, there are only cages and no channel, but a majority has at least eight-ring channels. The stacking sequences of layers, cages, or rings in zeolite frameworks are usually described using the "ABC-system" [16]. Detailed information about the different framework types is available on the website http//:www.iza-structure.org/database/. Thus, the zeolite frameworks can be classified in terms of structural subunits, pore openings, channel systems, framework densities, loop configurations, and/or coordination sequences. In fact, for each of the T-atoms in a given framework type, the coordination sequences and loop configurations are unique; this helps in detecting whether or not the framework of a new zeolite is novel [16].

The knowledge about zeolite framework types can only help in determining the general structural features (as discussed earlier in this section) of zeolites, but it fails to describe real materials because the effects of chemical composition of the zeolite framework, the nature of the extra-framework species, and the type of postsynthesis modification, which greatly influence the properties of a zeolite, have not been considered. Thus, for a comprehensive assessment of a zeolite material, it is necessary to see the effects of these factors on the properties of the zeolite.

The framework composition greatly affects the properties of zeolites. The substitution of some tetravalent Si^{+4} with the trivalent Al^{+3} in a neutral pure silica framework results in the formation of an anionic aluminosilicate framework with counterions such as Na^+ required to balance its charge. Many elements have been incorporated into the zeolite frameworks. By substituting P^{+5} cations into the zeolite framework in place of Si^{+4}, a new family of aluminophosphate materials could be prepared. Also, by varying the ratio of Si/Al in the zeolite framework from low to high, zeolites with varying properties have been produced. The extra-framework ions usually reside in the channels and cages of a zeolite framework. These include exchangeable cations which balance the negative charge of the framework, removable water molecules, and/or organic species. The nature of these cations can modify the properties of the zeolite. For example, in catalytic applications, the zeolite has to have acidic sites. This can be achieved by exchanging the extra-framework cations of the zeolite with NH_4^+ followed by calcining the zeolite. This removes NH_3, leaving behind protons attached to the framework oxygen atoms. The formation of stacking faults or intergrowth structures takes place when closely related framework structures are formed under very similar conditions. This may result in some properties of the intergrowth materials being different from those of the pure end members.

9.2.3 Methods of Synthesis of Zeolites

The pioneering and systematic work in the synthesis of zeolites is attributed to Barrer and Milton in the 1940s. Since then a large number of zeolites with new framework topologies, compositions, and properties have been successfully prepared through continued synthetic efforts. In 1948, Barrer synthesized the first zeolite with no natural counterpart [17] (KFI structure type as determined subsequently for zeolite ZK-5) [18, 19]. In the late 1940s, Milton and coworkers successfully synthesized zeolites A (LTA), X (FAU), and P (GIS) by hydrothermal crystallization of reactive alkali metal aluminosilicate gels at low temperatures (~100°C) and pressures (autogenous) under alkaline condition (pH typically higher than 12) [20, 21]. The large-scale synthetic methodologies of Milton and coworkers laid the foundations of zeolite technology. Overall, Milton's concept of hydrothermal crystallization of reactive gels has been followed with various additions and modifications for most of the molecular sieves, zeolite, and zeolite-type material syntheses since the late 1940s [1]. Hydrothermal synthesis refers to the formation of materials in a primarily aqueous solvent. The water performs the solubilization of the components of the reagent mixture to varying degrees. In the case of zeolites, it is the mixing of silicates and aluminates with the formation of a gel followed by aging of the hydrogel at 100°C or more for selected periods in order to facilitate crystallization.

Table 9.1 lists the components used in hydrothermal synthesis of zeolites and their role therein. While the synthetic procedure of zeolites seems simple, the

Table 9.1 Zeolite reaction mixtures: components and roles (Reprinted from Ref. [23] with kind permission of © Elsevier (2005))

SiO_2:TO_2:M_2O:R_2O:H_2O + Seed + Salt		
Component	Role	Comments
SiO_2, TO_2	Framework components	Si may be replaced by Ge
		T may be Si, Al, B, Fe, Ga, Ti, etc.
		Ratio Si/T affects crystallization rate
		Ratio Si/T affects framework charge and properties
M_2O	Mineralizer or base	Base (OH) is routine, but fluoride may be employed
		M is usually an alkali metal or a quaternary ammonium compound
		Na → K → Rb → Cs can alter product but also crystal morphology and size
		Concentration can affect product purity, crystallization rate, and framework composition (gradient)
R_2O	Template	Structure direction
		Structure blocking
		Framework stabilization (to overrun)
		Location of T-atoms
		Crystal morphology and size
H_2O	Solvent	Limited work with other solvents
		Alcohols/diols
		Ammonia
Seed	Seed	Structure directing (rare)
		Increase rate (overall) of crystallization
Salt	Direct/indirect	Direct addition
		Crystal size and morphology
		Indirect addition – through use, for example, of $Al_2(SO_4)_3$ as Al source
		Salts may act as framework stabilizers; cf. templates

synthetic chemistry of zeolites is rather complicated. Most zeolites are formed as metastable phases under hydrothermal conditions. Zeolite crystallization represents one of the most complex chemical problems in crystal nucleation and growth which involves complex reactions, such as solution-precipitation, polymerization-depolymerization, and nucleation-crystallization. A large number of variables affect the formation of a specific zeolite phase, such as batch composition, reactant sources, Si/A1 ratio, alkalinity, water content, inorganic cations, organic templates, solvents, temperature, aging, stirring, and seeding. Yu [22] describes many new routes that have been developed for the synthesis of zeolites. These include the solvothermal synthetic route, ionothermal synthetic route, F^- synthetic route, microwave-assisted hydrothermal synthetic route, microemulsion-based hydrothermal synthetic route, dry gel conversion synthetic route, and combinatorial synthetic route.

9.2.4 Methods of Characterization of Zeolites

There is a lot of productive activity going on in the field of zeolites. New zeolites are being developed for different applications, both conventional and novel. Characterization of zeolites is of paramount significance in the three main uses of zeolites, as cation exchangers, sorbents/hosts, and catalysts, because without knowing the full details about a zeolite system, it would be very difficult to understand it and put it to appropriate use. The task of characterization becomes more stupendous as a number of techniques have to be applied in order to get a balanced and informed view of the usually complex zeolite. Moreover, this multitude of techniques also accelerates the process of development of a zeolite, particularly as a catalyst, as different aspects come to the fore. Over the years, hectic research in the field of instrumentation has resulted in advanced techniques of analysis that have helped in characterization of molecular sieves in general and zeolites in particular. In general, the characterization of a zeolite has to provide information about (1) its structure and morphology, (2) its chemical composition, (3) its ability to sorb and retain molecules, and (4) its ability to chemically convert these molecules [24]. The techniques for zeolite characterization being currently employed include X-ray powder diffraction, electron-beam-based microscopy (transmission electron microscopy (TEM), scanning electron microscopy (SEM) with energy-dispersive X-ray spectroscopy (EDX)), infrared spectroscopy, magic angle spinning (MAS) nuclear magnetic resonance (NMR) spectroscopy (MASNMR), and physicochemical characterization techniques. Diffraction is a central technique for characterizing the structure of a zeolite. X-ray diffraction is commonly used to identify the presence and quantitatively determine the amount of crystalline zeolite present. Electron microscopy is useful for structural and morphological analysis and for chemical analyses at specific locations. The spectroscopic techniques of IR and NMR provide insight into acid sites and framework structure. Additional understanding of the acid sites can be achieved by various physicochemical characterization techniques along with the application of porosimetry and thermal analysis [25]. Though it is not possible to compress the discussion of these techniques and the work being done on their development and application in one book, but compact and knowledgeable works [24–28] exist in literature that provide useful insight regarding the various techniques for the characterization of zeolites.

9.3 Some Basic Facts About Ion Exchange

The process of ion exchange finds mention in the Bible and is regarded as a very old process known to man. During the Second World War, the scientific moorings of ion exchange began to take shape in the secret shroud of the Manhattan Project [29]. Ion exchange is a process in which solid particles containing cations or anions are contacted with an electrolyte solution to change the composition of the solution [30].

Depending upon the type of bonding involved, ion exchange is classified as an electrostatic sorption process, that is, coulombic attractive forces exist between ions and charged functional groups. Ion exchangers are solid materials that are able to take up charged ions from a solution and release an equivalent amount of other ions into the solution. The ability to exchange ions is attributable to the properties of the structure of the materials [31]. Ion exchange and adsorption are similar in that a substance is taken up by a solid in both the processes. Both adsorption and ion exchange can be grouped together as sorption processes for the purpose of study and application. However, while ion exchange is a stoichiometric process, adsorption is not. For every ion that is removed in ion exchange from the solution, another ion of the same sign is released into the solution. In adsorption, no such replacement of the solute takes place [32]. Ion exchange is a reversible reaction involving chemically equivalent quantities [31, 33]. It involves the redistribution of ions between two phases by diffusion, and chemical factors are less significant or even absent. The heat produced by ion exchange is usually very small to negligible [32]. Some of the advantages of ion exchange can be listed as high efficiency, recovery of valuable ions, large variety of ion exchange resins, and the principle that all ions or ionizable species can be removed from aqueous solutions. The disadvantages are that since suspended particles in the feed cause plugging, prefiltration is required; competing ions in waste streams cause undue interference, thereby affecting the efficiency of the ion exchanger; and organic ion exchangers have low-temperature resistance [31]. Some well-known applications of ion exchangers have been reported in literature as the treatment of wastewaters from industry (removal of NH_4^+ and nitrates), nuclear wastewater treatment (primarily the removal of cesium and strontium), and as builders in detergents (water-softening duty).

9.3.1 Equilibrium in Ion Exchange

The usual aim behind studying the equilibrium aspects of ion exchange in zeolites is to obtain a measure of preference the zeolite exhibits for one ion over either another one, or a group of other ions. The ion exchange equilibrium can be described in terms of any of the following: the ion exchange isotherms (namely, Langmuir, Freundlich, Dubinin–Raduskevish (D-R), Langmuir–Freundlich, Redlich–Peterson, and Sips isotherms), the separation factor, the selectivity coefficient, the distribution coefficient, and the thermodynamic equilibrium constant [34]. The expressions for these terms and their significance have been elaborated in many texts on mass transfer and unit operations. Essentially, the purpose of ion exchange thermodynamics is not to enable us to speculate about mechanism, but to enable us to predict the behavior of the system over ranges of conditions (temperature, pressure, concentration) which may be difficult to measure directly because of experimental limitations [35]. By making appropriate activity corrections to the experimental isotherm, the thermodynamic equilibrium can be calculated for *one set* of conditions. The beauty lies in the fact that this data can be used in principle to predict the behavior of the zeolite under *any other set* of conditions. The requirements to predict ion exchange equilibria are an adequate thermodynamic formulation, very accurate experimental data for at least *one set* of conditions, and accurate activity coefficient data for the dissolved salts in the external solutions for the range of conditions over which the prediction is to be made [35].

9.3.2 Kinetics of Ion Exchange

When designing an ion exchange system, it is desirable to have a definite assessment of the rate at which the reaction will occur. The required contact time will have a direct impact on the size of the actual plant. It is to be noted that when an exchanger particle is brought in contact with a solution, there is a static liquid film formed around it. The rate of ion exchange depends upon the rates of the following individual processes: (1) diffusion of the ion from the bulk of the liquid to the external surface of the exchanger particle, (2) diffusion of the ion through the hydrated film surrounding the particle, (3) diffusion of the ion across the film-particle interface, (4) inward diffusion of the ion through the solid to the site of exchange, (5) exchange of the ions, (6) outward diffusion of the released ions to the surface of the solid, (7) diffusion of the released ions across the film-particle interface, (8) diffusion of the released ion through the hydrated film surrounding the particle, and (9) diffusion of released ions from the hydrated film to the bulk of the liquid. Apart from these very obvious diffusion steps, there is the possibility of the formation of agglomerates by the individual crystallites. In such a case, the rate-determining steps may include other possible diffusion processes within the macropores which then exist between the individual crystallites within the individual crystallites and the crystal agglomerate. It is important to know that the rate of a particular ion exchange reaction may be determined by one or more of the several distinct mechanisms. The diffusion rates can be described by appropriate mass-transfer coefficients for equi-equivalent counterdiffusion through the solid and through the liquid, and in some instances at least it appears that the resistance to diffusion in the liquid phase may be controlling [33].The kinetics may be affected by the nature of the exchanger, the nature of the counterions, selectivity of the ion exchanger, temperature, particle size, etc. The actual rates of ion exchange may vary from a few seconds to several months to attain equilibrium.

The total available exchange capacity of an ion exchange material as described by the number of functional groups on it is termed as ion exchange capacity. This value is constant for a given exchange material. Its units are milliequivalents per gram (meq/g) based on the dry weight of material in a given form. The two capacity parameters required to characterize ion exchangers are the total static capacity and the dynamic capacity. The former is determined under static conditions, while the latter is determined by passing the solution through a bed of the exchanger (usually a packed bed or a in a column). The theoretical maximum capacity or the total capacity of the ion exchange bed can be calculated under dynamic conditions by multiplying the distribution coefficient with the mass of the ion exchange material in the column. However, in practice, this capacity can never be exploited owing to interference from coexisting counterions. It is thus the operating capacity or the breakthrough capacity of a column type exchanger that is of practical significance. It depends on the concentration of the ions being removed, the effects of interfering ions, ionic size, ionic valence, design of the column, and operating parameters. A plot between the concentrations of the removed ions in the effluent (ordinate) versus the effluent volume (abscissa) is drawn. The effluent volume corresponding to the point from where a sharp increase in effluent concentration of the ions being removed is observed is the breakthrough point (start of breakthrough). The extent of breakthrough increases up to a point where no ion exchange takes place, that is, the effluent contains the original concentration of the ions. This point marks the stage at which the ion exchanger is fully spent and must be regenerated or replaced. The breakthrough capacity or the operating capacity is the total liquid volume treated up to a particular percentage of breakthrough. It is always beneficial to have as steep a breakthrough curve as possible, in order to increase the extent of column utilization. The breakthrough capacity is of great interest in the design of a column type ion exchange system.

9.4 Environmental Applications of Zeolites as Inorganic Ion Exchangers

9.4.1 Water and Wastewater Treatment

A number of ion exchange materials are being extensively investigated in order to tackle the problem of removal of objectionable cations (heavy metals and ammonium) from water and wastewater. Among them, the zeolites have the potential to attain the objective of water treatment in a very cost-effective and efficient manner. Several experimental studies have been carried out over the years in various parts of the world, which have resulted in a better understanding of this particular application of zeolites. However, the application of zeolites is on the laboratory experiment scale. More work is needed for the application of zeolites at an industrial scale [36]. In this subsection of the chapter, an attempt has been made to present the salient features of some of the studies performed on zeolites for the removal of cations from water and wastewater.

9.4.1.1 Removal of Ammonium

Jorgensen et al. [37] studied a European type of clinoptilolite for the removal of NH_4^+ and investigated the dependence of capacity and efficiency on parameters, such as flow rate, pH, and intervals between regenerations. Further, they examined the performance of clinoptilolite in conjunction with a chemical treatment method for reducing the cost of the process and the amount of the total sludge and elution liquid generated for the overall process. It was observed that clinoptilolite worked by the process of ion exchange at low concentration of the counterion and via adsorption at high concentration of the counterion. The capacity increased with the contact time up to 120 min and the number of regenerations (up to three regenerations). The specific capacity was found to increase with the quantity of clinoptilolite used. Use of smaller particle sizes (more area of contact) resulted in higher uptake. However, the presence of other competing ions reduced the capacity. The combination of chemical precipitation-clinoptilolite-activated carbon gave a 90% removal efficiency of organics as well as ammonium both in the laboratory as well as on the pilot plant of 100 l/h capacity. In another study [38], Klieve and Semmens evaluated the samples of clinoptilolite, erionite, mordenite, and phillipsite for ammonium removal from wastewaters. Additionally, samples of clinoptilolite were pretreated in various ways to determine whether an improvement in ammonium removal performance could be realized. Total exchange capacities, capacities for ammonium removal from a synthetic waste, packed bed densities, and crushing strengths were measured. Phillipsite was found to have almost twice the weight capacity for ammonium removal from synthetic waste compared to that of clinoptilolite. The volumetric capacity was 26% better than that of clinoptilolite. However, the phillipsite sample due to its extreme friability could not be used for water treatment without being strengthened with a binder. Pretreatment of clinoptilolite with NaOH, HNO_3, and steam did not result in any improvement of the performance of clinoptilolite. However, heat pretreatment (600°C for 1 h) improved the zeolite's selectivity for ammonium significantly vis-à-vis the competing ions. Ammonium removal capacities were increased by approximately 17%, for heat-treated clinoptilolite samples, although the total exchange capacity of the zeolite was reduced somewhat. The phillipsite sample was very effective in ammonium removal, but erionite and mordenite behaved poorly. Hlavay et al. [39] studied the use of natural zeolites deposited at Tokaj Mountain, Hungary, for ammonia removal from synthetic and municipal wastewaters. It was observed that the clinoptilolite ion exchanger proved suitable to take up the ammonia even at a concentration level of 45 mg dm^{-3} NH_4^+-N. Optimal conditions including particle size and ion form of clinoptilolite as well as loading rate of the column were established, and taking into account the purity of the clinoptilolite, the highest possible ion exchange breakthrough capacity of 4.50 mg NH_4^+-N/g was achieved using synthetic wastewater in a column of 9.5 cm i.d. × 92 cm.

Czárán et al. [40] also investigated the cation-exchange properties of a Hungarian clinoptilolite containing rock from the Tokaj Mountains in order to prepare suitable cation containing forms for NH_4^+ removal and regeneration of the NH_4^+ form. The work included static laboratory experiments, dynamic measurements, and pilot plant studies. It was found that this zeolitic material was capable of removing 3–5 mg NH_3 per g rock. In another study [41], the suitability of Hector clinoptilolite as an exchange medium for fish farm water treatment in freshwater recycle systems was examined using small continuous fixed-bed columns with multications Na^+, NH_4^+, K^+, Ca^{2+}, and Cl^- as the co-ion. The most important criterion used to assess performance, ammonia uptake, was strongly influenced by ammonium ion concentration in the feed, and in the concentration range studied (1–5 mg NH_4^+-N/l), the exchange capacity for ammonium varied linearly with feed concentration. The ammonium uptake was maximized at a high-liquid flow rate, a high ammonium concentration, and a small zeolite particle size. Regeneration studies using sodium chloride solution at pH 12 revealed a substantial loss in exchange capacity following the first regeneration. The presence of calcium ion appeared to depress the capacity of the clinoptilolite after the first uptake/regeneration cycle. In an interesting work [42], a new concept for removal of ammonium from secondary effluent using ion exchange followed by bioregeneration was proposed using the zeolite as a carrier for the nitrifying biomass. As opposed to other hybrid ion exchange-biological systems which use multiple reactors, this system employed a single reactor. All the ammonium from the original effluent is concentrated in the zeolite by ion exchange and is released gradually during regeneration; nitrification is thus carried out in a small volume reactor in an almost batch mode where optimal conditions for nitrification can easily be maintained. The operational costs and production of large volumes of brine were claimed to be minimized.

The NH_4^+ uptake performances of the naturally occurring clinoptilolite and of New Zealand mordenite in the concentration range 0–200 mg/l were compared taking into account the influence of other cations (potassium, calcium, and magnesium) present in water [43]. In all cases, the anionic counterion present was chloride. It was observed that mordenite exhibited higher overall uptake concentrations at equilibrium compared with clinoptilolite at solution concentrations greater than 80 mg/l. The effect of the other cations upon uptake of ammonium ion was found to be relatively small. For both mordenite and clinoptilolite, the ammonium ion showed the highest uptake among all the cations. However, the clinoptilolite from New Zealand showed greater selectivity to ammonium than potassium, whereas the clinoptilolites from other regions exhibited a higher affinity to potassium than ammonium.

The ion exchange potential of Chinese clinoptilolite for ammonia removal from aqueous solution was assessed [44]. A batch study and a column study were carried out. From the batch study, the effect of such relevant parameters as contact time, pH, and initial ammonia concentration were revealed, whereas in the column study, operation and regeneration parameters and ion exchange capacity of clinoptilolite zrium. The optimum pH value of 6 was obtained, and the ammonia removal capacity of clinoptilolite increased with the increase of initial ammonia concentration. From the column study, it was clear that that flow velocity affected breakthrough capacity of clinoptilolite. Fifteen to twenty BV of 0.5 mol/l sodium chloride regeneration solution at pH 11–12 was sufficient for complete regeneration of clinoptilolite in column without loss of ammonia removal capacity. Leakage waters from waste dumps having a high content of ammonium (820 mg/l) and organic pollutants (1,033 mg/l) were treated using activated carbon and natural zeolite clinoptilolite [45]. Model and real solutions were used to test the removal efficiency of the zeolite. First, the organic content of the leakage waters was reduced by 0.04–2.5% (w/w) activated carbon that led to a reduction of total organic carbon in leakage water from 1,033 to 510 mg/l. Pretreatment of leakage water with activated carbon did not improve the exchange of ammonium ions on zeolite. Without pretreatment of leakage water, the exchange of ammonium ions amounted to 4.2 mg NH_4^+/g zeolite, and with the addition of activated carbon, it improved to only 5.7 mg NH_4^+/g zeolite. In the model solution of an equal concentration of ammonium as the real solution, 17.70 mg NH_4^+/g zeolite was exchanged. Remnant organic pollutants adsorbed to zeolite and prevented the exchange of NH_4^+ ions, which was also reduced due to the presence of K^+ and Ca^{2+} ions. Jha and Hayashi [46] chemically and mechanically modified the natural clinoptilolite available locally in Japan. The ammonium ion retention capacity of thus-obtained modified clinoptilolites was found to sharply increase with either increasing alkaline metal cation content or increasing specific surface area (decreasing particle size) of the clinoptilolite. The main mechanism of ammonium ion retention was found to be ion exchange, and Na^+ ions were observed to be more easily exchanged for ammonium ions. The maximum NH_4^+ retention capacities of natural clinoptilolite, clinoptilolite treated with NaOH solution for 72 h, and wet-milled clinoptilolite according to the Langmuir model were 0.89, 1.15, and 1.39 mmol/g, respectively.

Ramos et al. [47] studied the ammonium-exchange capacity of a natural chabazite. The Langmuir isotherm fitted quite well the equilibrium data obtained from a batch adsorber. The effects of the temperature and pH on the ammonium-exchange capacity of chabazite indicated that the capacity increased: with an increase in temperature from 15°C to 35°C and the pH from 3 to 6. The natural chabazite was modified by a hydrothermal treatment using NaCl and KCl solutions, and it was found that the modification influenced the ammonium-exchange capacity of the chabazite. The ammonium capacity of natural chabazite was compared with that of a natural clinoptilolite, and it was concluded that the chabazite capacity was 1.43 times higher than that of clinoptilolite. In a recent study [48], the behavior of the extra-framework cation system for ion exchange, in the form of a Greek heulandite, has been examined in detail. The sorption ability of the zeolite was measured using the ammonium acetate saturation method. It was observed that Ca^{2+} presented an unexpected extra-framework release and a surprisingly high degree of exchange (90%). The exchange of Mg^{2+} (57%) was also substantial, whereas the behavior of K^+ showed an expected rapid initial release. On analogy, the behavior of Na^+ was inferred to be similar. However, its lower concentration in the zeolitic material minimized its overall significance to some extent. The release of Ca^{2+} and Mg^{2+} was kinetically much slower, compared to that of alkali metal ions.

Malekian et al. [49] studied the ability of Iranian natural zeolites in millimeter and nanometer particle sizes to remove NH_4^+ from aqueous solutions with different Na^+ concentrations (0.03, 0.1, and 0.3 M). The effect of ionic strength on NH_4^+ released from Iranian zeolite was also evaluated, and several kinetic models were tested. It was shown that the initial NH_4^+ and Na^+ concentrations had a significant effect on the amount of NH_4^+ exchanged. Greater amounts of NH_4^+ were exchanged with a decrease in Na^+ concentration, leading to greater removal efficiency. The values of the coefficient of determination (R^2) were higher, and those of the standard error of estimate (SEE) were lower for the three parameter isotherm models (Redlich–Peterson and Langmuir–Freundlich) than for the two parameter isotherm models (Langmuir and Freundlich). The Elovich, power function, and pseudo-second-order models successfully described the NH_4^+ release data with R^2 in the range 0.866–0.958 and SEE in the range 0.043–0.194. The results of the study gave enough indication that the natural Iranian zeolite was a good NH_4^+ exchanger and as such could be used in water treatment and in agriculture as a slow release fertilizer.

9.4.1.2 Removal of Heavy Metal Cations

Joshi and Rao [50] reported the results of a study conducted by them to investigate the behavior of natural heulandite as an ion exchanger. They described the procedure for preparing the ion-exchanged samples with both monovalent (Li^+, Na^+, K^+, Cs^+, NH_4^+, and Ag^+) and divalent ions (Ba^{2+}, Sr^{2+}, Ca^{2+}, and Mg^{2+}). Heulandite crystals used in the investigation had a Si/Al ratio of 3.04 which was high. Zeolites with high Si/Al ratio (smaller ion exchange capacity) exhibit the weak field pattern. The anhydrous ion exchange capacity calculated from the unit cell contents was 3.45 meq/g. The water content was also rather high (≈16%). Thus, the following selectivity series were established for monovalent and divalent ions as $Ag^+ > NH_4^+ > Cs^+ > K^+ > Na^+ >> Li^+$ and series $Ba^{2+} > Sr^{2+} >> Mg^{2+}$, respectively. They predicted the use of heulandite in the removal of ammonium ions from wastewater streams and in the production of oxygen-rich air. Blanchard et al. [51] showed that the natural zeolite clinoptilolite can be used to remove NH_4^+ from drinking waters even though the water may contain heavy metal cations in small quantities. They had carried out a study to find the selectivity series for removing the various cations present in the water. The same was furnished as

$$Pb^{2+} > NH_4^+ > Cu^{2+}, Cd^{2+} > Zn^{2+}, Co^{2+} > Ni^{2+} > Hg^{2+}$$

Thus, it can be deduced that the zeolite had the highest affinity for lead followed by ammonium. The ion exchange kinetics was also studied, and rate equations were proposed. The regeneration conditions had been defined and optimized in order to achieve a good reproducibility of the results for several cycles. The optimized parameters of the batch study were tested on a pilot plant.

Semmens and Martin [52] studied the influence of pretreatment on the capacity and selectivity of clinoptilolite to remove metal ions. They attributed the inconsistencies observed by investigators in measured exchange capacities and selectivity data to the way the zeolite was treated prior to testing. Sodium chloride solution of varying concentrations was used to pretreat the clinoptilolite in order to remove potassium and calcium ions that were strongly held by the zeolite in preference to sodium. Prolonged exposure to high sodium concentrations was required to displace these ions from the zeolite. The presence of these cations definitely affects the performance of the zeolite as an ion exchanger. Thus, it was concluded that due attention must be paid to the conditioning procedure for zeolites as it affects zeolite performance. In another study [53], the use of Neopolitan yellow tuff containing phillipsite and chabazite for the removal of chromium from water by using ion exchange was reported. Though the results of the investigation were not very positive, but it contributed to an understanding of the interactions between high charge density cations and zeolite framework. Moreover, the low-cost and widespread occurrence of the yellow tuff and the irreversible nature of chromium ion exchange point to the potential lucrative use of Neopolitan zeolite.

Wark et al. [54] observed that when zeolites of types A, X, and Y were loaded with heavy metals from salt solutions via ion exchange, heavy metal hydroxides of low solubility precipitated onto the crystal structure because of the alkaline reaction of these zeolites in aqueous solutions. It was seen that the precipitation of hydroxides could be monitored by the pH value of the exchange batch, which was more acidic than the salt solution. The tendency of precipitation followed the sequence NaA > NaX > NaY, in line with the alkaline nature of each zeolite in an aqueous environment. To avoid the precipitation, the cation concentrations in the exchange solutions had to be limited, so that the solubility product of the corresponding hydroxide was not exceeded. Under these conditions, either the hydroxides were not precipitated or they were redissolved. Also, sulfidation of heavy metal ion-exchanged zeolites produced semiconductor particle-loaded zeolites of which the light absorption behavior was sensitive to the extent of the side reaction of hydroxide precipitation prior to the sulfide formation. Hernández et al. [55] studied the cation-exchange selectivity in Na^+- and H^+-exchanged phillipsite. For the Na^+-exchanged phillipsite, the selectivity sequence was $Ba^{2+} >> Pb^{2+} >> Cd^{2+} > NH_4^+ > Cu^{2+} \approx Zn^{2+} \approx K^+ > Na^+ >> Li^+$. The selectivity sequence for H^+ (acid treated) phillipsite was $Ba^{2+} \approx Pb^{2+} > Cd^{2+} \approx NH_4^+ > K^+ \approx Cu^{2+} \approx Zn^{2+} > H_3O^+ >> Li^+$. On regression of the data, a linear relationship was found between the product of the ionic radius of the cation and its valence and the logarithm of the selectivity coefficient. The general expression for both the zeolite variants obtained was

$$Rv = A + B \log K_s$$

where Rv is the product of the ionic radius of the cation and its valence, and A and B are constants. R^2 values of 0.9367 and 0.8367 were obtained for the Na^+- and H^+-exchanged phillipsite, respectively. The results exhibited the sequestration properties of phillipsite for heavy metal ions in dilute solutions. The decomposition behavior of zeolite A in freshwater was examined by Kuhm and Lortz [56], taking into consideration environmental factors. The zeolite NaA was used in heavily diluted suspensions. The release of Si and Al was studied in a closed system at constant pH levels in the range 4.5–9.5 using "synthetic" waters and those of natural origins. Drinking water at a pH range of 6–8 was used for open-system long-term trials, and the primary focus was on observing the changes in the individual zeolite crystallites using SEM, EDX, and electron diffraction analysis. The interesting find was that the zeolite had not only lost its crystalline structure but there was a change in its composition also: the extent of change depending on the environmental conditions, the time period of exposure, and the composition of the water used. The zeolite NaA completely transformed into relatively stable and insoluble Ca–Al–silicate–phosphate if Ca and phosphate were present in the water. Thus, it could be inferred that the NaA zeolite could possibly act to immobilize phosphate in natural inland waters contaminated with phosphate. Some of the different types of natural zeolites from United States and some synthetic zeolites have been assessed for their capacity to remove heavy metal cations from acid mine drainage and industrial wastewaters using both the batch and the column techniques [57]. Their effectiveness for cation removal, in the "as received form," beneficiated form, and after making the zeolites homoionic with strong Na^+ solutions, has been found to be in the following order:

$$\text{Erionite} > 13\text{X} > \text{Chabazite} > \text{Clinoptilolite} >> \text{Mordenite} \approx 3\text{A} \approx 4\text{A} \approx 5\text{A},$$

where 3A, 4A, 5A, and 13X are the synthetic zeolites. The selectivity series is as follows: $Ag^+ > Pb^{2+} > Cd^{2+} > Zn^{2+} > Cu^{2+} >> Ni^{2+} > Hg^{2+}$. The poor ion exchange performance of synthetic zeolites was attributed to the diffusion problem, and a method for solving this problem was proposed for zeolite 13X. Pansini et al. [58] have studied the ability of zeolite phillipsite to remove lead from water in detail. Equilibrium runs showed that Na-exchanged phillipsite displayed a very high selectivity for lead. Kinetic runs showed a fast lead uptake rate by sodium phillipsite ($k = 0.013$ 1/mequiv. min). Column run data supported the possibility of performing a massive lead removal from large volumes of water by using moderate quantities of phillipsite-containing materials, even in the presence of substantial amounts of interfering cations (sodium). Their investigation demonstrates that regeneration was effective since purging the exhausted bed with about 75 BV of regenerant solution results in the elution of about 70% of lead contained in the bed. Their conclusion was that the encouraging performance of Na^+ phillipsite and chabazite (from an earlier study) for lead removal implied that most of the Italian zeolitized tuff (total zeolite content about 70–80%) could be profitably used for removal purposes. Ćurković et al. [59] studied the removal of lead and cadmium by natural and modified Croatian zeolites. The pH value during the experiments was increased from 4.5 to 8 due to the hydrolysis of zeolites. At lead concentration of less than 2.5 mmol/l, removal efficiency was about 90% depending on the treatment of zeolites. At the same concentration of cadmium, removal efficiency was from 45% to 70%. It was clear that the lead was more selectively removed than cadmium. The exchange of the lead and cadmium increased with increasing temperature. They recommended these zeolites as ion exchangers and adsorbents for lead and cadmium removal from wastewaters.

Kang and Egashira [60] reported a study on three grades of Korean natural zeolites which contained clinoptilolite or both clinoptilolite and mordenite. These which were subjected to a hydrothermal process with 2 M NaOH solution at 103°C during different times ranging from 1 to 16 h and converted to zeolite Na-P as detected by X-ray diffraction and scanning electron microscopy. Zeolite Na-P was formed from both clinoptilolite and mordenite. During the reaction, a large amount of Si was dissolved into solution. The cation-exchange capacities of the products after 8 or 16 h reaction were higher by more than two times than those of the corresponding raw materials. A solid NaY zeolite has been used by Ahmed et al. [61] to remove lead and cadmium from aqueous solution by batch ion exchange under competitive and noncompetitive conditions. Equilibrium-exchange isotherms are presented for NaY treatment of lead and cadmium nitrate and chloride solutions at 293 K. In the Pb/NaY and Cd/NaY binary-exchange processes, Na-Y exhibited a greater preference for exchange with Pb, but selectivity in favor of both heavy metal ions decreased with increasing heavy metal concentration where the equilibria coincide for exchange with external nitrate and chloride solutions. The zeolite-exchange efficiency increased in the order $Ni^{2+} < Cu^{2+} < Cd^{2+} < Pb^{2+}$. It was found that for the ternary Pb/Cd/NaY-exchange process, Pb was preferred to a combination of Cd and Na at all loadings.

The removal of nickel and copper from aqueous solution by batch ion exchange with solid lithium-, sodium-, potassium-, rubidium-, and cesium-based Y zeolites under competitive and noncompetitive conditions has been studied by Keane [62]. Equilibrium-exchange isotherms were presented, and the effect of increasing the treatment temperature from 298 to 373 K on exchange selectivity was shown. The extent of transition metal removal was found to be strongly dependent on the nature of the outgoing alkali metal cation with the overall preference of the zeolite for exchange with both metals increasing in the order CsY < RbY < NaY < LiY. Copper removal was much greater than that of nickel for all the zeolite exchangers under identical experimental conditions. Townsend and Coker [35] have conducted a review of various aspects of ion exchange in zeolites with particular reference to their utility for predicting exchange behavior in zeolites and for the use of ion exchange techniques in detergency building, radionuclide separation, wastewater treatment, and the preparation of zeolitic catalysts and sorbents. They noted that how the basic understanding of the theory of ion exchange in zeolites which had been developed over the years, coupled with careful experimental measurements, had borne fruit in major up-to-date applications of these materials in processes of high commercial importance. They recommended the use of computational modeling for a better understanding of the ion exchange process.

Mier et al. [63] have studied multicomponent ionic exchange for heavy metal removal using Mexican clinoptilolite. The interactions of Pb^{2+}, Cd^{2+}, and Cr^{6+} competing for ion exchange sites in naturally occurring clinoptilolite have been discussed. While Pb and Cd were effectively removed within 18 h in batch reactors, with higher removal efficiencies (>95%) in the acidic pH range, the presence of Cr(VI), which can interact with these metals to form anionic complexes, significantly diminished the Pb and Cd removal efficiencies. At high pH values(>10), a decrease in the efficiency of the zeolite to remove lead was observed, which was due to the formation of anionic hydroxo-complexes with little affinity for cationic ion exchange sites. In a flow through column, lead was preferentially exchanged in comparison to cadmium but not in batch reactors because diffusion kinetics may have influenced the removal efficiency to a greater extent than equilibrium partitioning. Addition of phenol as an organic cocontaminant was observed to result in slightly hindered heavy metal removal in batch reactors due to the formation of organometallic complexes that could not penetrate zeolite-exchange channels. Overall, it was exhibited that zeolites hold a great promise to act as ion exchangers for heavy metal removal from industrial waters.

Cincotti et al. [64] investigated the Sardinian natural clinoptilolites to evaluate their heavy metals (copper, cadmium, lead, and zinc) uptake and ammonium ions. Both the "as received" as well as the homoionic zeolites were used. The corresponding behavior was quantitatively correlated using classical isotherms, whose parameters were estimated by fitting the equilibrium data. Breakthrough experiments of lead solutions were also performed. Fixed-bed runs were simulated using a mathematical model which included axial dispersion as well as a new approximate rate law for nonlinear adsorption and diffusion in spherical adsorbent particle based on an equivalent film resistance model.

Inglezakis et al. [65] generated equilibrium and kinetic data for Pb^{2+}, Cu^{2+}, Fe^{3+}, and Cr^{3+}exchange on natural clinoptilolite. The study investigated the ion exchange at $277 \pm 1°C$ and initial concentration of 10 meq/dm^3. Equilibrium was favorable for Pb^{2+}, unfavorable for Cu^{2+}, and sigmoid for Cr^{3+} and Fe^{3+}. Selectivity series deduced from equilibrium isotherms and confirmed by breakthrough point values was in the order $Pb^{2+} > Cr^{3+} > Fe^{3+} > Cu^{2+}$; the selectivity series order obtained by considering maximum exchange levels (MELs) and confirmed by operating capacity values was $Pb^{2+} > Cr^{3} > Cu^{2+} > Fe^{3+}$. Natural clinoptilolite from Caimanes deposit (Moa-Cuba) has been used for the nickel removal from aqueous solution by ion exchange, followed by elution of this nickel by using ammonia-rich solution [66]. The stability of clinoptilolite as an exchanger of Ni^{2+} and NH_4^+ cations under hydrothermal conditions and high ammonia concentrations was analyzed. No loss in the clinoptilolite-exchange capacity of Ni^{2+} and NH_4^+ ions during the Ni^{2+} removal-elution cycles was found. It was detected that the treatment of the clinoptilolite with NH_4OH concentrated solutions resulted in not only the Ni^{2+} elution by ion exchange but also a substantial increase in the Ni^{2+} ion removal capacity of the zeolite because the hydroxyl anions adsorbed (or occluded) on the zeolitic phase particles precipitated the Ni^{2+} as nickel hydroxide as revealed by the IR and XRD analyses. The possibility of nickel removal followed by its elution showed that the process was regenerable.

For the effective utilization of a natural zeolite as an ion exchanger, it is essential to have chemical models that help to describe accurately heavy metal-exchange equilibria. Originally, by analogy with the sorption process, empirical equations such as Freundlich or the Langmuir were adjusted to experimental data. Petrus and Warchoł [67] have studied the binary ion exchange equilibrium using the clinoptilolite as cation exchanger. The equilibrium zeolite loading data were generated for Na^+/Cu^{2+}, Na^+/Cd^{2+}, and Na^+/Pb^{2+} systems. The experimental data were predicted using Freundlich, Langmuir, and the thermodynamic equilibrium models. The Freundlich type isotherm was found to be inadequate as a model of ion exchange processes. Better results were obtained with the Langmuir and thermodynamic models. The thermodynamic model for ion exchange based on the Pitzer or Bromley equations used for calculation of the activity coefficients in liquid phase and the Wilson equation for the same in zeolite phase was excellent in predicting ion exchange equilibria between aqueous solutions and clinoptilolite. However, it was seen that sufficiently good results for binary systems were obtained assuming ideal behavior of the zeolite phase. The predicted values were in good agreement with the experimental values when the concentrations were replaced by activity coefficients in the calculations. It was found that the calculated binary interaction parameters were very useful for estimation of equilibrium constants for multicomponent solutions.

In another study, Inglezakis et al. [68] have examined the ion exchange of Pb^{2+}, Cu^{2+}, Fe^{3+}, and Cr^{3+} on natural Greek clinoptilolite in terms of selectivity toward the above heavy metals in single- and multicomponent solutions in batch systems. Further, the influence of clinoptilolite on solution acidity and the effect of acidity on the ion exchange process were also studied. The following order of selectivity for both single-multicomponent solutions at total concentration 0.01 N and acidity 2 was obtained as $Pb^{2+} > Fe^{3+} > Cr^{3+} \geq Cu^{2}$. On the other hand, selectivity in single metal solutions where acidity is not adjusted was in the following order: $Pb^{2+} > Cr^{3+} > Fe^{3+} \approx Cu^{2+}$. Papadopoulos et al. [69] have studied the removal of Ni^{+} from a wastewater stream originating from a metal finishing industry by ion exchange and by a combination of ion exchange and precipitation methods. The results showed that the individual application of the ion exchange led to the removal of nickel up to 74.8% while the metal uptake rates were higher in the first stage of the process. Higher reduction in nickel concentrations was obtained by using the combination of ion exchange and precipitation processes (from 94.2% to 98.3%). The optimum results were obtained by using wastewater at a pH of 10.5 for precipitation followed by the ion exchange, using pretreated clinoptilolite for a time period not exceeding 2 h.

Turan et al. [70] used clinoptilolite and sepiolite for the removal of lead from water in their study. Both batch and column experiments were conducted. The results revealed that the removal of lead ions, both by clinoptilolite and sepiolite, involved the contribution of both adsorption by ion exchange and precipitation. The contribution of precipitation was found to be pH dependent and significant compared to that of adsorption. Coker and Rees [71] have studied the kinetics of binary ion exchange of Ca^{2+} and Mg^{2+} for Na^{+} in quasi-crystalline and crystalline zeolite A. It was observed that the diffusion of the divalent cations was faster in crystalline NaA than in the quasi-crystalline material. The diffusion coefficient (D) for Ca^{2+}, DCa, was higher than that for Mg^{2+}. This was due to the larger hydration sphere of the Mg^{2+} ion. As the crystallinity of the zeolite increased, the DCa also increased; in fully crystalline NaA, the DCa: DMg was 12. They suggested that the quasi-crystalline coatings may be used atop crystalline materials (hierarchical pore systems) to tune ion exchange selectivity and/or kinetics.

Medvidović et al. [72] have investigated the removal of lead ions from aqueous solutions by clinoptilolite, using the column method with more successive service and regeneration cycles. The highest effectiveness of the column performance has been attained for the zeolite particle size of 0.6–0.8 mm, at the initial concentration of 212.5 mg Pb/L and the flow rate of 2 ml/min. Mendoza et al. [73] conducted experiments on NH_4Cl- and NaCl-modified clinoptilolite to generate experimental data for the removal of lead by ion exchange. Two types of isotherms were made to represent the ion exchange equilibrium – ion exchange isotherm based upon the constant of thermodynamic equilibrium for the ion exchange reaction and the Langmuir isotherm. A comparison between the two isotherms showed that both of them fitted the experimental data reasonably well, but the Langmuir isotherm was much simpler and easier to use. On the other hand, the fitting procedure of the isotherm based on the thermodynamic equilibrium constant was quite laborious and tedious, and additional thermodynamic information was also required to estimate the activity coefficients of the ions in solution.

Kurtoğlu et al. [74] have studied ion exchange kinetics and equilibria of lead on Na-pretreated clinoptilolite using batch method by means of electrometric and polarographic techniques. Results show that Pb/Na exchange was affected by H^+ uptake at lower Pb concentrations (<10^{-3} M) in the fast period whereas uptake of OH^- ions contributed to exchange reactions at >10^{-3} M. A comparison of equilibrium data showed that Pb^{2+} ions were removed by ion exchange mechanism from the solutions below 5×10^{-4} M while formation of surface complexes between Pb species and surface hydroxyl groups increased lead adsorption at higher concentrations. The equilibrium data for ion exchange and adsorption were reasonably well predicted by Langmuir and the Freundlich isotherms.

A high-surface-area clinoptilolite-iron oxide (Clin–Fe) system was synthesized by Doula [75] for the removal of Mn^{2+} ions from drinking water samples. The Clin–Fe system had a specific surface area equal to 151.0 m^2/g and was fully iron exchanged (Fe/Al = 1.23). Batch adsorption experiments were carried out to determine the effectiveness of the clinoptilolite and the Clin–Fe system for the removal of manganese from drinking water. The Mn adsorption capacity of clinoptilolite was 7.69 mg/g, whereas, for Clin–Fe system, it was 27.12 mg/g. The main factors that contributed to difference in adsorption capacity of the two solids were due to new surface species and the negative charge of Clin–Fe system. The release of counterions (i.e., Ca^{2+}, Mg^{2+}, Na^+, and K^+) and the dissolution of Si and Al from the framework of the two zeolitic materials were also examined. It was found that for the most of the samples, the Clin–Fe system released lower concentrations of Ca, Mg, and Na and higher concentrations of K than clinoptilolite, while the dissolution of Si/Al was limited. After treatment with the Clin–Fe system, the water samples were seen to significantly lower the hardness, and it seemed that, except the adsorption of Mn^{2+} ions, the new system acted simultaneously as a water-softening material as well.

Argun [76] presented a detailed study of equilibrium and kinetics of the ion exchange process for removing Ni^{2+} from aqueous solution using a common, naturally occurring clinoptilolite. The removal efficiency of Ni^{2+} was found to depend upon operational parameters such as the amount of clinoptilolite, contact time, initial Ni^{2+} concentrations, and the pH of the solution. The Langmuir, D–R, and Freundlich isotherms were used to model the ion exchange of Ni^{2+} on clinoptilolite. The kinetics data has been modeled by a pseudo-second-order kinetics equation. Also discussed are the thermodynamic parameters, including changes in Gibbs free energy, entropy, and enthalpy, for the ion exchange of Ni^{2+} on clinoptilolite, and it was revealed that the ion exchange process was spontaneous and exothermic under natural conditions. The maximum removal efficiency obtained was 93.6% at a pH = 7 and with a 45 min contact time (for 25 mg^{l-1} initial concentration and a 15 g/l solid-to-liquid ratio).

NaY zeolite was tested for its uptake capacity for Fe^{3+} and Zn^{2+} [77] by carrying out experiments in a fixed-bed column using particles of size 0.180 mm at 30°C and pH values of 3.5 and 4.5 for the two ions, respectively. The influence of flow rate on the breakthrough curves at feed concentrations of 1.56 meq/l for Fe^{3+} and 0.844 meq/l for Zn^{2+} was investigated in order to minimize the diffusional resistances. The column equilibrium data has been represented by Freundlich and Langmuir isotherms. The Langmuir model described well the experimental breakthrough curves for feed concentrations from 0.1 up to 3.5 meq/l for Fe^{3+}. The Freundlich equation modeled the Zn^{2+} dynamic isotherm quite well. To investigate the kinetics of adsorption of the Fe^{3+} and Zn^{2+} ions, batch reactor studies were carried out. The second-order kinetic model best described the Fe^{3+} ion exchange, and the Langmuir kinetic model appropriately described the Zn^{2+} ion exchange.

Calvo et al. [78] studied the removal of cations found in acid mine drainage (Pb^{2+}, Cu^{2+}, Zn^{2+}, H^{+}) and municipal wastewater (NH_4^{+}) by using ion exchange on three types of zeolitic tuffs (clinoptilolite, heulandite, and stilbite) from Ecuador and Brazil. The clinoptilolite sample CLI-1 showed exceedingly good exchange capacities for Pb^{2+} and NH_4^{+} as received, and also a very high exchange capacity for Cu^{2+} and Zn^{2+} when treated with 2 M sodium chloride. These values were much higher than those reported in the literature for other clinoptilolite ores. The general order of cation-exchange capacity as inferred from breakthrough curves showed that the clinoptilolite-rich tuff was the most effective and that the stilbite was the least effective. It was observed that these zeolitic tuffs could be used in an economically affordable, environmentally compatible, and highly selective and effective way for removing cations found in acid mine drainage and municipal and agricultural wastewaters, in those parts of the world where zeolite deposits are in abundance.

9.4.2 Builders for Detergents

The extensive use of sodium tripolyphosphate (STPP) as a builder or water softener in detergents resulted in increased levels of phosphates in natural water bodies promoting eutrophication, that is, the phosphate served as a fertilizer leading to large growths of algae on the surface of the water, which depleted the dissolved oxygen in the water, thus annihilating precious flora and fauna. STPP basically acts by forming strong complexes with Ca^{2+} and Mg^{2+} ions in solution, thereby preventing their precipitation with the surfactant, as is well known that zeolites are excellent inorganic ion exchangers. Thus, those zeolites that can selectively exchange Ca^{2+} and Mg^{2+} ions from hard water with sodium were investigated for their use in detergents as builders. Some of these have established themselves as able replacements for STPP, while research is going on to find still better zeolites for this particular application. Detergents, both domestic and commercial, are used with waters having a wide variation in calcium and magnesium concentrations. Their compositions most often are complex, and the washing conditions around the globe also vary to a large extent [9]. Added to these is the fact that the large hydration shells surrounding the Ca^{2+} and Mg^{2+} ions make it difficult for them to enter the pores of the zeolite without expunging a number of their associated water molecules. In this, Mg^{2+}, due to its higher charge density, has a larger and more tightly held hydration shell than Ca^{2+}. Also, the rate of dissociation of water molecules from the hydration shells is slower for Mg^{2+} than Ca^{2+} [35]. Studies regarding the application of zeolites as builders for detergents have been designed toward developing an in-depth understanding of the cation-exchange process, with an ability to formulate accurate predictive models.

Schwuger et al. [79] explored the physicochemical background to the substitution of phosphates in laundry detergents, with special emphasis on combinations of zeolite A with other detergent ingredients, especially water-soluble cobuilders and optical brighteners. It was demonstrated that the addition of small amounts of polycarboxylates was very efficient in reducing calcium that was bound with the soil on the fabric itself and was aiding the soil to adhere to the surface of the fabric. NaZSM-5 zeolite was studied for its monovalent and divalent ion exchange characteristics [80]. It was found that all sodium was not replaced by the divalent cations. This incomplete exchange was attributed to divalent ions being unable to exchange into sites where the A1 atoms are more than a certain distance apart. By comparing the isotherm cutoffs with the distribution of Al–Al distances predicted by Monte Carlo simulations, they concluded that the charge on the divalent ions can only be satisfied by 2 AI atoms closer than 12 Å. The isotherms for the exchange of the Na^{+} ions in ZSM-5 samples by Ca^{2+}, Sr^{2+}, and Ba^{2+} at 298 and 338 K were found to be reversible. These isotherms clearly demonstrated that divalent exchange in these zeolite samples was completely different in behavior from that found for monovalent exchange.

Watling and Rees have studied the effect of Si/Al ratio [81] and the chemical nature of the exchange sites on ion exchange [82], by monovalent and divalent cations. The ideas propounded by McAleer [80] were extended to zeolite EU-1 in [81]. There are only a few studies that have looked at the effect of changing the chemical nature of the exchange site; nearly all others have studied ion exchange in aluminosilicates where exchange sites are $(AIOSi)^{-}$. In [82], ion exchange in EU-1 samples in which the exchange sites were predominantly SiO^{-} and $(GaOSi)^{-}$ was studied. It was concluded that di-univalent ion exchange isotherms were influenced not only by the chemical nature of the exchange site but also by the location of these sites. The zeolite first used in detergents was the LTA-type zeolite NaA, which possessed high ion exchange capacity, good selectivity for calcium, and acceptable Ca–Na ion exchange kinetics, and, more importantly, it was easy to manufacture and was low in cost. The Ca–Na ion exchange has thus been extensively studied for this zeolite. Some of the pertinent studies carried on this zeolite are mentioned [83–86]. A new development in detergent technology was the introduction of zeolite "maximum aluminum P" or MAP with a Si/Al ratio of 1:1, as a builder in detergents. The MAP has a GIS typology. The advantages of MAP include more rapid and more selective exchange of calcium, particularly at low temperatures; it provides enhanced stability to sensitive and expensive performance components, especially bleaches; and it extends the range of possible formulations to higher surfactant levels by its ability to carry larger amounts of nonionic surfactants. These advantages are due to the flexibility of its framework and the careful engineering of its particles [87]. It was much better in performance than 4A. It was the first zeolite tailor-made for use as a detergent builder.

Adams et al. [88] have discussed the Ca–Na ion exchange over MAP zeolite and compared the results with those obtained previously by Ca–Na–zeolite A. Zeolite MAP shows greater selectivity for calcium than does zeolite A over much of the exchange system. The high selectivity for calcium shown by MAP and the unusual Z-shaped selectivity plot are attributed to the formation of coexisting sodium- and calcium-rich zeolite phases. Allen et al. [89] performed ion exchange studies on MAP. The exchanges of Na by Li, K, Rb, Cs, Mg, Ca, Sr, and Ba in MAP were examined. When cation exchanges were shown to be reversible, their isotherms were analyzed to produce thermodynamic data. Where possible, ion exchange isotherms were constructed and compared with those of similar processes in other zeolites. Thermal analyses provided supporting data, and structural studies provided further information on the cation-exchange mechanisms taking place inside the zeolite phase. The shape of the isotherms, coupled with the Gibbs energies of exchange calculated from the Kielland plots, facilitated the construction of a selectivity series for the zeolite MAP as Ba > Sr > Ca >> Na > K > Rb > Cs > (Mg, Li). Many other investigations have been carried out to study the various aspects of the application of zeolites to detergents.

9.4.3 Nuclear Waste

A large number of ion exchange materials are available for the treatment of radioactive liquids. These include natural inorganic ion exchangers (clays, vermiculite, and zeolites), natural organic ion exchangers (proteins, polysaccharides, and carbonaceous materials), modified natural ion exchangers (both organic and inorganic), synthetic inorganic ion exchangers (zeolites, titanates, and silicotitanates, and transition metal hexacyanoferrates), synthetic organic ion exchangers (polystyrene divinylbenzene, phenolic resins, acrylic resins), composite ion exchangers, and ion exchange membranes. In nuclear power plants, ion exchange materials are being used in primary coolant water purification, treatment of primary effluents, treatment of steam generator blow-down demineralization, liquid waste and drainage water treatments, treatment of fuel storage pond water, boric acid purification for recycling, and condensate polishing for nuclear power plants having boiling water reactors [34]. Over the years inorganic ion exchangers have evolved as an increasingly important complement or replacement for conventional organic ion exchangers in the nuclear industry operations, particularly in liquid radioactive waste treatment and spent fuel reprocessing processes. They have also made a mark in containing fallout from nuclear mishaps. The inorganic ion exchangers have greater thermal stability, radiation stability, and greater selectivity for certain species such as Cs and Sr; they have wider operating temperature range; and they may exhibit advantages with regard to immobilization and eventual disposal vis-à-vis the organic ion exchangers. However, they also have some disadvantages like relatively low ion exchange capacities, low mechanical strength and abrasion resistance, noncontrollable pore size, and limited chemical stability in many solutions. In order to apply ion exchange on a particular waste, it is absolutely necessary that the waste conforms to some criteria. In reference [34], IAEA recommends such criteria, which are as follows: (1) In columns with downward flow of liquid, the concentration of total suspended solids in the waste should be low (less than 4 mg/l); in upward flow, a prefiltration of the stream may be necessary to prevent the escape of fine particular material. (2) It is recommended that the waste should have a low total dissolved salt content (1–2 g/l) to prevent the rapid exhaustion of the ion exchange capacity. (3) The radionuclides should be present in a suitable ionic form. (4) The waste should contain very small quantities of organic contaminants such as oils and greases. (5) Strong oxidizing compounds such as nitrates should not be there, as they can cause a catalytic degradation of organic media in the presence of some metallic ions. The ion exchange processes can be applied in the form of batches, in columns, in continuous loops and as part of, or in combination with membrane processes [34, 35]. In this section, we shall restrict the discussion to the application of natural and synthetic zeolites in the treatment of radioactive nuclear wastes and in containing radioactive fallout, in the light of some relatively recent literature.

In order to appreciate the application of ion exchange to separate radioactive components of nuclear waste, it is important to know the nature of this waste. It contains $^{137}Cs^{+}$ and $^{90}Sr^{2+}$ as the major components, together with appreciable quantities of competing ions, such as Ca^{2+}, Mg^{2+}, Na^{+}, and a range of anions. The composition of this waste varies considerably, depending on its source in the nuclear power plant. It may be alkaline with a very dilute concentration of nuclides as in the case of nuclear pond water, or it may be highly acidic and of higher concentration as in the case of recycling spent fuel rods by dissolving them in concentrated nitric acid [35]. Liquid radioactive waste (LRW) produced as a result of the operation, repair, and disposal of nuclear energy installations can be classified in different ways. First, the source of LRW can be considered as a criterion. In this case, one should mention decontamination LRW, LRW from waste nuclear fuel (WNF) storage basins, and LRW from reactor coolant liquids. All of these LRW types differ in chemical and radiochemical composition. The radiochemical composition changes significantly over time because of the decay of the short-lived radionuclides. In this process, the fraction of long-lived Cs and Sr isotopes increases continuously. Regarding chemical composition, LRW is divided into low-salt, salt content up to 20 mg/l (e.g., LRW from WNF storage basins), and high-salt LRW, salt content higher than 20 mg/l (e.g., decontamination LRW). Concerning activity values, LRW is divided into high active (higher than 10 Ci/l), middle active (from 10 down to 10^{-2} Ci/l), low active (from 10^{-2} down to 10^{-5} Ci/l), and very low active (from 10^{-6} down to 10^{-9} Ci/l). LRW of activity lower than 10^{-9} Ci/l is conventionally not processed but dumped so they will be diluted to below accepted concentration limits [90].

The pH, temperature, and other factors decide whether the nuclides will exist in solution as cations, neutral species, or anions. Owing to the complexity of nuclear waste, the prediction of the effectiveness of a zeolite in removing nuclear waste is usually done by carrying out laboratory tests on actual waste and not from binary-exchange data or with simulated waste samples. The zeolites can be operated in a regenerative mode or a nonregenerative mode. In the former case, the regeneration of the zeolite bed is done by using a concentrated salt solution to elute the nuclide. But there is always a problem to fix the nuclide. In the nonregenerative mode, the saturated zeolite is transformed into concrete, glass, or ceramic bodies and stored indefinitely [2]. Zeolites score over other ion exchangers in that they are compatible with concrete and glass matrices used to solidify and contain the ion exchanger phase once radioactive [35]. Also, zeolites exhibit stability to radiation as mentioned earlier and can withstand very high values without any change in structure or ion exchange properties. Natural zeolites, additionally, have the advantages of being cheap and abundant.

The literature in the field of nuclear waste treatment using zeolites extending back to the late 1940s has been reviewed thoroughly by Dyer [91].The sorption capabilities of some 20 natural zeolites have been studied for Sr, Cs, and other radionuclides from nuclear waste effluents (americium, cobalt, krypton, radium, thorium, tritium, and uranium). The main reason for this has been the unsuitability of organic ion exchange resins for this application due to their low radiation and thermal stability. Komarneni et al. [92] studied the effect of radiation by gamma rays on ion exchange and selective properties as well as the leaching behavior of Ionsiv-IE-95 zeolite and also investigated the effect of gamma radiation on the leaching behavior of a natural pollucite containing a small percentage of Na and water substituted for Cs. No measurable difference in the Cs-exchange capacities of the irradiated and unirradiated Ionsiv-IE-95 samples was observed. There was no statistically significant difference between the irradiated and unirradiated Ionsiv-IE-95 samples in their selective Cs and Sr sorption properties. Leaching studies of the Cs-loaded Ionsiv-IE-95 were conducted under irradiation. The results showed that there was a significant increase in the desorption of the exchangeable Cs and Na ions from the Ionsiv-IE-95 upon irradiation. There was also a slight increase in the amount of Al released into solution from Ionsiv-IE-95 by gamma irradiation. The desorption of exchangeable Cs and Na and the release of Al were attributed to a decrease in pH of the irradiated leachates by the formation of acidic species. It was further observed that the irradiated pollucite when leached in deionized water exhibited a higher leachability for Cs and Na than the unirradiated pollucite. The higher leachability of the irradiated pollucite was also attributed to the generation of a lower pH. Dyer and Keir [93] reported the separation of 90-strontium from daughter nuclide 90-yttrium in a synthetic radioactive waste solution using clinoptilolite, ferrierite, phillipsite, and a synthetic mordenite (Zeolon 900). It was demonstrated that ferrierite had comparable effectiveness as clinoptilolite in radioactive waste treatment. All the four zeolites studied in the experiment produced a clean and harmless effluent. Chelishchev [94] observed that the substitution of clinoptilolite for ion exchange resins increased the volume of purified low-level radioactive sewage water up to 5–10 times and its substitution for quartz sand in rapid waterworks and individual filters reduced the specific radioactivity down to n.10^{-9} cu/1. The concentration of ^{137}Cs on the solid phase was found to increase by a factor of at least 1,000. Perona [95] used binary interaction parameters for the Wilson equation to model solid-phase activity coefficients for a five-component system in chabazite comprising Ca, Cs, Mg, Na, and Sr. Paschoa and Filho [96] discussed the principal forms of low-level radioactive waste management in the developing and newly industrialized countries. The wastes produced by the thorium and uranium mining and milling activities have been discussed, taking into account their importance, and wastes produced from the military nuclear weapon complex have also been mentioned. Sinha et al. [97] have made an exhaustive comparative study of the treatment of radioactive liquid waste containing cesium by locally available synthetic zeolites such as 13X, AR-1, 4A, and ZSM-5 using pure aqueous solutions as well as from actual radioactive liquid wastes. The kinetics of the exchange process was studied in batch experiments with pure CsCI solutions spiked with radioactive ^{137}Cs. Equilibrium studies were carried out under stirred conditions, using various zeolites and isonormal solutions of Cs and Na, at a constant total ionic concentration of 0.1 N. The adsorption isotherms indicated a high selectivity of the zeolite AR-I for Cs, which was further confirmed by drawing Kielland-type plots. The breakthrough curves indicated that AR-1 was the best zeolite for effluent that contained low dissolved solids, while none of the other zeolites were found to be promising for the treatment of waste streams with high dissolved solids.

In another investigation [98], Iranian natural zeolites (clinoptilolite) from three different locations and their sodium-exchanged forms were studied for cation exchange of Cs^+, Ni^{2+}, Sr^{2+}, and Ba^{2+} from radioactive or municipal wastewaters. It was observed that the affinity of the zeolites for these cations was relatively high. Elizondo et al. [99] studied the removal of Cs^+ and Sr^{2+} from liquid radioactive waste of low activity by using natural clinoptilolite. They recommended that the clinoptilolite having grain size 0.25–1.0 mm be used and the used solutions should have a pH = 8 and flow velocity should not exceed five column volumes per hour for effective cation removal. Marinin and Brown [90] have thoroughly studied the removal of strontium from low-level LRW using natural zeolite-clinoptilolite, modified natural clinoptilolite, synthetic zeolites, new synthetic crystalline materials selective to strontium, ion exchange resins, and modified fiber sorbents. Comparison of materials was made in terms of batch distribution coefficient K_d, sorption selectivity in the presence of competing ions, kinetics of radionuclide sorption, and the approximate cost of the sorption materials used in the study. In another interesting study, Kazemian et al. [100] investigated the ion exchange behaviors of an Iranian natural clinoptilolite and its modified forms as well as a relevant synthetic zeolite P for removal of cerium and thorium from nuclear wastewaters. Results of column experiments showing the effect of parameters such as particle size, pH, temperature, and time were exhibited, together with the calculations for distribution coefficient, cation-exchange capacity, and some thermodynamic parameters. Ion exchange isotherms and breakthrough curves were plotted. A comparison of all the types of zeolites used in the study was made based upon the above data reduction.

Keheyan et al. [101] have studied the sorptive behavior of radioactive waste on Armenian zeolites. The effects of temperature, acidity, basicity, specific activity, electron, and gamma irradiation on sorption have been reported. The model experiments were utilized to design a laboratory plant for evaluating the radioactive waste from the nuclear reactor. A detailed description of the technologies being employed in radioactive waste management practices in India has been presented by Raj et al. [102] which shows the complexity involved in the safe and efficient management of radioactive waste. Osmanioglu [103] carried out a study to assess the Turkish natural zeolites from different formations for the removal of radioactive nuclides: ^{137}Cs, ^{60}Co, ^{90}Sr, and ^{110m}Ag as generated in liquid wastes emanating from industrial applications, research, and medicine. It was deduced from the study that Turkish zeolites possessed high cation-exchange capacities and they had good selectivity. Decontamination costs of liquid waste streams could be decreased by using natural zeolite as sorbent instead of other expensive chemical compounds. In quite a recent study [104], the sorption of cesium with zeolites was investigated from an alkaline nuclear waste containing sodium aluminate and sodium nitrate. Isotherms at different temperatures and for different sodium concentrations as well as the kinetics were investigated. Breakthrough curves have been reported both for laboratory and pilot plant columns. For a high-salt-content alkaline solution, a capacity of 1 mg of Cs per gram of zeolite was found in fixed-bed column operation at room temperature. Yeritsyan et al. [105] investigated the performance of Armenian clinoptilolite for treating radioactive wastewater from the Armenian nuclear power plant. They reported a significant decrease in the radioactivity of the treated water.

In addition to their applicability in treating radioactive wastewaters, zeolites are actively employed in combating radioactive fallout from accidents and tests. Their selectivities to cesium and strontium, the two major radionuclides in any outbreak of radioactivity, confer on them high effectiveness at low cost to deal with nuclear mishaps. Nishita and Haug [106] observed that by adding clinoptilolite to soils contaminated with strontium, there was dramatic decrease in the uptake of ^{90}Sr by plants. In the USA, after the Three Mile Island nuclear accident, more than 1.5 million gallons of water was treated by a 60/40 mixture by volume of IE-96 and LTA zeolites to immobilize 340,000 Ci of fission products [107]. The addition of clinoptilolite in the soils of Bikini Atoll inhibited the uptake of ^{137}Cs from them to the plants [108]. It was shown in a study [109], conducted on Swedish reindeer in the aftermath of the Chernobyl nuclear disaster, that zeolite supplements in diet accelerated the excretion of ^{137}Cs ingested by taking contaminated food. In another study [110], it was detected that ^{137}Cs in soils was not taken up by the plants after treating the soil with a zeolite. Chelitshchev [111] reported the use of some 50,000 t of clinoptilolite from Ukraine, Georgian Republic, and regions of Russia together with many thousands of tons from Bulgaria to prevent the release of radioactive ^{137}Cs from the burning reactor to the atmosphere. The cation exchange of natural zeolites has attracted a great deal of attention for use as barriers to nuclear waste repositories [112]. Very recently, sandbags of zeolite were dropped into the seawater near the Fukushima nuclear plant to adsorb radioactive cesium that was present there in high levels.

9.5 Conclusions

The importance of zeolites as inorganic ion exchangers has been highlighted, with special reference to their use in environmental applications. Though it was not possible to discuss in this text the large body of literature available in this field, nevertheless, an attempt has been made to provide a panoramic window to the type of research work that has been carried out over the years by selecting a modest number of research papers. It was observed that substantial progress has been made in the application of zeolites to water and wastewater treatment, detergent building, and nuclear waste treatment and fallout. Studies have been performed on both natural and synthetic zeolites. Most of the work is related to equilibrium and kinetic aspects of zeolite ion exchange. Some investigations have been directed to study the performance of zeolites for multiple cation exchanges, where competition among cations makes the process quite complex.

The preceding sections of this chapter clearly point to the fact that the level of the basic understanding of the theory of ion exchange supported by meticulous experimentation has resulted in the application of zeolites to areas of vital import to modern civilization. The use of zeolites and related materials for environmental applications has emerged as the most rapidly growing area in zeolite ion exchange. However, the continuing advances in synthesis and measurement techniques are opening new avenues for enhancing our understanding and capabilities in the field of zeolite ion exchange. The need of the hour is to apply the new and powerful computational techniques to carry out modeling and simulation of equilibrium and kinetics in ion exchange, in order to predict and design accurately, efficient processes concerned with the application of zeolites. The vast quantities of natural zeolites throughout the world and the rapidly growing family of synthetic zeolites are irresistible inducements to carry out research for utilizing these invaluable minerals in novel applications for the benefit of mankind.

Acknowledgment The author wishes to gratefully acknowledge and thank all the researchers in the field of zeolite exploitation and development whose work provides the basis for this review chapter as well as the publishers of such work.

References

1. Flanigen EM, Broach RW, Wilson ST (2010) Introduction. In: Kulprathipanja S (ed) Zeolites in industrial separation and catalysis. Wiley-VCH Verlag GmbH & Co, KGaA, Weinheim
2. Mumpton FA (1999) La roca magica: uses of natural zeolites in agriculture and industry. Colloquium paper. Proc Natl Acad Sci USA 96:3463–3470
3. Valdés MG, Pérez-Cordoves AI, Díaz-Garcia ME (2006) Zeolites and zeolite-based materials in analytical chemistry. Trends Anal Chem 25(1):24–30
4. Bailey S, Olin T, Bricka M, Adrian D (1999) A review of potentially low-cost sorbents for heavy metals. Water Res 33:2469–2479

5. Elizalde-González M, Mattusch J, Einicke W, Wennrich R (2001) Sorption on natural solids for arsenic removal. Chem Eng J 81:187–195
6. Rožić M, Cerjan-Stefanović Š, Kurajica S, Vančina V, Hodžić E (2000) Ammoniacal nitrogen removal from water by treatment with clays and zeolites. Water Res 34:3675–3681
7. Bowman RS (2003) Applications of surfactant-modified zeolites to environmental remediation. Micropor Mesopor Mater 61:43–56
8. Ackley M, Rege S, Saxena H (2003) Application of natural zeolites in the purification and separation of gases. Micropor Mesopor Mater 61:25–42
9. Dyer A (2007) Ion-exchange properties of zeolites and related materials. In: Čejka J, van Bekkum H, Corma A, Schüth F (eds) Introduction to zeolite science and practice, 3rd Revised edn. Elsevier B. V, Amsterdam/The Netherlands
10. Drews TO, Tsapatsis M (2005) Progress in manipulating zeolite morphology and related applications. Curr Opin Colloid Interface Sci 10:233–238
11. McCusker LB, Liebau F, Engelhardt G (2003) Nomenclature of structural and compositional characteristics of ordered microporous and mesoporous materials with inorganic hosts (IUPAC recommendations 2001). Micropor Mesopor Mater 58(11):3–13
12. Baerlocher C, McCusker LB, Olson DH (2007) Atlas of zeolite framework types, 6th edn. Elsevier, Amsterdam
13. Meier WM, Olson DH (1971) Zeolite frameworks. Adv Chem Ser 101:155–170
14. Barrer RM (1979) Chemical nomenclature and formulation of compositions of synthetic and natural zeolites. Pure Appl Chem 51:1091–1100
15. McCusker LB, Liebau F, Engelhardt G (2001) Nomenclature of structural and compositional characteristics of ordered microporous and mesoporous materials with inorganic hosts. Pure Appl Chem 73:381–394
16. McCusker LB, Liebau F, Baerlocher C (2001) Zeolite structures. In: van Bekkum H, Flanigen EM, Jacobs PA, Jansen JC (eds) Stud Surf Sci Catal 137. Elsevier Science B.V, Amsterdam/The Netherlands
17. Barrer RM, Marcelli C (1970) Hydrothermal chemistry of silicates. Part XV. Synthesis and nature of some salt-bearing aluminosilicates. J Chem Soc A 2735–2745
18. Kerr GT (1966) Chemistry of crystalline aluminosilicates. III. The synthesis and properties of zeolite ZK-5. Inorg Chem 5:1539–1541
19. Kerr GT (1963) Zeolite ZK-5: a new molecular sieve. Science 140:1412
20. Milton RM (1959) US Patent 2,882,243
21. Milton RM (1959) US Patent 2,882,244
22. Yu J (2007) Synthesis of zeolites. In: Čejka J, van Bekkum H, Corma A, Schüth F (eds) Introduction to zeolite science and practice, 3rd Revised edn. Elsevier B. V, Amsterdam/The Netherlands
23. Casci JL (2005) Zeolite molecular sieves: preparation and scale-up. Micropor Mesopor Mater 82:217–226
24. Jentys A, Lercher JA (2001) Techniques of zeolite characterization. In: van Bekkum H, Flanigen EM, Jacobs PA, Jansen JC (eds) Stud Surf Sci Catal 137. Elsevier Science B.V, Amsterdam/The Netherlands
25. Bradley SA, Broach RW, Mezza TM, Prabhakar S, Sinkler W (2010) Zeolite characterization. In: Kulprathipanja S (ed) Zeolites in industrial separation and catalysis. Wiley-VCH Verlag GmbH & Co, KGaA, Weinheim
26. Breck DW (1974) Zeolite molecular sieves. Wiley, New York
27. Karge HG, Weitkamp J (eds) (2004) Molecular sieves – science and technology, characterization 1, vol 4. Springer, Berlin
28. Karge HG, Weitkamp J (eds) (2007) Molecular sieves – science and technology, characterization II, vol 5. Springer, Berlin
29. Lucy CA (2003) Evolution of ion-exchange: from Moses to the Manhattan Project to modern times. J Chromatogr A 1000:711–724
30. McCabe WL, Smith JC, Harriot P (2005) Unit operations of chemical engineering, 7th edn. McGraw Hill International Edition, Singapore

31. Inglezakis VJ, Poulopoulos SG (2007) Adsorption, ion exchange and catalysis. Elsevier B.V, Amsterdam/The Netherlands
32. Helfferich F (1962) Ion exchange. Dover, New York
33. Treybal RE (1981) Mass transfer operations, 3rd edn. McGraw-Hill International Editions, Singapore
34. IAEA (2002) Application of ion exchange processes for the treatment of radioactive waste and management of spent ion exchangers, Technical reports series no. 408, International Atomic Energy Agency (IAEA), Vienna
35. Townsend RP, Coker EN (2001) Ion exchange in zeolites. In: van Bekkum H, Flanigen EM, Jacobs PA, Jansen JC (eds) Stud Surf Sci Catal 137. Elsevier Science B.V, Amsterdam/ The netherlands
36. Fu F, Wang Q (2011) Removal of heavy metal ions from wastewaters: a review. J Environ Manag 92:407–418
37. Jorgensen SE, Libor O, Graber KL, Barkacs K (1976) Ammonia removal by use of clinoptilolite. Water Res 10:213–221
38. Klieve JR, Semmens MJ (1979) An evaluation of pretreated natural zeolites for ammonium removal. Water Res 14:161–168
39. Hlavay J, Vigh G, Olaszi V, Inczédy J (1982) Investigations on natural Hungarian zeolite for ammonia removal. Water Res 16:417–420
40. Czárán E, Mészáros-Kis Á, Domokos E, Papp J (1988) Separation of ammonia from wastewater using clinoptilolite as ion exchanger. Nucl Chem Waste Manag 8:107–113
41. Dryden HT, Weatherley LR (1989) Aquaculture water treatment by ion exchange: continuous ammonium ion removal with clinoptilolite. Aquacult Eng 8:109–126
42. Lahav O, Green M (1998) Ammonium removal using ion exchange and biological regeneration. Water Res 32(7):2019–2028
43. Weatherley LR, Miladinovic ND (2004) Comparison of the ion exchange uptake of ammonium ion onto New Zealand clinoptilolite and mordenite. Water Res 38:4305–4312
44. Du Q, Liu S, Cao Z, Wang Y (2005) Ammonia removal from aqueous solution using natural Chinese clinoptilolite. Sep Purif Technol 44:229–234
45. Farkaš A, Rožić M, Mikočević BZ (2005) Ammonium exchange in leakage waters of waste dumps using natural zeolite from the Krapina region, Croatia. J Hazard Mater B117:25–33
46. Jha VK, Hayashi S (2009) Modification on natural clinoptilolite zeolite for its NH_4^+ retention capacity. J Hazard Mater 169:29–35
47. Ramos RL, Rocha JEM, Piña AA, Mendoza MSB, Coronado RMG, Davila PA, Barron JM (2010) Removal of ammonium from aqueous solution by ion exchange on natural and modified chabazite. J Environ Manag 91:2662–2668
48. Kantiranis N, Sikalidis K, Squires C, Papastergios G, Filippidis A (2011) Extra-framework cation release from heulandite-type rich tuffs. J Environ Manag 92:1569–1576
49. Malekian R, Koupai JA, Eslamian SS, Mousavi SF, Abbaspour KC, Afyuni M (2011) Ion-exchange process for ammonium removal and release using natural Iranian zeolite. Appl Clay Sci 51:323–329
50. Joshi MS, Rao PM (1983) Cation sieve properties of natural heulandite. J Colloid Interface Sci 95(1):131–134
51. Blanchard G, Maunaye M, Martin M (1984) Removal of heavy metals from waters by means of natural zeolites. Water Res 18(12):1501–1507
52. Semmens MJ, Martin WP (1988) The influence of pretreatment on the capacity and selectivity of clinoptilolite for metal ions. Water Res 22(5):537–542
53. Pansini M, Colella C, De' Gennaro M (1991) Chromium removal from water by ion exchange using zeolite. Desalination 83:145–157
54. Wark M, Lutz W, Schulz-Ekloff G, Dyer A (1993) Quantitative monitoring of side products during high loading of zeolites by heavy metals via pH measurements. Zeolites 13:658–662
55. Hernández JEG, Notario del Pino JS, Martín MMG (1994) Cation-exchange selectivity in Na^+- and in H^+-phillipsite. Appl Clay Sci 9:239–249

56. Kuhm P, Lortz W (1994) Examination of the "decomposition behavior" of zeolite A in fresh-water, particularly taking into consideration environmentally relevant conditions. In: Weilkamp J, Karge HG, Pfeifer H, Hölderich H (eds) Stud Surf Sci Catal 84 Elsevier Science B.V, Amsterdam/The Netherlands
57. Mondale KM, Carland RM, Aplan FF (1995) The comparative ion exchange capacities of natural sedimentary and synthetic zeolites. Miner Eng 8(4/5):535–548
58. Pansini M, Colella C, Caputo D, de Gennaro M, Langella A (1996) Evaluation of phillipsite as cation exchanger in lead removal from water. Micropor Mater 5(357):364
59. Ãrković L, Štefanović SC, Filipan T (1997) Metal ion exchange by natural and modified zeolites. Water Res 31(6):1379–1382
60. Kang SJ, Egashira K (1997) Modification of different grades of Korean natural zeolites for increasing cation exchange capacity. Appl Clay Sci 12:131–144
61. Ahmed S, Chughtai S, Keane MA (1998) The removal of cadmium and lead from aqueous solution by ion exchange with Na-Y zeolite. Sep Purif Technol 13:57–64
62. Keane MA (1998) The removal of copper and nickel from aqueous solution using Y zeolite ion exchangers. Colloid Surface A 138:11–20
63. Mier MV, Callejas RL, Gehr R, Cisneros BEJ, Alvarez PJJ (2001) Heavy metal removal with Mexican clinoptilolite: multi-component ionic exchange. Water Res 35(2):373–378
64. Cincotti A, Lai N, Orrù R, Cao G (2001) Sardinian natural clinoptilolites for heavy metals and ammonium removal: experimental and modeling. Chem Eng J 84:275–282
65. Inglezakis VJ, Loizidou MD, Grigoropoulou HP (2002) Equilibrium and kinetic ion exchange studies of Pb^{2+}, Cr^{3+}, Fe^{3+} and Cu^{2+} on natural clinoptilolite. Water Res 36:2784–2792
66. Iznaga IR, Gómez A, Fuentes GR, Aguilar AB, Ballan JS (2002) Natural clinoptilolite as an exchanger of Ni^{2+} and NH_4^+ ions under hydrothermal conditions and high ammonia concentration. Micropor Mesopor Mater 53:71–80
67. Petrus R, Warchol J (2003) Ion exchange equilibria between clinoptilolite and aqueous solutions of Na^+/Cu^{2+}, Na^+/Cd^{2+} and Na^+/Pb^{2+}. Micropor Mesopor Mater 61:137–146
68. Inglezakis VJ, Loizidou MD, Grigoropoulou HP (2003) Ion exchange of Pb^{2+}, Cu^{2+}, Fe^{3+}, and Cr^{3+} on natural clinoptilolite: selectivity determination and influence of acidity on metal uptake. J Colloid Interface Sci 261:49–54
69. Papadopoulos A, Fatta D, Parperis K, Mentzis A, Haralambous KJ, Loizidou M (2004) Nickel uptake from a wastewater stream produced in a metal finishing industry by combination of ion-exchange and precipitation methods. Sep Purif Technol 39:181–188
70. Turan M, Mart U, Yüksel B, Çelik MS (2005) Lead removal in fixed-bed columns by zeolite and sepiolite. Chemosphere 60:1487–1492
71. Coker EN, Rees LVC (2005) Kinetics of ion exchange in quasi-crystalline aluminosilicate zeolite precursors. Micropor Mesopor Mater 84:171–178
72. Medvidović NV, Perić J, Trgo M (2006) Column performance in lead removal from aqueous solutions by fixed bed of natural zeolite–clinoptilolite. Sep Purif Technol 49:237–244
73. Mendoza MSB, Ramos RL, Davila PA, Rubio LF, Coronado RMG (2006) Comparison of isotherms for the ion exchange of Pb(II) from aqueous solution onto homoionic clinoptilolite. J Colloid Interface Sci 301:40–45
74. Kurtoğlu AE, Atun G (2006) Determination of kinetics and equilibrium of Pb/Na exchange on clinoptilolite. Sep Purif Technol 50:62–70
75. Doula MK (2006) Removal of Mn^{2+} ions from drinking water by using clinoptilolite and a clinoptilolite–Fe oxide system. Water Res 40:3167–3176
76. Argun ME (2008) Use of clinoptilolite for the removal of nickel ions from water: kinetics and thermodynamics. J Hazard Mater 150:587–595
77. Ostroskia IC, Barros MASD, Silvab EA, Dantas JA, Arroyoa PA, Lima OCM (2009) A comparative study for the ion exchange of Fe(III) and Zn(II) on zeolite NaY. J Hazard Mater 161:1404–1412
78. Calvo B, Canoira L, Morante F, Bedia JMM, Vinagrea C, Gonzáleza JEC, Elsenc J, Alcantara R (2009) Continuous elimination of Pb^{2+}, Cu^{2+}, Zn^{2+}, H^+ and NH_4^+ from acidic waters by ionic exchange on natural zeolites. J Hazard Mater 166:619–627

79. Schwuger MJ, Liphard M (1989) Fundamentals of phosphate substitution in detergents by zeolites – cobuilders and optical brighteners. In: Karge HG, Weitkamp J (eds) Zeolites as catalysts, sorbents and detergent builders. Elsevier Science Publishers B.V, Amsterdam
80. Andrew M, McAleer AM, Rees LVC, Nowak AK (1991) Ion exchange and aluminum distributions in ZSM-5 zeolites. Zeolites 11:329–336
81. Watling TC, Rees LVC (1994) Ion exchange in zeolite EU-I: part 1. The effect of Si/Al ratio. Zeolites 14:687–692
82. Watling TC, Rees LVC (1994) Ion exchange in zeolite EU-I: part 2. The effect of the chemical nature of the exchange site. Zeolites 14:693–696
83. Barrer RM, Rees LVC, Ward DJ (1963) Thermochemistry and thermodynamics of ion exchange in a crystalline exchange medium. Proc R Soc Lond A 237:180–197
84. Barri SAI, Rees LVC (1980) Binary and ternary cation exchange in zeolites. J Chromatogr 201:21–34
85. Franklin KR, Townsend RP (1985) Multicomponent ion exchange in zeolites. Part 1 – Equilibrium properties of the sodium/calcium/magnesium-zeolite system. J Chem Soc Faraday Trans I 81:1071–1086
86. Franklin KR, Townsend RP (1985) Multicomponent ion exchange in zeolites. Part 2 – Prediction of equilibria over a range of solution concentrations. J Chem Soc Faraday Trans I 81:3127–3141
87. Adams CJ, Araya A, Carr SW, Chapple AP, Franklin KR, Graham P, Minihan AR, Osinga TJ, Stuart JA (1997) Zeolite map: the new detergent zeolite. In: Chon H, Ihm S-K, Uh YS (eds) Stud Surface Sci Catal 105 Elsevier Science B.V, Amsterdam/The Netherlands
88. Adams CJ, Araya A, Cunningham KJ, Franklin KR, White IF (1997) Measurement and prediction of Ca–Na ion-exchange equilibrium in maximum aluminium P (MAP), a zeolite with the GIS framework topology. J Chem Soc Faraday Trans 93(3):499–503
89. Allen S, Carr S, Chapple A, Dyer A, Heywood B (2002) Ion exchange in the synthetic gismondine, zeolite MAP. Phys Chem Chem Phys 4:2409–2415
90. Marinin DV, Brown GN (2000) Studies of sorbent/ion-exchange materials for the removal of radioactive strontium from liquid radioactive waste and high hardness ground waters. Waste Manag 20:545–553
91. Dyer A (2000) Application of natural zeolites in the treatment of nuclear wastes and fallout. In: Cotter-Howells JD, Campbell LS, Valsami-Jones E, Batchelder M (eds) Environmental mineralogy: anthropogenic influences, contaminated land and waste management, vol 9. The Mineralogical Society, London
92. Komarneni S, Palau GL, Pillay KKS (1983) Radiation effects on a zeolite ion exchanger and a pollucite. Nucl Chem Waste Manag 4:335–338
93. Dyer A, Keir D (1984) Nuclear waste treatment by zeolites. Zeolites 4:215–217
94. Chelishchev NF (1994) Use of natural zeolites for liquid radioactive wastes treatment (Russian experience). In: Karge HG, Weitkamp J (eds) Stud Surf Sci Catal 98 Elsevier Science B.V, Amsterdam/The Netherlands
95. Perona JJ (1993) Model for Sr-Cs-Ca-Mg-Na ion exchange equilibria on chabazite. AIChE J 39(10):1716–1720
96. Paschoa AS, Filho AT (1995) Radioactive waste management in developing and newly industrialized countries. Appl Radiat Isot 46(6/7):707–715
97. Sinha PK, Panicker PK, Amalraj RV, Krishnasamy V (1995) Treatment of radioactive liquid waste containing caesium by indigenously available synthetic zeolites: a comparative study. Waste Manag 15(2):149–157
98. Faghihian H, Marageh MG, Kazemian H (1999) The use of clinoptilolite and its sodium form for removal of radioactive cesium, and strontium from nuclear wastewater and Pb^{2+}, Ni^{2+}, Cd^{2+}, Ba^{2+} from municipal wastewater. Appl Radiat Isot 50:655–660
99. Elizondo NV, Ballesteros E, Kharisov BI (2000) Cleaning of liquid radioactive wastes using natural zeolites. Appl Radiat Isot 52:27–30

100. Kazemian H, Modarres H, Mobtaker HG (2003) Iranian natural clinoptilolite and its synthetic zeolite P for removal of cerium and thorium from nuclear wastewaters. J Radioan Nucl Chem 258(3):551–556
101. Keheyan Y, Khachatryan S, Christidis G, Moraetis D, Gevorkyan R, Sarkisyan H, Yeritsyan H, Nikoghosyan S, Sahakyan A, Kekelidze N, Akhalbedashvili L (2005) Sorption behavior of Armenian zeolites. Fusion Sci Technol 48:230–233
102. Raj K, Prasad KK, Bansal NK (2006) Radioactive waste management practices in India. Nucl Eng Des 236:914–930
103. Osmanlioglu AE (2006) Treatment of radioactive liquid waste by sorption on natural zeolite in Turkey. J Hazard Mater B137:332–335
104. Marrocchelli A, Pietrelli L (2007) Cesium adsorption with zeolites from nuclear high salt content alkaline wastes. Solvent Extr Ion Exc 7:159–172
105. Yeritsyan HN, Sahakyan AA, Harutunyan VV, Nikoghosyn SK, Hakhverdyan EA, Grigoryan NE (2007) Natural zeolites and application in liquid waste treatment. In: Wiedemann H, Tsakanov V (eds) Brilliant light in life and material sciences. Springer, Dordrecht
106. Nishita H, Haug RM (1972) Influence of clinoptilolite on Sr90 and Cs137 uptakes by plants. Soil Sci 114(2):149–157
107. Hofstetter KJ, Hintz C (1986) Water cleanup systems. ACS Symp Ser 293:228–238
108. Robinson WL, Stone GR (1988) Bikini atoll rehabilitation committee summary report no. 6, (BARC, Berkeley, CA), Appendix A, A1–A48
109. Forberg S, Jones B, Westermark T (1989) Can zeolites decrease the uptake and accelerate the excretion of radiocaesium in ruminants. Sci Total Environ 79(1):37–41
110. Shenbar MA, Johanson KJ (1992) Influence of zeolite on availability of radiocaesium in soil to plants. Sci Total Environ 113(3):287–295
111. Chelishchev NF (1995) Use of natural zeolites at Chernobyl. In: Ming DW, Mumpton FA (eds) Natural zeolites'93 occurrence, properties, use. Int Comm Natural Zeolites, Brockport
112. Vanniman DT, Bish LD (1995) The importance of zeolite in the potential high-level radioactive waste repository at yucca mountain, Nevada. In: Ming DW, Mumpton FA (eds) Natural zeolite'93 occurrence, properties, use. Int Comm Natural Zeolites, Brockport

Chapter 10
Ion Exchange Materials and Environmental Remediation

Mu. Naushad and Zeid A. AL-Othman

Abstract The presence of fluoride, pesticides, radionuclides, organic wastes, and heavy toxic metal ions in the aquatic environment has been of great concern to engineers, environmentalists, and scientists because of their increased discharge, toxic nature, and adverse effects on receiving water. Several technologies such as thermal treatment, precipitation, extraction, adsorption, and ion exchange have been proposed for the removal of these pollutants. Among these techniques, ion exchange is one of the most common and effective treatment methods. Composite materials are one of the most important classes of engineered materials, as they offer several outstanding properties as compared to conventional materials for the separation of these hazardous materials. Composite materials have shown their versatility in the separation science. The present chapter showed the application of ion exchange materials for the environmental monitoring.

10.1 Introduction

Groundwater pollution has been an issue of concern for the environmentalists since the pollution could be hardly reformed. The amount of water on planet earth is estimated to be approximately 1,388 million billion cubic meters. The major part of this total amount, 1,348 million billion cubic meters (97.3%), is constituted by the salt water in the oceans. Only 37.5 billion cubic meter (2.7%) water occurs in the form of freshwater. Of this total freshwater, 28,200 thousand billion cubic meter (2.04%) is in the form of polar ice and glaciers, 8,450 thousand billion cubic meter (0.61%)

M. Naushad (✉) • Z.A. AL-Othman
Department of Chemistry, College of Science, King Saud University,
Building-5, Riyadh, Saudi Arabia
e-mail: shad123@gmail.com

Inamuddin and M. Luqman (eds.), *Ion Exchange Technology II: Applications*,
DOI 10.1007/978-94-007-4026-6_10,

as groundwater, and 127 thousand billion cubic meters in the form of lakes, rivers, etc. [1]. As we know, life began in water and life is nurtured with water. There are some organisms such as anaerobes, which can survive without oxygen. But no organism can survive for any length of time without water. It is a universal solvent, and it provides the ionic balance and nutrients, which support all forms of life [2]. But water is continuously polluted by many sources mainly fluoride, pesticides, radionuclides, organic wastes, and heavy toxic metal ions.

10.1.1 Pollution by Fluorides

Fluoride toxicity in the water is a major environmental problem in many regions of the world. There are mainly two sources of fluoride. It can appear naturally in water, or it can appear as a toxic waste. It is found in many common household products, namely, toothpaste, dietary supplements, vitamins, chrome-cleaning agents, insecticides, and rodenticides. When the concentration of fluoride exceeds (>1.5 mg/L) in drinking water, it becomes harmful to the human health [3]. A low dose of fluoride is deemed responsible for inhibiting dental caries, while a higher daily dose is linked to permanent tooth and skeletal fluorosis [4]. Fluoride has several mechanisms of toxicity. Ingested fluoride initially acts locally on the intestinal mucosa. It can form hydrofluoric acid in the stomach which leads to gastrointestinal irritation or corrosive effects. The gastrointestinal tract is the earliest and most commonly affected organ system. Once absorbed, fluoride binds calcium ions and may lead to hypocalcemia. Fluoride has direct cytotoxic effects and interferes with a number of enzyme systems. It disrupts oxidative phosphorylation, glycolysis, coagulation, and neurotransmission [5].

10.1.2 Pollution by Pesticides

The term "pesticide" is a composite term that includes all chemicals that are used to kill or control pests. In agriculture, this includes herbicides (weeds), insecticides (insects), fungicides (fungi), nematocides (nematodes), and rodenticides (vertebrate poisons). The uses of pesticides for the control of a wide variety of insectivorous and herbaceous pests diminish the quantity and quality of food products. The concept of pesticides is not new. Around 1,000 B.C.E., Homer referred to the use of sulfur to fumigate homes, and by 900 C.E., the Chinese were using arsenic to control garden pests. The use of pesticide has grown because not only our exploding population is supplied with food, but crops and food are grown for export to other countries. The United States has become the largest producer of food products in the world, partly owing to our use of modern chemicals (pesticides) to control the insects, weeds, and other organisms that attack food crops. We have learned that pesticides can potentially harm the environment and our own health. Water plays an important role here because it is one of the main ways that pesticides are transported from the areas where they are applied to other locations. Pesticide contamination of groundwater is a subject of national importance because groundwater is used for drinking water by about 50% of the nation's population. This especially concerns people living in the agricultural areas where pesticides are most often used, as about 95% of that population relies upon groundwater for drinking water. There is increasing anxiety about the importance of small residues of pesticides, often suspected of being carcinogens or disrupting endocrine activities, in drinking water and food. Over the last 50 years, many human illnesses and deaths have occurred as a result of exposure to pesticides.

Pesticide contaminates land and water when it escapes from production sites and storage tanks and runs off from fields, sprayed aerially and sprayed into water to kill algae. Pesticides can enter the human body through inhalation of aerosols, oral exposure by consuming food and water, and dermal exposure by direct contact of pesticides with skin. The effects of pesticides on human health are more harmful based on the toxicity of the chemical and the length and magnitude of exposure [6]. Farm workers and their families experience the greatest exposure to agricultural pesticides through direct contact with the chemicals. But every human contains a percentage of pesticides found in fat samples in their body. Children are most susceptible and sensitive to pesticides due to their small size and under development. The chemicals can bioaccumulate in the body over time. Exposure to pesticides can range from mild skin irritation to birth defects, tumors, genetic changes, blood and nerve disorders, endocrine disruption, and even coma or death. Some pesticides, including aldrin, chlordane, DDT, dieldrin, endrin, heptachlor, hexachlorobenzene, mirex, and toxaphene, are considered POPs.

10.1.3 Pollution by Radioactive Materials

Radioactive materials contaminate groundwater from natural and industrial processes. Radionuclides are radioactive chemicals that occur in either discharge of industrial wastes or erosion of natural rock or soil deposits containing radioactive minerals. The presence of high levels of radionuclides in groundwater can increase the risk of cancer. The small level of radiation is not any more harmful than sunlight, but just as excessive quantities of sunlight can be dangerous. The hazards to the people and the environment from radioactive contamination depend on the nature of the radioactive contaminant, level of contamination, and the extent of the spread of contamination. Even a very low level of radiation can be life threatening when in long exposure to it. High levels of contamination may pose major risks to people and the environment. People can be exposed to potentially lethal radiation levels, both externally and internally, from the spread of contamination following an accident involving large quantities of radioactive material. The biological effects of internally deposited radionuclides depend greatly on the activity and the biodistribution and removal rates of the radionuclide which in turn depend on its chemical form.

The biological effects may also depend on the chemical toxicity of the deposited material, independent of its radioactivity. Some radionuclides may target specific organs and have much lower removal rates. For instance, the thyroid gland takes up a large percentage of any iodine that enters the body. If large quantities of radioactive iodine are inhaled or ingested, the thyroid may be impaired or destroyed, while other tissues are affected to a lesser extent. Radioactive iodine is a common fission product; it was a major component of the radiation released from the Chernobyl disaster, leading to nine fatal cases of pediatric thyroid cancer and hypothyroidism. On the other hand, radioactive iodine is used in the diagnosis and treatment of many diseases of the thyroid precisely because of the thyroid's selective uptake of iodine. Radioactive contamination can enter the body through ingestion, inhalation, absorption, or injection. For this reason, it is important to use personal protective equipment when working with radioactive materials. Radioactive contamination may also be ingested as the result of eating contaminated plants and animals or drinking contaminated water or milk from exposed animals. The amount of injury caused by radioactive waste materials depends on its physical half-life and the rate, how quickly it is absorbed and then excreted by an organism. The most sensitive regions of the human body appear to be those which have many actively dividing cells, such as the skin, gonads, intestine, and tissues that grow blood cells (spleen, bone marrow, lymph organs). Radioactivity is toxic because it forms ions when it reacts with biological molecules. These ions can form free radicals, which damage proteins, membranes, and nucleic acids. Radioactivity can damage DNA (deoxyribonucleic acid) by destroying individual bases (particularly thymine), breaking single and double strands, cross-linking different DNA strands, and cross-linking DNA and proteins. Damage of DNA can lead to cancers, birth defects, and even death. Radioactive contamination is typically the result of a spill or accident during the production or use of radionuclides, an unstable nucleus which has excessive energy. Contamination may occur from radioactive gases, liquids, or particles. For example, if a radionuclide used in nuclear medicine is accidentally spilled, the material could be spread by people as they walk around. Radioactive contamination may also be an inevitable result of certain processes, such as the release of radioactive xenon in nuclear fuel reprocessing. Nuclear fallout is the distribution of radioactive contamination by a nuclear explosion. Radioactive waste comes from a number of sources. The majority of waste originates from the nuclear fuel cycle and nuclear weapons reprocessing. However, other sources include industrial and medical wastes as well as naturally occurring radioactive materials (NORM) that can be concentrated as a result of the processing or consumption of coal, oil and gas, and some minerals. Industrial source waste can contain alpha, beta, neutron, or gamma emitters. Gamma emitters are used in radiography, while neutron emitting sources are used in a range of applications, such as oil well logging. Naturally occurring radioactive materials (NORM) are available in the environment and vary with location. NORM may be found in oil, gas, and water production facilities and accumulated during the drilling and processing of crude oil, natural gas, and well waters inside pipes, around valves and inside tanks. Accumulated NORM may produce radioactive pipe scale and sludge that leave sites and equipment contaminated. Industrial waste like sludge and scales are samples of products that might contain NORM. Radium-226 and radium-228 are natural groundwater contaminants that usually occur in trace quantities. At high exposure levels, radium-226 and radium-228 can cause bone cancer in humans and are believed to cause stomach, lung, and other cancers as well. Uranium is a naturally occurring radioactive contaminant found in both groundwater and surface water. At high exposure levels, uranium is believed to cause bone cancer and other cancers in humans. EPA also believes that uranium can be toxic to the kidneys. Gross alpha emitters occur naturally as radioactive contaminants, but several come from man-made sources. They may occur in either groundwater or surface water. At high exposure levels, alpha emitters are believed to cause cancer in humans. Beta and photon emitters are primarily man-made radioactive contaminants associated with operating nuclear power plants, facilities that use radioactive material for research or manufacturing, or facilities that dispose of radioactive material. Some beta emitters occur naturally. Beta and photon emitters primarily occur in surface water. At high exposure levels, beta and photon emitters are believed to cause cancer in humans.

10.1.4 Pollution by Organic Waste

Organic waste is a major component of municipal solid waste. Most originates from household waste, but commercial, institutional, and industrial waste can also contain significant proportions of organic waste, e.g., market waste. Organic pollution occurs when an excess of organic matter, such as manure or sewage, enters the water. When organic matter increases in a pond, the number of decomposers will increase. These decomposers grow rapidly and use a great deal of oxygen during their growth. This leads to a depletion of oxygen as the decomposition process occurs. A lack of oxygen can kill aquatic organisms. As the aquatic organisms die, they are broken down by decomposers which lead to further depletion of the oxygen levels. With increase in the global population and the rising demand for food and other essentials, there has been a rise in the amount of waste being generated daily by each household. This waste is ultimately thrown into municipal waste collection centers from where it is collected by the area municipalities to be further thrown into the landfills and dumps. However, either due to resource crunch or inefficient infrastructure, not all of this waste gets collected and transported to the final dumpsites. If at this stage the management and disposal is improperly done, it can cause serious impacts on health and problems to the surrounding environment. Waste that is not properly managed, especially excreta and other liquid and solid waste from households and the community, is a serious health hazard and leads to the spread of infectious diseases. Unattended waste lying around attracts flies, rats, and other creatures that in turn spread disease. Normally, it is the wet waste that decomposes and releases a bad odor. This leads to unhygienic conditions and thereby to a rise in the health problems. In particular, organic domestic waste poses a serious threat, since they ferment, creating conditions favorable to the survival and growth of microbial pathogens.

Direct handling of solid waste can result in various types of infectious and chronic diseases, with the waste workers and the rag pickers being the most vulnerable. Exposure to hazardous waste can affect human health, children being more vulnerable to these pollutants. In fact, direct exposure can lead to diseases through chemical exposure as the release of chemical waste into the environment leads to chemical poisoning. Disposal of hospital and other medical waste requires special attention since this can create major health hazards. This waste generated from the hospitals, health-care centers, medical laboratories, and research centers such as discarded syringe needles, bandages, swabs, plasters, and other types of infectious waste is often disposed with the regular noninfectious waste. Certain chemicals if released untreated, e.g., cyanides, mercury, and polychlorinated biphenyls, are highly toxic and exposure can lead to disease or death. Some studies have detected excesses of cancer in residents exposed to hazardous waste. The unhygienic use and disposal of plastics and its effects on human health has become a matter of concern. Colored plastics are harmful as their pigment contains heavy metals that are highly toxic. Some of the harmful metals found in plastics are copper, lead, chromium, cobalt, selenium, and cadmium. In most industrialized countries, color plastics have been legally banned.

10.1.5 Pollution by Heavy Metals

The presence of heavy metals in the aquatic environment has also been of great concern because of their toxicity at lower concentrations. Some metal ions have cumulative effect capable of being assimilated and stored in the tissues of organism, causing significant adverse physiological effect. The water pollution caused by heavy metal ions from various industrial effluents has gained a great worldwide attention. Toxic metals are very often discharged by a number of industrial process which leads to the contamination of freshwater. Domestic and municipal wastewater have also contributed toward the contamination of freshwater by lead. Nonbiodegradability and persistence of these pollutants in the environment is responsible for health hazards [7, 8]. Toxic metals have an adverse effect on human health. Water pollution due to wastewater originating from industrial and mining wastewaters is one of the most significant environmental problems. Even at a very low level, heavy metal ions can cause serious health effects, including reduced growth and development, cancer, organ damage, nervous system damage, and, in extreme cases, death. The most common heavy metal toxins are Cr, Se, Al, As, Cd, Pb, and Hg. Once emitted from their sources, they have the property of accumulating in the environment for many years. They enter the human body through breathing, drinking, and skin absorption. They also can accumulate in the bodies of animals and humans before they even cause damage. The sources of heavy metal ions are diverse and specific to each element. Aluminum is mostly found in canned food, cookware, antacids, aluminum foils, and drinking water. Arsenic is encountered through air pollution and in drinking water, meat from commercial farms, and wood preservatives. Cadmium originates mainly from cigarette smoke, air pollution, batteries, fish, paint, and food grown in cadmium-laden soil. The major sources of lead are gasoline, house paints, lead pipes, mines, and batteries. "Silver" dental fillings constitute the most important source of mercury. Other sources of mercury are from gold mining and fish. In general, the increased use of coal favors exposure to major heavy metals. The heavy metal ions form complexes with proteins, in which carboxylic acid (–COOH), amine ($-NH_2$), and thiol (–SH) groups are involved. These modified biological molecules lose their ability to function properly and result in the malfunction or death of the cells. When metals bind to these groups, they inactivate important enzyme systems, or affect protein structure, which is linked to the catalytic properties of enzymes. This type of toxin may also cause the formation of radicals, dangerous chemicals that cause the oxidation of biological molecules. Some diseases associated with heavy metal ions are as follows:

- Chromium(III) is an essential nutrient for humans and shortages may cause heart conditions, disruptions of metabolisms, and diabetes. But the uptake of too much chromium(III) can cause health effects as well, for instance, skin rashes. Chromium(VI) is known to cause various health effects. When it is a compound in leather products, it can cause allergic reactions, such as skin rashes. Chromium(VI) can cause nose irritations and nosebleeds. Other health problems that are caused by chromium(VI) are lung cancer, respiratory problems, kidney and liver damage, and even death.
- Overexposure of selenium may produce accumulation of fluid in the lungs, garlic breath, bronchitis, pneumonitis, bronchial asthma, nausea, chills, fever, headache, sore throat, shortness of breath, conjunctivitis, vomiting, abdominal pain, diarrhea, and enlarged liver. Selenium is an eye and upper respiratory irritant and a sensitizer. Overexposure may result in red staining of the nails, teeth, and hair.
- Aluminum has been associated with Alzheimer's and Parkinson's disease, senility, and presenile dementia.
- Arsenic exposure can cause cancer, abdominal pain, and skin lesions.
- Cadmium is therefore a potential pollutant in the environment. When excessive amounts of cadmium are ingested, it replaces Zn at key enzymatic sites, causing metabolic disorders, kidney damage, renal dysfunction, anemia, hypertension, bone marrow disorders, cancer, and toxicity to aquatic biota [9].
- Lead and mercury may cause the development of autoimmunity, in which a person's immune system attacks its own cells. This can lead to joint diseases and ailments of the kidneys, circulatory system, and neurons. At higher doses, lead and mercury can cause irreversible brain damage.

Due to many difficulties that arise from nitrate, fluoride, pesticides, radionuclides, organic wastes, and heavy toxic metal ions, several technologies such as thermal treatment, precipitation, extraction, membrane adsorption, and ion exchange have been proposed for the removal of these pollutants. Among these techniques, ion exchange is one of the most common and effective treatment methods.

10.2 Ion Exchange Process and Its Applications for Environmental Monitoring

Ion exchange is basically the exchange of ions between two electrolytes or between an electrolyte solution and a complex. Ion exchangers are of three types:

1. Cation exchangers: Those which carry exchangeable cations
2. Anion exchangers: Those which carry exchangeable anions
3. Amphoteric ion exchangers: Those which are capable of exchanging both cation and anion

A typical ion exchange reaction may be represented as follows:

$$AX^- + B_{(aq)} \rightleftarrows BX^- + A_{(aq)} \tag{10.1}$$

where A and B are the exchangeable ions or counterions, respectively, and X is the structural unit of ion exchanger. Bar indicates the exchanger phase, and "aq" means that the species is in the aqueous phase.

A typical cation exchange reaction is as follows:

$$2NaX^- + CaCl_{2(aq)} \rightleftarrows CaX^-{}_2 + 2NaCl_{(aq)} \tag{10.2}$$

The reaction (10.2) occurs in softening of water by ion exchange.

An anion exchange reaction is as follows:

$$2X^-Cl + Na_2So_4(aq) \rightleftarrows X^-{}_2S0_4 + 2NaCl_{(aq)} \tag{10.3}$$

Depending upon the number of functional groups attached, ion exchangers may be monofunctional or bifunctional. The acidity and basicity of an exchanger depends upon the nature of its functional group. A strongly acidic ion exchange resin has $-SO_3$ as a functional group. On the other hand, a strongly basic ion exchange resin has $-N(CH_3)_3$ as a functional group.

Ion exchange is a reversible process. When an exchanger with counterion "A" is placed in a solution of another counterion "B," then the ion exchange process may be written as

$$\bar{A}^+ + B^+ \rightleftarrows B^+ + \bar{A}^+ \tag{10.4}$$

where the bar refers to the exchanger phase and the unbarred quantities refer to the solution phase.

Since the reaction reaches equilibrium, expression in terms of concentrations can be written as

$$K_A^B = \frac{[A^+][\bar{B}^+]}{[\bar{A}^+][B^+]} \tag{10.5}$$

The fundamental principle of an ion exchange process is based on the following facts:

1. Donan exclusion—The ability of the resin to exclude ions under most conditions but not undissociated substances
2. Equivalence of exchange
3. Difference in migration rates of absorbed substances down a column
4. Ionic mobility restricted to the counterions and the exchangeable ions only
5. Miscellaneous—Swelling, surface area, and other mechanical properties

Ion exchange is a process of nature occurring throughout the ages from even before the dawn of human civilization. The phenomenon of ion exchange is not of a recent origin. The earliest of the references were found in the Holy Bible, which says "Moses" succeeded in preparing drinking water from brackish water by an ion exchange method [10]. Aristotle [11] affirmed that the seawater loses part of its salt content when percolated through certain sands. In 1623, Francis Bacon and Hales described a method for removing salts by filtration and desalination from seawater. Lowritz purified sugar beet juice by passing it through charcoal in 1790. In the starting of nineteenth century, chemists were quite aware about ion exchange and were busy in new researches. Gazzari (1819) discovered that clay retained dissolved fertilizer particle. In 1826, Sprengel stated that the humus frees certain acids from soils. Fuchs (1833) pointed out that the lime frees potassium and sodium from some clay. By the middle of the nineteenth century, sufficient experimental observations and information had been collected, but the principle of ion exchange had not yet been discovered. Thompson (1845) and Way in (1850) laid the foundation of ion exchange by base exchange in soil. They observed that when soils are treated with ammonium salts, ammonium ions are taken up by the soil and an equivalent amount of calcium and magnesium ions are released. During 1850–1855, the agrochemist Way demonstrated the following mechanism to be one of the ion exchange methods involving the complex silicates present in the soil. As described by Way, the process observed by Thompson could be formulated as

$$Ca-Soil + (NH_4)_2SO_4 \leftrightarrow NH_4-Soil + CaSO_4 \tag{10.6}$$

The ion exchange process is reversible and aluminosilicates (zeolites) are responsible for the exchange in soil, established by Eichorn (1850). The first synthetic aluminum-based ion exchanger was prepared by Harms and Rumpler in 1903 to purify the beet syrup. According to Lamberg and Wiegner, the materials responsible for the phenomenon were mainly clays, zeolites, gluconites, and humic acids. Lamberg also obtained analcite ($Na_2O.Al_2O_3.4SiO_2.2H_2O$) by leaching the mineral leucite ($K_2O.Al_2O_3.4SiO_2$) with solution of sodium chloride and found that this transformation could be reversed by treating analcite with a solution of potassium chloride, thereby exchanging Na^+ again by K^+. The first application of synthetic zeolite for collection and separation of ammonia from urine was made by Follin and Bell [12]. Due to the limitations in the applications of neutral and synthetic silicates in various industrial applications and in an attempt to meet the demands of the industries, Adam and Holms [13] in 1935 laid the foundation of organic exchangers when they observed that the crushed phonograph records exhibit ion exchange properties. This led investigators to develop synthetic ion exchange resins. These resins were developed and improved by the former I.G. Farben Industries in Germany followed by the manufacturers in USA and UK, which proved very effective for separations, recoveries, the ionization catalysis, etc. An ion exchange may be organic or inorganic, depending upon the nature of the matrix of which it is made up. More interest in ion exchange materials took place in the twentieth century due to their use in the field of nuclear research. At that time, there was need of some new materials that were stable at high temperatures and in presence of intense radioactive radiations. Amphlett summarized the work on ion exchangers up to 1963 in the classical book "Inorganic Ion Exchangers." The later work up to 1970 has been condensed by Vasely and Pekarek [14]; Clearfield [15, 16], Alberti [17, 18] and Watton [19, 20] have also worked on different aspects of synthetic inorganic ion exchangers. Synthetic ion exchanger may be classified in the following categories:

- Synthetic zeolites
- Polybasic acid salts
- Hydrous oxides
- Metal ferrocyanides
- Heteropolyacids
- Various insoluble materials

Zeolites are microporous, aluminosilicate minerals commonly used as commercial adsorbents. The term zeolite was originally coined in 1756 by Swedish mineralogist Axel Fredrik Cronstedt, who observed that upon rapidly heating the material stilbite, it produced large amounts of steam from water that had been adsorbed by the material. Zeolites are widely used in industry for water purification, as catalysts, and in nuclear reprocessing. Their biggest use is in the production of laundry detergents. They are also used in medicine and agriculture. Zeolites have a porous structure that can accommodate a wide variety of cations, such as Na^+, K^+, Ca^{2+}, Mg^{2+}, and others. These positive ions are rather loosely held and can readily be exchanged for others in a contact solution. Some of the more common mineral zeolites are analcime, chabazite, clinoptilolite, heulandite, natrolite, phillipsite, and stilbite. An example mineral formula is $Na_2Al_2Si_3O_{10}{\bullet}2H_2O$, the formula for natrolite.

Natural zeolites form where volcanic rocks and ash layers react with alkaline groundwater. Zeolites also crystallize in postdepositional environments over periods ranging from thousands to millions of years in shallow marine basins. Naturally occurring zeolites are rarely pure and contaminated to varying degrees by other minerals, metals, quartz, or other zeolites. For this reason, naturally occurring zeolites are excluded from many important commercial applications where uniformity and purity are essential. Synthetic zeolites hold some key advantages over their natural analogs. The synthetics can be manufactured in a uniform and pure state. It is also possible to manufacture desirable zeolite structures which do not appear in nature. Zeolite A is a well-known example. Since the principal raw materials used to manufacture zeolites are silica and alumina, which are among the most abundant mineral components on earth, the potential to supply zeolites is virtually unlimited. Synthetic zeolites have some limitations too.

The main limitations of synthetic zeolites are:

- They have a relatively high cost compared to natural zeolites.
- They have a limited chemical stability at extreme pH ranges (either high or low).
- Their ion specificity is susceptible to interference from similar sized ions.

Synthetic zeolites are widely used as catalysts in the petrochemical industry, for instance, in fluid catalytic cracking and hydrocracking. Zeolites confine molecules in small spaces which causes changes in their structure and reactivity. The hydrogen form of zeolites is powerful solid-state acids and can facilitate a host of acid-catalyzed reactions, such as isomerisation, alkylation, and cracking. The specific activation modality of most zeolitic catalysts used in petrochemical applications involves quantum-chemical Lewis acid site reactions. Catalytic cracking uses a furnace and reactor. First, crude oil distillation fractions are heated in the furnace and passed to the reactor. In the reactor, the crude meets with a catalyst such as zeolite. It goes through this step three times, each time getting cooler. Finally, it reaches a step known as separator. The separator collects recycled hydrogen. Then, it goes through a fractionator and becomes the final item.

Acidic salts of multivalent metals form by mixing acidic oxides of metals belonging to IV, V, and VI groups of the periodic table. Acid salts of the quadrivalent metals are the most studied groups of this class. They are extremely insoluble. Their composition is nonstoichiometric and depends on the condition under which they are precipitated. The materials which have been so far synthesized includes the phosphates, arsenates, molybdates, tungstates, antimonates, silicates, vanadates, and tellurates of zirconium, titanium, thorium, tin, cerium, chromium, iron, niobium, tantalum, etc.

Hydrous oxides are metal oxides which exhibit amphoteric ion exchange behavior depending on the basicity of the central metal ions as well as the metal oxygen (M–O) and oxygen hydrogen (O–H) bonds of the hydrous oxides. The hydrous oxides of some metal ions have also been the well-established materials for ion exchange purpose. Mixed oxides can be prepared in which second cations of higher charge than the parent cation is introduced into the structure, the resulting net positive charge being balanced by the presence of anions other than oxides and hydroxides. Examples of such materials include $Zn(OH)_2$ in which Zn^{2+} is partly replaced by Al^{3+}, and $Al(OH)_3$ containing Si^{4+}, Ti^{4+}, or Zr^{4+}. The general formulas are $Zn_{n-1}Al_n(OH)_2X_n$ and $Al_{n-1}M_n(OH)_3X_{n-1}$, where M^{4+} is a tetravalent oxide and x is a monovalent anion. Quadrivalent metal oxides are also commonly used as inorganic ion exchangers such as SnO_2, SiO_2, ThO_2, and ZrO_2. Actually, these materials do not possess simple oxide formula as given above unless they are ignited at a high temperature. They are found to contain varying amount of water, which is not present as water of hydration since on heating, it is lost continuously over a range of temperature. Consequently, these oxides are usually described as hydrous oxides. Inoue et al. [21–23] and De et al. [24–26] have done important work on hydrous oxides. The parent acids of the compounds considered in this section are 12-heteropolyacids of the general formula $H_nXY_{12}O_{40}.nH_2O$, where X may be phosphorus, arsenic, silicon, germanium, and boron, and Y different elements such as molybdenum, tungsten, and vanadium. The basic physicochemical and ion exchange properties of these compounds have been widely described in many earlier papers and surveyed in the book by Amphlett [27]. The heteropolyacids exhibit high affinity to heavy alkali metals, thorium, and silver. The size of univalent ions of these elements is suitable for their retention in the crystal lattice of heteropolyacids. Ammonium molybdophosphate is very selective for large monovalent cations in acidic condition [28].

Insoluble materials have been prepared by precipitation from metal salt solution with Na_2S or H_2S. The ion exchange properties of insoluble sulfides (e.g., Ag_2S, SnS, CuS, PbS, FeS, NiS, As_2S_3, Sb_2S_3) have been investigated. The sulfides are selective toward cations forming insoluble sulfides. The exchange reaction occurs through metathetical reactions in which the metal of sulfide is displaced by appropriate ion from the solution. The organic ion exchange resins are beads of highly polymerized cross-linked organic materials containing acidic or basic ionogenic groups. The backbone is generally styrene; divinylbenzene is frequently used. To improve the ion exchange capacity and selectivity, some naturally occurring organic ion exchangers are modified. Cellulose-based cation exchangers may be modified by the introduction of phosphate, carbonic, or other acidic functional groups. The largest groups of ion exchangers available today are synthetic organic resins in a powdered or bead form. The main advantages of synthetic organic ion exchange resins are their high capacity, wide applicability, wide versatility, and low cost relative to some synthetic inorganic materials. But they also have some limitations. The main limitations are their limited radiation and thermal stabilities. Since organic ion exchangers were found to be unstable at high temperature and strong radiation, inorganic ion exchangers were taken as alternatives. However, the inorganic ion exchangers also have their own limitations, e.g., nonsuitable for column operation and low chemical stability. In order to overcome the above limitations of organic resins and inorganic ion exchangers, many investigators have introduced organic–inorganic composite ion exchangers consisting of inorganic materials and organic binding matrices [9, 29, 30]. In literature, various methods of preparing these hybrid materials have been reported [31–33]. Composite materials, formed by mixing organic polymers and inorganic particles, possess all the good properties of both the constituents and an enhanced utility. The combination of organic and inorganic precursors yields hybrid materials that have mechanical properties not present in the pure materials. The organic group can be reactive which implies that it is able to form an organic network as well as inorganic network. In designing composite materials, scientists and engineers have ingeniously combined various metals, ceramics, and polymers to produce a new generation of extraordinary materials that encompass a wide variety of applications. Most composites have been created to improve combination of mechanical characteristics such as stiffness, toughness, and high temperature strength. Thus, the synthesis of polymeric–inorganic composite has received a great deal of attention because it provided new material with special mechanical, chemical, electrochemical, optical, and magnetic properties [34, 35]. There has been considerable interest in the use of these materials to remove different ions by both surface adsorption and ion exchange. Their high uptake levels of pollutants can be accounted by their large surface area and high ion exchange capacities. These materials exhibit excellent water treatment properties. Many composite materials are composed of just two phases; one is termed the matrix, which is continuous and surrounds the other phase, often called the dispersed phase. In most cases, composites consist of a bulk material (the matrix) and a reinforcement of some kind. So in another way, today's most man-made composites can be divided in to four main groups depending upon the matrix material as (1) polymer matrix composites (PMC), (2) metal matrix composites (MMC), (3) ceramic matrix composites (CMC), and (4) carbon and graphite matrix composites.

The need for treatment of water is becoming more urgent for drinking purpose, even in developed countries. The increasing populations of developing countries are encroaching more on previously undisturbed watersheds. The daily demand of drinking water of a man is normally 7% of his body weight. Thus, it is vital for the healthy growth of a person. But the same water may become a hazard, a threat to the continuation of life if it gets polluted with harmful or toxic substances. Usually, the groundwater is considered as less polluted as compared to the surface water due to the reduced exposure to the external environment. But lack of sanitation and improper waste management have a potential to spoil the purity of the groundwater leading to increased pollution levels; hence, it has been reported that about 40% or even more disease outbreaks are attributed to be water borne in nature [36]. Safe drinking water has always been a concern to both those who consume it and to those who provide it. Ion exchange methods can be used for elimination of these pollutants. Ion exchangers have a varying degree of preference for each metal ion depending on exchange type, operating conditions, and the metal being removed. Usually, the only pretreatment required ahead of the ion exchange is pH adjustment [37]. A pH of about 4–7 is usually optimum because a lower pH will prevent the metal from loading onto the exchanger and higher pH could cause precipitation of the heavy metal hydroxide. Once the capacity of the ion exchange is exhausted, the exchanger is regenerated to remove the heavy metal ions and return the exchanger to its usable form. Regeneration process is accomplished by passing strong acids and some inorganic salt solutions. Argonne National Laboratory recently developed a new ion exchange treatment by which water contaminated with uranium, plutonium, americium, and other contaminants (e.g., lead, mercury, cadmium, and zinc) can be cleaned to meet regulated discharge and drinking water standards. Contaminated water is passed through a column of diphosphonic ion exchange resin (Diphonix) to remove the contaminants. The Diphonix resin is commercially available and can be regenerated for reuse.

Khan et al. have reported polyaniline Sn(IV) tungstoarsenate [9], polypyrrole Th(IV) phosphate, polyaniline Sn(IV) arsenophosphate [38], and polystyrene Zr(IV) tungstophosphate [39]. Ion exchange kinetics on styrene supported zirconium (IV) tungstophosphate was made by Khan et al. [40]. Beena Pandit et al. have synthesized such type of ion exchange materials, i.e., o-chlorophenol Zr(IV) tungstate and p-chlorophenol Zr(IV) tungstate [41]. Polyaniline Zr(IV) tungstophosphate has been synthesized by Gupta et al. [42] which was used for the selective separation of La^{3+} and UO_2^{2+}. Chanda et al. reported polyacrylic acid-coated SiO_2 as a new ion exchange material. Islam et al. [43] prepared polycinnamamide thorium (IV) phosphate, and it was used for the separation of Pb^{2+} metal ion. The adsorption of lead from aqueous solution by polycinnamamide thorium (IV) phosphate was found to occur readily. Adsorption of lead was found to follow first order kinetics. The effect of other cations was also studied and it was found that the cations reduced the lead adsorption in the order of calcium > magnesium > potassium > sodium. The percentage removal was found to be 81%, and with increase in pH, the percentage removal gradually increased, and almost 99% removal was achieved at pH 10. Polyacrylamide thorium (IV) phosphate, a fibrous ion exchanger, has also been synthesized by Islam et al. [44] from acrylamide, thorium nitrate, and phosphoric acid by coprecipitation method. It was used for the separation of toxic metal ions. An inorganic–organic adsorber consisting of an inorganic ion exchanger-modified polyacrylonitrile (PAN) binding matrix was prepared by Nilchi et al. [45]. It was used for separation of cesium and cobalt from radioactive wastes. Ion exchange adsorbers are widely used for radioisotope separation as well as for the removal of hazardous fission products from aqueous waste prior to discharge to the environment. Humic acid (HA) was immobilized onto amine-modified polyacrylamide/bentonite composite which was prepared by direct intercalation polymerization technique, and the product (HA-Am-PAA-B) was used as an adsorbent for the removal of copper(II) ions from aqueous solutions [46] and cationic dyes malachite green (MG), methylene blue (MB), and crystal violet (CV) from aqueous solutions [47]. A new adsorbent having carboxylate functional group at the chain end was synthesized by graft copolymerization of acrylamide onto banana stalk (*Musa paradisiaca*) using ferrous ammonium sulfate/H_2O_2 redox initiator system. The efficiency of the adsorbent in the removal of cobalt [Co(II)] from water was investigated using batch adsorption technique. The adsorbent exhibits very high adsorption potential for Co(II), and under optimum conditions, more than 99% removal was achieved [48]. Removal of uranium [U(VI)] from aqueous solutions with humic acid-immobilized zirconium-pillared clay (HA-Zr-PILC) was investigated using a batch adsorption technique by Anirudhan et al. [49]. Apart from their application in sorption processes for removal of environmental contaminants, the hybrid nanocomposite also helps to avoid interference of phosphate and natural organic matter (NOM), in a process that, in conjunction with another hybrid material, enables for cost-effective real-time sensing of trace toxic metals in natural waters using pH as a surrogate parameter [50]. Use of CIX in anaerobic bioreactors is particularly attractive for high-strength biological wastewater where unforeseen fluctuations in feed compositions are rather common [51]. Sen Gupta et al. studied a new energy-efficient hybrid ion exchange-nanofiltration (HIX-NF) process to desalinate brackish and seawater [52]. N-butylacetatezirconium(IV) iodate cation exchanger has been established to recover toxic metals from environmental wastes [53]. Parachlorophenol anchored tin antimonate, an inorgano–organic ion exchanger, has been used for the selective separation of heavy metals like Bi(III) and Cu(II) [54]. The adsorption behavior of thorium from aqueous solutions by a composite adsorbent has been investigated by Akyil et al. [55]. Polyacrylamide Al(III) phosphate, polyacrylamide Ce(IV) phosphate, and polyacrylamide Zr(IV) phosphate have been used for the removal of fluoride from drinking water [56]. A solid phase extractor, nanoscale diboron trioxide/titanium dioxide composite material was synthesized and used for separation and/or preconcentration of trace cadmium ion from various environmental samples [57]. Unsaturated polyester Ce(IV) phosphate was prepared by plastic waste bottles, and its application was performed for removal of malachite green dye from water samples [58]. A novel composite adsorbent, hydroxyapatite/magnetite (HAp/Fe_3O_4), has been studied for the purpose of removing lead ions from aqueous solution [59]. The polypyrrole/sawdust (PPy/SD) and polyaniline/sawdust (PAn/SD) were used as effective adsorbents for the removal of heavy metals, anions, color, and COD (chemical oxygen demand) from mill paper wastewater [60]. Recently, Nabi et. al. have also synthesized and characterized some new composite cation exchange materials, namely, poly-o-toluidine stannic molybdate [61], acetonitrile stannic(IV) selenite [62], acrylonitrile stannic(IV) tungstate [63], EDTA-Zr(IV) phosphate [64], EDTA-Zr(IV) iodate [65], and cellulose acetate Zr(IV) molybdophosphate [66]. The practical applicability of poly-o-toluidine stannic molybdate was demonstrated for the separation of Hg^{2+} from thermal power plant effluents and of Pb^{2+} from lead storage battery industry wastes. Acetonitrile stannic(IV) selenite was used for the quantitative binary separations of metal ions, namely, Pb^{2+}–Th^{4+}, Ni^{2+}–Th^{4+}, Ni^{2+}–Zn^{2+}, Cu^{2+}–Ce^{4+}, Al^{3+}–Bi^{3+}, and Al^{3+}–Zn^{2+}. Acrylonitrile stannic(IV) tungstate was conveniently utilized for the removal and isolation of heavy toxic metal ions released from wastewater stream. EDTA-Zr(IV) phosphate was found to be highly selective toward Th(IV), and its selectivity was examined by achieving some important binary separations such as Cd(II)–Th(IV), Ni(II)–Th(IV), Hg(II)–Th(IV), Zn(II)–Th(IV), Pb(II)–Th(IV), and Al(III)–Th(IV) by column means, indicating its utility in environmental pollution control in one way or other. The selectivity of EDTA-Zr(IV) iodate was assessed by examining its use for some important binary separations such as Mg^{2+}–Pb^{2+}, Cd^{2+}–Pb^{2+}, Hg^{2+}–Pb^{2+}, Zn^{2+}–Pb^{2+}, Cu^{2+}–Pb^{2+}, and Al^{3+}–Pb^{2+} on a packed column of this material. Cellulose acetate Zr(IV) molybdophosphate was offered a variety of technological opportunity for quantitative determination and separation of Cr(III) from a synthetic mixture of metal ions and Ca(II) from commercially available vitamin and mineral formulation, namely, Recovit, and organic–inorganic composite materials were used in the quantitative separation of $Fe^{3+}$and Zn^{2+} contents of a commercially available pharmaceutical sample, namely, Fefol-Z. Pb(II)- and Fe(III)-selective sensors and their application were also performed by Naushad in the analysis of rock sample, pharmaceutical sample, and water sample [67, 68]. Inamuddin et al. determined the ion exchange kinetic parameters for the poly-o-methoxyaniline Zr(IV) molybdate and polyaniline Ce(IV) molybdate composite cation exchangers [69, 70] and adsorption thermodynamics of trichloroacetic acid herbicide on polypyrrole Th(IV) phosphate composite cation exchanger [71]. More recently, Naushad et al. synthesized polyaniline Sn(IV) molybdate [72] and poly-o-toluidine Zr(IV) tungstate [73]. Polyaniline Sn(IV) molybdate was used for the quantitative separation of Pb^{2+} from a standard reference material (Rompin Hematite) of lead. Poly-o-toluidine Zr(IV) tungstate was found to be highly selective for Hg^{2+} metal ion. On the basis of selectivity studies, some important binary separations were also achieved to explore the potential application of this composite cation exchanger in environmental analysis of toxic heavy metals.

10.3 Conclusion

In fact, ion exchange materials are one of the most important classes of engineered materials, as they offer several outstanding properties as compared to conventional materials. These materials have found increasingly wider utilities in the area of separation science, chemical sensors, chromatography, fabrication of selective materials, and electrical and optical applications. So, we can say that these materials have enormous role for the environmental monitoring.

Acknowledgment The Authors extend their appreciation to the Deanship of Scientific Research at King Saud University for funding the work through the research group project No RGP-VPP-043.

References

1. Raj B (2001) Water quality and corrosivity of ground water of northwestern part of Bankura District, West Bengal. J Environ Pollut 8:329–332
2. C.G.W.M (1990) About groundwater protection. Wright State University, Dayton
3. Tsai WT (2009) Environmental hazards and health risk of common liquid perfluoro- n-alkanes, potent greenhouse gases. Environ Int 35:418–424
4. Mahramanlioglu M, Kizilcikli I, Bicer IO (2002) Adsorption of fluoride from aqueous solution by acid treated spent bleaching earth. J Fluorine Chem 115:41–47
5. Islam M, Mishra PC, Patel R (2011) Fluoride adsorption from aqueous solution by a hybrid thorium phosphate composite. Chem Eng J 166:978–985
6. Maroni M, Jarvisalo J, Ferla FL (1986) The WHO-UNDP epidemiological study on the health effects of exposure to organophosphorus pesticides. Toxic Lett 33:115–123
7. Bakkaloglu I, Butter TJ, Evison LM et al (1998) Screening of various types biomass for removal and recovery of heavy metals (Zn, Cu, Ni) by biosorption, sedimentation and desorption. Water Sci Technol 6(38):269–277
8. Yetis U, Ozcengiz G, Filiz B, Ergen N, Erbay A, Dolek A et al (1998) Heavy metal biosorption by white-rot fungi. Water Sci Technol 38:323–330
9. Khan AA, Alam MM (2003) Synthesis, characterization and analytical applications of a new and novel 'organic–inorganic' composite material as a cation exchanger and Cd(II) ion-selective membrane electrode: polyaniline Sn(IV) tungstoarsenate. React Funct Polym 55:277–290
10. The second book of Moses, Exodus, Chapter15, verse 25
11. Aristotle BA, works, 7 (about 330, B.C.), 933 b
12. Folin O, Bell R (1917) Applications of a new reagent for the separation of ammonia. I. The colorimetric determination of ammonia in urine. J Biol Chem 29:329–337
13. Adams BA, Holmes EL (1935) Synthetic ion exchange resins. J Soc Chem Ind 54:I–6
14. Vesely V, Pekarek V (1972) Synthetic inorganic ion-exchangers—I: hydrous oxides and acidic salts of multivalent metals. Talanta 19:219–262
15. Clearfield A, Styne JA (1964) The preparation of crystalline zirconium phosphate and some observations on its ion exchange behaviour. J Inorg Nucl Chem 26:117–129
16. Clearfield A, Laudis AM, Medina AS et al (1973) Crystalline zirconium phosphates. J Inorg Nucl Chem 35:1099–1108

17. Alberti G, Allulli S (1968) Ion Exchange in fused salts: III. Preparation, thermal stability and ion exchange properties of amorphous zirconium phosphate completely converted to the salt form. J Chrom 32:379–383
18. Alberti G, Costantino U (1970) Crystalline insoluble acid salts of tetravalent metals: X. Fibrous thorium phosphate, a new inorganic ion-exchange material suitable for making (support-free) inorganic sheets. J Chrom A 50:482–486
19. Walton HF (1970) Ion exchange. Anal Chem Ann Rev 42:86R
20. Walton HF (1974) Ion exchange. Anal Chem 46:398R
21. Inoue Y, Yamazaki H (1980) Studies of the hydrous titanium oxide ion exchanger. V. The stability of the hydrogen form in air. Bull Chem Soc Jpn 53(4):811–816
22. Inoue Y, Tsuji M (1976) Studies of hydrous titanium oxide ion exchanger. II. The equivalence of the exchange adsorption of cations and the dissociation constant. Bull Chem Soc Jpn 49:111–116
23. Inoue Y, Tsuji M (1978) Studies of the hydrous titanium oxide ion exchanger. IV. The effect of radiation and heal treatment. Bull Chem Soc Jpn 51(3):794–799
24. De AK, Chawdhury ND (1978) Synthetic inorganic-ion exchangers hydrous tungsten (VI) oxide-synthesis, physicochemical properties and ion-exchange separations. Chromatographia 11(10):586–592
25. De AK, Das SK (1978) Synthetic inorganic ion-exchangers hydrous cerium(IV) oxide-synthesis, properties, and ion-exchange separations. Sep Sci Technol 13(5):465
26. De AK, Das SK, Charaborty P (1981) Synthetic inorganic ion exchangers—XVIII hydrous Th(IV) oxide: synthesis, properties and some ion exchange separations. J Inorg Nucl Chem 43:1909–1913
27. Amphlett CB (1964) Inorganic ion exchangers. Elsevier, Amsterdam
28. Van J, Smit R (1965) Ion exchange on ammonium molybdophosphate—III: preparation and properties of coarse ammonium heteropolyacid salts. J Inorg Nucl Chem 27:229–232
29. Inamuddin, Khan SA, Siddiqui WA et al (2007) Synthesis, characterization and ion-exchange properties of a new and novel 'organic–inorganic' hybrid cation-exchanger: Nylon-6,6, Zr(IV) phosphate. Talanta 71:841–847
30. Khan AA, Alam MM (2004) New and novel organic–inorganic type crystalline 'polypyrrolel/ polyantimonic acid' composite system: preparation, characterization and analytical applications as a cation-exchange material and Hg(II) ion-selective membrane electrode. Anal Chim Acta 504:253–264
31. Chujo Y (1996) Organic-inorganic hybrid materials. Curr Opin Solid State Mater Sci 1:806–811
32. Sanchez C, Ribot F (1994) Design of hybrid organic-inorganic materials synthesized via sol-gel chemistry. New J Chem 18:1007–1047
33. Judeinstein P, Sanchez C (1996) Hybrid organic-inorganic materials: a land of multi-disciplinarity. J Mater Chem 6:511–525
34. Coetzee CJ, Qureshi M, Varshney KG (1991) Inorganic ion exchange chemical analysis. CRC Press, Boca Raton
35. Alexander PW, Dimitrakopoulos T, Hibbert DB (1997) Electroanalysis 9:813–817
36. Daniela C, Scagliarini M (2005) Modelling the effect of salinity on the multivariate distribution of a water quality index. J Math Stat 1(4):268–272
37. Kamel MM, Ibrahm MA, Ismael AM et al (2004) Adsorption of some heavy metal ions from aqueous solutions by using kaolinite clay. Ass Univ Bull Environ Res 7:101–110
38. Khan AA, Niwas R, Vershney KG (1999) Synthesis and ion exchange behaviour of polyaniline Sn(IV) arsenophosphate: a polymeric inorganic ion exchanger. Coll Sur A Phys Eng Asp 150:7–14
39. Khan AA, Niwas R, Vershney KG (1980) Preparation and properties of styrene supported zirconium(IV) tungstophosphate: a Hg(II) selective inorganic-organic ion exchangers. Indian J Chem 37A:464–472
40. Khan AA, Niwas R, Alam MM (2002) Ion exchange kinetics on styrene supported Zirconium (IV) tungstophosphate: an organic-inorganic type of cation exchanger. Ind J Chem Technol 9:256–260

41. Pandit B, Chudasma U (2001) Synthesis, characterization and application of an inorgano organic material: *p*-chlorophenol anchored onto zirconium tungstate. Bull Mater Sci 24:265–271
42. Saiqa Ikram (2000) Ph.D. Thesis, D.C.E., Delhi, p 84
43. Islam M, Patel R (2009) Removal of lead (II) from aqueous environment by a fibrous ion exchanger: polycinnamamide thorium (IV) phosphate. J Hazard Mater 172:707–715
44. Islam M, Patel R (2008) Polyacrylamide thorium (IV) phosphate as an important lead selective fibrous ion exchanger: synthesis, characterization and removal study. J Hazard Mater 156:509–520
45. Nilchi A, Atashi H, Javid AH et al (2007) Preparations of PAN-based adsorbers for separation of cesium and cobalt from radioactive wastes. App Radiat Isot 65:482–487
46. Anirudhan TS, Suchithra PS (2010) Humic acid-immobilized polymer/bentonite composite as an adsorbent for the removal of copper(II) ions from aqueous solutions and electroplating industry wastewater. J Ind Eng Chem 16:130–139
47. Anirudhan TS, Suchithra PS (2009) Adsorption characteristics of humic acid- immobilized amine modified polyacrylamide/bentonite composite for cationic dyes in aqueous solutions. J Environ Sci 21:884–891
48. Shibi IG, Anirudhan TS (2005) Adsorption of Co(II) by a carboxylate- functionalized polyacrylamide grafted lignocellulosics. Chemosphere 58:1117–1126
49. Anirudhan TS, Bringle CD, Rijith S (2010) Removal of uranium(VI) from aqueous solutions and nuclear industry effluents using humic acid-immobilized zirconium-pillared clay. J Environ Radioact 101:267–276
50. Sarkar S, Chatterjee PK, Cumbal LH et al (2011) Hybrid ion exchanger supported nanocomposites: sorption and sensing for environmental applications. Chem Eng J 166:923–931
51. Mitra IN, SenGupta AK, Kugelman IJ (1998) Water Sci Technol 38:369–376
52. Sarkar S, SenGupta AK (2008) A new hybrid ion exchange-nanofiltration (HIX- NF) separation process for energy-efficient desalination: process concept and laboratory evaluation. J Memb Sci 324:76–84
53. Nabi SA, Shahadat M, Bushra R et al (2010) Development of composite ion-exchange adsorbent for pollutants removal from environmental wastes. Chem Eng J 165:405–412
54. Chithra PG, Raveendran R, Beena B (2008) Parachlorophenol anchored tin antimonate -an inorgano-organic ion-exchanger selective towards heavy metals like Bi(III) and Cu(II). Desalination 232:20–25
55. Kaygun AK, Akyil S (2007) Study of the behaviour of thorium adsorption on PAN/zeolite composite adsorbent. J Hazard Mater 147:357–362
56. Sundaram CS, Meenakshi S (2009) Fluoride sorption using organic–inorganic hybrid type ion exchangers. J Coll Int Sci 333:58–62
57. Kalfa OM, Yalçınkaya O, Turker AR (2009) Synthesis of nano B_2O_3/TiO_2 composite material as a new solid phase extractor and its application to preconcentration and separation of cadmium. J Hazard Mater 166:455–461
58. Khan AA, Ahmad R, Khan A et al (2010) Preparation of unsaturated polyester Ce(IV) phosphate by plastic waste bottles and its application for removal of Malachite green dye from water samples. Arabian J Chem. doi:10.1016/j.arabjc.2010.10.012
59. Dong L, Zhu Z, Qiu Y et al (2010) Removal of lead from aqueous solution by hydroxyapatite/magnetite composite adsorbent. Chem Eng J 165:827–834
60. Ghorbani M, Esfandian H, Taghipour N et al (2010) Application of polyaniline and polypyrrole composites for paper mill wastewater treatment. Desalination 263:279–284
61. Nabi SA, Bushra R, Naushad M et al (2010) Synthesis, characterization and analytical applications of a new composite cation exchange material poly-o-toluidine stannic molybdate for the separation of toxic metal ions. Chem Eng J 165:529–536
62. Nabi SA, Bushra R, Al-Othman ZA et al (2011) Synthesis, characterization, and analytical applications of a new composite cation exchange material acetonitrile stannic(IV) selenite: adsorption behavior of toxic metal ions in nonionic surfactant Medium. Sep Sci Technol 46:847–857

63. Nabi SA, Naushad M, Bushra R (2009) Synthesis and characterization of a new organic–inorganic Pb^{2+} selective composite cation exchanger acrylonitrile stannic(IV) tungstate and its analytical applications. Chem Eng J 152:80–87
64. Nabi SA, Naushad M, Bushra R (2009) Synthesis, characterization and analytical applications of a new hybrid cation exchanger EDTA-Zirconium phosphate. Ads Sci Technol 27:423–434
65. Nabi SA, Ganie SA, Naushad M (2008) Synthesis, characterization and analytical applications of a new hybrid cation exchanger EDTA-Zirconium iodate. Ads Sci Technol 26:463–468
66. Nabi SA, Naushad M (2008) Synthesis, characterization and analytical applications of a new composite cation exchanger cellulose acetate-Zr(IV) molybdophosphate. Coll Surf A Phys Eng Asp 316:217–225
67. Naushad M (2008) PVC based membrane of Ti(IV) iodovanadate for Pb(II) determination. Sens Transduc J 95:86–96
68. Naushad M (2008) A new ion-selective electrode based on aluminium tungstate for Fe(III) determination in rock sample, pharmaceutical sample and water sample. Bull Mater Sci 31:957–965
69. Al-Othman ZA, Inamuddin, Naushad M (2011) Determination of ion-exchange kinetic parameters for the poly-*o*-methoxyaniline Zr(IV) molybdate composite cation-exchanger. Chem Eng J 166:639–645
70. Al-Othman ZA, Inamuddin, Naushad M (2011) Forward ($M^{2+}-H^{+}$) and reverse ($H^{+}-M^{2+}$) ion exchange kinetics of the heavy metals on polyaniline Ce(IV) molybdate: a simple practical approach for the determination of regeneration and separation capability of ion exchanger. Chem Eng J 171:456–463
71. Al-Othman ZA, Inamuddin, Naushad M (2011) Adsorption thermodynamics of trichloroacetic acid herbicide on polypyrrole Th(IV) phosphate composite cation-exchanger. Chem Eng J 169:38–42
72. Al-Othman ZA, Naushad M, Nilchi A (2011) Development, characterization and ion exchange thermodynamics for a new crystalline composite cation exchange material: application for the removal of Pb^{2+} ion from a standard sample (Rompin Hematite). J Inorg Organometal Poly Mater 21:547–559
73. Al-Othman ZA, Naushad M, Inamuddin (2011) Organic–inorganic type composite cation exchanger poly-o-toluidine Zr(IV) tungstate: preparation, physicochemical characterization and its analytical application in separation of heavy metals. Chem Eng J 172:369–375

Chapter 11
Metal Recovery, Separation and/or Pre-concentration

Cláudia Batista Lopes, Patrícia Ferreira Lito, Simão Pedro Cardoso, Eduarda Pereira, Armando Costa Duarte, and Carlos Manuel Silva

Abstract Metals are essential for the existence of life. Due to their chemical, physical, electrical and mechanical properties, they found a large number of applications, their use being intrinsically associated with the development of society. Besides their natural occurrence in the ecosystem, the application of metallic compounds in several industrial and agricultural sectors gives an inevitable rise to their release and dispersion into the environment. Accordingly, metallic ions must be separated from water and industrial water or effluents prior to final discharge whenever toxic or radioactive, or recovered in the case of valuable and precious species.

In this chapter, the ion exchange of metallic ions solutions with the aim to recover, separate or pre-concentrate them is reviewed and discussed. After a general introduction on the topic, the discussion is divided into several sections according to the nature of the ionic species: precious, radioactive, priority pollutants and other metals. The main parameters of the sorption process are generally analysed, such as the influence of pH, temperature and initial concentration on the equilibrium and kinetic performance, as well as sorbent regeneration details and modelling.

11.1 Introduction

Metals are a natural part of our environment; they are present in all of earth's compartments (air, water and soil), and life has evolved in this natural milieu in such a way that metals should be present in appropriate levels and combinations, since

C.B. Lopes • E. Pereira • A.C. Duarte
CESAM, University of Aveiro , Campus Universitário de Santiago,
Aveiro 3810-193, Portugal

P.F. Lito • S.P. Cardoso • C.M. Silva (✉)
CICECO/Department of Chemistry, University of Aveiro,
Campus Universitário de Santiago, 3810-193 Aveiro, Portugal
e-mail: carlos.manuel@ua.pt

Inamuddin and M. Luqman (eds.), *Ion Exchange Technology II: Applications*,
DOI 10.1007/978-94-007-4026-6_11, © Springer Science+Business Media B.V. 2012

whether they can be essential for a healthy life or extremely hazardous for all kinds of life depends on this.

There are a number of metals (Table 11.1), such as cobalt (Co), chromium (Cr), copper (Cu), iron (Fe), manganese (Mn), molybdenum (Mo), nickel (Ni), selenium (Se), tin (Sn), vanadium (V) and zinc (Zn), that are essential to life and are used as micronutrients by humans, plants and/or microorganisms [1]. For example, iron is mostly used as haemoglobin, which transports oxygen to the brain and muscles. Iron deficiency, or anaemia, occurs when our body does not have enough iron and causes one to become chronically tired. Cobalt is another metal that our bodies need. It is a component of vitamin B12 which humans need in their diet. However, when the concentration of the micronutrients in the environment becomes higher than is required by living organisms, they can become toxic and are considered to be contaminants [2]. Other metals, such as lead (Pb), mercury (Hg) and cadmium (Cd), are toxic substances that have no known metabolic or beneficial function on organisms [1], and their accumulation over time in the bodies of living organisms can cause serious damage. These metals become an environmental concern when their concentrations in the environmental compartments begin to affect human health and the ecosystems, and so it is desirable to measure and control their concentrations.

The ingestion of potentially toxic metals/metalloids may occur via respiration, food and drinking water. These potentially toxic metals may have no effects on an organism if they are excreted without any bioaccumulation or interference with biochemical and organ functions [3]. Conversely, the bioaccumulation of potentially toxic metals/metalloids has severe impacts on human health, which are often long term and manifest themselves in many ways, such as decrease of mental acuity, loss of motor control, critical organ dysfunction, cancer, chronic illness and concomitant suffering, incapacitation, and even death [3].

Due to their chemical, physical, electrical and mechanical properties, metals have a large number of applications, and their uses are intrinsically associated with the development of society as we know it, representing one of the principal applications of natural resources for the benefit of mankind. Metals are fundamental for the existence of life, and also for its progression through time.

Metals are good conductors, malleable, ductile and lustrous and mix easily, allowing the formation of alloys. Almost everything around us is made from metals. Some possess high structural strength per unit mass, making them useful materials for carrying large loads or resisting impact damage. Their strength and flexibility have led to their recurrent use in high-rise building and bridge construction, as well as in the manufacturing of vehicles, pipes, tools and railroad tracks, among others. Iron and aluminium (Al), which are the most abundant metals in the earth's crust, are also the two most commonly used structural metals. The good conductivity of metals makes them valuable in electrical appliances and for carrying an electric current over a distance with little energy lost. Copper is widely used on electricity distribution. Additionally, the thermal conductivity of metals is useful for containers to heat materials over a flame. Precious metals, such as gold (Au), silver (Ag), palladium

Table 11.1 Essential micronutrients for optimal functioning of biological processes and organs in humans and non-essential heavy metals (Adapted from Ref. [3] with kind permission of © Springer (2002))

Essential metal/metalloid micronutrients (a few mg or μg per day):	As, Co, Cr, Cu, Fe, Mn, Mo, Ni, Se, Sn,V, Zn
Non-essential heavy metals	Be, Cd, Hg, Pb, Sb, Ti
Macronutrients (100 mg or more per day):	Ca, Cl, Mg, P, K, Na, S

(Pd) and platinum (Pt) may have industrial applications (e.g. catalytic use) but are better known for their uses in art, jewellery and coinage. The high reflectivity of some metals is important in the construction of mirrors, including precision astronomical instruments. Some other metals have specialized uses, like the radioactive metals, such as uranium (U) and plutonium (Pu), which are used in nuclear power plants to produce energy via nuclear fission. Metals are also the key to making different coloured paints, photoreactive eye glasses, or to convert solar energy into electric energy, in photovoltaic solar cells, etc. Nowadays, with the booming of technology and the aerospace industry, metals with high conductivity and large strength to weight ratios are at top demand.

As result of so many applications in the industrial, construction and agricultural sectors, there is inevitably some loss and dispersion of metals back into the environment – the so-called anthropogenic sources. However, metal releases into the environment can also be of natural origin, and assessing the effects of inputs of metals into the environment as results of human activities is not a trivial task due to the very large natural inputs from natural igneous rocks, weathering and soils, sedimentary rocks, metamorphic rocks, erosion of ore-bearing rocks, wind-blown dust, volcanic activity, forest fires, geysers and thermal springs [3, 4]. Yet, excessive levels of metals can be introduced into the environment from anthropogenic activities. Alloys, batteries and electrochemical cells, biocides, ceramic and glass, chemicals, pharmaceuticals and dental, coatings, electrical equipment and apparatus, fertilizers, fossil fuel combustion, mining, smelting and metallurgy, nuclear reactor, paints and pigments, petroleum refining, pipes, sheets and machinery, plastics, pulp and paper, rubber, semi-conductors and super-conductors, tanning and textiles, wood preservative treatment, and solid waste disposal, both industrial and domestic, are some examples of uses of potentially toxic metals through which they may be introduced into the environment directly as a result of industrial processes, via effluent discharge or emissions, or indirectly through runoff or waste disposal [3].

The amount of metals/metalloids in the earth's ecosystems has been growing with our planet's increasing populations and their requirements for natural resources and metal-based goods. The continuous demand for metals has originated metal-rich wastewaters and several metal contamination problems, all over the world. Today, metal pollution has accomplished high visibility and is garnering public concern, since population is less and less tolerant of any risks to their health, safety and quality of life due to discharges from local factories or even distant sources [5]. Metals have become of utmost importance due to our ever-growing population and economy, their demand will continue as long as life as we know continues. The impacts that metal pollution has on public health and ecosystems have forced the regulatory entities to drive for higher environmental standards and reduction in metal releases. All over the world, industries are forced to reduce the metal contents in water and industrial water to an acceptable level [6].

Metals in water and in industrial water or effluents must be recovered prior to final discharge not only to protect the environment and the living beings from metal's toxic effects but also to recover metals with economical value. Several processes are reported in literature to remove and recover metals from solutions (e.g. ion exchange, precipitation, flotation, electrochemical techniques, reverse osmosis, solvent extraction and adsorption) [7–9].

Chemical precipitation is commonly employed for removal of most of the metals. Hydroxide (OH^-) and sulphide (S^{2-}) are the common precipitants, but in some cases, carbonate (CO_3^{2-}) has also been used [10]. Precipitation is accompanied by flocculation or coagulation, and one of the major disadvantage of this technique is the production of a large amount of sludge which contains metal ions [6], and in some cases, the effluent discharge limits are not achieved [11]. For example, an effluent that contains 0.1 $g\,dm^{-3}$ of Cu(II), Cd(II) or Hg(II) compounds originates 10, 9 and 5 times larger amounts of sediments [6].

Ion exchange is one of the most effective and common treatments used for removal of metals. The economic feasibility of this process improves, if it is also used for the recovery of valuable metals [10]. Moreover, by ion exchange, either all ions can be removed from solutions or separated, allowing the complete deionization or selective removal of ionic contamination or of a valuable metal [6]. A number of materials can be used as exchangers, but the most known and used include zeolites, weak and strong anionic and cationic resins, chelating resins and microbial and plant biomass [10]. However, further developments on new types of ion exchangers are in progress, and consequently, the range of application of the ion exchange technique has been remarkably extended. The ion exchange is highly pH-dependent, since the solution pH has a strong impact on the metal speciation and on the interaction between counter ions [10]. Generally, the metals bind better at higher pH, due to the less competition with protons for the sorption sites. The performance of the ion exchange process is determined by the operating and wastewater conditions (pH, temperature, other ionic species and chemical background, e.g. oxidants, particles, solvents and polymers) [10].

This chapter is designed to discuss the various aspects of metal recovery, separation and/or pre-concentration by ion exchange. This technology allows not only reducing metal contents in waters and wastewaters, dropping the risk of metal pollution and its consequences, but recovering and subsequently recycling of valuable metals as well. Ion exchange may also be used to pre-concentrate metals; in some cases, metal concentrations are too low that a previous pre-concentration step is a requisite, for instance in metal analysis.

11.2 Groups and Classification

The metal/metalloids considered in this chapter were subdivided into several classes: the precious (Au, Ag, Pt and Pd), the radioactive (U, Ra, Cs and Sr), the priority pollutants (Pb, Cd and Hg) and other metals (Cu, Zn, Ni, Cr, Co, Fe, Mn, Ca, Mg, Mo, Se, Sr, V, Zr, Ti and the metalloid As). This subdivision arises from the fact that many of the studies available in the literature have grouped, themselves, the metals by their properties, applications and/or toxicity, and so this global appreciation on metal recovery, separation and/or pre-concentration issue has followed, in part, the bibliographic approach available.

In the following, each group is described in detail. Nonetheless, more emphasis has been given to the last two groups owing to the extensive research and number of publications on the topic; especial attention is devoted to Pb, Cd and Hg, given their priority hazardous character.

11.3 Gold, Silver, Platinum and Palladium

Precious metals are broadly applied in several industrial processes owing to their attractive physical and chemical properties [12, 13]. For instance, they found applications as catalysts, in electronic devices, in jewellery and, more recently, in energetic and environmental fields as well as in medical instruments [12, 14–17].

Silver and gold are two of the most economically interesting metals [18]. Gold, in particular, is by far a specially required metal. Beyond its great demand as a safe investment, the relevance of gold has been increasing in industrial applications as well [19]. Silver is widely used as raw material in several industries. Its exceptional malleability, ductility, electrical and thermal conductivity, photosensitivity and antimicrobial properties make silver particularly suitable for photographic films, electrical contacts, batteries catalysts, mirrors and electroplating [18, 20–23].

Platinum and palladium are also widely consumed materials. These precious metals are commonly found not only in health care, equipment and jewelleries but also in diverse chemical processes [17, 24, 25]. For instance, they are used as effective catalysts for industries like ammonia oxidation, petroleum reforming and petrochemical, and pharmaceutical [26–29].

The high value and scarcity of precious metals, associated with their increasing industrial demand, have prompted their effective recovery from waste aqueous solutions, electrical and electronic wastes, and spent catalysts [26, 30–34]. For instance, substantial amounts of spent catalysts containing imbedded platinum and other precious metals are produced by industries, which are replaced periodically by fresh ones [34]. The recovery of precious metals from spent catalysts, which are considered environmentally hazardous waste, is therefore a quite viable alternative to their expensive production from natural supplies [24, 26, 35–37]. Furthermore, significant amounts of precious metals are discharged from industries into the environment, and due to their high toxicity, their removal becomes essential. In view of that, numerous research studies have been focused on the development of methods and materials for selective recovery and pre-concentration of precious metals.

The main technologies for recovery of precious metals from its solutions include ion exchange, solvent extraction, adsorption, electrodeposition, cementation, reductive exchange, precipitation, membrane processes and electrolytic recovery [12]. Nonetheless, precipitation, ion exchange and adsorption have been considered the most appropriate technologies available for metal removal [12, 13, 19, 32, 38–43]. However, most of these processes may be only successfully applied for the treatment of relatively high concentrated effluents.

Adsorption has been a subject of interest mainly because of the availability of new materials especially suitable for the treatment of large volumes of very dilute wastes [22, 32]. Activated carbon has been the most commonly used adsorbent for heavy metal removal, but recently, more environmental-friendly and cost-effective materials have been investigated. For instance, interest has been paid to several agricultural and other natural by-products, such as flax [44], banana and orange peels [45], tea waste [46] and bagasse fly ash [43, 47]. Furthermore, Qu et al. [19] used two kinds of cotton fibre/chitosan composite adsorbents to recover Au(III) ions from aqueous solution; cotton fibres were selected as supporting materials since cellulose is their major constituent, which is the most abundant biopolymer in nature and exhibits excellent hydrophilicity, biocompatibility and biodegradability properties. Both fibres were shown to have strong affinity for Au(III) in aqueous solution and exhibited 100% selectivity for the Au(III) in the presence of coexistent Ni(II), Cd(II), Zn(II), Co(II) and Mn(II) in binary mixtures. The effect of solution pH on the sorption capacity of both fibres was investigated in the range of 1–8. The adsorption capacities for Au(III) were found to increase with increasing pH from 1 to 3; below pH 2, a small amount of Au(III) was adsorbed onto both fibres because of the competition between Cl^- and $AuCl_4^-$ (which is the predominant complex of gold below pH 3) on the positively charged sites of both fibres. At pH higher than 3.0, the adsorption capacities decreased with increasing pH, therefore the optimum pH value around 3, probably due to electrostatic attraction between $R\text{-}NH_3^+$ and $AuCl_4^-$. The decrease of the Au(III) adsorption was ascribed to the hydrolysis reaction of $AuCl_4^-$ at higher solution pH and to the presence of hydrolyzed chloro-gold complexes such as $AuCl_3(OH)^-$, which together with the deprotonation of amine lead to a decrease in electrostatic attractions between the negatively charged Au(III) anions and the positively charged adsorption sites of both fibres.

Notwithstanding, ion exchange is considered the most economically viable alternative for the separation and purification of metal ions at trace concentrations [31]. Its high efficiency and easy phase separation makes ion exchange the most preferential method. Besides, the recovery of precious metals from their ores and wastes is frequently subjected to a leaching step prior to the final metal recovery, using, in general, cyanides, thiosulphate and chloride solutions as leaching agents [31, 34, 41, 48–50]. Precious metals form anionic complexes with those agents, which may significantly complicate their recovering from solutions [41, 50, 51]. In contrast to activated carbons, such complexes can show a high affinity for anion exchange resins, due to their anionic nature.

A large variety of well-established ion exchange materials has been available for the recovery of metal ions from aqueous solutions. Table 11.2 compiles the most frequently used materials for the removal of precious metals together with the corresponding experimental conditions (namely pH, temperature and initial solution concentration) and major results found in literature. The uptake of Au(III) ions from aqueous solution has been investigated using, for instance, the commercial strong acidic ion exchange resin Purolite C-100 [39]; the weak base anion exchange resins Amberlite IRA93, DiAnion CR20 and Lewatit MP 62 [50], Lewatit MP 64 [41] and Purolite A100 [39]; and the strong base anion exchange resins 717 [12], Amberlite IRA400 [50], Amberlite IRA410 [50, 52], Amberlite IRA416 [53], Amberlite IRA910 [50, 53], Dowex G51 [50, 52], Dowex G55G, Dowex 21K [50, 52], Purolite A500/2788 [50], Purolite A600U [50] and Purolite A400 [39].

Concerning Pt(IV) and Pd(II) ions, their recovery has been investigated using anion exchange resins such as the 717 anion exchange resin [12], AV-17-9 [24], AM-2B [24], AN-251 [24], Amberlite IRA416 [53], Amberlite IRA910 [53], Lewatit MP 600 WS [24], Lewatit MP 500 A [13], Purolite A100 [24], Duolite S37 [33], Duolite A6 [13], Duolite A7 [13], Duolite A30B [13], Varion AP [13], Amberlyst A21 and Amberlyst A29 [54], and Dowex 1X10 [55], and chelating resins like Amberlite IRC718, Duolite ES346 and Duolite C467 [43]. The obtained results have demonstrated the potential of the above resins to successfully uptake precious ions from their aqueous solutions. For instance, Liu et al. [12] obtained around 100% removal for Ag, Au and Pd after 30 min of contact time and around 80% for Pt, with a commercial 717 anion exchange resin at pH 2 HCl medium in batch experiments (Fig. 11.1). In fact, the sorption rates for Ag and Au were found to be particularly fast being only 5 min sufficient to remove nearly 100%, which reveal the exceptional affinity of 717 anion exchange resin for the recovery of these precious metals.

In general, the performance of the investigated resins has been established to be strongly dependent on the operating conditions, such as contact time, pH of the solution, initial concentration and presence of competing ions. Besides, as expected, the chemical and physical structure of the selected resin has been demonstrated to play an important role on the uptake efficiency. Kononova et al. [24] compared the performance of five commercial resins with different physical and chemical structures, namely AV-17-8, Lewatit MP 600 WS, Purolite A100, AM-2B and AN-251, for the recovery of platinum(II,IV) from chloride solutions. The obtained isotherms for an initial HCl concentration of 0.5 mol L^{-1} are shown in Fig. 11.2. It can be observed that the four investigated resins exhibit distinct sorption behaviour for platinum ion exchange. For instance, under the experimental conditions tested, the strong basic anion exchanger Lewatit MP 600 WS exhibited the highest ion exchange capacity (i.e. almost 0.50 mmol g^{-1}), whereas the maximum capacities obtained for

Table 11.2 Materials most frequently used for the removal of precious metals together with the corresponding experimental and major results found in literature

		Operation conditions			Results						
Metal	Material	pH	T (°C)	C_0 (mg/L)	Batch	Column	Kinetics	Isotherms	Regeneration	Others	Ref.
Ag	Dithiooxamide acid-formaldehyde resin and thiourea-formaldehyde resin	0–6	N.A.	Equil: 1000; column: 200	x	x	x	x	x	Capacity; separation; competitive ions	[58]
Ag	Clinoptilolite	2–8	23	50–300	x		x	x	x	Capacity; removal efficiency %; desorption efficiency %	[22]
Pt	Resins AV-17-9, Lewatit MP 600 WS, Purolite A100, AM-2B, AN-251	N.A.	20	0.25–1 mM	x		x		x	Recovery %; distribution coefficient; saturation degree	[24]
Ag	Clinoptilolite	2–6	20	10–500	x		x	x	x	Capacity	[65]
Ag	11A tobermorites	N.A.	25	3 mM	x		x	x		Capacity	[62]
Pd	Amberlyst A29 and Amberlyst A21	0–8	25	100	x	x	x	x		Capacity; weight and bed distribution; pre-concentration	[54]
Pd	Polyaniline (PANI)	1–12	R.T.	1×10^{-3}, 1	x	x	x			Preconcentration and recovery; competitive ions; distribution coefficient; removal %	[25]
Pt	Acrylic polymers	1–5	25	164–375	x		x	x		Removal %; capacity; thermodynamics	[31]
Ag, Au, Pd, Pt	Anion exchange resin 717	0–4	R.T.	0.40, 5.0	x	x				Adsorption rate %; competitive ions	[12]
Au	Cotton fiber/chitosan composites	1–8	5–35	4.08 mM	x		x	x		Selectivity; thermodynamics; competitive ions	[19]
Pt, Pd	Vinylbenzyl chloride-divinylbenzene resin	N.A	N.A.	Kin: 10–30; equil: 10–300	x		x	x	x	Capacity; desorption efficiency %; selectivity; competitive ions	[57]
Au, Pt, Pd	Glycine modified crosslinked chitosan resin	1–11	30	50–500	x		x	x	x	Capacity; removal %; desorption efficiencies	[17]
Pt	Duolite S37	N.A.	N.A.	143,1		x	x			Fraction extracted; distribution coefficients; capacity; removal %; weight distribution	

										coefficient; bed distribution coefficient	[33]
Pd	Amberlite IRC718, Duolite ES346 and Duolite C467	0–7	25	0.0011 M	x		x	x		Capacity; recovery factor	[43]
Au	Lewatit MP 64	N.A.	20, 40, 60	20, 40, 80, 160	x	x	x	x	x	Fractional attainment of equilibrium; thermodynamics; elution %	[41]
Ag, Au	Resins derived from bisthiourea	1–11 and 1–5	28	Ag: 1×10^{-2} M; Au: 5×10^{-3} M	x	x	x	x	x	Capacity; rate constant of intraparticle diffusion	[18]
Pt, Pd	Chitosan derivatives	2	20	Batch: 20–40; column: 50–200	x	x	x	x	x	Capacity; selectivity; competitive ions; desorption efficiency	[48]
Au, Ag, Pd, Pt	Carbon from flax shive	Au: 2; Ag: 5; Pt: 5, 6.5	25	N.A.	x		x	x	x	Rate of adsorption	[32]
Pd	Duolite A6, Duolite A7, Duolite A30B, Varion AP, and Lewatit MP 500 A	N.A.	25	100	x	x	x		x	Recovery factors; weight and bed distribution coefficients; capacity	[13]
Au	Dowex G51, Dowex 21K and Amberlite IRA410	11	23–25	20 ppm		x	x		x	Elution efficiency %; recovery %	[52]
Pt, Pd	Dowex 1X10	4	60	5×10^3–30×10^3		x	x			Recovery %; separation; interfering elements	[55]
Au	Anion exchanger resins	8, 9	23–25	10, 10.87 ppm	x		x	x		Competitive ions; elution %	[50]
Ag	Strong acidic, strong basic, weak basic resins	N.A.	23	100 ppm	x	x	x	x	x	Capacity	[39]
Au, Pd	Amberlite IRA416 and Amberlite IRA910	N.A.	22–60	Pd: 61.3, 86.8; Au: 84.3, 121.7	x	x	x	x	x	Separation; capacity; adsorption %; breakthrough capacities; elution percentages	[53]
Au, Pt, Pd	Polymeric resins	3–12	R.T.	10	x		x			Capacity	[49]
Pd, Au, Ir, Os, Pt, Ru, Rh	1,2,4,5-tetrazine-modified resin	N.A.	R.T.	10^{-3} M	x		x		x	Competitive ions; sorption %; desorption %; extraction %	[56]

N.A. – Not available
R.T. – Room temperature

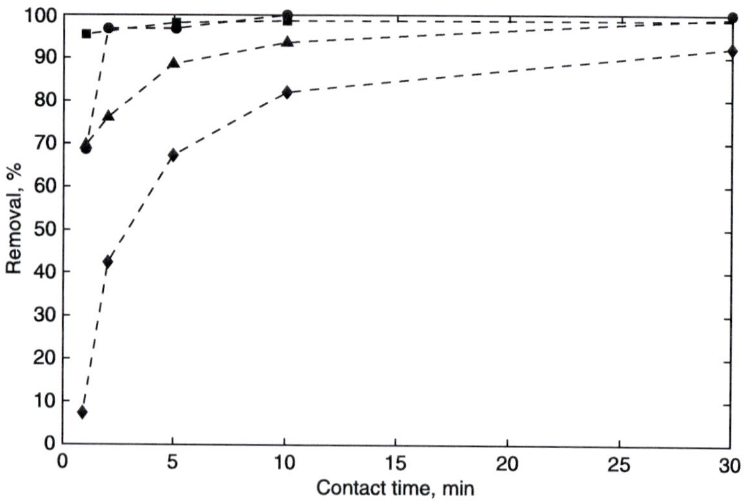

Fig. 11.1 Effect of contact time on metal removal with commercial 717 anion exchange resin in pH 2.0 HCl medium. Symbols: (♦) Pt; (■) Ag; (▲) Pd; (•) Au. *Curves* are *lines* to guide the eyes (Reproduced from Ref. [12] with kind permission of © Elsevier (2009))

AM-2B, Purolite A100 and AV-17-8 were quite similar (between 0.40 and 0.45 mmol g^{-1}). However, the shape of the isotherms of the last three exchangers is different. The weak basic resin AN-251 was shown to exhibit the lower exchange capacity for platinum(II,IV).

An effort has been made by researchers towards the development of new resin structures suitable to recover precious metals. Accordingly, a novel 1,2,4,5-tetrazine-modified ion exchange resin was successfully synthesized by Topp and Grote [56] through reaction of an aminomethylated macroporous polystyrene-DVB copolymer with 3-bromo-6-phenyl-1,2,4,5-tetrazine, and its anion exchange properties towards precious metal ions were successfully confirmed. Cortina et al. [49] developed new polymeric sorbents by impregnation of Amberlite XAD2, using Aliquat 336, Alamine 336, LIX79, Cyanex 923, Cyanex 47 and tributylphosphate as extractants, and demonstrated their effectiveness for the recovery of precious metals. Chassary et al. [48] prepared chitosan derivatives by glutaraldehyde cross-linking, by polyethyleneimine grafting, and by thiourea grafting, for the recovery of platinum group metals. Four resins were synthesized by Atia [18] through polymerization of bisthiourea with formaldehyde to be applied on the uptake of silver ions from processed photo films. Parodi et al. [57] synthesized a vinylbenzyl chloride-divinylbenzene resin by reaction with sodium salt of diethylmalonate followed by reaction with 3-aminopropyl imidazol to recover platinum and palladium from HCl solutions. Ramesh et al. [17] synthesized a novel glycine-modified cross-linked chitosan resin and investigated its ability

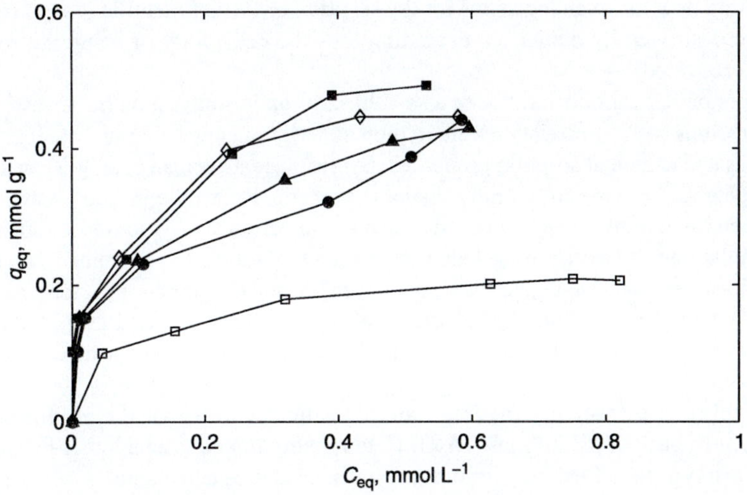

Fig. 11.2 Isotherms of platinum(II,VI) on several anion exchangers for C_o(HCl) = 0.5 mol L^{-1}. Symbols: (■) Lewatit MP 600 WS; (◊) AM-2B; (•) Purolite A100; (▲) AV-17-8; (□) AN-251. *Curves* are *lines* to guide the eyes (Reproduced from Ref. [24] with kind permission of © Elsevier (2010))

for removal of Au(III), Pt(IV) and Pd(II) from aqueous solution. Neagu et al. [31] synthesized two acrylic sorbents with distinct morphological structures bearing aminoethylenamine and thiol groups for platinum uptake from chloride solutions. Krishna et al. [25] investigated the applicability of polyaniline resin (PANI) for pre-concentration and recovery of palladium from aqueous solution. Recently, Çelik et al. [58] synthesized a new dithiooxamide acid-formaldehyde chelating resin and a thiourea-formaldehyde resin for the separation and concentration of silver ions.

Most ion exchange resins, however, can generate hazardous products in some processes (e.g. in pyrolysis and incineration), which must be eliminated in order to avoid secondary pollution sources [19, 59]. Accordingly, increasing attention has been focused on natural materials extracted for instance from agriculture wastes and seafood by-products, which have biodegradable character. Moreover, the substitution of petroleum derivatives by environmental-friendly resources is of great importance for a sustainable society [30, 60, 61]. Cox et al. [32] investigated the sorption of gold, silver, palladium and platinum chlorides onto a carbonaceous sorbent prepared from flax shive by treatment with hot sulphuric acid. Their results pointed out the ability of the prepared sorbent to effectively remove precious metals from dilute solution by ion exchange with functional groups on the sorbent. The sorption of the metal followed the order Au(III) > Pd(II) > Ag(I) > Pt(II) ≈ Pt(IV). This study revealed that after uptake, except for Pt(IV), the loaded ions were reduced by the sorbent to the metal with oxidation of the sorbent creating new ion exchange sites for the sorption of further metallic ions to replace those destroyed by oxidation, extending thus the possibility of using the sorbent to be recycled.

Inorganic materials have been also subject of major interest on the investigation of precious metals recovery owing to their superior chemical and mechanical resistances and excellent sorption properties. For instance, Coleman et al. [62] evaluated the potential of two structurally distinct synthetic 11 Å tobermorites, a bioactive layered calcium silicate hydrated ion exchanger, as carrier for Ag^+ ions in antimicrobial formulations, by performing batch experiments. During the exchange reactions, Ag^+ ions were found to replace labile interlayer Na^+ and Ca^{2+} ions in 11 Å tobermorites. The obtained equilibrium Ag^+ uptake levels exceed those reported for palygorskite and montmorillonite clays and were similar to those presented for diverse natural zeolites.

Zeolites are crystalline, hydrated aluminosilicates with open three-dimensional structures built of $[SiO_4]^{4-}$ and $[AlO_4]^{5-}$ tetrahedra linked to each other by sharing all the oxygens to form regular intracrystalline cavities and channels of molecular dimensions [63]. This type of materials gained a significant interest among scientists not only due to their valuable properties such as ion exchange capability but also due to their large availability in many countries, which provide to the local industries some promising benefits since they are able to treat wastewater contaminated with heavy metal at low cost [64]. As a result, zeolites have been the most frequently studied inorganic materials due to their superior mechanical, sorption and ion exchange properties [22, 63]. Clinoptilolite is the most abundant natural zeolite occurring in large sedimentary deposits with sufficiently high purity in many parts of the world. This material has been reported to be effective in the removal of precious metals from aqueous solution. For instance, Çoruh et al. [22] investigated the efficiency of two natural clinoptilolites collected from Çankırı-Çorum and Manisa-Gördes regions of Turkey to uptake silver ions from aqueous solutions. Results revealed the potential of both clinoptilolites as effective materials to recover Ag(I) ions from aqueous solutions and wastewaters. Their removal efficiency was found to strongly depend on the pH, which could improve significantly the sorption rate and capacity for Ag(I) ions. The capacity of these two materials was found to increase when the initial pH of the solution was raised from 2 to 4 and reached a maximum in the pH range of 6–8. The adsorption capacities of Çankırı-Çorum and Manisa-Gördes clinoptilolites were found to be 31.44 and 22.57 mg g^{-1}, respectively. Lihareva et al. [65] evaluated the sorption capacity of natural and preliminary sodium-exchanged clinoptilolite from a Bulgarian deposit in Eastern Rhodopes Mountain towards silver(I) uptake. The preliminary treatment of clinoptilolite with sodium chloride was shown to enhance its sorption ability in comparison with the natural material, especially at low Ag^+ concentrations. Maximum uptake values of 43% and 83.6% were obtained for the natural material and for its Na-form, respectively. The relation between the Ag^+ uptake and the amount of released metal ions (Ca^{2+}, K^+, Mg^{2+} and Na^+) revealed that silver removal occurred by ion exchange for both forms up to the Ag^+ concentration of 500 mg L^{-1}. For a concentration of 5,000 mg L^{-1}, both ion exchange and sorption mechanisms were detected. The adsorption capacity of both forms increased rapidly with increasing pH from 1.96 to 2.43, and after the initial pH = 3.24, it remained almost constant up to pH = 6.0.

11.4 Uranium, Radium, Cesium and Strontium

Radioactive materials have been extensively used in industrial and medical applications, military and nuclear power programmes and research activities [66]. Concerning medical applications, the use of these materials is of great importance in diagnostic procedures, therapies and research. The use and production of radionuclides inevitably results in a considerable amount of radioactive wastes. Additionally, large amounts of highly active waste have been stored in tanks for more than half a century, mainly in US and Russian sites where nuclear weapons have been produced [67]. Such radioactive waste needs to be carefully managed in order to protect both human health and environment safety, given its inherent radiological, biological, chemical and physical hazard potential [66].

Uranium is a natural radioactive element generally occurring in soils, rocks, black sand and seawater [68–70]. Natural uranium consists of a mixture of three isotopes, ^{234}U, ^{235}U and ^{238}U, which may be dissolved in contact with water, being frequently found in groundwater at trace concentrations [69, 71]. Uranium is one of the most important heavy metals given its strategic importance in nuclear power plants and in the production of nuclear weapons [72–75]. Moreover, uranium found applications in the electron microscope, glass and pottery glaze, copper mining industry, milling industry, etc. [76].

Uranium resulting from nuclear and industrial wastes may contaminate surface water and groundwater, which comprises a serious environmental problem given its heavy metal toxicity and radioactivity. Even though natural uranium has very low levels of radioactivity, exposure to enriched uranium certainly originate a radiological health hazard [68]. The uptake of uranium by human beings embraces risks to humanity, particularly at relatively high concentration [77, 78], causing severe toxicological effects and harmful diseases such as lung, pancreatic and liver cancer, kidney damage and eventually death [79]. In view of that, discharge of uranium from nuclear industries and the resulting surface and groundwater contamination must be globally monitored. According to the standards of World Health Organization, the U(VI) concentration in water should not exceed 0.05 mg L^{-1} [78, 80], which enforces the uranium removal from wastewater before it is discharged into the environment. Beyond the environmental concerns, its recovery, concentration and purification from geological samples and nuclear wastes fuels are extremely important for its application in nuclear and atomic energy industries [73].

Radium is a naturally occurring radioactive material and is found as a daughter nuclide of both uranium and thorium [81–83]. Ra^{2+} ions are present in the rocks subsurface being leached by highly mineralized formation waters or brines, contaminating ores, oil and gas, and drinking water. Moreover, the use of subsurface materials like coal or phosphates in several industries produces significant levels of Ra^{2+} on the surface.

Superficial Ra^{2+} ions could be inhaled and ingested with food, giving rise to internal exposure to its radioactive emissions and to those of generated daughter radionuclides [81–83]. Ra^{2+} has a similar metabolic behaviour as Ca^{2+} in the human body, following parallel accumulation pathways to the bones, which contributes considerably to the internal dose of radioactivity. Furthermore, its gaseous daughter nuclide radon is retained in the lungs, where it decays to Pb and Po. Such harmful effects incited an effective Ra^{2+} separation from other desirable non-toxic ions.

Cesium is an alkaline metal naturally present as ^{133}Cs in ores and in the soil, occurring mainly as an aluminosilicate of cesium [84]. Cesium finds widespread applications as, for instance, in photoelectric cells, hot-cathode arcs, optical instruments, in the production of glasses and ceramics, and more recently as catalysts [84, 85]. Concerning medical applications, a cesium chloride solution has been recently used in the development of a method for the isolation of plasmid DNA [86]. Besides, radioactive cesium isotopes can be applied in some cancer treatments. Furthermore, cesium is generally associated with the operation of fuel reprocessing plants, nuclear power industry and nuclear tests [84, 87, 88].

Strontium is naturally present in the environment in both the earth's crust and in seawater [89]. The minerals strontianite and celestite are the most economically important naturally occurring strontium compounds, being the first ones widely applied in industry as, for instance, in the manufacture of X-ray absorbing glass for cathode ray tubes, in electrolytic zinc process, or in electroceramics [89–92]. Additionally, strontium found application in optical materials, in the production of the red flame colour of pyrotechnic devices, such as fireworks and signal flares, as oxygen eliminator in electron tubes and in the production of glass for colour television tubes [93].

^{137}Cs, ^{134}Cs and ^{90}Sr are the main radioactive products present in high volumes of liquid waste [85, 87, 88, 94–97]. Such waste has been stored within stainless-steel lined underground tanks (e.g. 65 million gallons in Hanford reservation, USA), whose integrity is problematic due to corrosion (wastes are highly acidic or basic). Actually, a number of old storage tanks are known to be leaking into the surrounding environment, contaminating the soil and groundwater with radioisotopes [98, 99]. Furthermore, cesium and strontium have been accidentally released into the environment as a result of fallout from nuclear arms tests or accidents in nuclear plants [85, 96, 97].

Cesium and strontium are strongly sorbed by a number of soil particles, sediments and suspended particles, and owing to their high solubility, they can migrate through the groundwater. This reality has raised public health concerns owing to their relatively long half-life (of about 30 years), extremely high radiotoxicity and ionizing radiation [97, 100–102]. Besides, in the same way as radium, ^{90}Sr is

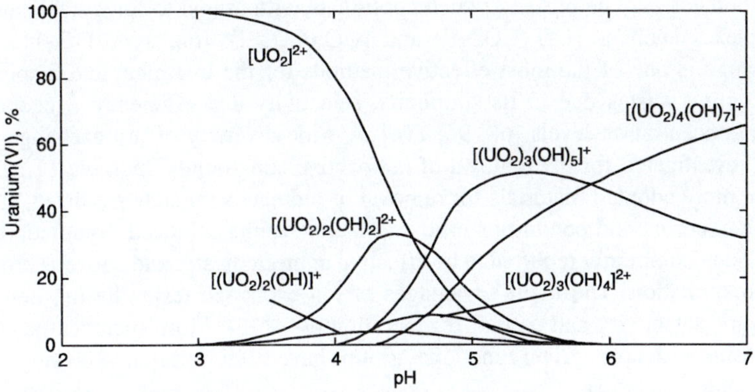

Fig. 11.3 Distribution of uranium species *versus* pH for 10^{-3} M uranium(VI) concentration (Reproduced from Ref. [109] with kind permission of © Elsevier (1995))

chemically similar to calcium and could be incorporated in human bones, increasing the probability of serious diseases such as leukaemia.

Accordingly, the removal of radioactive metals is a serious concern. An effort has been made to remove radionuclides from industrial wastewaters and radioactive wastes, and also to reduce their level on underground water. The most commonly used techniques for selective removal of radionuclides from aqueous streams include once more chemical precipitation, membrane separation, solvent extraction, adsorption and ion exchange [67, 69, 73, 80, 87, 96]. In the particular case of radioactive metals, most of these techniques exhibit disadvantages related with the production of toxic sludge, which requires further disposal [103].

Adsorption has been recognized as an important method for the purification of contaminated solutions and for the disposal of radioactive wastes [78, 80, 104–107]. A variety of organic and inorganic materials have been explored. Nonetheless, activated carbon is still the most adopted and effective adsorbent, given its large specific surface area, high thermal, chemical and mechanical stability, and strong resistance to high radiation. For instance, Mellah et al. [73] investigated the removal of uranium(VI) from aqueous solution onto commercial activated carbon under batch conditions, measuring an adsorption capacity of 28.30 mg g^{-1} at 293.15 K. However, uranyl ions may be present in different hydrolyzed forms, depending on the solution pH and concentration, which may affect its adsorption process. According to calculations of the uranium speciation (see Fig. 11.3), the predominant species present in aqueous solution in the pH range of 1–4.5 is the uranyl ions UO_2^{2+}, while at pH higher than 3, the hydrolysis products $[UO_2(OH)]^+$ and $[(UO_2)_2(OH)_2]^{2+}$ are readily formed [108, 109].

At pH value higher than 4.5, the uranyl hydrolysis becomes more intense and cationic species like $[(UO_2)_3(OH)_5]^+$ appear. Furthermore, in systems open to atmospheric CO_2 or in neutral or slightly alkaline groundwater, carbonates and bicarbonates may be present, which coordinate with uranyl to form stable anionic complexes such as $[UO_2(CO_3)_2]^{2-}$ and $[UO_2(CO_3)_3]^{4-}$ [68, 104, 110–113]. Ion exchange is one of the most effective methods for the treatment and disposal of radioactive wastes due to its simplicity, selectivity and efficiency, especially at trace concentration levels [68, 96, 114]. A wide diversity of ion exchangers has been investigated for the removal of radioactive compounds. In Table 11.3, some of the more adopted materials for removal of radioactive metals are listed, as well as the experimental conditions used and major results achieved. Synthetic resins have been commonly reported to be effective in groundwater and aqueous effluents decontamination. The main advantages of ion exchange resins lie on their high capacity, selectivity and ease of regeneration [74, 115]. More specifically, chelating resins with appropriate functional groups have been revealed to be very promising materials [115].

A number of commercially available resins have been used for radioactive ions removal. For instance, uranium(VI) uptake has been carried out using (a) the anion exchange resins Varion AP [110], PANSIL [110], Lewatit MP 62 [68], Eichrom Diphonix [116], Amberlite IRA910U [113], Amberlite IRA67 [68], Purolite A520E [111, 116], Purolite D3696 [116], Purolite A530E [111], Purolite A500 [111], Purolite A850 [111], Dowex 21K [116] and Dowex 1X8 [111, 116]; (b) the cationic resins Amberlite IRN77 [110] and Amberlite IR118H [117]; and (c) the chelating resins Bio-Rad Chelex 100 [116] and Dowex A [113]. Once more, the removal performance of those ion exchange resins has been recognized to depend on the operating conditions, such as contact time, pH, initial concentration and existence of competing ions, as well as on the chemical and physical structure of the exchanger. Riegel et al. [68] compared the capability of two weakly basic anion exchanger resins (Amberlite IRA67 and Lewatit MP 62) with respect to the uranium(VI) sorption from drinking water under distinct solution pH values. It may be concluded from their isotherm results (Fig. 11.4) that the behaviour of both exchangers towards uranium showed a strong pH dependence, which is very pronounced for the weaker basic exchanger Lewatit MP 62 owing the strong increase/decrease of the share of protonated amino groups in the pH range evaluated. Furthermore, the resin Amberlite IRA67 was demonstrated to be more effective for uranium uptake under those specific conditions. For both resins, the Langmuir isotherm was shown to fit equilibrium data accurately for all the pH values studied.

Chanda and Rempel [118] investigated the uranium uptake using a commercial weak base polyethyleneimine resin gel-coated on silica. The obtained resin revealed high capacity and fast kinetics. Besides, by using this gel-coated resin, uranium uptake from dilute aqueous solutions of UO_2SO_4 at pH 4–5 was found to be several times higher than that using a conventional bead-form resin Dowex IX8. The advantage of the gel-coating process was related to the increase of the accessibility of the inner sorption sites, enhancing their attainable capacity as well as the rate of attainment of equilibrium.

Moon et al. [119] investigated the pre-concentration of radium isotopes from natural waters using a new MnO_2 resin, which was evaluated over wide ranges of

Table 11.3 Materials most frequently used for the removal of radioactive metals together with the corresponding experimental and major results found in literature

		Operation conditions			Results						
Metal	Material	pH	T (°C)	C_0 (mg/L)	Batch	Column	Kinetics	Isotherms	Regeneration	Others	Ref.
U	Humic acid-immobilized zirconium-pillared clay	2–8	30	50–200	x		x	x	x	Removal %; capacity	[80]
U	Clinoptilolite	3–9	20, 25	0.01–500	x		x	x		Removal efficiency %; distribution coefficient	[108]
Cs	Clinoptilolite-rich tuffs	3–9	30, 40, 60	7.9–3946.5	x	x	x	x		Capacity; thermodynamics	[87]
U	Amazon kaolinite functionalized	1–11	R.T.	1.25×10^{-2}–3×10^{-1} mM	x			x		Capacities; thermodynamics; adsorbed amount %	[127]
U	Ethylenediamino tris (methylenephosphonic) acid graphed on polystyrene resin	1.5–6	25	0.25–4.5 mM	x		x	x	x	Capacity; extraction yield %; rate of adsorption; recovery %; thermodynamics	[250]
U	Tamarind fruit shell	2–9	30	10–100	x		x	x	x	Removal %	[78]
U	Sulphonated phenol-formaldehyde resin	0.739–3.250	25	0.05 M	x		x	x	x	Capacity; distribution coefficient	[104]
Cs	Clinoptilolite, chabazite, and natural and synthetic mordenite	3–9	25	0.001–0.1 N	x		x	x		Distribution coefficients; selectivity; removal %; fractional attainment of equilibrium	[132]
U	Natural sepiolite	2–9	25–45	100–500	x		x	x		Capacity; thermodynamics; removal %; distribution constant	[103]
Cs	Layered sulphide material	0.7–12	R.T.	1 and 55–550 ppm	x		x	x	x	Distribution coefficients; selectivity; competitive ions; removal %; capacity	[130]
U	Titanium-rich sand	1.5–2	4, 22	250	x		x			Capacity; distribution coefficient; thermodynamics	[69]

(continued)

Table 11.3 (continued)

		Operation conditions			Results						
Metal	Material	pH	T (°C)	C_0 (mg/L)	Batch	Column	Kinetics	Isotherms	Regeneration	Others	Ref.
Cs	Potassium copper hexa-cyanoferrate on porous activated carbon substrate	N.A.	R.T.	2×10^{-3} M	x		x			Capacity; competitive ions; distribution coefficients	[88]
U	Natural zeolite coated with manganese oxide	3.2–8.1	20	50–120		x	x		x	Column capacity; competitive ions	[105]
Cs, Sr	Zeolite A	2–8	25, 40, 60	50, 100, 150 and 100–1000	x	x	x	x		Removal %; column capacity; thermodynamics	[96]
U	Eichrom Diphonix, Bio-Rad Chelex 100, Purolite A520E, Purolite D3696, Dowex 21K, Dowex 1X8	5, 8	N.A.	5–25		x	x			Distribution coefficients; competitive ions	[116]
U	Lewatit MP 62 and Amberlite IRA67	IRA: 7–7.8; MP: 6.9–9.6	20	1–2	x	x	x	x		Mass transfer coefficient	[68]
U	Analogue of mesolite	1–9	30, 40, 50	0.001 N	x		x	x	x	Sorption %; thermodynamics	[66]
U	Manganese oxide coated zeolite	2–11	20, 35, 55	50–100	x		x	x		Capacity; thermodynamics; competitive ions	[138]
U	Amberlite IRA910U and Dowex A	2.7, 3.9	25	12	x	x	x	x		Capacity; competitive ions; distribution coefficient	[113]
U	Activated carbon	1–8	20–70	100–300	x		x	x		Capacity; thermodynamics; removal %; distribution constant	[73]
U	Purolite A530E, A500, A520E, and A850, Dowex 1X8, and WBR109	2, 8.5	N.A.	~0.06 mM	x	x		x	x	Capacity; sorption affinity; competitive ions; distribution coefficient	[111]
U	Cork biomass	5,2	22	1.9×10^{-5}–3.3×10^{-4} M	x		x	x		Capacity	[147]

^{226}Ra	Ionizable (Thia)calix[4] crowns	8–13	22–24	2.9×10^{-8} M	x				x	Selectivity coefficients; competitive ions; extraction percentage	[81]
U	Amberlite IRN77, and anionic Varion AP and PANSIL	Equil: 4–5; column: CER: 5.1–5.8; AER: 7.1–7.3; PAN: 5.5–6.5	23	Equil: 1–2000; column: 10	x	x	x	x		Capacity; conditional distribution coefficient; geochemical modelling	[110]
U	Amberlite IR118H	N.A.	20, 40, 60	1×10^{-2}–3×10^{-1} M	x		x	x		Thermodynamics; distribution constant	[117]
^{137}Cs	Hanford pristine sediment	N.A.	20–65	10^{-1}–10^{-9} M	x	x	x	x	x	Distribution coefficient; thermodynamics	[126]
Ra	MnO_2 resin	1–13	20	20000 dpm/ml	x	x	x			Preconcentration; distribution coefficient; recovery %	[119]
^{137}Cs	Nb-substituted titanosilicate	7–12	R.T.	340–350	x		x	x		Selectivity; distribution constant; uptake; recovery %	[146]
^{137}Cs	Clinoptilolite	N.A.	25, 40, 60, 80	10^{-6}–10^{-1} M	x	x	x	x		Distribution coefficient	[137]
^{85}Sr, ^{134}Cs, ^{57}Co	Sodium titanosilicate and clinoptilolite	1–13	R.T.	Tracer	x		x	x		Distribution coefficient; maximum processing capacity; competitive ions	[145]
Cs	Micaceous subsurface sediments from the Hanford site	5.5–6.9 and 7.5–9	30	10^{-9}–10^{-1} M	x		x	x	x	Capacity; competitive ions; conditional equilibrium constant	[97]
Ra	Bentonite	7–11	R.T.	6.0×10^{-7} M	x	x		x	x	Distribution coefficients; diffusion coefficients; capacity; desorption ratio %	[125]

(continued)

Table 11.3 (continued)

		Operation conditions			Results						
Metal	Material	pH	T (°C)	C_0 (mg/L)	Batch	Column	Kinetics	Isotherms	Regeneration	Others	Ref.
Cs	Zeolite NaY	7.5±0.5	R.T.	0.2–2 mM	x		x	x	x	Capacitiy	[142]
Cs, Sr	Phillipsite	N.A.	25	0.1 N	x			x		Selectivity; thermody namic; capacity	[136]
^{137}Cs, ^{89}Sr	Titanosilicate analogue of pharmacosiderite	1–13	R.T.	2×10^{-8}–3.7×10^{-7} M	x					Selectivity factor; distribution coefficients; capacity	[133]
UO_2^{2+}, U^{4+}	Zeolite X	1.5–3.5	R.T.	0.001–0.1 N	x		x			Capacity	[141]
Cs, Sr	Pharmacosiderite	N.A.	R.T.	5×10^{-3}–0.1 M	x		x	x		Distribution coefficient; competitive ions; selectivity	[94]
Ra, Ba	Zeolites 3A, 5A and Y	3	R.T.	1.5 kBq/cm^3	x					Percentage sorption; competitive ions; selectivity	[140]
U	Polyethylenemine resin	1–6	R.T.	0.1, 0.4 mM	x		x	x		Capacity	[118]

N.A. - Not available
R.T. - Room temperature

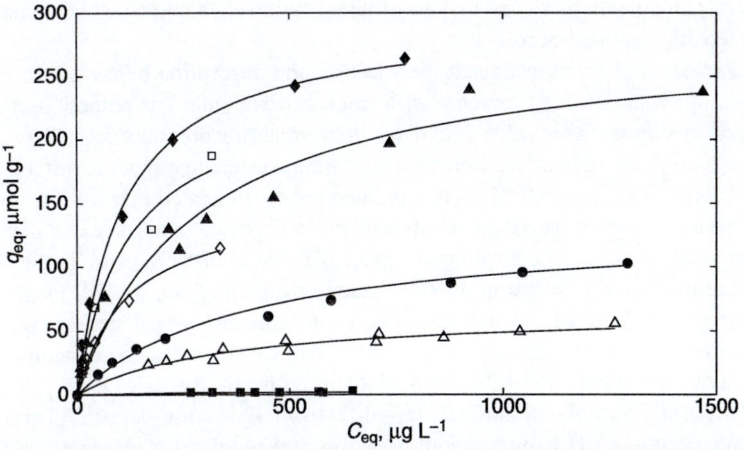

Fig. 11.4 Isotherm of uranium(VI) species onto Amberlite IRA67 and Lewatit MP 62; *lines*: Langmuir isotherms. Symbols: (♦) IRA 67, pH=7.3; (□) IRA 67, pH=7.8; (▲) MP 62, pH=6.9; (◊) MP 62, pH=7.1; (•) MP 62, pH=7.2; (Δ) MP 62, pH=7.6; (■) MP 62, pH=9.6 (Reproduced from Ref. [68] with kind permission of © Elsevier (2008))

pH, reaction times and salt concentrations. The sorption of ^{133}Ba was found to be highly dependent on pH, with most favourable range from 4 to 8, since the surface layers of the Mn oxides apparently become more positively charged under acidic conditions, which prevents diffusion of positively charged alkaline earth species into the exchanger; at higher pH, carbonate complexation was assumed to inhibit removal. Atun and Ortaboy [104] used a sulphonated phenol-formaldehyde resin to capture uranium(VI) from aqueous solution and estimated a sorption capacity of 0.29 mol g^{-1}. Kadous et al. [72] developed a new chelating polymeric sorbent using polystyrene resin grafted with ethylenediamino tris(methylenephosphonic) acid for uranium(VI) ions recovery and obtained an U(VI) sorption capacity of 41.76 mg g^{-1} under optimum conditions.

The use of anion exchange resins has been successfully implemented to recover uranium in mining industry and in water purification where the concentration of natural uranium is considerable. Moreover, strong base anionic resins are especially suitable to remove uranium from industrial effluents where anionic uranyl carbonate complexes are formed [68, 112, 120, 121]. However, organic exchangers may be easily decomposed by highly ionizing radiation. Besides, due to their weak acidic properties, these sorbents may be not effective in removing radionuclides from highly acidic medium.

Inorganic ion exchange materials are in general more advantageous than their organic counterparts in the treatment of nuclear waste effluents owing to their high radiation stability and remarkable selectivity in the presence of large excess of competing counter ions [67, 88, 96, 99, 122–124]. Besides, natural materials such as zeolites, metal hexacyanoferrates and several soil constituents, including clay minerals, are inexpensive and may be efficient cation exchangers, which make them an interesting subject of research.

Tachi et al. [125] investigated the sorption and desorption behaviour of radium on bentonite and purified smectite with batch experiments. The authors used crude sodium bentonite Kunigel V1 and purified smectite obtained from Kunipia F (from Kunimine Industries, Japan) with cation exchange capacities of 60.1 and 110.2 meq/100 g, respectively. The obtained results indicated that Ra sorption was dominated by ion exchange and evidenced the occurrence of surface complexation, which could increase Ra sorption at higher pH. Liu et al. [126] and Zachara et al. [97] studied the Cs^+ sorption on subsurface sediments from the US Department of Energy Hanford Site, which has been contaminated by accidental leakages of high-level nuclear waste containing $^{137}Cs^+$. The cation exchange capacity of the sediment ranged between 4.26 and 8.25×10^{-5} eq g^{-1} of dry soil. Donat [103] used a commercial sample of natural sepiolite from Eskisehir deposit (Turkey) to recover uranium(VI) from aqueous solution and obtained a maximum sorption capacity of 34.61 mg g^{-1}. Tykva et al. [69] reported the uptake of uranium(VI) from aqueous solution on a Ti-rich black sand, collected from the Egyptian black beach sand of the Abou-Khashaba area near Rosetta north of Egypt. Anirudhan et al. [80] studied the removal of U(VI) from aqueous solutions using humic acid-immobilized zirconium-pillared clay under batch conditions and attained a maximum removal of 97.6±2.1 and 94.7±3.3% for an initial concentration of 50 and 100 mg L^{-1}, respectively, at pH 6; the uranium sorption capacity was found to be 132.68±5.04 mg g^{-1}. Guerra et al. [127] applied a natural and a modified Amazon kaolinite clay samples in the U(VI) uptake from aqueous solution. The Amazon kaolinite functionalized with diethylenetriamine moieties was obtained by anchoring N-[3-(trimethoxysilyl)propyl] diethylenetriamine onto the Amazon kaolinite surface by heterogeneous route. The maximum sorption was determined to be 8.37×10^{-3} and 13.87×10^{-3} mmol g^{-1} for the natural and modified samples, respectively, at 298 K.

Metal hexacyanoferrates have been demonstrated to be strongly selective for cesium over other radionuclides and alkali cations [96, 128, 129]. However, the particles prepared are frequently small or may even be colloidal, being the powder-like resultants frequently unsuitable for practical application. Accordingly, in order to improve their mechanical properties, several investigators have reported the use of a support material. Recently, Wang et al. [88] synthesized and characterized a potassium copper hexacyanoferrate on porous activated carbon substrate for cesium removal. The maximum sorption capacity obtained was about 0.46 mmol g^{-1}, and it was not significantly reduced in the presence of a large amount of competing cations. Besides, the activated carbon was found to have no significant effect on selectivity.

Manos and Kanatzidis [130] investigated the efficiency of a layered metal sulphide (KMS-1) for Cs^+ uptake. The KMS-1 exhibited high exchange capacity for Cs^+ under a very wide pH range (0.8–12). For neutral pH, the capacity obtained was 226(4) mg g^{-1}. Besides, very fast sorption kinetics of Cs^+ was observed, showing a removal capacity superior to 90% for relatively low concentrations (ca. 1 ppm of

Cs^+) within only 5 min, using 10 mg of solid. Actually, this result diverges from those reported for related layered manganese oxides, being efficient only under acidic conditions. Such distinction may be ascribed to the higher mobility of K^+ ions in KMS-1 in comparison to that of interlayer cations in the manganese oxides. Furthermore, KMS-1 was also found to be effective for cesium uptake from solutions containing large excess of competitive cations.

Zeolites have been gaining increasing interest as decontamination agents for radioactive waste solutions owing to their well-known high selectivity and ion exchange capacity, even for trace concentrations, extremely high radiation and thermal resistance, and possibility of regeneration and radionuclides recovery [63, 66, 67, 105, 131–133]. Furthermore, zeolite structures can be modified by the inclusion of novel functional groups in order to improve their selectivity for specific metallic ions [132, 134, 135]. Natural zeolites, like clinoptilolite and mordenite, have been widely studied. For example, Borai et al. [132] investigated the ability of four zeolite minerals including natural clinoptilolite, natural chabazite, natural mordenite and synthetic mordenite to remove cesium from low-level radioactive liquid waste. The natural clinoptilolite and mordenite were from Japan (Akita prefecture and Miyagi prefecture, respectively), synthetic mordenite was from Norton Company (USA) and chabazite sample was AW-500 (Linde corporation). The natural chabazite was found to have the highest distribution coefficients and capacity for cesium, being the obtained apparent capacities 1.27, 2.07, 1.93 and 1.67 mmol g^{-1} for natural clinoptilolite, chabazite and mordenite, and synthetic mordenite, respectively. Furthermore, a novel impregnated zeolite material was prepared by van Leeuwen et al. [83] by loading Calix[4]arene-bis-2,3-naphtho-crown-6 onto synthetic mordenite in order to combine the high removal uptake of mordenite with the high Cs selectivity of Calix[4]arene. The obtained results pointed out that the impregnation process leads to an improvement of the distribution coefficients of Cs^+ ion (from 0.52 to 27.63 L g^{-1}). Adabbo et al. [136] compared the ion exchange selectivity of sedimentary (from northern outskirts of Naples, Italy), hydrothermal and synthetic phillipsites for Cs and Sr as a function of framework composition. The sedimentary and synthetic phillipsite samples exhibited good selectivity for Cs and moderate selectivity for Sr, whereas the hydrothermal phillipsite showed a lower selectivity for Cs and higher selectivity for Sr. Such results were attributed to the field strength of the anionic zeolite framework and to the ion charge density. The experimental cation exchange capacities for sedimentary, hydrothermal and synthetic samples were 3.55, 4.71 and 3.60 meq g^{-1}, respectively. This work revealed the potential of phillipsite for simultaneous removal of Cs and Sr from wastewaters. Abusafa and Yucel [137] tested different cationic forms (Na^+, K^+, NH_4^+ and Ca^{2+}) of a natural clinoptilolite from a sedimentary deposit in Bigadiç (Turkey) on the removal of ^{137}Cs from aqueous solutions. The equilibrium exchange capacities were found to range between 0.79 and 1.52 meq g^{-1}. Han et al. [138] and Zou et al. [105] performed experiments of uranium(VI) removal from aqueous solution using a natural zeolite from Xinyang (China) coated with manganese oxide, under batch and column conditions, respectively. The capacity of the coated zeolite from batch experiments was shown to strongly depend on the initial pH and

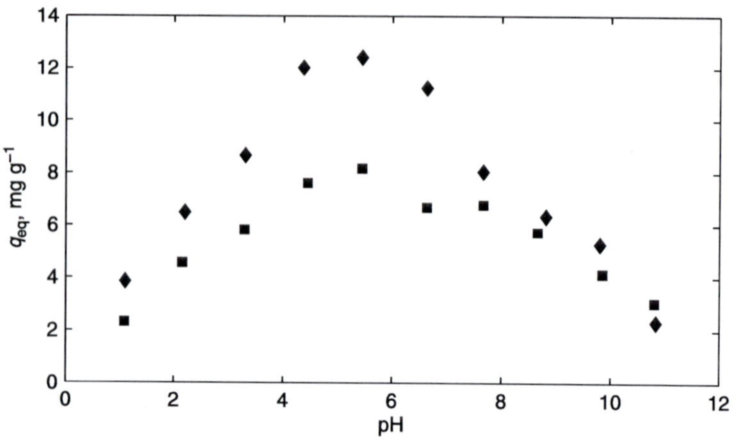

Fig. 11.5 Effect of the initial pH upon uranium(VI) sorption by the coated zeolite in 0.01 mol L^{-1} $NaNO_3$ solution with log P_{CO_2} = −3.5. Symbols: initial uranium(VI) concentration: (♦) 100 mg L^{-1}; (■) 50 mg L^{-1} (Reproduced from Ref. [138] with kind permission of © Elsevier (2007))

U(VI) concentration and temperature. The solution pH was found to influence both the solubility and speciation of metal ions on solution, as discussed previously (Fig. 11.3), and the overall load of the exchanger. Figure 11.5 illustrates the uranium(VI) uptake for both initial concentrations of 50 and 100 mg L^{-1} for a solid/solution ratio of 5.0 g L^{-1}, in 0.01 mol L^{-1} $NaNO_3$ with $\log P_{CO_2} = -3.5$ was low under acidic conditions but increased as pH raised from 2 to 6; at pH 6, the uranium uptake reached a maximum followed by a monotonous decline until pH 11. Such behaviour was attributed to the increase of dissolved carbonate as pH raises, which then increases the concentration of soluble uranium(VI)-carbonate complexes. Besides, both studies revealed the effectiveness of the coated zeolite for uranium(VI) removal from aqueous solutions and showed its better ability in comparison to the original new material.

Camacho et al. [108] studied the sorption of uranium(VI) from groundwater under batch experiments on a natural clinoptilolite zeolite from Southeast of Bitter Creek in Sweetwater County. It has been shown that clinoptilolite can neutralize both acidic and low basic water solutions through its alkalinity and ion exchange with U(VI). A uranium removal efficiency of 95.6% and maximum capacity of 0.7 mg g^{-1} were measured at pH 6.0. Cortés-Martínez et al. [87] compared the sorption behaviour of cesium using two Mexican clinoptilolite-rich tuffs (from Oaxaca City and Parral Chihuahua regions) considering both batch and continuous systems. Both zeolitic materials have good properties for the sorption of cesium from aqueous solutions, when ion exchange has been the mechanism responsible in both cases. The capacities at 30°C were 104.32 mg g^{-1} and 140.6 mg g^{-1} for the zeolitic materials from Oaxaca and Chihuahua, respectively.

The natural zeolites generally contain impurities, which can seriously compromise their performance [132, 139]. Besides, effective materials for highly basic or acidic waste solutions containing high salt concentrations and complexion agents are largely required, which thus motivates the development of synthetic inorganic exchangers [94, 123, 124, 133]. For instance, samples of zeolites 3A, 5A and Y were used by Jurado-Vargas et al. [140] to analyse the influence of other cations during the sorption of radium, barium and mixtures of both elements. Their results allowed them to suggest the predominance of the ion exchange mechanism for the uptake of both metals in all the materials studied. The carrier-free ^{226}Ra was established to be completely retained in the three zeolites, regardless their structure and composition, while the presence of barium in solution influenced radium exchange, independently of the counter ions initially present in each structure. A selectivity of zeolite Y towards barium was observed in the presence of radium. Besides, the competitive sorption of H^+ by the zeolites was found to be also present in all the studied materials. Olguín et al. [141] evaluated the capacity of zeolite X to uptake UO_2^{2+} and U^{4+} from aqueous solutions and found that U^{4+} was more easily retained than UO_2^{2+}, being the maximum equilibrium uptake values 1.86 and 2.86 meq g^{-1}, for UO_2^{2+} and U^{4+} solutions, respectively. Gu et al. [142] investigated the effect of amorphization on the Cs ion exchange and retention capacity of zeolite NaY. With that purpose, crystalline zeolite NaY and Cs exchanged zeolite NaY were transformed into the amorphous state by thermal treatment. The ion exchange capacity of the amorphized zeolite NaY was observed to be significantly reduced as compared with that of the original crystalline zeolite, which was ascribed to the loss of exchangeable cations and/or the closure of access to exchangeable sites. Besides, the Cs-exchanged zeolite NaY presented a slightly higher thermal stability than the unexchanged zeolite NaY. El-Kamash et al. [143] evaluated the possibility of solidifying exhausted synthetic zeolite A, loaded with ^{137}Cs and/or ^{90}Sr radionuclides, in ordinary (Portland) cement, given the low cost, high density, durability and simple handling of Portland cements. The obtained results confirmed that the presence of zeolite A in the cemented wastes has enhanced their mechanical characteristics, such as mechanical strength and setting times, towards the safety requirements and reduce considerably the radionuclide leach rates. Sharma and Tomar [66] synthesized and investigated the sorption behaviour of an analogue of mesolite for the removal of uranium(VI), thorium(IV) and europium(III) from aqueous waste. The measured sorption for U(VI) and Th(IV) was smaller than that for Eu(III), being the maximum uptake capacities for these metals registered at pH 5. The strong pH dependency and increment with temperature observed suggested adsorption as well as ion exchange mechanisms for U(VI), Th(IV) and Eu(III) uptake. El-Kamash [96] synthesized a zeolite A in order to assess its ability for the removal of cesium and strontium ions from aqueous solutions, adopting batch and fixed bed column operations and investigated the effect of various process parameters like pH, initial ion concentration, temperature, and bed depth and flow rate. The results confirmed the effectiveness of the synthetic zeolite A for cesium and strontium ions removal. The total metal ion uptake and the overall bed capacity were shown to decrease with increasing flow rate and to increase with increasing initial ion concentrations and bed depth. The assessed experimental capacities ranged between 2.51–2.54 and 4.90–5.51 mmol g^{-1} for Cs^+ and Sr^{2+}, respectively.

Notwithstanding their high selectivity and resistance to radiation damage, zeolites can in general only be applied in a limited pH range due to the aluminium and silicon dissolution in extreme acid and alkaline media and may exhibit low selectivity in the presence of high concentration of competitive ions [67, 98, 144]. Such stringent requirements have driven the design and development of new synthetic inorganic materials for the selective removal of radionuclide elements from nuclear solutions. Crystalline titanosilicates with an open and rigid framework structure have attracted considerable attention for this purpose especially because of their high thermal, radiation and chemical stabilities and unique selectivity for certain elements [67, 145].

Several reports have revealed the potential of the titanosilicate pharmacosiderite to preferentially remove cesium and strontium from neutral and alkaline solutions with relatively high sodium concentrations. Behrens and Clearfield [94] synthesized sodium and potassium titanosilicates analogues of the mineral pharmacosiderite with three-dimensional tunnel structures and studied their ion exchange behaviour for the selective removal of strontium and cesium from wastewater solutions. Their results indicated that potassium and sodium titanosilicate forms are selective for both Sr^{2+} and Cs^+ in the presence of ppm levels of competitive Ca^{2+}, Mg^{2+}, K^+ and Na^+ in slightly acidic to neutral solutions. Besides, those materials were presumed to be effective for removal of ppm levels of Sr^{2+} or Cs^+ from ^{90}Sr and ^{137}Cs-contaminated waters. Afterwards, Dyer et al. [133] reported the synthesis of different cationic forms of a titanosilicate analogue of the mineral pharmacosiderite and their application for the removal of trace cesium (^{137}Cs) and trace strontium (^{89}Sr) from aqueous solution. High selectivity factors were observed for trace strontium exchange against sodium, potassium, magnesium and calcium ions. However, the efficiency of the synthesized material for strontium removal was found to considerably decrease for low pH values. Furthermore, a decline in selectivity was also observed in dilute sodium and potassium ion solutions probably due to hydrolysis, which may partially convert the exchanger to the hydrogen form. The overall efficiency for trace cesium in the presence of sodium and potassium was lower than for strontium, whereas a significantly high selectivity was obtained in the presence of magnesium and calcium.

Bortun et al. [122] investigated the capability of distinct synthetic sorbents, including layered and framework titanosilicates, for cesium and strontium removal from contaminated groundwater and wastewater and evaluated their selectivity to alkali and alkaline earth metal cations in individual and complex solutions. The experiments were carried out under the concentration range 0.005–1.0 M and for distinct pH values. Several exchangers were found to have high selectivity for cesium and strontium.

Several commercial and under development inorganic materials were also evaluated by Sylvester and Clearfield [99] for the removal of strontium and cesium from a simulated groundwater found in the Hanford waste storage area. A sodium titanosilicate (from Texas A&M University) and two titanium silicate pharmacosiderites (from AlliedSignal) were found to be promising materials for strontium and cesium removal. Besides, the ion exchange kinetics for those titanosilicate materials was

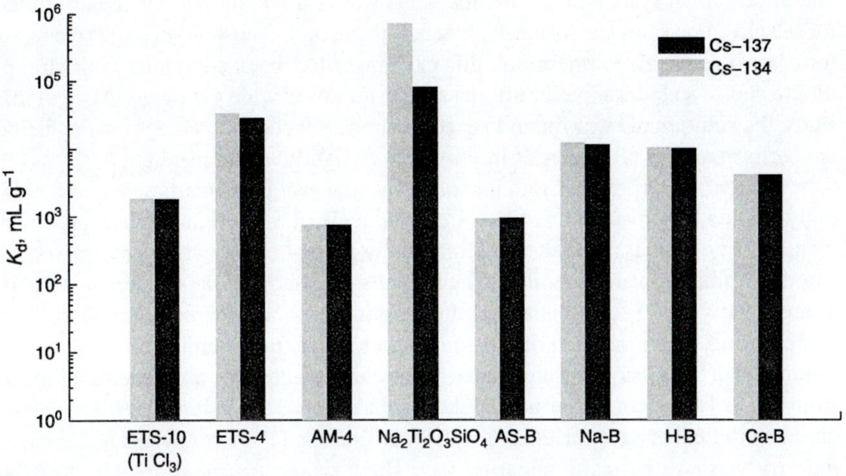

Fig. 11.6 Distribution coefficients of isotope cesium ions on several inorganic materials at initial pH = 7.10 (Reproduced from Ref. [67] with kind permission of © Royal Society of Chemistry (2003))

particularly fast. Moller et al. [145] performed the uptake of ^{85}Sr, ^{134}Cs and ^{57}Co in a sodium titanosilicate and investigated the effect of crystallinity on its selectivity. The synthesized material revealed a high selectivity for cesium in slightly acidic solutions and for strontium in neutral and alkaline media. However, competing cations K^+ and Na^+ exerted a strong interference at very high pH values, which may complicate the removal of radioactive cesium from highly alkaline nuclear waste solutions. Furthermore, the crystallinity was found to have a significant effect on the selectivity, and therefore, materials of lower crystallinity could be more efficient in the treatment of nuclear waste solutions. Tripathi et al. [146] studied the selectivity of Nb-substituted crystalline titanosilicate with sitinakite topology for radioactive ^{137}Cs and ^{89}Sr. The Nb-substituted titanosilicate showed significantly higher uptake values for ^{137}Cs, while for ^{89}Sr, lower values were obtained. The authors attributed this discrepancy mainly to the differences in coordination environments of cesium and strontium in the titanosilicate channels. Accordingly, the selectivity could be enhanced by a higher coordination number of cesium or strontium. Al-Attar et al. [67] evaluated the sorption properties of four synthetic titanosilicates, ETS-10, ETS-4, AM-4 and $Na_2TiO_3SiO_2 \cdot H_2O$, a layered manganese oxide and its M-exchanged form (M=H^+, Na^+, Ca^{2+}), and antimonysilicalite sample towards ^{134}Cs and ^{137}Cs, under batch conditions. Figure 11.6 depicts the obtained distribution coefficients for a solution with initial pH=7.10. The results demonstrated an important distinction in the efficiency of the eight exchangers evaluated, which was explained based on their structural differences and ion exchange properties. It was clear that $Na_2TiO_3SiO_2 \cdot H_2O$ was the best exchanger for cesium uptake. Actually, the one-dimensional system of this titanosilicate was found to have a size ideally suited for selective uptake of Cs^+ ions, which makes it one of the most promising exchangers for cesium removal. Furthermore, this exchanger has been compared with clinoptilolite and revealed a superior affinity for cesium over wide pH range. Also, in this study, the solution pH was found to exert a crucial effect on metal ions uptake, being the performance of all materials improved in a slightly acidic media.

Recently, biosorption of radionuclides by biomass products has been increasingly studied. Psareva et al. [147] evaluated the effect of an HCl solution pre-treatment of cork biomass, which is a biologically neutral cheap waste product, on its sorption affinity towards uranium. Their results revealed that the conversion of cork biomass into the H-form enhanced the sorption capacity for uranium more than twice, owing to the increase of the concentration of strong acidic and weak acidic groups at surface. The measured capacity of cork biomass towards uranium ranged from 0.1 to 1.7 meq g^{-1}. Anirudhan and Radhakrishnan [78] studied a biomaterial-based cation exchanger for the adsorption of uranium(VI) from water and nuclear industry wastewater, using tamarind fruit shell, which is a waste material of the tamarind pulp industry that causes disposal problems. The authors prepared a new cation exchanger with carboxylate functional groups at the end of the chains by graft copolymerization of hydroxyethylmethacrylate onto tamarind fruit shell and found a maximum U(VI) uptake of 100.79 mg g^{-1}.

11.5 Lead, Cadmium and Mercury

Lead, cadmium and mercury are among the metals whose presence in the environment causes great concern not only due to their persistent character in the environment and biota but also due to their bioaccumulation in organisms and bioamplification along the food chain. Because of their injurious effects, these metals were classified in 2001 by the European Union as 'priority hazardous substances' (Decision No 2455/2001/EC), and thus, they will be subject to cessation or phasing out of discharges, emissions and losses within an appropriate timetable that shall not exceed 20 years. However, the list of priority substances was revised in 2008, and the final Directive (2008/105/EC) confirmed mercury and cadmium and their compounds as priority hazardous substances, whereas lead and its compounds have been excluded. Nonetheless, because of their deleterious effect on the aquatic environment, lead compounds are included in the List II of Directive 2006/11/EC, and their discharges into the aquatic environment are subjected to prior authorization which specifies emission standards. Nowadays, there is a global agreement that metals like lead, cadmium and mercury have severe and harmful effects in ecosystems and organisms, and their discharges into the environment compartments should be eradicated.

Lead occurs in a wide variety of minerals. Globally, it has been used in construction, ceramics, ammunition, pigments, petrol additives, production of lead-acid batteries, solder and alloys [148]. The organo-lead compounds tetraethyl and tetramethyl lead have been used extensively as antiknock and lubricating agents in petrol, although

Table 11.4 Guideline values recommended by the World Health Organization [149], levels usually found, and maximum concentration levels established by the United States Environmental Protect Agency (USEPA) and by the European Union (EU) in drinking water for Pb, Cd and Hg

Metal	WHO guideline value (mg L^{-1})	Occurrence levels in drinking water usually less than (mg L^{-1})	Maximum concentration level in drinking water (mg L^{-1})	
			USEPA	EU
Pb	0.01	5	0.015	0.010
Cd	0.003	1	0.005	0.005
Hg	0.006	0.0005	0.002	0.001

their use for these purposes in many countries is phased out. Owing to the decreasing use of lead-containing additives in petrol and of lead-containing solder in the food processing industry, its concentration in air and food is declining, and drinking water is becoming the major source of total intake [149]. The main cause of lead presence in drinking water is not its dissolution from natural sources but its dissolution from household plumbing systems containing lead in pipes, solder, fittings or the service connections to homes [149]. Lead accumulates in the skeleton and interferes with calcium metabolism, both directly and by interfering with vitamin D metabolism, being also toxic to the central and peripheral nervous systems.

Cadmium is widely distributed in the earth's crust but is particularly associated with zinc and non-ferrous ores and is produced commercially only as a by-product of zinc smelting [4]. It has been used in quantity since about 1950 as stabilizers and pigments in plastics, in electroplating, in solders and other alloys. It is also usually present in cement and phosphate fertilizer, as well as in fossil fuels [4, 148], though its main use has been in Ni-Cd batteries. Cadmium releases into the environment are mainly in wastewater and via diffuse pollution caused by contamination from fertilizers and local air pollution. Contamination in drinking water may also be caused by impurities in the zinc of galvanized pipes and solders and some metal fittings. Cadmium and its compounds are toxic (acute and chronic) for mammals and are listed as probable human carcinogens [150], being also very toxic to aquatic life forms, especially in freshwater. Cadmium accumulates in soil and plants, and in water it is adsorbed to particulate matter. In humans, cadmium accumulates predominantly in the liver and kidneys [150].

Mercury is one of the most hazardous contaminants in natural environment. Globally, it has been used in dentistry, measuring and control equipment, batteries and lamps [148]. The chloro-alkaline industry also used large amounts of this metal. Mercury has also been used in pesticides and biocides on grain and in paper industry. Currently, it is found in old electrical appliances, amalgam in teeth and in lights (tubes, energy-saving light bulbs and headlamps on cars) [150]. Mercury causes damage on inner organs and is very toxic to aquatic organisms, which, together with mammals, accumulate the metal. Both short-term and long-term exposure to mercury in humans may result in central nervous system effects.

Depending upon the toxicity of each metal, different public health standards have been established for each one (Table 11.4).

Due to their well-known harmful effects on human health and to an increasing concern on environmental protection, in the last years, a number of studies have focused on reducing Pb, Cd and Hg content in water and industrial waters. Depending on the metal content, pH, temperature, type of ion exchangers and presence or not of other ions, different removal efficiencies and sorption capacities can be achieved. Table 11.5 displays the main results, found in literature, for several ion exchangers used to recover, separate or pre-concentrate Pb, Cd and/or Hg, under different operating conditions.

A vast variety of ion exchangers have been studied. In the last 10 years or so, particular attention has been paid in the application of resins, composites, clays, zeolites, several types of biomaterials and more recently nanomaterials.

Nowadays, resins are still one of the most commonly used ion exchangers for metal recovery, separation and/or pre-concentration. Several studies available in literature reported the use of this type of materials.

In 1999, Rivas et al. [151] reported the binding capacity of a commercial poly(4-vinylpyridine) hydrochloride resin for several metals. This resin evidenced a high ability and selectivity to remove Hg(II) from aqueous solutions in comparison to other metals such as Cd(II), Zn(II), Pb(II), Co(II), Cr(III) and U(VI). These authors suggested that the retention of Hg(II) occurs by exchange of the chloride anion by $[HgCl_4]^{2-}$ and/or $[Hg(NO_3)_4]^{2-}$. Later, Bessbousse et al. [152] also reported the use of poly(4-vinylpyridine) for Hg(II) removal from aqueous solutions but incorporated in a novel metal-complexing membrane. This poly(vinyl alcohol)/poly(4-vinylpyridine) (PVA/P_4VP) complexing membrane has a retention capacity of 450 mg g^{-1}. In other study, Monteagudo and Ortiz [153] tested a number of commercial resins (Dowex XZS-1, Dowex XVS and Amberlite 252) to eliminate inorganic Hg from mine wastewaters. Among them, Dowex XZS-1, a strongly acidic polystyrene-divinylbenzene cationic resin with $-SO_3H$ functional groups, has the most pronounced selectivity for Hg in either of its two common states (Hg^+ and Hg^{2+}). Kocaoba [154] also investigated the sorption capacity of a commercial resin, Amberlite IR120, a strong cation exchange resin, for Pb and Cd uptake. In their study, optimum conditions were determined in a batch system, and it was demonstrated that this resin possesses good capacity and efficiency for the removal of both metals from aqueous solutions. Thiol (-SH) based chelating polymeric resins (an alkyl thiol resin, thiomethyl resin, and an aryl thiol resin, Duolite GT73) have been investigated for the sorption of Cd and other trace metals by Saha et al. [155]. These resins are polymeric solids with organic mainframes containing active functional groups capable of interacting with metal ions by forming coordinate bonds, so the sorption is mainly due to complex formation within the resin, which distinguishes itself from conventional ion exchange [155]. Duolite GT73 is a macroporous ion exchange resin based on a cross-linked polystyrene matrix. The functionality of this resin is attributed to the thiol group, with a minor proportion of sulphonic acid groups. In the study of Saha et al. [155], Duolite GT73 showed slightly better performance than the thiomethyl resin for the removal of Cd and other metals, probably due to the existence of a relatively greater number of thiol groups in its matrix. The authors reported that the Cd sorbed on sodium form thiol resin is in +2 state, which

Table 11.5 Materials most frequently used for the removal of priority pollutants together with the corresponding experimental and major results found in literature

Metal	Material	Operation conditions			Results						Ref.
		pH	T (°C)	C_0 (mg/L)	Batch	Column	Kinetics	Isotherms	Regeneration	Others	
Pb [(a)]	Versatic 911 coated on silanized silica gel	2.5–7.5	N.A.	50–400	x	x		x		Distribution coefficient; thermodynamics; competitive anions; multi-component synthetic mixtures	[180]
Pb	*Moringa oleifera* bark	2–8	25	20–200	x		x	x	x	Competition with other ions	[206]
Pb, Hg	Clays	2	20, 40, 60	N.A.	x		x	x		Capacity; removal %; thermodynamics	[191]
Cd [(a)]	Wool powders	3–9	R.T.	N.A.	x				x	Distribution coefficient; capacity	[249]
Cd	*Hydrilla verticillata*	1.5–7	20–50	10–100	x			x	x	Sorption %; thermodynamics; competitive ions	[201]
Cd	Polyaniline Ce(IV) molybdate composite	N.A.	25	0.01 M	x	x			x	Distribution coefficient; selectivity; binary mixtures	[177]
Cd, Pb [(a)]	Natural muscovite	2–10	20	0–100	x		x	x		Capacity; removal %	[252]
Cd, Hg, Pb [(a)]	Hybrid inorganic/organic adsorbents, hybrid alumina adsorbents	1–6	R.T.	0.1 M	x		x			Capacity; distribution coefficient	[179]
Hg	ETS-4 titanosilicate	2–10	4, 20, 40	0,05	x		x			Capacity; removal %; thermodynamics	[178]
Hg [(a)]	Carbonaceous chemically prepared from rice husk	1.5–6	25–45	100–1,500	x		x	x		Capacity	[213]
Hg	Polyaniline/humic acid nanocomposite	3–11	24 ± 1	0–150	x		x	x		Capacity; removal %; competitive ions	[220]

(continued)

Table 11.5 (continued)

Metal	Material	Operation conditions			Results						
		pH	T (°C)	C_0 (mg/L)	Batch	Column	Kinetics	Isotherms	Regeneration	Others	Ref.
Hg	1,5-diphenylcarbazide doped magnetic Fe_3O_4 nanoparticles	6–8	R.T.	2	x		x		x	Capacity; pre-concentration factor	[219]
Pb	Purolite C100	N.A.	25	26,5	x	x			x	Binary mixtures; distribution coefficient; separation factors; recovery %	[161]
Pb [(a)]	Negatively charged hybrid materials	N.A.	N.A.	0.05 M	x					Capacity; conductometric analysis	[175]
Pb, Cd [(a)]	Natural zeolite	1–10	N.A.	10–100	x		x			Removal %; capacity	[186]
Cd	ETS-4 titanosilicate	4	20	0.4–0.8	x		x	x		Capacity; removal %; modelling	[174]
Cd	ETS-4 titanosilicate	4–8	20	0,85	x		x	x		Capacity; removal %; modelling	[172]
Cd	ETS-10 titanosilicate	4–8	20	0,85	x		x	x		Capacity; removal %; modelling	[173]
Hg	Poly(vinyl alcohol)/poly (4-vinylpyridine) (PVA/P4VP) complexing membrane	2,5	15, 25, 35, 45, 55, 65	70–3,500	x		x	x	x	Amount of metal adsorbed; retention ratio; retention efficiency; thermodynamics	[152]
Hg	Moss (*Drepanocladus revolvens*)	2–8	20–50	10–400	x		x	x	x	Biosorption %; desorption efficiency; thermodynamics	[205]
Hg	Lichen (*Xanthoparmelia conspersa*) biomass	2–8	20–50	10–400	x		x	x	x	Biosorption %; desorption efficiency; thermodynamics	[194]

Hg	ETS-4 titanosilicate	4–5	21	0.050–0.250	x		x	x		Removal %; capacity	[176]
Pb [(a)]	Chitosan and chitosan entrapped in polyacrylamide	N.A.	10, 20, 40	1×10^{-4}– 8×10^{-3} M	x		x	x	x	Thermodynamics; capacity; ion selectivity	[216]
Cd	1-(2-pyridylazo)-2-naphthol modified SiO_2 nanoparticles	3.4–10	N.A.	N.A.	x		x		x	Recovery %; capacity; pre-concentration factor; coexisting ions	[221]
Pb	ETS-10 titanosilicate	5	25	0–15 mM	x	x		x		Breakthrough curves; modelling	[189]
Cd, Hg, Pb [(a)]	Rice husk	2–7	32	50–200	x	x	x	x	x	Capacity; break-through curve	[212]
Hg	Eucalyptus bark	2–10	20, 30, 40	25–300	x		x	x		Capacity; thermodynamics	[204]
Hg	Poly(hydroxyethyl methacrylate) grafted coconut coir pith	2–8	N.A.	10–600	x		x	x	x	Capacity; removal efficiency %; thermodynamics	[211]
Pb, Cd	Amberlite IR120 and dolomite	1–8	20, 40, 60	5–100	x		x	x		Removal %; thermodynamics	[154]
Pb, Cd [(a)]	ETS-10 titanosilicate	N.A.	25, 40, 60	5–15 mM		x			x	Breakthrough curves; competitive ions	[188]
Cd, Pb [(a)]	Chelating resins	3–9	30–50	N.A.	x		x	x		Uptake %; capacity; thermodynamics	[156]
Hg, Pb [(a)]	Chitosan	5–7	N.A.	N.A.	x					Metal chelation; recovery	[215]
Pb, Cd	ETS-10 and Al-substituted ETAS-10 titanosilicates	~10	25, 40, 50	0–4,000	x		x	x		Capacity	[190]
Cd	*Sargassum muticum*	4,5	N.A.	50, 250	x	x			x	Elution efficiency; breakthrough curves	[198]
Cd [(a)]	*Sargassum fluitans* brown seaweed	5	N.A.	N.A.		x				Breakthrough curves; modelling	[199]

(continued)

Table 11.5 (continued)

Metal	Material	Operation conditions			Results						
		pH	T (°C)	C_0 (mg/L)	Batch	Column	Kinetics	Isotherms	Regeneration	Others	Ref.
Cd, Pb [a]	Polyethylenimine methylenephosphonic acid (PEIMPA)	2.4–5.9	N.A.	N.A.	x			x		Liquid–solid extraction; recovery %	[169]
Hg	1,5-Diphenylcarbazide (DPC) functionalized sol–gel silicas (DPCSG)	1–8	28	N.A.	x		x	x	x	Capacity; sorption %; distribution ratios; competitive ions	[170]
Hg	Industrial minerals: atapulgites, zeolites, magnetites, kaolin, vermiculite and bentonite	N.A.	N.A.	0.010–10	x				x	Sorption %; desorption %	[183]
Pb, Cd [a]	ETS-10 titanosilicate	5	25	N.A.	x		x	x		Capacity; selectivity; binary mixtures	[187]
Cd [a]	Microcapsule containing chelating resin	1–7	25	N.A.	x	x	x			Distribution coefficients; sorption %; breakthrough curve	[160]
Hg	Amine–mercaptan chelating resin	1–9	28	N.A.	x	x	x		x	Capacity; selective separation; binary mixtures	[159]
Pb	Zeolites and activated carbon	2–12	23, 35, 45	50	x		x	x		Removal %; thermodynamics; competitive ions	[184]
Pb [a]	Natural zeolite	5	23	0.55–8.9 mM	x			x		Uptake degree	[185]
Cd [a]	Lewatit TP 207	4–7	25	N.A.	x			x		Separation factor; selectivity	[158]
Cd, Pb, Hg [a]	Wheat bran	1–9	25	Pb: 30; Cd: 2; Hg: 500	x			x		Percent adsorbed; selectivity coefficient; capacity; preconcentration	[209]

Cd [(a)]	Metal fix chelamine resin and cation exchanger Dowex 50WX4	4	N.A.	0,1		x			x	Pre-concentration; recovery %	[162]
Hg	Natural zeolite	N.A.	N.A.	0.0028–0.0246 mg/kg	x		x			Capacity	[181]
Hg	Resins Purolite S920, Rohm and Haas GT73; biosorbent *Azolla filiculoides*	4, 6	25	0–0.50 mM	x		x	x		Capacity; distribution coefficient; metal separation	[203]
Cd, Hg, Pb [(a)]	Modified silica gel phases: amino ($SiNH_2$) and chloro-(SiCl)	N.A.	N.A.	0.1 M	x					Percentage recovery; distribution coefficient	[167]
Hg	*Ulva lactuca* seaweed	3.5, 5.5, 7	25	N.A.	x	x		x	x	Removal efficiency %; capacity; breakthrough time	[196]
Hg, Pb [(a)]	Water-soluble N-benzoylthiourea modified PAMAM polymers	1–9	20	0.5 mM	x					Retention %	[168]
Hg, Pb, Cd [(a)]	Thiol functional organoceramic composite	3, 4, 5	25	150–1800	x	x	x	x	x	Capacity; selectivity; breakthrough curve	[166]
Hg, Pb, Cd [(a)]	Olive mill solid residues	4–5	N.A.	N.A.	x		x	x	x	Capacity; regeneration %	[208]
Pb, Hg, Cd [(a)]	Polyacrylamide-grafted iron(III) oxide	2–10	30, 40, 50, 60	Pb:100–40 Hg, Cd:50–200	x		x	x	x	Removal %; thermodynamics; cationic competition	[165]
Pb [(a)]	Anionic resin Duolite A7 and cationic resin Purolite C106	6	N.A.	4		x			x	Selective removal, separation and recovery; breakthrough curves	[157]

(continued)

Table 11.5 (continued)

Metal	Material	Operation conditions			Results						
		pH	T (°C)	C_0 (mg/L)	Batch	Column	Kinetics	Isotherms	Regeneration	Others	Ref.
Cd [(a)]	Thiomethyl resin and Duolite GT73	2, 3, 4	20	0.2–4 mM	x	x	x	x	x	Capacity; diffusion coefficient; breakthrough curves; mixtures	[155]
Hg	Amberlite 252, Dowex XVS/M, and Dowex XZS1/G	N.A.	20	70–90	x	x	x	x	x	Breakthrough curves	[153]
Hg	Dithiocarbamate-incorporated monosize polystyrene based microspheres	3–7.5	R.T.	10–100			x		x	Capacity; competitive sorption	[164]
Hg, Cd, Pb [(a)]	*Pseudomonas fluorescens* 4F39	1–10	30	0.85 mM	x					Metal accumulation	[207]
Pb, Cd, Hg [(a)]	Poly(4-vinylpyridine) hydrochloride resin	N.A.	20–35	7–40	x			x	x	Sorption and elution selectivity; mixtures	[151]
Pb, Cd, Hg [(a)]	Poly(acrylaminophosphonic-carboxyl-hydrazide) chelating ion exchange fibre	1–7	25	0.015 M	x					Binding capacity; distribution coefficient	[163]
Pb, Cd, Hg [(a)]	Tree barks	0–10	N.A.	10, 100, 1,000	x	x				Metal removal; retention capacity	[202]
Pb, Cd, Hg [(a)]	*Escherichia coli* and *Zoogloea ramigera*	6	R.T.	10–100	x	x	x			Metal binding; multiple cycles	[195]
Pb, Cd, Hg [(a)]	Erionite, zeolite 13X, chabazite, clinoptilolite, mordenite, zeolite 3A, zeolite 4A, zeolite 5A	3.6–5.3	N.A.	10^{-2} M		x				Capacity, breakthrough curves	[182]

N.A. - Not available
R.T. - Room temperature

suggests that sorption predominantly takes place by ion exchange. Sharaf et al. [156] also tested two new chelating resins for the removal of Cd and Pb from water. These resins, containing methylthiourea and phenylthiourea groups, showed high capacities for Cd, because of the high stability of the complexes formed [156]. On the contrary, the sorption of Pb(II) on both resins was very low due to its large size [156]. In 2000, Pagano and co-workers carried out an investigation for the optimization of a new process for the separation and recovery of Pb/Fe species from automobile battery manufacturing wastewater. The process involved the use of a commercial weak anion resin (Duolite A7) as selective sorbent for the ferric species and a weak cationic resin with carboxylate functionality (Purolite C106) for the lead species [157]. The Pb/Fe separation is feasible and efficiently carried out by adopting proper operating conditions, and both metals are quantitatively recycled to the production lines after recovery from the loaded resins after regeneration. In 2004, Valverde et al. [158] studied the possibility of removing metal ions from industrial liquid streams using Lewatit TP 207, a weak acid resin. According to the measured separation factors, Lewait TP-207 is more selective for other metals than for cadmium, following, at 298 K, the order Cu(II)>Zn(II)>Cd(II). Atia et al. [159] investigated the recovery of Hg(II) using a chelating resin with amine–mercaptan groups and found that Hg(II) has high affinity towards the resin (621.8 mg g^{-1}). The interaction between the metal ion and the resin was dependent upon the acidity of the medium as well as on the type of the conjugate anion, and the interaction was explained to proceed via complex formation or ion exchange mechanism. In the same year, Singh and Srivastava [160] synthesized microcapsules containing 2-(p-sulphophenylazo)-1,8-dihydroxy-3,6-naphthalenedisulphonate chelate resin for efficient recovery of Cd(II) ions from water samples. They found that the sorption rate of Cd(II) onto the microcapsules was about four times higher than that for Zn(II), Co(II), Ni(II) and Cu(II). Badaway et al. [161] studied a simple solid-phase extraction procedure based on cation exchange with Purolite C100, which is a high-capacity premium-grade bead-form conventional gel polystyrene sulphonate resin. The procedure relies upon a column technique, and the results proved that Purolite C100 can be employed for the separation and recovery of Pb(II) ions from synthetic binary mixtures.

In addition to recovery, removal and separation of metals, resins are also used to pre-concentrate them. Pohl and Prusisz [162] studied the chelating resin metalfix chelamine and the strong cation exchanger Dowex 50WX4 for the off-line pre-concentration of Cd and other metals prior to measurements by inductively coupled plasma atomic emission spectrometry. The procedure was applied for the determination of traces of Cd and other metals such as, Co, Cu, Ni and Zn, in beer. For the analysis, mineralized and not mineralized beer samples were treated with Dowex 50W-X4 in order to assess their total contents and the fraction of the cationic labile species of the analytes, respectively. With this procedure, the sensitivity increased allowing the determination of the selected metals at concentrations of the order of 1 ng mL^{-1}. The accuracy of the entire procedure was verified by the recovery test, performed on spiked samples of beer, and the proposed method provides full recovery for Cd [162].

Besides resins, a number of inorganic, hybrid and composites materials have also been prepared and tested for metal recovery, separation and/or pre-concentration.

Examples are the studies of Liu et al. [163], Denizli et al. [164], Manju et al. [165], Nam et al. [166], Osman et al. [167], Rether and Schuster [168], Abderrahim et al. [169], Khan et al. [170], Lopes et al. [171], Barreira et al. [172], Bessbousse et al. [152], Camarinha et al. [173], Ferreira et al. [174], Liu et al. [175], Lopes et al. [176], Alam et al. [177], Lopes et al. [178], Mahmoud et al. [179] and Mandal and Ghosh [180].

Liu et al. [163] prepared a chelating fibre containing carboxyl, amino, phosphonic and hydrazide groups. The authors reported that this fibrous sorbent exhibits high capacities for Pb(II) (1.55 mmol g^{-1}), Cd(II) (1.36 mmol g^{-1}), Hg(II) (1.25 mmol g^{-1}) and other metals, and so it might be an effective tool for removal and pre-concentration of such species from natural waters and industrial effluents. Denizli et al. [164] used dithiocarbamate-incorporated monosize polystyrene-based microspheres for selective uptake of Hg(II) from aqueous solutions under distinct experimental conditions. The authors reported that they may effectively be used for specific removal of Hg(II) ions because of their high sorption rates and capacity (33.2 mg g^{-1}). Other type of sorbent was synthesized by Manju et al. [165] by introducing carboxylate functional group into polyacrylamide-grafted hydrous iron(III) oxide (PGHyFeO-COOH). The results shown that this new sorbent is very effective for Pb(II), Hg(II) and Cd(II) recovery.

Nam et al. [166] and Alam et al. [177] published the application of composite materials for the removal of metals from water. Nam et al. [166] investigated the sorption of Hg by a thiol functional organoceramic composite (SOL-AD-IV), which showed high selectivity and capacity (726 mg g^{-1}) for Hg(II) at low concentrations, while Alam et al. [177] synthesized a polyaniline Ce(IV) molybdate composite, which demonstrated to be an excellent ion exchange material in terms of selectivity for Cd(II).

Other authors, such as Osman et al. [167], Khan et al. [170], and Mandal and Ghosh [180], focused on the use of silica gels chemically modified. Osman et al. investigated the properties of two silica gel phases modified with 2-hydroxy-1-naphthaldehyde; Khan et al. studied 1,5-diphenylcarbazide functionalized sol–gel silicas for the solid-phase extraction of Hg(II) from aqueous solution by a batch equilibrium technique; Mandal and Ghosh developed a selective method for the chromatographic separation of Pb(II) with Versatic 911 (liquid cation exchanger) coated on silanized silica gel.

Hybrid sorbents, such as alumina/phenylazoformic acid 2-phenylhydrazide and novel negatively charged hybrid materials containing sulphonic acid group, have also been investigated as sorbents for separation and recovery of environmental hazardous substances, such as Pb, Cd and/or Hg [175, 179].

Minerals, clays, natural and synthetic zeolites and other zeolitic-type materials (e.g. titanosilicates) offer a substantial potential for a variety of industrial applications as molecular sieves, ion exchangers, catalysts, detergents builders [181] and on the removal of cations from acid mine drainage and industrial wastewaters [182]. Generally, this kind of materials has large availability, low cost and high sorption capacities [183], so great attention has been given to them in the removal and recovery of metals.

Several authors have been investigating the ion exchange capacities of natural (e.g. erionite, chabazite, clinoptilolite, mordenite, dolomite, Semnan natural zeolites) and synthetic zeolites (e.g. 3A, 4A, Linde type A (5A) and Faujasite (13X))

[181, 182, 184–186]. More recently, other studies reported the huge capacity of synthetic titanosilicates (ETS-10 and ETS-4), to uptake Pb, Cd [172–174, 187–190] and Hg [176, 178] from solutions. Other minerals, such as clays [191], atapulgite, bentonite, bahia, bentonite, paraíba, caulim magnetite, vermiculite [183] and muscovite [192], have been evaluated for their capacity to remove metal ions from aqueous solutions.

In 1995, Mondale and co-authors investigated the ion exchange ability of several natural and synthetic zeolites through different metal ions. They found a wide variation in the ion exchange capacities of these zeolites, though, in general, they followed the order Erionite > 13X > Chabazite > Clinoptilolite >> Mordenite ≈ 3A ≈ 4A ≈ 5A. Similar results were found by Payne and Abdel-Fattah [184], who studied the uptake of Pb(II) ions. Perić et al. [185] studied the sorption of Pb(II) and other metals (Cu and Zn) by ion exchange on natural Croatian zeolitic tuff (ca. 50% clinoptilolite content) and found that the removal efficiency is higher for Pb(II) and Cu(II) than for Zn(II) ions. In the same year, Chojnacki et al. [181] reported the application of natural zeolites from clinoptilolite group for Hg removal from industrial effluents from copper smelter and refinery. Their study concluded that the dominant separation mechanism was ion exchange and the maximum capacity was 1.21 meq g^{-1}. Later, Kocaoba [154] investigated the performance of a natural zeolite, dolomite, for Pb and Cd removal and find out that it provides good capability and efficiency. More recently, Mousavi and Asghari [186] studied the potential of Semnan natural zeolites to remove metal cations from industrial wastewaters and verified the following sequence of affinity: Pb > Cu > Zn > Cd.

Besides zeolites, there are a variety of microporous frameworks that can occur naturally or can also be synthesized. Titanosilicates are an important class of zeolitic materials which have attracted a considerable attention during the last decades since they are stable, have ion exchange properties and remarkable selectivity and consist of a variety of framework structures. ETS-10 is a microporous zeolitic titanosilicate constituted from SiO_4 tetrahedra and TiO_6 octahedra, possessing a high cation exchange capacity because of the presence of tetravalent Ti in an octa-coordinated state, which generates two negative charges balanced by ion exchangeable alkali cations Na^+ and K^+ [188]. Essays of several authors [173, 187–190] have demonstrated that ETS-10 exhibits fast sorption kinetics and high equilibrium capacity towards Pb(II) and Cd(II). ETS-4 is another of the main members of a class of microporous heteropolyhedra transition-metal silicates. The frameworks of this material and mineral zorite are identical and are highly disordered [193]. Its framework comprises corner-sharing SiO_4 tetrahedra, TiO_5 pentahedra and TiO_6 octahedra [176], and like ETS-10, each titanium ion has an associated −2 charge, which is neutralized by extra-framework cations usually Na^+ and K^+. The studies of Barreira et al. [172] and Ferreira et al. [174] have focused on the removal of Cd(II) by ETS-4. Both articles presented a large ion exchange capacity and high removal efficiencies. Lopes and co-workers [176, 178] have been investigating the ion exchange ability of ETS-4 to remove Hg(II) ions from aqueous solutions and found that it has a great potential for wastewater purification and reported a maximum capacity of 246 mg g^{-1}. With only a few milligrams of material and for an initial concentration of 50 µg L^{-1}, the authors achieved efficiencies between 60% and 99% at 21°C and pH = 4–5 (Fig. 11.7).

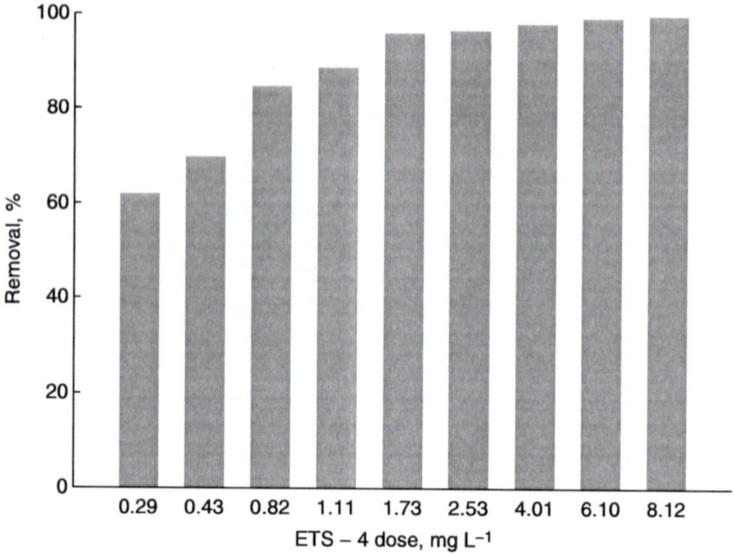

Fig. 11.7 Effect of ETS-4 dose on the Hg^{2+} removal in solution with initial concentration of 50 μg L^{-1}, at 21°C and pH = 4–5 (Reproduced from Ref. [176] with kind permission of © Elsevier (2009))

Clay minerals constitute alternative materials for the ion exchange of metal ions from aqueous solutions. Similar to zeolites, clay minerals are also important inorganic components in soil whose sorption abilities come from their high surface area and exchange capacities resultant from the negative charge of the solid structure [64]. Hristodor et al. [191] investigated two natural clays, from Valea Chiuoarului and R zoare areas, as sorbents for Pb(II) and Hg(II). From their results, the uptake of metal ions in diluted solutions is quite high, especially for Pb(II) ions. Melamed and Benvindo da Luz [183] tested several minerals, which included atapulgites, zeolites, magnetites, kaolin, vermiculite and bentonite to remove Hg(II) from liquid effluents. Among them, magnetite, vermiculite and zeolite displayed better removal efficiencies. Results also indicated that vermiculite and zeolites may be used as mercury filters while magnetite would have a better application as *in situ* soil amendment at high contaminated sites [183].

Biological materials, known as biosorbents, are widely used and implemented for metal removal from various matrices and may include fungi [194], algae [195–201], plants and trees [202–206], microorganisms [195, 207], agriculture wastes [208–214] and several biopolymers [215, 216]. These biomasses have been reported to bind a variety of metals to different extents. Some of them have shown high metal binding capacity and are able to collect the majority of metals with no specific priority, while others can even be specific for certain types of metals [217].

In 1996, Xie et al. [195] used *Escherichia coli* (ATCC 23724) and *Zoogloea ramigera* 115 (ATCC 25935) as biosorbents for removal and recovery of Cd and other metal ions from wastewater. Zeroual and collaborators [196] investigated the Hg biosorption onto non-living protonated biomass of *Ulva lactuca*. Other authors [197] studied the biosorption of Hg(II), Cd(II) and Pb(II) ions onto microalgae *Chlamydomonas reinhardtii* and reported that this biosorbent is suitable for the development of an efficient metal removal and recovery system from aqueous media. The microalgae exhibited higher affinity for Pb(II) ions when compared to Hg(II) and Cd(II): the sorption capacities observed were 96.3 mg g^{-1} for Pb(II), 72.2 mg g^{-1} for Hg(II) and 42.6 mg g^{-1} for Cd(II). Another algae specie, *Sargassum muticum*, was also used as biosorbent of Cd(II) from wastewaters by Lodeiro et al. [198]. Naja and Volesky [199] studied the biosorption of Cd(II) and other metals by *Sargassum fluitans*, a brown seaweed biomass, pre-loaded with Ca^{2+}, and reported that the biosorption occurred by ion exchange accompanied by the release of Ca^{2+} into the liquid phase. According to them, the carboxylic, the sulphonate and phosphonate moieties of the biomass were quantitatively established as being responsible for the sequestering of metals [199]. Romera et al. [200] studied the sorption capacity of six different algae (green, red and brown) in the recovery of Cd, Pb and other metals from aqueous solutions and reported that the brown algae (*Fucus spiralis*) produced the lowest metal concentration levels in solution. Haung et al. [201] examined the biosorption of Cd(II) onto *Hydrilla verticillata* in aqueous solutions. *H. verticillata* is a perennial submerged macrophyte, growing in rivers, lakes, pools and drain. The monolayer capacity of Cd(II) was 50 mg g^{-1} at pH 6 and 20°C.

Although algae are among the most explored biomass, other biomaterials like agricultural wastes have also been investigated. Gaballah and Kilbertus [202] used raw and chemically treated tree barks for decontamination of synthetic solutions and industrial effluents containing Cd, Pb, Hg and other metals. The decontamination was possible using European barks from oak (*Quercus robur*), beech (*Fagus sylvatica*), piecea (*Picea abies*) and Scotch pine (*Pinus sylvestris*), and the percentage of metal ions removal from the solutions was higher than 70%. Another bark, eucalyptus *camaldulensis* bark, a forest solid waste, was investigated by Ghodbane and Hamdaoui [204] as a novel material for Hg(II) removal from aqueous media. Its sorption behaviour was dependent on usual operating variables such as sorbent dosage, ionic strength, stirring speed, temperature, solution pH, contact time and initial metal concentration. Reddy et al. [206] investigated *Moringa oleifera* bark to recover Pb(II) from aqueous solution. Results showed that it is an effective and efficient biosorbent for Pb(II) ions uptake and can be considered a viable alternative to activated carbon, ion exchange resin and other synthetic sorbents adopted for this purpose. Pagnanelli et al. [208] investigated the use of olive mill solid residue, another agricultural waste, to sorb different metals (Hg, Pb, Cu, Zn and Cd). Farajzadeh and Monji [209] studied the sorption characteristics of wheat bran towards metal cations and found that the equilibrium was achieved in about 10 min for all species, and the measured capacities were 70 mg g^{-1} for Hg(II), 62 mg g^{-1} for Pb(II) and 21 mg g^{-1} for Cd(II). Minamisawa et al. [210] investigated the biosorption capacity of several biomaterials (chitosan, coffee, green tea, tea, yuzu, aloe and Japanese coarse tea) towards Cd(II) and Pb(II) ions, and high capacities have been observed for all of them at pH 4 and 6.7, though coffee offered the best results. Krishnani et al. [212] examined the biosorption mechanism of nine different metals onto a biomatrix from rice husk, a lignocellulosic waste from agro-industry. Batch and column sorption studies were performed and indicated that the developed biomatrix, obtained by partial alkali digestion of rice husk, could be used to remove as many as eight different metal ions (Ni, Zn, Cd, Mn, Co, Cu, Hg and Pb) in single and mixed metal solutions effectively. Recently, El-Shafey [213] also used a sorbent chemically prepared from rice husk to remove Hg(II) and Zn(II) from aqueous solution. Sousa et al. [214] applied green coconut shells for the removal of toxic metals (Pb(II), Cd(II) and others) from aqueous effluents in percolation columns. The breakthrough curves for multiple elements gave rise to the following order of sorption capacities: Cu(II) > Pb(II) > Cd(II) > Zn(II) > Ni(II).

Another biosorbent, moss (*Drepanocladus revolvens*), was investigated by Sari and Tuzen [205] for the removal of mercury from aqueous solution, which measured a capacity of 94.4 mg g^{-1}. Additionally, these researchers concluded that the *D. revolvens* biomass can be used as alternative biosorbent for such treatment because of the following advantages: it is natural and low cost, possesses high capacity and offers a reasonably rapid sorption rate and good reusability without considerable loss of activity.

The use of sorbents containing natural polymers has also received great attention, in particular polysaccharides such as chitin and its derivate chitosan. Chitin is the second most abundant natural biopolymer after cellulose and the most abundant amino-polysaccharide. It can be found in the shells of crustaceans, shells and skeletons of molluscs and krill, on the exoskeletons of some arthropods and in the cell walls of some fungi [218]. Chitosan, a copolymer that is primarily composed of β-(1→4)-linked 2-amino-2-deoxy-D-glucopyranose units, and residual 2-acetamido-2-deoxy-D-glucopyranose units, is a chemical derivative produced by alkaline deacetylation of chitin. It is being also found naturally in some fungal cell walls [218]. Because of its large number of NH_2 groups, chitosan is well known as an excellent biosorbent for metal cation removal in near-neutral solutions and consequently has received considerable interest not only due to its excellent metal-binding capacities but also because of its low cost. In literature, numerous studies can be found related with chitin and chitosan properties. The works of Gamage and Shahidi [215] and Akkaya and Ulusoy [216] are only two examples of the application of chitosan on the uptake of metal ions from water. The former one reported that chitosan served as an effective coagulating agent to remove metal ions [Hg(II), Fe(II), Ni(II), Pb(II), Cu(II) and Zn(II)] from industrial wastewater, especially at pH 7, while Akkaya and Ulusoy [216] found that both chitosan and chitosan entrapped in polyacrylamide showed a higher affinity to Pb(II) ions.

Recently, nanomaterials are emerging as a new class of sorbents for recovery, separation and/or pre-concentration of metals. Kaur and Gupta [219] developed a new analytical method using 1-(2-pyridylazo)-2-naphthol modified SiO_2 nanoparticles as solid-phase extractant for the pre-concentration of trace amounts of Cd(II) in different water samples, and the sorption capacity of this material was 60.6 μmol g^{-1}.

Zhai et al. [219] also reported a solid-phase extraction and pre-concentration of trace Hg(II) from aqueous solution using magnetic nanoparticles doped with 1,5-diphenylcarbazide (DPC-NF-SiO_2), whose maximum sorption capacity was 220 μmol g^{-1}. Zhang et al. [220] published the removal of mercury by a polyaniline/humic acid nanocomposite (PANI/HA). This PANI/HA nanocomposite exhibits remarkable enhancement of material stability and Hg removal efficiency; the maximum sorption capacity of 671 mg g^{-1} was measured at ionic strength of 0.20 M and pH 5.0.

11.5.1 Effect of pH

Once more, the pH of the medium is one of the most important parameters affecting ion exchange: it determines the surface charge of the sorbent, the degree of ionization and speciation of the components to be separated, the ionic strength of the media, and consequently the interaction between sorbent and metal ions. Due to the protonation and deprotonation properties of the ion exchanger groups, its metal uptake behaviour may be directly affected by the pH value. Owing to its importance, numerous researchers have investigated the influence of pH on the performance of the various types of ion exchangers mentioned previously.

Monteagudo and Ortiz [153] found that the Dowex XZS-1 resin operates over a very broad pH range (1–12) for the removal of Hg(II) from mine wastewater. Valverde et al. [158], on their study concerning the possibility of removing metal ions from industrial aqueous streams using Lewatit TP 207, verified that the operating resin capacity with respect to Cd(II) ions increased with increasing pH from 4 to 7. Atia et al. [159] studied the effect of the acidity of the medium on the uptake of Hg(II) by chelating resin with amine–mercaptan groups. According to these authors, the highest uptake value was recorded at pH 5 (natural), suggesting that this may be due to the presence of the lone pair of electrons on nitrogen and sulphur atoms (or deprotonation of –SH group) suitable for coordination with the metal ion to form the corresponding resin–metal complex. The higher affinity of Hg(II) towards the –SH sites may be attributed to the soft base nature of the S atom which is capable to interact with it. The reducing uptake in the acidic media, observed by Atia et al. [159], may be attributed to the protonation of the lone pair of nitrogen along with the lower ionization of the SH group which hinders the complex formation. The uptake of Hg(II) beyond the natural pH (pH >5) is doubtful to be attributed only to the interaction of the free Hg(II) cations with resin active sites but also to the formation of metal hydroxide species such as soluble $Hg(OH)^+$ and/or insoluble precipitate of $Hg(OH)_2$ [159]. Singh and Srivastava [160] verified that in the pH interval from 1 to 7, the sorption rate of Cd(II) ion by microcapsule incorporated 2-(p-sulphophenylazo)-1,8-dihydroxy-3,6-naphthalenedisulphonate chelate resin was very high. Kocaoba [154] also studied the effect of pH on the sorption capacity of a strong cation exchange resin, Amberlite IR120, towards Pb(II) and Cd(II), and observed that the sorption efficiency increased by incrementing solution pH from 1 to 8, both for Pb(II) and

Cd(II). Different results were found by Sharaf et al. [156] for Cd with two new chelating resins (methylthiourea- and phenylthiourea-supporting resins). Their study depicts that the sorption process is more favourable near neutral conditions, which is also reflected on the relatively low acidity of the chelating polymers.

The effect of pH on the efficiency of inorganic, hybrid and composite materials has been also investigated by several authors. Liu et al. [163] found that the removal of Hg(II) by a chelating fibre containing carboxyl, amino, phosphonic and hydrazide groups is not pH dependent. According to them, such insensitiveness is attributed to the presence of multi-groups in the chelating fibre. At low pH values, the protonation of amino groups results in the decrease of the affinity of the nitrogen atom to Hg(II) ion, but the phosphonic groups of the chelating fibre, with good ionization character, can complex with Hg(II) in strong acidic solution. Denizli et al. [164] showed that dithiocarbamate-incorporated microspheres exhibited a low affinity for Hg(II) in acidic condition (pH <4.0), somewhat higher affinity between pH 4.0 and 5.0, and increased affinity above pH 5.0. The sorption amounts of Hg(II) ions incremented with increasing pH and reached almost a plateau around pH 6.0. Such pH growing favours complex formation between dithiocarbamate groups and Hg(II) [164]. Manju et al. [165] found that the percentage removal of metal ions by PGHyFeO-COOH increased when pH was raised up to a certain value and then started to decrease. The maximum removal of Pb(II) was observed at an optimum pH of 6.0 (95.7%), whereas the maximum removal of Hg(II) and Cd(II) was observed at pH 5.0, 84.9% and 81.0%, respectively. Below and above this pH, a decreasing trend was observed. The possible active sites on PGHyFeO-COOH for specific sorption under acid pH includes H^+ ions in –COOH functional groups [165]. Hence, the influence of pH may be explained by the exchange of H^+ ions from peripheral –COOH groups. At very low pH, the competition between M^{2+} (M=Pb, Hg, Cd) and the higher concentration of H^+ ions favours H^+; on the other hand, by increasing pH up to the optimum value enhances metal sorption apparently due to the hydrolysis of the counter ions, since the hydroxy complex $[M(OH)]^+$ is sorbed preferentially to the uncomplexed cations [165]. Khan et al. [170] found that the extraction of mercury with 1,5-diphenylcarbazide doped sol–gel silica (DPCSG) was sensitive to pH changes, since below pH 4, the protonated ligand sites of DPCSG exhibited no reactivity towards the positively charged mercuric ions, while the extraction increased and attained its maximum value at pH 6. The extensive extraction at pH 6.0 may be attributed to the availability of deprotonated active sites responsible for the coordination of the mercury ion [170]. Bessbousse et al. [152] verified that the Hg(II) retention ratio of a membrane containing P_4VP increased between pH 1 and 3 and reached its maximum value (99%) at pH 3. The solution pH strongly influences the protonation degree of the P_4VP ligands, whose ionization is significantly reduced at pH <3.

The effect of pH on zeolites and zeolitic materials performances was also investigated by several authors. Payne and Abdel-Fattah [184] found that the application of chabazite to remove Pb(II) ions requires pH values between 4 and 11, while clinoptilolite may operate between 3 and 11. Kocaoba [154] investigated the performance of dolomite to remove Pb(II) and Cd(II) from aqueous solutions and

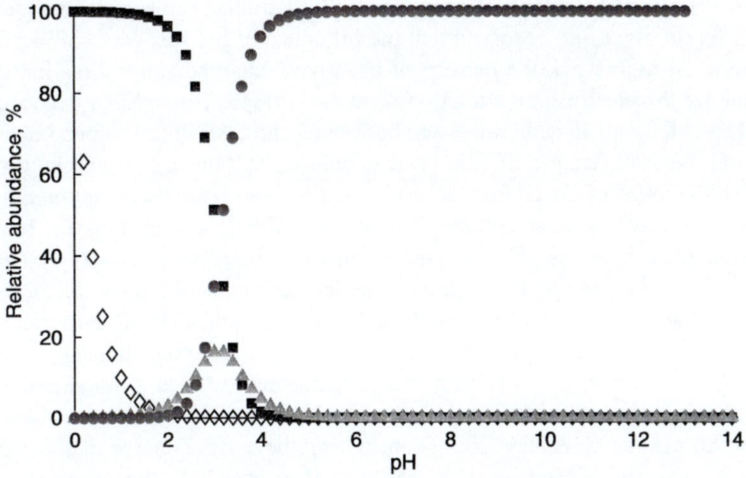

Fig. 11.8 Speciation diagram of mercury in aqueous solution at 298 K, calculated with the Ka values of Hg^{2+} hydrolysis ($K_{a1} = 10^{-35}$, $K_{a2} = 10^{-27}$). Symbols: (◊) H_3O^+; (■) Hg^{2+}; (▲) $Hg(OH)^+$; (•) $Hg(OH)_2$

observed that the sorption efficiency increased from 53% to 99% in the case of Cd(II) and from 65% to 100% in the case of Pb(II), when pH changes from 1 to 8. Such increments are explained by the significant protonation of the sorbent at low pH, i.e. the competitive sorption between H^+ protons and free metal ions prevails [154]. Mousavi and Asghari [186] verified that the removal percentages by Semnan natural zeolites increased for all metals studied after pH 4, attaining 99.9% at pH 10. Camarinha et al. [173] observed that Cd(II) removal by ETS-10 increases with increasing pH. The authors reported that the amount of metal sorbed is much higher at pH 6 (3.6×10^3 eq m^{-3}) than at pH 4 (1.6×10^2 eq m^{-3}). The same conclusion has been drawn by Barreira et al. [172] for the removal of Cd(II) by ETS-4. Lopes et al. [178] reported that changes in pH have a significant impact on Hg(II) ions uptake by microporous titanosilicate ETS-4. These authors observed that the removal increased with increasing pH up to a maximum at pH = 4–6. It was suggested that the efficiency of the ETS-4 is lower at low pH values (< 3) due to the competition between H^+ and free Hg^{2+} ions, which is the mercury predominant specie in solution (Fig. 11.8) for exchange with ETS-4 counter ion, Na^+. Furthermore, at very low pH, the removal decreased drastically, which may also be a consequence of the partial collapse of the ETS-4 structure [178]. On the contrary, the increasing pH reduces H^+ concentration, and consequently, the ion exchange of mercury species is favoured [178].

Under the context of ion exchange, pH affects essentially the biosorption by the following aspects: metal ions solubility and speciation, ionization of sorbent active sites and competition between counter ions and H^+ [200]. Tüzün et al. [197] verified that the biosorption of Hg(II), Cd(II) and Pb(II) ions on algae biomass was pH dependent, and the maximum sorption was obtained at pH 6 for Hg(II) and Cd(II) and at pH 5 for Pb(II) ions. Romera et al. [200] studied the sorption capacity of six different algae and observed that the influence of pH was very similar for all of them. At higher pH, the density of negatively charged active sites increased, promoting the electrostatic attraction of metal cations. The optimal pH value for Cd(II) was 6 for all algae biomass and between 3 and 5 for Pb(II), depending on the type of biomass. Huang et al. [201] also concluded that the pH is a crucial variable in the biosorption of Cd(II) from aqueous solution onto *Hydrilla verticillata*. At low values, there was a clear competition between Cd(II) and protons for biomass sorption sites, while as pH increased up to 5–7, the removal efficiency attained 96%. When pH arose 7, the authors observed cadmium hydroxide precipitation. Ghodbane and Hamdaoui [204] observed that the sorption of Hg(II) onto eucalyptus bark was once more strongly pH dependent. The maximum sorption capacity takes place at pH 8, and beyond this value, the sorbent exhibits the same maximum value. It has been suggested that the variation of the Hg removal by eucalyptus bark with respect to pH can be elucidated by considering the surface charge of the material and the speciation of Hg (Fig. 11.8). At low pH values (2–4), the number of negatively charged surface sites decreased, which did not favour the uptake of the metal cations Hg^{2+} and $Hg(OH)^{+}$. Additionally, the presence of a large excess of H^{+} ions competing with metal counter ions also reduces exchange efficiency. The improvement of mercury sorption in the pH range 2–6 is due to the less competition of protons to reaction sites, to an increase of the $Hg(OH)^{+}$ species concentration and to the modification of the sorbent surface charge [204]. Anirudhan et al. [211] studied the sorption capacity of two materials, coconut coir pith (CP) and coconut coir pith grafted with polyhydroxyethylmethacrylate and with a carboxylate functional group at the chain end (PGCP-COOH), for Hg(II) uptake using initial concentrations of 10 and 25 mg L^{-1} over the pH range 2.0–8.0. According to the results plotted in Fig. 11.9, the maximum uptake percentage attained by both sorbents was measured in the pH range 5.5–8.0 and significantly decreases by reducing the pH values until 2.0, suggesting that the metal ions interact with both materials by a chelating mechanism [211].

The effect of pH on the biosorption of Hg(II) ions onto *D. revolvens* biomass was evaluated by Sari and Tuzen [205], who observed that the efficiency increased from 70% to 91.5% as pH increased from 2 to 5, passing through a maximum of 96% at pH 5.5. The authors proposed a similar explanation to Ghodbane and Hamdaoui [204] for the biosorption trend and mechanism at different pHs. Reddy et al. [206] found that the equilibrium removal of *M. oleifera* bark was minimum at pH 2 (23.3%) and reached maximum value (96.5%) at pH 5; above this pH, a slow decrease in the sorption of Pb(II) was reported.

The effect of pH on Pb, Cd and Hg recovery, separation and/or pre-concentration by nanomaterials was also examined by some researchers. Kaur and Gupta [221] detected a strong influence on the recovery of Cd(II) by 1-(2-pyridylazo)-2-naphthol modified SiO_2 nanoparticles. The quantitative recovery of Cd(II) was obtained for pH between 9.2 and 10.0. Zhai et al. [219] also investigated the effect of pH on the performance of DPC-NF-SiO_2 through Hg(II) ions and found that below pH 4, the protonated ligand sites had no reactivity to the positively charged metal ions;

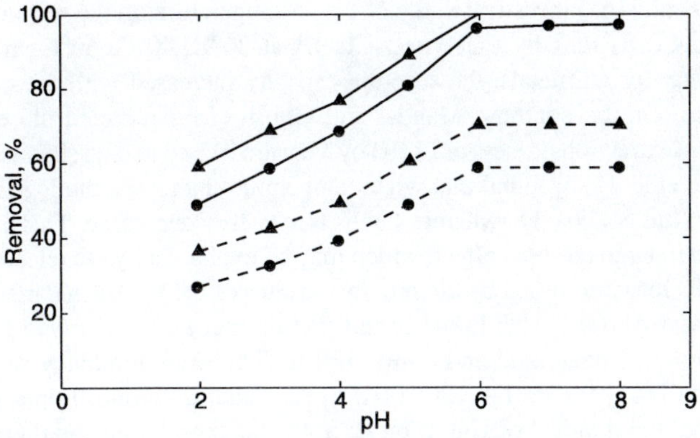

Fig. 11.9 Effect of pH on the removal of Hg(II) with CP and PGCP-COOH. Symbols: initial concentration (▲) 10 mg L^{-1}; (•) 25 mg L^{-1}; *full lines*: PGCP-COOH; *dashed lines*: CP. *Curves* are *lines* to guide the eyes (Reproduced from Ref. [211] with kind permission of © Elsevier (2008))

increasing pH, the sorption greatly increased and attained quantitative extraction (>95%) at pH about 6–8. Zhang et al. [220] observed that Hg(II) removal by PANI/HA nanocomposite is strongly affected by this parameter also. The optimal pH range for Hg(II) removal was from 4 to 7.5, whereas alkaline and more acidic solutions inhibited the sorption. This significant impact could be explained by changes in both the physico-chemical properties of PANI and aqueous Hg(II) speciation [220].

In most cases, Pb, Cd and Hg recovery, separation and/or pre-concentration by different types of sorbents (resins, inorganic materials, zeolites, biomaterials and nanomaterials) are strongly affected by pH. The extent of that influence is determined by the type of sorbent and type of the metal ions. Overall, in a particular pH range, most metal sorption is enhanced with pH, increasing to a certain plateau value or going through a maximum followed by a reduction on further pH increase.

11.5.2 Effect of Temperature

It is well known that depending on the structure and/or surface functional groups of a particular sorbent, temperature has an impact on its sorption capacity to certain extent, since temperature variations alter equilibrium and transport properties [222]. A number of studies have been performed concerning its influence on Pb, Cd and Hg recovery, separation and/or pre-concentration, reinforcing the role played by temperature on this issue. Moreover, thermodynamic parameters like Gibbs free energy ($\Delta G°$), enthalpy ($\Delta H°$) and entropy ($\Delta S°$) provide valuable information about the process and allow addressing the possibility and feasibility of a certain separation.

Manju et al. [165] investigated the effect of temperature on the sorption of Pb(II), Hg(II) and Cd(II) ions by PGHyFeO-COOH at 30°C, 40°C, 50°C and 60°C and observed that for all metals, the sorption capacity increased with the increasing of temperature. On the contrary, Mandal and Ghosh [180] reported the endothermic nature of the extraction process of Pb(II) by Versatic 911 coated on silanized silica gel. Bessbousse et al. [152] found that when temperature increased the Hg(II), uptake at equilibrium did not change, whereas Gibbs free energy decreased. This phenomenon was related to large entropic effect, which may be explained by the release of protons when Hg(II) ions are sorbed by the protonated sites of the PVA/P_4VP membrane.

Payne and Abdel-Fattah [184] found that temperature improved the sorption of Pb(II) ions by chabazite and clinoptilolite. The same tendency was observed by Kocaoba [154] for Pb(II) and Cd(II) uptake onto dolomite. Lopes et al. [178] reported that the removal of Hg(II) by ETS-4 is an exothermic and thermodynamically favourable process, being weakly affected by temperature. Hristodor et al. [191] studied the sorption of Pb(II) and Hg(II) ions on original and Na-exchanged clays and observed that the increase of temperature is beneficial on the amount of ion metal retained from the solution.

Khambhaty et al. [223] reported that the temperature is an important parameter in metal biosorption by microorganisms. Tüzün et al. [197] verified that the biosorption capacity of microalgae for Hg(II), Cd(II) and Pb(II) ions did not significantly change for temperatures in the range tested (5–35°C). On the other hand, Huang et al. [201] observed a slow decrease in the sorption of Cd(II) with the rise of temperature and suggested it was due to the increasing tendency to desorb from the interface to the solution. This study also indicated that the sorption onto *H. verticillata* was exothermic. Ghodbane and Hamdaoui [204] observed that the biosorption of Hg(II) onto eucalyptus bark increased with temperature, pointing out the endothermic nature of the process. The authors suggest that this enhancement might be due to the interaction between sorbate ions and sorbent and creation of some new active sites at higher temperatures [204]. Sari and Tuzen [205] verified that the removal of Hg(II) ions by *D. revolvens* biomass decreased from 98% to 82% when temperature increased from 20°C to 50°C.

11.5.3 Effect of Initial Concentration

Several studies have been undertaken in order to establish the effect of initial metal concentration on their uptake by different types of sorbents such as resins, inorganic materials, titanosilicates, clays and microalgae and eucalyptus bark biomasses.

Atia et al. [159] studied the uptake of Hg(II) by an amine–mercaptan resin and observed that it increases monotonous with the concentration until reaching a saturation value (3.10 mmol g^{-1}). Sharaf et al. [156] found that the uptake percentage of Cd(II) by two new chelating resins (methylthiourea- and phenylthiourea-supporting resins) increases with the rising of the initial concentration of metal up to a plateau. The loading capacities were 3.2 and 3.0 mmol g^{-1} for the methylthiourea resin and phenylthiourea resin, respectively. In the case of Hg(II) and dithiocarbamate-incorporated monosize PS microspheres, the sorption increased first by incrementing the initial concentration of metal and then reached a plateau for a value about 40 mg L^{-1}, for which the active binding sites of the microspheres got saturated [164]. Bessbousse et al. [152] found that the retention ratio of Hg(II) ions onto a membrane containing poly(4-vinylpyridine) decreased with increasing initial concentration of Hg(II). For small initial concentrations (70 mg L^{-1}), the values of the retention ratio were close to 100%, indicating complete removal of the Hg(II) ions present in solution; at higher concentration (3,500 mg L^{-1}), that ratio decreased due to the saturation of the membrane active sites.

Lopes et al. [176] investigated the effect of the initial concentration in the system Hg(II)/ETS-4 and detected a noticeable variation in the amount of Hg(II) removed per gram of ETS-4 in the range of concentrations studied. Hristodor et al. [191] reported that the degree of metal removal from aqueous solution by the clays decreases steeply with the increase of the initial metal ion concentration, once more due to the saturation of the solid.

Tüzün et al. [197] verified that the biosorption capacity of microalgae biomass increased first with rising initial metal concentration and reached a saturation value, 200 mg L^{-1} for Hg(II) and Cd(II) and 300 mg L^{-1} for Pb(II). In the biosorption of Hg(II) onto eucalyptus bark, Ghodbane and Hamdaoui [204] noticed that an increase in initial metal concentration leads to an increase in the sorption capacity of eucalyptus bark.

11.5.4 Competition Effects

The presence of foreign ions influences the sorption of metal ions because of competition for the active sites, complex formation, oxidation–reduction of targeted ions by foreign ions, and precipitation [170]. In most cases, metal recovery, separation and/or pre-concentration are applied to wastewater where more than one type of metal is encountered. As consequence of competitive processes, it is expected that the desired uptake could be lower than in individual solutions.

Saha et al. [155] conducted an investigation on the selectivity of thiomethyl and Duolite GT73 resins towards the desired metal ion, using a solution containing a mixture of Cu, Cd, Ni and Zn ions, and found that the removal followed the order Cu(II) > Cd(II) > Zn(II) > Ni(II). Atia et al. [159] studied the selective separation of Hg(II) from binary mixtures. The results obtained reveal that it may be selectively separated from Ca(II), Mg(II), Pb(II), Cd(II), Cu(II) and Zn(II) at pH 2.7 by an amine–mercaptan chelating resin. The microcapsules incorporated 2-(p-sulphophenylazo)-1,8-dihydroxy-3,6-naphthalenedisulphonate chelate resin synthesized by Singh and Srivastava [160] showed relatively large separation factors for Cd/Co, Cd/Ni, Cd/Cu and Cd/Zn. Denizli et al. [164] conducted an investigation on competitive sorption of Cu(II), Cd(II), Pb(II) and Hg(II) from their mixture by dithiocarbamate-incorporated monosize PS microspheres. The authors found that the amounts of

Hg(II) sorbed are higher than those measured for Cu(II), Cd(II) and Pb(II), in both weight and molar basis. The capacities reported are 1.92 mg g^{-1} for Cd(II), 3.76 mg g^{-1} for Cu(II), 6.17 mg g^{-1} for Pb(II), and 22.92 mg g^{-1} for Hg(II) ions. Such order of affinity has been interpreted on the basis of Pearson's theory of hard and soft acids and bases. Sulphur-containing ligands, which are usually considered soft bases, can strongly coordinate soft acids like Hg(II) ions. Denizli et al. also reported that the equilibrium concentrations of each metal ion decreased by 12–23% in the presence of other ions, since they might share the same binding sites on the microspheres or inhibit the metal binding of adjacent sites. Manju et al. [165] evaluated the competition among Pb(II), Hg(II) and Cd(II) for the sorption sites of PGHyFeO-COOH and reported the following loadings: Pb(II) = 14.6 mg g^{-1}, Hg(II) = 9.76 mg g^{-1} and Cd(II) = 8.31 mg g^{-1}. According with the authors, for ions of the same valence, the sorbent prefers the metal with higher atomic number. Khan et al. [170] verified that the presence of Se(IV), Cd(II), Y(III), Ni(II) and Pb(II) reduced the extraction of Hg(II) by 1,5-diphenylcarbazide doped sol–gel silica (DPCSG) in less than 2.7%, while Mn(II), Cu(II), Fe(II) ascorbate and EDTA reduced the extraction to high degree and $S_2O_3^{2-}$ completely stopped the removal. The results obtained suggest that DPCSG can be used for Hg separation from bimetal solutions of Hg(II) with Se(IV), Cd(II), Y(III), Ni(II) and Pb(II). Payne and Abdel-Fattah [184] analysed the competition between Pb(II) ions and cations, such as potassium, and found that its presence reduced Pb(II) removal by the natural zeolites chabazite and clinoptilolite. Lv et al. [187] conducted an investigation on the competitive sorption characteristics of binary and ternary metal ions Pb(II), Cu(II) and Cd(II) on microporous titanosilicate ETS-10 and concluded that the ETS-10 displays high selectivities towards one metal in two- or three-component systems according to the affinity order Pb(II) > Cd(II) > Cu(II). These researchers suggest that such behaviours are determined by the hydrated ionic radii and the hydration energy of the metals species [187]. Recently, Huang et al. [201] studied the influence of competitive ions on the sorption of Cd(II) onto *H. verticillata* and found that it increased in the order of Pb(II) < Ni(II) < Cu(II) < Pb(II) + Cu(II) < Pb(II) + Ni(II) < Ni(II) + Cu(II) < Pb(II) + Ni (II) + Cu(II) (Fig. 11.10). The biosorption reduction of cadmium in the presence of other cations could be attributed to the difference in their class behaviour on the basis of their covalent indices. Pb(II) is classified as a class *b* ion, while Cu(II) and Ni(II) are classified as borderline ions [201]. The covalent index of a metal ion is related with its potential to form covalent bounds with biological ligands [224–226].

Zhai et al. [219] investigated the effect of common electrolytes on the sorption of Hg(II) by DPC-NF-SiO_2 nanocomposite and found that 1,000-fold K^+, Na^+, Cl^-, NO_3^- and HCO_3^- and 400-fold Ca^{2+} and Mg^{2+} did not interfere with Hg(II) removal, indicating that this functional sorbent can be suitably used in high ionic strength solutions. Zhang et al. [221] studied the effects of Cl^-, SO_4^{2-} and PO_4^{3-} on mercury sorption onto PANI/HA nanocomposite and observed that neither the presence of SO_4^{2-} nor the PO_4^{3-} has any significant impact on the removal, though the effect of Cl^- is notable. The authors suggested that Cl^- competed favourably with OH^- for Hg(II) to form Hg–Cl complexes which possess less affinity to the PANI/HA, causing a substantial decrease of the separation.

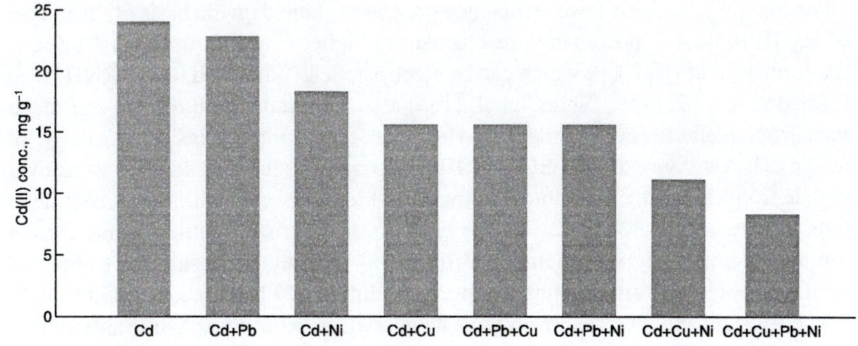

Fig. 11.10 Effect of competitive ion on the sorption of Cd(II) onto *Hydrilla verticillata* (Reproduced from Ref. [201] with kind permission of © Springer (2010))

11.5.5 Sorbent Regeneration

The regeneration of sorbents is a very important step in metal recovery, separation and/or pre-concentration processes, in order to assure their economical viability for practical use. Many researchers have been conducting investigation on this issue.

Rivas et al. [151] showed the easy regeneration of a Hg(II)-loaded poly (4-vinylpyridine) hydrochloride resin with 4 M $HClO_4$, under competitive and non-competitive conditions. Bessbousse et al. [152] found that a similar material, a poly(vinyl alcohol)/poly(4-vinylpyridine) complexing membrane, could be regenerated several times, without noticeable loss of efficiency, by treatment with 0.5 M HNO_3. Monteagudo and Ortiz [153] regenerated successfully the Hg(II)-loaded Dowex XZS-1 resin with 0.5 M HCl. The authors reported that the regeneration is possible due to increased competition between Hg(II) and hydronium ions and formation of $[HgCl_4]^{2-}$ complex. Thiomethyl and Duolite GT73 resins loaded with Cd(II) were regenerated by elution with hydrochloric acid (1 M and 4 M), sulphuric acid (1 M and 4 M) and acidified thiourea. The maximum Cd(II) desorption ratios were 49% for thiomethyl resin and 71% for Duolite GT73 using 4 M HCl [155]. Pagano et al. [227] also used HCl as regeneration agent for Purolite C106, a weak cation resin with carboxylate functionality, used for separation and recovery of Pb/Fe species from automobile battery manufacturing wastewater. Such regeneration was easily and efficiently carried out with 1 M HCl for quantitative recovery of the lead. In 2004, Pohl and Prusisz [162] used the chelating resin Metalfix Chelamine and the strong cation exchanger Dowex 50WX4 for the off-line pre-concentration of cadmium and other metals. The elution of cadmium ions from the solid was performed with 1 and 2 M solutions of hydrochloric and nitric acids, which permitted the recovery with efficiency higher than 90%; the best results were obtained using 2 M HNO_3 and 1 M HCl. Atia et al. [159] conducted a research on the recovery of Hg(II) from an amine–mercaptan resin using HCl, HNO_3 and H_2SO_4 with different concentrations. The last ones originated high elution ratios (>90%), whereas HCl attained only 32%. This lower efficiency may be explained on the basis of the uptake of Hg(II) through ion exchange mechanism at higher concentration of HCl due to the formation of $[HgCl_3]^-$, which can be electrostatically attached on the RNH^+ sites formed at low pH [159]. Sharaf et al. [156] also achieved a high recovery of metal ions from methylthiourea- and phenylthiourea-supporting resins in the range of acidic pH. A mixture of 2 M HCl and 10% thiourea was the best solution to recover metals from the resins since thiourea forms stable complexes with the sorbed metal ions. Badawy et al. [161] studied the behaviour of various eluting agents for the removal of Pb(II) ions sorbed onto Purolite C100. Quantitative results were obtained with nitric acid and ammonium acetate. The elution of Pb(II) reached 95.6% with 2 M nitric acid and increased with increasing acid concentrations, while ammonium acetate was found to be an efficient elute in the 1–4 M range. In the study of Denizli et al. [164], the dithiocarbamate-incorporated microspheres loaded with Hg(II) ions were placed within the desorption medium containing 1 M HNO_3. The results obtained indicated that the exchanger regeneration by strong acids is feasible, as the desorption ratio was very high (up to 99%). Chelate-forming interactions between incorporated dithiocarbamate groups and Hg(II) ions are weaker at decreasing pH. The sorption capacity of the recycled solid can still be maintained at 98% at the fifth cycle [164]. Manju et al. [165] tested two solutions for the desorption of Pb(II), Hg(II) and Cd(II) from PGHyFeO-COOH: 0.5 M HCl and 0.2 M NaCl. The authors reported that the Na^+ ions were easily displaced by the divalent metal ions during the sorption step of experiments indicating an ion exchange process. Notwithstanding, efficiencies of 98.3% for Pb(II), 93.1% for Hg(II) and 90.1% for Cd(II) were obtained by using 0.2 M NaCl, concluding that this solution was therefore suitable for the regeneration of the spent PGHyFeO-COOH. The small fraction of sorbed metals not recoverable presumably represents ions bound through strong interaction and so the capacity is reduced in subsequent cycles [165]. Lv et al. [188] used 1 M $NaNO_3$ and 0.1 M EDTA-Na_2 solutions to regenerate an ETS-10 bed. As the structure of ETS-10 is known to be unstable at pH <4, conventional reagents such as acids cannot be used [188]. The efficiency of EDTA-Na_2 solution was much better than that of $NaNO_3$ solution. Three consecutive sorption–regeneration cycles were performed with EDTA-Na_2, and the recovery efficiency was maintained at about 94%, though the sorption capacity of Pb(II) ions has lowered [188].

Tüzün et al. [197] studied the desorption of biosorbed Hg(II), Cd(II) and Pb(II) ions from *C. reinhardtii* biomass with 0.1 M HCl and attained efficiencies from 87% to 98%. The authors also found that the whole process is reversible, namely the algae biomass can be used repeatedly for metal ions removal from wastewater. Lodeiro et al. [198] studied 11 types of eluting solutions at different concentrations to find out the most appropriate to treat Cd(II)-loaded biomass. It was pointed out the high efficiency of chloride and nitrate calcium salts as desorbing agents for the Cd(II) previously bound to protonated *S. muticum* alga, causing no structural damage to the biomass or even reinforcing it. HNO_3 and HCl were observed to provide similar desorption capacity but were more aggressive for the stability of the algae. Nevertheless, acid desorption contributed to sweep soluble biomass material, which could block the fixed bed columns employed in the cyclic process [198].

Sari and Tuzen [205] successfully recovered Hg(II) from *D. revolvens* biomass (99%) using 1 M HCl, while Huang et al. [201] verified that Cd(II) ions sorbed on *H. verticillata* were desorbed effectively by about 85%, 79% and 85% employing 0.10 M EDTA, HCl and HNO_3, respectively. Zhai et al. [219] used various concentrations of HCl or HNO_3 solutions to elute Hg(II) ions from the composite magnetic particles (DPC-NF-SiO_2) under study, and better results were accomplished with HNO_3.

Summarily, research pointed out that the majority of sorbents can be eluted and regenerated by acids such as HCl, HNO_3 and H_2SO_4. Chelating agents such as EDTA can also be applied, especially when exchangers are not resistant to acidic conditions.

11.5.6 Equilibrium Isotherms

The distribution of metal ions between solution and exchanger is a measure of the position of equilibrium and can be expressed by one or more isotherms [228]. Equilibrium isotherms are usually classified into empirical equations and mechanistic models (see Chap. 3). The mechanistic models are based on physical/chemical mechanisms of metal ion sorption being able not only to represent but also to explain and predict the experimental behaviour. However, the application or empirical extension of models such as Langmuir, Freundlich, Langmuir–Freundlich, Redlich–Peterson, Tempkin and Dubinin–Radushkevich (D–R) are more often used. Among them, the most common are the monolayer Langmuir and the multilayer Freundlich equations.

Monteagudo and Ortiz [153] applied the Langmuir model to the equilibrium data of the removal of Hg from mine wastewater by commercial resins (Dowex XZS1, Dowex XVS and Amberlite 252) and determined maximum sorption capacities ranging from 3.88 to 4.39 meq g^{-1}. Valverde et al. [158] investigated the ion exchange equilibrium of H^+/Cu^{2+}, H^+/Zn^{2+} and H^+/Cd^{2+} in aqueous solutions with the weak acid resin Lewait TP-207 and found that the equilibrium data were satisfactorily correlated using the Langmuir-type empirical equation. Atia et al. [159] reported that the amine–mercaptan resin displays higher uptake capacity (3.10 mmol g^{-1}) as well as higher binding constant (2.0) towards Hg(II) over other metal ions (Cd(II), Zn(II), Cu(II), Pb(II), Ca(II), Mg(II)), according to the Langmuir equation. Kacaoba [154] verified the sorption of Pb(II) and Cd(II) by the cation exchanger Amberlite IR120 is supported by Freundlich and Langmuir isotherms. Moreover, following Langmuir isotherm, the maximum concentrations in the resin are 0.0236 meq g^{-1} for Pb(II) and 0.0104 meq g^{-1} for Cd(II) [154]. Sharaf et al. [156] also found that Langmuir and Freundlich equations fitted quite well the equilibrium data obtained for the removal of Cd(II) by methylthiourea- and phenylthiourea-supporting resins. The maximum capacities of both materials were quite similar, 3.97 mmol g^{-1} for the methylthiourea resin and 3.87 mmol g^{-1} for the phenylthiourea resin. In 2002, Manju et al. [165] adopted the Langmuir model to the equilibrium data of Pb(II), Hg(II) and Cd(II) on PGHyFeO-COOH at 30°C, 40°C, 50°C and 60°C. At 30°C, the sorption capacities of PGHyFeO-COOH were around 211.4 mg g^{-1} for Pb(II), 155.0 mg g^{-1} for Hg(II) and 147.2 mg g^{-1} for Cd(II). Bessbousse et al. [152] modelled the equilibrium of Hg(II) onto PVA/P_4VP membrane using Langmuir, Freundlich and Dubinin–Radushkevich isotherms. The last two models were not obeyed, whereas the Langmuir equation provided excellent fitting with the experimental data, supporting the hypothesis that sorption is limited by saturation of the membrane sites. According to these researchers, the sorption capacity of the PVA/P_4VP membrane, given by the Langmuir equation, is 455 mg g^{-1}. Mandal and Ghosh [180] used Langmuir and Freundlich models to fit the sorption data of Pb(II) by Versatic 911 coated on silanized silica gel. The results obtained suggested the applicability of the Langmuir isotherm and indicated the formation of a monolayer at the surface of the sorbent corresponding to 33.8 mg g^{-1}. Payne and Abdel-Fattah [184] reported that the sorption of Pb(II) ions by synthetic zeolites 13X and 5A is best fitted by the Langmuir isotherm, while the Freundlich expression represents better the ion exchange of Pb(II) ions with natural zeolites chabazite and clinoptilolite. Perić et al. [185] performed an extensive analysis and screening of isotherms for the equilibrium data of Pb(II) and other metals on natural Croatian zeolitic tuff. Equations like Langmuir–Freundlich, Redlich–Peterson, Toth, Dubinin–Radushkevich, modified Dubinin–Radushkevich and Lineweaver–Burk have been tested. The best fitting results were obtained with models, which assume that the ionic species bind first to the energetically most favourable sites, followed by multilayer sorption. Kocaoba [154] found out that sorption of Pb(II) and Cd(II) ions onto dolomite is supported by Freundlich and Langmuir isotherms. Lv et al. [187] observed that the equilibrium behaviour of metal species with a stronger affinity towards ETS-10 can be described by Langmuir equation, while Choi et al. [190] verified that the data for both Pb(II) and Cd(II) onto ETS-10 and ETAS-10 were well fitted by Langmuir–Freundlich isotherm. In other study on Cd(II) uptake by ETS-10, both Langmuir and Langmuir–Freundlich isotherms provided accurate representation over the range of the experimental conditions points, though Langmuir–Freundlich model performed slightly better [173]. Barreira et al. [172] reported that the equilibrium data for Cd(II) removal by ETS-4 at pH 6 were accurately represented by the Langmuir–Freundlich isotherm. In other essay, Lopes et al. [176] found that the equilibrium results for Hg(II) on titanosilicate ETS-4 are well described by Langmuir, Freundlich and Dubinin–Radushkevich isotherms, although the second one offered slightly better fittings. The biosorption equilibria of Hg(II), Cd(II) and Pb(II) ions onto *C. reinhardtii* biomass were well represented by the Freundlich equation [197], while the biosorption data of Cd(II) and Pb(II) onto six different algae were quite well optimized by Langmuir model [200]. Huang et al. [201] applied Langmuir and Freundlich models to describe the equilibrium isotherms of Cd(II) sorption onto *H. verticillata* with success. The Dubinin–Radushkevich isotherm was also applied and the mean free energy determined (11.2 kJ mol^{-1}) indicated that the cadmium sorption onto *H. verticillata* might be carried out via ion exchange mechanism. Ghodbane and Hamdaoui [204] reported that the biosorption equilibrium of Hg(II) onto eucalyptus bark is much better fitted

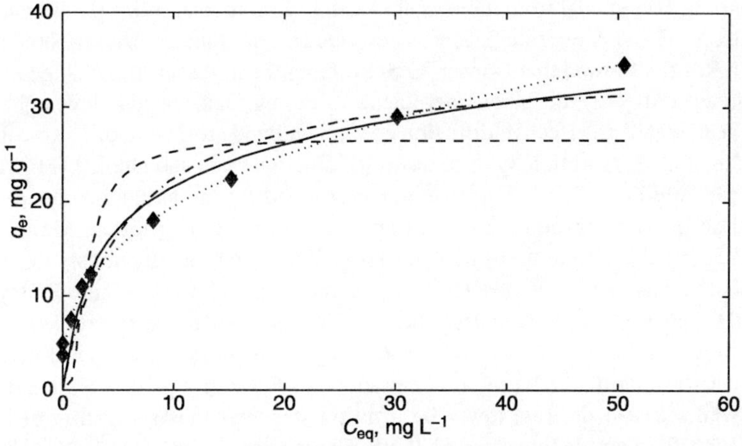

Fig. 11.11 Sorption isotherms of Pb^{2+} on *Moringa oleifera* bark. Symbols: experimental data; *full line*: Temkin model; *dashed line*: Dubinin–Radushkevich model; *dotted line*: Freundlich model; *dash-dotted line*: Langmuir model (Reproduced from Ref. [206] with kind permission of © Elsevier (2010))

by the Langmuir model than Freundlich model. The capacities obtained were 33.1, 33.2 and 34.6 mg g^{-1} at 20°C, 30°C and 40°C, respectively. In the study performed by Reddy et al. [206], the biosorption equilibrium data of Pb(II) on *Moringa oleifera* were analysed using Langmuir, Freundlich, Dubinin–Radushkevich and Temkin isotherm models. As Fig. 11.11 points out, the best results are accomplished with the Freundlich model. Equilibrium sorption data of metal ions, such as Cd(II), Pb(II), Hg(II) and others, onto a biomatrix prepared from rice husk are well represented by both Langmuir and Freundlich isotherms, probably due to the real heterogeneous nature of the surface sites involved in the metal uptake [212]. Recently, Zhang et al. [220] used both Langmuir and Freundlich equations to describe the equilibrium of Hg(II) on PANI/HA nanocomposite. The former model correlated with the equilibrium experimental data, suggesting that the sorption occurs in a monolayer. The theoretical maximum adsorption capacities for Hg(II) at three ionic strengths (0.02, 0.20 and 1.00 M) were about 380, 671 and 493 mg g^{-1}.

11.5.7 Kinetic Models

As shown in Chap. 3, several models have been employed to describe ion exchange kinetics. Although the models based on the Nernst–Planck (NP) and Maxwell–Stefan (MS) equations are the most appropriate for this purpose, empirical kinetic models based on the capacity of the sorbent, such as the Lagergren's first-order equation and Ho's second-order expression, are the most widely used to describe the sorption process.

The Hg(II) removal from mine wastewater by commercial resins (Dowex XZS-1, Dowex XVS and Amberlite 252) was described by the squared driving force model [153]. Results showed that Dowex XZS-1 offered the highest ion exchange rate and the greatest effective diffusion coefficient. Adopting Fick's second law, Saha et al. [155] calculated an effective diffusion coefficient inside resins, in order to compare the rate of Cd(II) sorption by thiol resins (thiomethyl resin and Duolite GT73) from aqueous solutions. The kinetics studies performed by the authors evidenced that metal sorption by these resins is intraparticle diffusion controlled. Manju et al. [165] adopted a simple first-order reversible kinetic model, previously developed by Bhattacharya and Venkobachar [229], to model metal sorption onto PGHyFeO-COOH. The process was well described by this model and the intraparticle diffusion was shown to be the main rate-limiting step. Bessbousse et al. [152] observed that the sorption of Hg(II) by a membrane with P_4VP groups was quite slow and followed a first-order rate law. The limiting step was the association of Hg(II) ions with the complexing sites of P_4VP. Chojnacki et al. [181] verified that the sorption of Hg(II) on Sokornit zeolite was fast, followed a first-order reversible mechanism and equilibrium. Lv et al. [187] investigated the competitive sorption kinetics of binary systems Cu(II)/Pb(II) and Cd(II)/Pb(II) and ternary systems Cd(II)/Cu(II)/Pb(II) and observed that the uptake of Pb(II) ions by ETS-10 was rapid, essentially attaining equilibrium within 5 min under the experimental conditions used. Moreover, the kinetic data of metal species with a stronger affinity towards ETS-10 were well represented by the pseudo-second-order model, and all correlation coefficients were close to unity. Choi et al. [190] verified that the uptake rates for both Pb(II) and Cd(II) by ETS-10 and ETAS-10 were extremely rapid and fitted to the pseudo-second order. Barreira et al. [172], Ferreira et al. [174] and Camarinha et al. [173] successfully described the kinetic of Cd(II) uptake by ETS-4 or ETS-10 with a Nernst–Planck based model, which combines both film and intraparticle diffusion control, where the convective mass transfer coefficient and the self-diffusivities of Cd(II) and Na^+ ions are the unique parameters. Lopes et al. [176] reported that the removal of Hg(II) from aqueous solution by ETS-4 is well predicted by the pseudo-second-order kinetic model; moreover, film diffusion is the rate-controlling step. In other work, the same authors reported that the ion exchange kinetic of Hg(II) uptake was also successfully described by a Nernst–Planck based model with both intraparticle and film diffusion resistances [171]. Pagnanelli et al. [208] found that Cd(II) biosorption by olive mill solid residue takes place within 2 h, while for Hg(II) and Pb(II), a 4-h period is necessary to reach the equilibrium conditions, probably because of the large ionic radius and the related problems of diffusion in the biological matrix. The biosorption equilibrium of Hg(II), Cd(II) and Pb(II) ions onto *C. reinhardtii* followed a second-order kinetics model [197]. Ghodbane and Hamdaoui [204] reported that the biosorption of Hg(II) onto eucalyptus bark was uniform and rapid. The authors investigated the mechanism of sorption by modelling the kinetic data using pseudo-first-order and pseudo-second-order kinetic equations, and intraparticle diffusion model. They have concluded that among the kinetic models studied, the pseudo-second-order equation was the best applicable to describe the process. In the case of Pb(II) sorption onto *Moringa oleifera* bark, the pseudo-first-order, pseudo-second-order and intraparticle diffusion models were analysed by Reddy et al. [206]. The best kinetic correlation was provided by the second-order kinetics, although intraparticle diffusion could not be ignored. Studies conducted by Krishnani et al. [212] on the kinetics of metal ions (II) removal revealed that the majority of them (Cd(II), Pb(II), Hg(II), etc.) were removed within the first 90–120 min contact with a biomatrix prepared from rice husk. More recently, El-Shafey [213] found that Zn(II) sorption onto a carbonaceous sorbent chemically prepared from rice husk was fast, reaching equilibrium within 2 h, while Hg(II) sorption was slow attaining equilibrium within 120 h. The experimental data for both metals were found to follow pseudo-second-order model [213]. Zhai et al. [220] reported that the sorption of Hg(II) ions onto DPC-NF-SiO_2 was very rapid initially and gradually became slower with lapse of time. About 95% of total Hg(II) was sorbed within the first 5 min. Zhang et al. [220] verified that the kinetic of Hg(II) ions removal by PANI/HA nanoparticles was initially rapid and reached equilibrium after approximately 200 min. Moreover, the pseudo-second-order equation was the model that furthered the best fit of the experimental data, suggesting chemical sorption as the rate-limiting step of the process and no involvement of a mass transfer in solution [220].

Summarily, the pseudo-second-order equation fitted the kinetic data very well in a large quantity of publications in the literature for Pb, Cd and/or Hg recovery, separation and/or pre-concentration.

11.5.8 Column Experiments

Investigations conducted under batch operation provide useful information on the sorption capacity of the sorbents and kinetic parameters, which are the basis for fixed bed implementation. The percolation column is one of the most effective configurations for cyclic sorption–desorption operation, and some works have been published regarding Pb, Cd and/or Hg recovery, separation and/or pre-concentration in fixed bed.

Monteagudo and Ortiz [153] studied the removal of Hg(II) from mine wastewater with Dowex XZS-1 cationic resin in a semi-continuous operation, subjecting the resin to a series of recycle runs using a ion exchange loading step followed by an acid regeneration step. The mine wastewater was pumped through a column of fresh resin (20 g of dry resin) at 0.3 $dm^3\ min^{-1}$ flow rate. The breakthrough curves did not change after consecutive regeneration cycles, showing the effectiveness of the Dowex XZS1 in the semi-continuous operation. Saha et al. [155] used thiomethyl and Duolite GT73 resins in mini-columns to predict the metal sorption performance and selectivity towards the metal ion. About 1 g of dry sorbent was loaded into a mini-column, where a Cd(II) feed concentration of 1 mM (pH 4) was pumped through at a flow rate of ~10 BV h^{-1}. Both thiol resins showed a similar trend for Cd(II) sorption (5% breakthrough capacity for cadmium was 0.03 mM g^{-1} for both resins; 5% breakthrough capacity was calculated based on the total amount of metal removed when the outlet metal concentration reached 5% of the feed concentration). Singh and Srivastava [160] tested the microcapsule in columns in order to obtain the breakthrough curve of Cd(II). The authors used 1 g of microcapsule resin packed into a glass column, and a feed solution containing 0.01 M Cd(II) + 0.1 M HCl was fed to the column at a constant flow rate of 0.5 mL min^{-1}. The breakpoint volume (5% breakthrough) was estimated to be 268 mL, for a bed volume of 223 mL, and the breakthrough capacity was evaluated to be 2.68 mmol g^{-1}. Lv et al. [188] examined the breakthrough curves of Pb(II) in an ETS-10 fixed bed at various operating conditions. The results obtained showed that the sorbed amount of Pb(II) ions per unit mass solid particles was about 1.68 mmol g^{-1}. This amount was not markedly affected by the operating conditions because of the rapid sorption rate of Pb(II) ions on ETS-10. In other study, the breakthrough curves of Pb(II) sorption on ETS-10 particles were simulated using a two-phase homogeneous diffusion model [189], and a good agreement between the numerical simulation and the experimental results was observed. Zeroual et al. [196] investigated the biosorption of Hg(II) on protonated *U. lactuca* in a packed biomass column and found the flow rate affects directly the efficiency of the treatment, as the breakpoint was 2.7 h for 12 mL min^{-1} and 1.18 h for 20 mL min^{-1}, which evidences mass transfer limitations and/or hydrodynamic deviations to ideal behaviour. Lodeiro et al. [198] used the protonated alga *Sargassum muticum* in column studies to remove Cd(II) ions from wastewaters and concluded that it could compete with commercial biosorbents for the removal. The system tested was able to operate during 605 h for sorption and 66 h for desorption, maintaining a Cd(II) effluent concentration lower than 0.02 mg L^{-1} during at least 2 h in the 11 cycles performed. After an acid treatment, the invasive algae *S. muticum* was able to support an uninterrupted use during, at least, 28 days subjected to continuous Cd(II) sorption–desorption cycles in a fixed bed column, with no apparent diminution of its performance, despite the 27% weight loss found for the whole process [198]. Sousa et al. [214] investigated the application of green coconut shells as sorbent for the removal of Pb(II), Cd(II) and other metals from aqueous effluents under percolation operation. They determined optimum operating conditions for a bed height of 10 cm, and according to the breakthrough curve obtained, they found that larger quantities of effluent can be treated for individual ions with respect to multiple ions.

11.6 Other Metals

Besides the precious, radioactive and 'priority' pollutants reported in the previous sections, many other metals are released into the environment by industrial activities such as mining and metal processing, causing a serious worldwide problem. Once more, numerous attempts have been made to treat industrial wastewater in order to recover, separate or pre-concentrate metals, not only to reduce the risk of human and environmental contamination but also to recycle metals with economical value.

Most of the studies available in literature are regarded to copper (Cu), zinc (Zn), nickel (Ni), chromium (Cr), cobalt (Co), iron (Fe), arsenic (As) and manganese (Mn) due to their toxicity and/or economical importance. Other metals such as calcium (Ca), magnesium (Mg), molybdenum (Mo), selenium (Se), vanadium (V), zirconium (Zr) and titanium (Ti) are also focused but in much less extent.

Copper is an essential nutrient as well as a water contaminant. It has many commercial applications. It is used to make pipes, valves and fittings and is present in alloys and coatings. Copper toxicity inhibits natural growth and development of plants and affects negatively self-purification of water and living organisms in the water environment. The application of ion exchange to different types of wastewater has been widely used to remove and recovery copper ions [6]. Nickel is an important contaminant in industrialized areas. It is used in steel and other alloys, electroplating and batteries. The combustion of oil and coal results in a significant contribution of Ni to the atmosphere [4]. For this metal, the ion exchange has also been successfully applied. Zinc is an essential trace element, and due to its dissolution from pipes, its concentration in tap water can be much higher than the established limits. According to literature, the most appropriate method to remove zinc from wastewater is extraction [6]. Yet, this method has several economical disadvantages, and ion exchange appears as a promissory alternative. Chromium is another metal usually present in industrial wastewater. The main source of water contamination is the industries of surface metal treatment and tannery. These industrial wastewater is characterized by low pH caused by the presence of free acids [6]. Besides these metals, many others can cause environmental and public health problems, and many researchers have been exhaustively studying their removal, recovery and separation with very satisfactory results.

Table 11.6 displays the main results of several researchers on the recovery, separation or pre-concentration of other metals by ion exchange under different operating conditions.

The ion exchangers used for these metals are almost the same cited before and include a vast variety of resins, composites, zeolites and other microporous materials, numerous types of biomaterials and more recently some nanomaterials.

Resins are probably the most adopted ion exchangers. Several types of commercial resins, such as Amberlite, Lewatit, Versatic, Purolite, Indion, Amberlyst, Dowex, Cyanex, Duolite, have been intensely employed. In 2001, Yalçin et al. [229] developed a combined treatment scheme for the recovery of Cu and Cr from acidic and cyanide-containing alkaline wastewater of electroplating industries. Acid bath wastes were treated with Dowex 50X8, and the retained Cu was eluted with H_2SO_4 [230]. The cyanide-containing alkaline bath waste was firstly oxidized with excessive hypochorite, then neutralized and recovered with the Amberlite IRC718 resin [230]. Deepatana and Valix [231] investigated the recovery of Ni and Co from organic acid complexes using a chelating aminophosphonate Purolite S950 resin. The authors reported that the sorption of the various Ni and Co complexes were quite low, 16–18 and 5.4–9 mg g^{-1} of resin, respectively. Another Purolite-type resin (Na-form of strong-acid Purolite NRW100) was studied by Juang et al. [232], also to recover Ni(II) from a simulated electroplating waste solution containing $NiSO_4$, NH_4Cl,

Table 11.6 Materials most frequently used for the removal of other metals together with the corresponding experimental and major results found in literature

		Operation conditions			Results						
Metal	Material	pH	T (°C)	C_0 (mg/L)	Batch	Column	Kinetics	Isotherms	Regeneration	Others	Ref.
Cu	Bromomethylated poly(2,6-dimethyl-1,4-phenylene oxide) hollow fibre membrane	4,75	R.T.	0–5.0×10^{-3}	x	x	x	x	x	Capacity; desorption performance; breakthrough curve	[243]
Cr	Amberlyst 15 (H$^+$)	1.7–2.9	20–60	1.92–19.2 mM	x		x	x		Capacity; thermodynamics	[240]
Co, Cu, Cd	Three wool powders	3–9	R.T.	10^{-3}–10^{-6} M	x				x	Distribution coefficient; capacity	[248]
Zn, Hg	Carbonaceous sorbent chemically prepared from rice husk	1.5–6	25–45	Zn: 25–300; Hg: 100–1,500	x		x	x		Capacity	[213]
Mg, Ca, Mn, Cr, Fe,Co, Ni, Cu, Zn, Cd, Hg, Pb	Hybrid inorganic/organic adsorbents – hybrid alumina adsorbents	1–6	R.T.	0.1 M	x		x			Capacity; distribution coefficient	[179]
Zn	PVA/EDTA resin	N.A.	25, 30, 35	10–100	x		x	x	x	Capacity; removal efficiency %; thermodynamics	[238]
Pb, Ni, Cd, Zn, Cu	Green coconut shells	N.A.	N.A.	N.A.		x			x	Maximum capacity; breakthrough curves	[214]
As	*Paecilomyces variotii* biomass	2–9	N.A.	0.5–2.5	x	x	x	x	x	Capacity; breakthrough curves	[250]
As, Cd, Cu, Pb	Natural Muscovite	2–10	20	0–100	x		x	x		Capacity; removal %	[251]
Cu	Fresh fungal biomass of *Agaricus bisporus*	1–6	20	30–100			x	x		Capacity; thermodynamics	[252]

As, Fe, Cr	Purolite ArsenXnp, Purolite D3235, Purolite D3442, Purolite D3687, and Fuji PEI-CS-07	N.A.	N.A.	N.A.	x	x				Removal %	[237]
Cr	Indion 790	0.5–5	R.T.	$0.5–5\times10^3$	x	x	x	x	x	Sorption performance and capacity; extraction %	[239]
Pb, Cu	Negatively charged hybrid materials	N.A.	N.A.	0.05 M	x					Capacity; conducto-metric analysis	[175]
Pb, Cd, Hg, Cu, Ni, Zn, Cr	Natural zeolites	1–10	N.A.	10–100	x		x			Metal removal %; capacity	[186]
Ni, Zn, Mn	Hyperbranched poly(ethy-lenimine) (HBPEI)	1–9	N.A.	0–120	x			x		Sorption behaviour; capacity; recovery %	[242]
Cu, Ni	Amberlite IRC748	2–6	25–45	4.73–18.9 M	x	x	x	x	x	Capacity; breakthrough curve	[236]
Cu, Cd	Chelating resin, cross-linked poly(glycidyl methacrylate-aspartic acid)	1–5	25	0–12 mM	x		x	x	x	Capacity; desorption %; competitive adsorption and selectivity	[253]
Cr	Lewatit MP 64	N.A.	N.A.	700		x	x		x	Breakthrough curves; laboratory scale; pilot scale	[254]
Fe, Cu, Zn, Ni	Amberlite IRC718	2–7	25	15–30 mg/50 mL	x		x			Metal-ion uptake	[235]
Ni, Zn, Cd, Mn, Co, Cu, Hg, Pb	Rice husk	2–7	32	50–200	x	x	x	x	x	Capacity; breakthrough curve	[212]
Cu, Zn	Amberlite XAD4	N.A.	25	5–50 mM	x		x	x		Sequential extraction; batch separation	[234]

(continued)

Table 11.6 (continued)

Metal	Material	Operation conditions			Results						
		pH	T (°C)	C_0 (mg/L)	Batch	Column	Kinetics	Isotherms	Regeneration	Others	Ref.
Cd, Pb, Zn	Chelating resins	3–9	30–50	N.A.	x		x	x		Uptake %; capacity; thermodynamics	[156]
Pb, Cu, Cd	ETS-10 titanosilicate	N.A.	25, 40, 60	5–15 mM		x			x	Breakthrough curves; competitive ions	[188]
Hg, Fe, Ni, Pb, Cu, Zn	Chitosan	5–7	N.A.	N.A.	x					Metal chelation and recovery	[215]
Cd, Co, Cu, Fe, Ni, Pb, Zn	Polyethylenimine methylenephosphonic acid (PEIMPA)	2.4–5.9	N.A.	N.A.	x			x		Liquid–solid extraction; recovery (%)	[169]
Ni, Co	Purolite S950	N.A.	N.A.	15–2,000	x			x	x	Capacity; regeneration efficiency	[231]
Ca, Mg	Dowex 50WX4, Diaion WT01S and Dowex 1X4	4, 6.5	N.A.	10×10^{-3}		x				Fractionation analysis	[233]
Cu, Zn, Cd	*Sargassum fluitans* brown seaweed biomass	5	N.A.	N.A.		x				Breakthrough curves; modelling	[199]
Pb, Mg, Ni, Fe, Ca	Amberlite IRC718	1–6	N.A.	300	x	x				Capacity; distribution coefficient; metal separation	[255]
Ni	Purolite NRW100	0.5–6	15–45	0.85–11.9 mM	x		x	x		Amount of metal exchanged; thermodynamics	[232]
Cu, Co, Mn, Zn	ETS-10 and ETAS-10 titanosilicates	N.A.	25, 40, 50	1–30 mM	x		x	x		Capacity; distribution coefficient; thermodynamics	[245]
Zn, Cd, Hg	Polymer (3-n-propyl (4-methylpyridinium) silsesquioxane chloride)	N.A.	25	N.A.	x			x		Capacity; sorption equilibria modelling	[256]

Pb, Cu, Cd	ETS-10 titanosilicate	5	25	N.A.	x		x	x		Capacity; selectivity; binary mixtures	[187]
Fe, Ti	Polysulphone cation-exchange membrane (SA_3S)	N.A.	N.A.	0.01, 0.001 M						Metal flux; dialysis	[241]
Cd, Co, Ni, Zn, Cu, Zr	Microcapsule containing chelating resin	1–7	25	1.0×10^{-2} M	x	x	x			Distribution coefficients; adsorption %; breakthrough curve	[160]
Cu, Cd, Zn	Lewatit TP 207	4–7	25	N.A.	x			x		Separation factor; selectivity	[158]
Cd, Co, Cu, Ni, Zn	Metal fix Chelamine resin and cation exchanger Dowex 50WX4	4	N.A.	0,1		x			x	Pre-concentration; recovery %	[162]
Zn, Cu, Pb	Natural zeolite	5	23	Zn:0.57–10 mM Cu: 0.65–10 mM Pb: 0.55–8.9 mM	x			x		Uptake degree	[185]
Cu, Ni, Cd, Pb, Fe, Hg, Cr	Wheat bran	1–9	25	Cr:20; Pb:30; Ni:20; Fe:10; Cd:2; Cu:10; Hg: 500	x			x		Percent adsorbed; selectivity coefficient; capacity; pre-concentration	[209]
Co, Cu, Hg, Ni, Pb, Zn	Water-soluble N-benzoylthiourea modified PAMAM polymers	1–9	20	0.5 mM	x					Retention %	[168]
Cr	Resins 1200H, 1500H and Amberlite IRN97H	2–9	25	10–30	x		x	x		Removal %	[257]

(continued)

Table 11.6 (continued)

		Operation conditions			Results						
Metal	Material	pH	T (°C)	C_0 (mg/L)	Batch	Column	Kinetics	Isotherms	Regeneration	Others	Ref.
Hg, Pb, Cd, As, Cr	Thiol functional organoceramic composite	3, 4, 5	25	150–1,800	x	x	x	x	x	Capacity; selectivity; breakthrough curve	[166]
Cr, Mn, Fe, Co, Ni, Cu, Zn, Cd, Hg, Pb	Modified silica gel phases: amino ($SiNH_2$) and chloro-(SiCl)	N.A.	N.A.	0.1 M	x					Percentage recovery; distribution coefficient	[167]
Hg, Pb, Cu, Zn, Cd	Olive mill solid residues	4–5	N.A.	N.A.	x		x	x	x	Capacity; regeneration %	[208]
Cu, Cr	Dowex 50X8, Dowex 1X8, Amberlite IRC50, and Amberlite IRC718	N.A.	N.A.	N.A.		x				Breakthrough curves	[230]
Pb, Fe	Duolite A7 and Purolite C106	Fe: 3; Pb: 6	N.A.	Fe: 2; Pb: 4		x			x	Selective removal; separation and recovery; breakthrough curves	[157]
Cu, Cd, Ni, Zn	Thiomethyl resin and Duolite GT73	2–4	20	0.2–4 mM	x	x	x	x	x	Capacity; diffusion coefficient; breakthrough curves	[155]
Ni, Hg, U, As, Cu, Cd, Co, Cr, Pb	*Pseudomonas fluorescens* 4F39	1–10	30	0.85 mM	x					Metal accumulation	[207]
Cu, Cd, Zn, Hg, Pb, Cr, U	Poly(4-vinylpyridine) hydrochloride resin	N.A.	20–35	7–40	x			x	x	Adsorption and elution selectivity	[151]

Cu, Pb, Zn, Co, Ni, Hg, Cd, Mn, Cr, Ag	Poly(acrylamino-phosphonic-carboxyl-hydrazide) chelating ion exchange fibre	1–7	25	0.015 M	x			Binding capacity; distribution coefficient	[163]
As, Cd, Cr, Cu, Fe, Ni, Hg, Pb, Zn	Tree barks	0–10	N.A.	10, 100, 1,000	x	x		Removal (%); capacity	[202]
As, Au, Cd, Co, Cr, Cu, Hg, Mo, Ni, Pb, Se, Sr, V, Zn	*Escherichia coli* and *Zoogloea ramigera*	6	R.T.	10–100	x	x	x	Metal binding; multiple cycles	[195]
Ag, Pb, Cd, Zn, Cu, Ni, Hg	Erionite, zeolite 13X, Chabazite, Clinop-tilolite, Mordenite, zeolite 3A, zeolite 4A, and zeolite 5A	3.6–5.3	N.A.	N.A.		x		Capacity; breakthrough curves	[182]

N.A. - Not available
R.T. - Room temperature

NaH_2PO_4 and citrate. Pohl and Prusisz [233] applied several resins (the strong acidic Dowex 50WX4, the weak acidic Diaion WT01S and the strong basic anion exchanger Dowex 1X4) to study the operational fractionation of Ca and Mg in bee honeys, fruit juices and tea infusions. Chen and Huang [234] analysed the separation of Cu and Zn in the citric acid leachate of industrial print wiring board sludge by D_2EHPA-modified Amberlite XAD4 resin. It has been found that it preferentially remove Zn in the practical leachate, leaving only Cu ion in solution. Noureddine et al. [235] studied the sorption properties of the commercial Amberlite IRC718 towards the divalent metal ions Fe(II), Cu(II), Zn(II) and Ni(II),and found that the resin exhibited higher capacities and more pronounced sorption towards Fe(II) and that the metal ion uptake follows the order: Fe(II) > Cu(II) > Zn(II) > Ni(II). Lin et al. [236] used another Amberlite type resin (Amberlite IRC748), which has iminodiacetic acid (IDA) functional ligand, to remove Cu(II) and Ni(II) from synthetic aqueous solutions. The obtained values for the saturated amounts of exchange at 25°C were 1.13 mol Cu(II) kg^{-1} and 1.06 mol Ni(II) kg^{-1}. Fernández-Olmo et al. [237] conducted an investigation on the removal of As(III), Fe(III) and Cr(III) at trace levels from HF solutions by means of specialty ion exchange resins. From the eight resins tested, Fuji PEI-CS-07 showed the highest capacity among the commercially available resins and Purolite D3235 and D-3687 from among the non-commercially ones. Zhang et al. [238] investigated the poly(vinyl alcohol)/EDTA (PVA/EDTA) resin containing hydroxyl, carboxyl and amine functional groups for removing the target pollutant Zn(II) from aqueous solutions. Sahu et al. [239] and Mustafa et al. [240] studied the removal of Cr(III) by ion exchange resins. The former researchers used the Indion 790 resin, which is a macroporous strongly acidic cation exchange of sulphonated polystyrene group, while the latter ones used Amberlyst 15 (H^+), a macroporous strong cation exchange resin with a polystyrene-DVB structure with SO_3^{2-} functional groups. Mustafa et al. [240] reported that Amberlyst 15 (H^+) provided fast kinetics and high exchange capacity towards Cr(III) ions in the acidic medium and that the Cr(III) sorption was the result of exchange between H^+ and $Cr(OH)^{2+}$ ions in the solution. Sahu et al. [239] also reported fast kinetics and high exchange capacity towards the Cr(III) ions. According to them, the resin extracted up to 92% of Cr(III) in 15 min from a 500 mg L^{-1} Cr(III) solution at pH 2.7 but less at higher concentrations.

Other types of ion exchangers like inorganic and hybrid materials, composites and membranes have been also investigated for metal recovery, separation and/or pre-concentration. Liu et al. [163] prepared a poly(acrylaminophosphonic-carboxyl-hydrazide) chelating ion exchange fibre by the amination and phosphorization reaction of the hydrazine-modified polyacrylonitrile fibre and reported that this sorbent exhibits high capacities for Cu(II), Pb(II), Zn(II), Co(II), Ni(II), Hg(II), Cd(II), Mn(II), Cr(III) and Ag(I). Osman et al. [167] studied metal sorption, solid-phase extraction and pre-concentration properties of two silica gel phases chemically modified with 2-hydroxy-1-naphthaldehyde. The authors referred in this essay that the two new chemically modified phases were highly selective for Cu(II), Zn(II), Cd(II), Hg(II) and Pb(II). Kır et al. [241] investigated the effect of chelating agent on the separation of Fe(III) and Ti(IV) from binary mixture solution by a charged polysulphone cation exchange membrane (SA_3S). Schulte-Bockholt and Schuster [242] conducted an investigation on the removal, enrichment and recovery of Ni(II), Zn(II) and phosphate from phosphation rinsing water with liquid-phase polymer-based retention technique. In that study, they used hyperbranched polyethylenimine (HBPEI) as complexing and ion exchange agent. The recovery of Ni(II) and Zn(II), accomplished in a pilot unit, constantly reached values over 99%, and starting from an initial concentration of ca. 20 mg L^{-1}, an enrichment up to 1,000 mg L^{-1} of both ions was achieved. In 2010, Mahmoud and co-workers [179] synthesized, characterized and examined three hybrid inorganic/organic sorbents based on alumina and phenylazoformic acid 2-phenylhydrazide for their metal sorption properties. The main purpose of their essay was to study and explore the combined hybrid characters of inorganic/organic sorbents for the selective removal and pre-concentration of metals via static and dynamic solid-phase techniques from industrial wastewater and drinking tap water samples. These hybrid sorbents were successfully implemented for the selective removal of Pb(II), Cu(II), Fe(III) and Cr(III) from industrial wastewater samples with recoveries of 91–99 ± 2–3%, as well as 98–99 ± 1–3% for the selective pre-concentration of Pb(II), Cu(II) and Cr(III) from drinking tap water samples. Recently, Cheng et al. [243] developed a novel hollow fibre cation exchange membrane from bromomethylated poly(2,6-dimethyl-1,4-phenylene oxide) which exhibited good performance for the removal of Cu(II), and its capacity can reach 69.1 mg g^{-1}.

As described in the previous section, natural zeolites are abundant and low cost resources. They are crystalline hydrated aluminosilicates with a framework structure containing which sorbed water and alkali and alkaline earth cations. Due to their high cation exchange ability and molecular sieve properties, natural zeolites have been widely used as sorbents in separation and purification processes in the past decades [244]. Mondale et al. [182] investigated the ion exchange capabilities of several natural and synthetic zeolites towards different metal ions and reported that the approximate order of effectiveness removal from pristine solutions was Ag(I) > Pb(II) > Cd(II) > Zn(II) > Cu(II) >> Ni(II) > Hg(II). Perić et al. [185] studied the uptake of Zn, Cu and Pb from aqueous solutions by ion exchange on natural zeolitic tuff. The authors reported that the efficiency of removal is higher for Pb and Cu than for Zn ions. This method is the most efficient process when dealing with small pollution concentrations. Choi et al. [245] used nano-sized large-pore titanosilicate materials, ETS-10 and ETAS-10 with different Al_2O_3/TiO_2 ratio, for the removal of heavy metal ions, in particular, Cu(II), Co(II), Mn(II) and Zn(II). The monolayer capacities on ETS-10 and ETAS-10 at 298 K were 1.55 and 1.62 mmol g^{-1} for Cu(II), 1.30 and 1.32 mmol g^{-1} for Zn(II), 0.98 and 1.11 mmol g^{-1} for Mn(II), and 0.91 and 0.89 mmol g^{-1} for Co(II), respectively. Moreover, the authors suggest that ETS-10 and its variant ETAS-10 would be highly prospective and efficient in removing heavy metal ions from wastewater compared to other types of conventional sorbents including zeolites.

A vast array of biological materials, especially bacteria, algae, yeasts and fungi, have received increasing attention for metal removal and recovery due to their good performance, low cost and large available quantities [217]. Their effectiveness is due to the presence of a variety of functional sites including carboxyl, imidazole, sulphydryl, amino, phosphate, sulphate, thioether, phenol, carbonyl, amide and hydroxyl moieties [217]. Xie et al. [195] used *Escherichia coli* (ATCC 23724) and *Zoogloea ramigera* 115 (ATCC 25935) as biosorbents for removal and recovery of a large amount of metals, e.g. Cd(II), Co(II), Cr(III), Cu(II), Hg(II), Ni(II), Pb(II), Se(II), Sr(II), V(II) and Zn(II), from wastewater. These authors also reported that generally, most biomass exhibits a good ability to bind cations with little evidence of selectivity. Gaballah and Kilbertus [202] used raw and chemically treated tree barks for the decontamination of synthetic solutions containing 10,100 and 1,000 mg L^{-1} of As, Cd, Cu, Cr, Fe, Pb, Hg, Ni and Zn. Several European barks from the following species, oak (*Quercus robur*), beech (*Fagus sylvatica*), piecea (*Picea abies*) and Scotch pine (*Pinus sylvestris*), were included in their study. Gaballah and Kilbertus [202] reported that this alternative can be used for the decontamination of industrial effluents or for the polishing of pre-treated effluents by conventional technologies, being possible to decrease the effluents' metal ions content to ppb level. López et al. [207] studied the accumulation of metals by *Pseudomonas fluorescens* 4F39 and observed that the affinity series decreased in the order Ni > Hg > U > As > Cu > Cd > Co > Cr > Pb. Moreover, they emphasized that the accumulation of metals on the surface of microbial cells is a consequence of the net negatively charged surface and is influenced by the chemistry of the cell wall, physico-chemical characteristics of the environment, such as pH, and the sequence of metal hydrolysis. Among biological sorbents, agricultural materials have a highlight role due to being easily and widely produced. Food crops, like sugar cane, rice, corn, wheat and many others, are being cultivated all over the world, and the parts other than fruit, grain and juice are available for biosorption experimentation [227]. Farajzadeh and Monji [209] investigated the sorption characteristics of wheat bran towards metal cations and found capacities of 93 mg g^{-1} for Cr(III), 70 mg g^{-1} for Hg(II), 62 mg g^{-1} for Pb(II), 21 mg g^{-1} for Cd(II), 15 mg g^{-1} for Cu(II) and 12 mg g^{-1} for Ni(II). Their results also evidenced the selectivity of the wheat bran for metal cations compared with sodium and potassium ions and showed that more than 82% of studied cations, except Ni(II), were removed from aqueous solution in single step. Rice husk, another waste from agro-industry, was investigated by Krishnani et al. [212] for the uptake of several metals (Ni, Zn, Cd, Mn, Co, Cu, Hg and Pb) from single and mixed solutions. These authors reported maximum capacities, obtained from the Langmuir isotherm, of 0.094 mmol g^{-1} for Ni, 0.124 mmol g^{-1} for Zn, 0.149 mmol g^{-1} for Cd, 0.151 mmol g^{-1} for Mn, 0.162 mmol g^{-1} for Co, 0.172 mmol g^{-1} for Cu, 0.18 mmol g^{-1} for Hg and 0.28 mmol g^{-1} for Pb. Algae are other type of bisorbents widely investigated to sequester metals from industrial effluents or to recover precious metals from processing solutions. However, the term algae refers to a large and diverse assemblage of organisms which contain chlorophyll and carry out oxygenic photosynthesis [246], and according to their type, (green, red or brown), algal biosorbents may depict distinct biosorption capacities. Romera et al. [200] measured the capacity of green, red and brown algae to recover Cd, Ni, Zn, Cu and Pb from aqueous solutions. The brown algae achieved the lowest metal concentration levels in solution, and the best results were obtained with *Fucus spiralis*. Davis et al. [247] and Naja and Volesky [199] studied the effectiveness of *Sargassum fluitans* brown seaweed biomass to remove toxic metals, e.g. Cu(II), Zn(II) and Cd(II), from aqueous solutions. Grimm et al. [247] also investigated a brown alga, *Fucus vesiculosus*, together with other two biomaterials, birch wood *Betula* sp. and terrestrial moss *Pleurozium schreberi*, for removal of copper ions from diluted aqueous solutions. The maximum sorption capacities, in decreasing order, is as follows: marine alga (23.4 mg g^{-1})>moss (11.1 mg g^{-1})>birch wood sawdust (4.9 mg g^{-1}). These researchers also verified that the capacities of the studied biosorbents seem to correlate well with their amounts of surface acidic groups. Grimm et al. [247] concluded that alga and moss biomasses can conveniently be utilized as potential biosorbents in removal of Cu(II) ions from aqueous media, on contrary to sawdust from birch wood which was found less effective for this purpose. Recently, Naik et al. [248] investigated metal binding properties of novel wool powders and noted that the wool fibres exhibited potential for the removal and recovery of toxic chemical and metal ions; nonetheless, the slow kinetics of binding has limited their widespread application.

11.6.1 Effect of pH

As mentioned previously, the pH of solution is probably the most important parameter on metal recovery, separation and/or pre-concentration by ion exchange, and so its influence has been widely analysed by researchers.

Noureddine et al. [235] verified that metal ion uptake by Amberlite IRC718 increased with pH and approached a maximum at about pH 6.0. This effect was more pronounced with Fe(II), whereas the sorption of other metal ions like Cu(II), Zn(II) and Ni(II) exhibited only a slight increase. This behaviour may be explained by the nature of the chelating group [235]. Sahu et al. [239] verified that the extraction of chromium by the Indion 790 resin from 500 mg L^{-1} Cr(III) solutions of various pH remained almost constant (92%) within the pH range 0.5–3.5, when the resin was loaded to about 30% of its capacity. Therefore, the equilibrium constant for ion exchange is large and the distribution coefficient remains sufficiently high even at low pH. However, Cr(III) uptake decreased beyond pH 3.5, and at pH 5.0, only 65% was extracted due to the hydrolysis of Cr(III) ion to form $Cr(OH)_3$, $Cr(OH)SO_4$ [239]. Zhang et al. [238] observed that, in general, the sorption of Zn(II) onto a PVA/EDTA resin was affected significantly by pH values, as its sorption incremented from 0.51% to 99.7%, when the solution pH varied from 1.0 to 8.0. At pH 1, there was nearly no exchange at all, but increasing it from 4 to 6, the sorbed Zn(II) increased from 37.0% to 99.7% [238]. At higher pH values, the authors observed a slight decrease, since Zn(II) could be suffering hydrolysis to form $Zn(OH)^+$ species, which promotes a reduction of the solid load, due to the diminution of the formal charge of the metallic ion.

Liu et al. [163] reported that the sorption of the chelating fibre for Cu(II) is greatly influenced by the pH value: the degree of removal is constant and equals

99% above pH 3.0, and decreases rapidly in the range 1–3, while Cu(II) ion is scarcely sorbed below pH 1.0. Cheng et al. [243] verified that the capacity of a novel hollow fibre cation exchange membrane is also strongly pH dependent. These authors reported that at pH 1.8, the sorption capacity is about 28.8 mg g^{-1}; after pH increases, the maximum capacity leaps to 37.4 mg g^{-1} at pH 2.5. However, when pH increases further, the capacity initially holds almost constant and then reduces quickly to 32.6 mg g^{-1} at pH 4.8. Such variation can be explained by the fact that H^+ can prevent the sorption of Cu(II) due to competition in an acid medium, and OH^- can reduce the concentration of Cu(II) in an alkaline medium.

López et al. [207] observed that accumulation of metals by *Pseudomonas fluorescens* 4F39 was pH dependent. The cell wall contains amines, amides and carboxylic functional groups that are protonated or deprotonated, depending on the pH of the aqueous medium. Increasing pH increases the negative charge at the surface of the cells until all relevant functional groups are deprotonated, which favours electrochemical attraction and sorption of cations. Furthermore, the increase in metal uptake with increasing pH may be the result of more efficient competition of cations with H^+ for binding sites on bacteria [207]. These researchers grouped metal cations into those whose accumulation increased as pH increased, with a maximum accumulation at the pH before precipitation (Ni, Cu, Pb, Cd, Co), and those whose maximum accumulation was not associated with precipitation (Cr, As, U, Hg). Farajzadeh and Monji [209] investigated the effect of pH on the sorption characteristics of wheat bran towards metal cations and observed that the sorbed cations percentage is intensively decreased at pH 1 relative to pH 5, due to competition between proton ions with target cations (Fig. 11.12). Krishnani et al. [212] also considered the effect of pH but on the binding of metal ions to lignocellulosic biomatrix of rice husk and verified that more metal cations were bound at higher pH with maximum uptake taking place at pH between 5.5 and 6. According to the surface complexation theory, such trend can be explained on the basis of a decrease in competition between proton and metal species for the surface sites and by the decrease in positive surface charge [212]. Romera et al. [200] verified that sorption capacity depended on the pH and found that the optimum pH value for recovery of Cd, Ni and Zn was 6 for the six algae under study. The optimum sorption pH for Cu ranged from 4 to 5 and for Pb from 3 to 5.

Again, it is demonstrated that metal recovery, separation and/or pre-concentration is strongly affected by pH, independently of the type of sorbent used (resins, inorganic materials, zeolites and biomaterials). The extent of that influence is determined by the system.

11.6.2 Sorbent Regeneration

Another important characteristic of a sorbent is whether it can be regenerated. Deepatana and Valix [231] investigated the elution efficiency of both Ni and Co complexes from Purolite S950 resins using citric, malic and lactic acids (0.01 and 0.1 M).

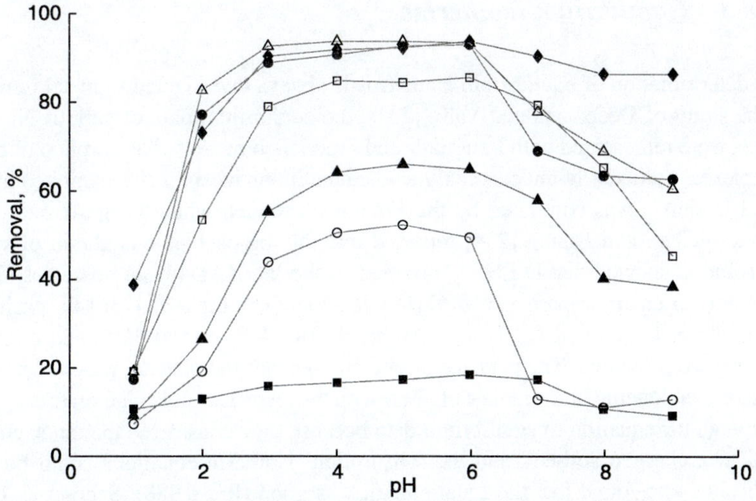

Fig. 11.12 Influence of pH on the removal of metal ions by the bran of wheat. Symbols: (♦) Hg; (●) Fe; (Δ) Pb; (□) Cd; (▲) Cr; (■) Ni; (◊) Cu. *Curves* are *lines* to guide the eyes (Reproduced from Ref. [209] with kind permission of © Elsevier (2004))

They reported high efficiencies for both complexes (82–98%), which indicates that the process involves weak interactions, where the nature of metal complexes and metal speciation have little effect on the overall desorption efficiency from Purolite S950 [231]. Lin et al. [236] compared solutions of EDTA, citrate and HCl as elution agents of Cu(II) and Ni(II) from Amberlite IRC748 and registered that its ability decreases in the order EDTA > HCl > citrate. This trend is related with the fact that stability constants of Cu(II) or Ni(II) with resin ligands (1:1 complex) are larger than that with citrate, though being much smaller than that with EDTA in aqueous solutions. Additionally, since HCl is a strong acid, the concentration of H^+ overlaps that originated by citric acid in solution; therefore, the equilibrium elution using citric acid is the poorest. Zhang et al. [238] investigated the desorption and regeneration potential of PVA/EDTA resin using an acid treatment (0.01–0.15 M HCl). The authors reported a high desorption efficiency (95%) of Zn(II) from the exchanger using acid treatment.

Grimm et al. [247] studied the reusability of the biosorbents alga, moss and sawdust by performing five consecutive sorption–desorption tests after using dilute HCl solution as elute. It was concluded that the desorption of saturated biomasses could be performed in 1 h with diluted HCl as eluting agent, without damage of the biosorbents and/or loss of performance after those cycles.

To make the sorption process more economical and viable, it is indispensable to regenerate the loaded sorbent for reusing in multiple cycles. In the whole, several methods are presented in the literature for elution and regeneration of metal-loaded sorbents using some elutes, usually acids.

11.6.3 Equilibrium Isotherms

The determination of equilibrium isotherms is always investigated in metal removal. In the study of Deepatana and Valix [231], the equilibrium data of various Ni complexes were represented with Langmuir and Freundlich models. The sorption of bulky complexes, particularly under weakly acidic conditions where resins exhibited higher metal loadings, was correlated by the Freundlich model, which suggests multilayer sorption. Chen and Huang [234] reported that the ion exchange isotherm of single ion solution showed that D_2EHPA-modified Amberlite XAD4 resin possesses higher selectivity to Zn in comparison with Cu ion. In the study of Lin et al. [236], the exchange isotherms of Cu(II) and Ni(II) onto Amberlite IRC748 resin at 25°C and three pH values were presented. The isotherm eventually reaches the plateau, which represents the saturated loading of the metals of interest on the resin. Lin and collaborators applied the Langmuir equation to equilibrium data because they considered the ion exchange resin as a charged sorbent, and the 'adsorption' isotherm equations could be thus applicable, and found that the Langmuir fit was good ($R^2>0.988$). Sahu et al. [239] reported that the sorption of Cr(III) on Indion 790 followed a Freundlich isotherm, and the high value of the Freundlich constant confirmed strong chemical interaction with the resin; the maximum loading was found to be 86.9 mg g^{-1} in a 500 ppm Cr(III) solution [239]. The equilibrium data for the exchange of Zn(II) on PVA/EDTA resin was tested with various equations (Langmuir, Freundlich, Tempkin and Dubinin–Radushkevich), among which three models were found to be suitable for the Zn(II) adsorption (Langmuir, Tempkin and Dubinin–Radushkevich isotherms) [238].

In another work, the sorption of Cu(II) ions on a novel hollow fibre cation exchange membrane was studied by Cheng et al. [243]. The equilibrium data were fitted by the Langmuir and Freundlich equations, and it was observed the former found much better results.

Perić et al. [185] studied the uptake of Zn, Cu and Pb from aqueous solutions by ion exchange on natural zeolitic tuff and used the adsorption isotherm of Langmuir–Freundlich, Redlich–Peterson, Toth, Dubinin–Radushkevich, modified Dubinin–Radushkevich and Lineweaver–Burk. The best correlation of the experimental points was accomplished by models that assume that the ionic species bind first on energetically most favourable sites, with multilayer sorption taking place subsequently. In the study of Choi et al. [245], the equilibrium data of Cu(II), Co(II), Mn(II) and Zn(II) removal by ETS-10 and ETAS-10 have been fitted by Langmuir, Freundlich and Langmuir–Freundlich isotherms. The authors found that Langmuir–Freundlich isotherm achieved best representation.

Krishnani et al. [212] applied successfully both Langmuir and Freundlich isotherms to the sorption of components from liquid phase onto rice husk and verified that the Langmuir equation gives rise to better fitting. The capacity taken from the Langmuir model increases in the order: Ni(II) < Zn(II) ≈ Cd(II) ≈ Mn(II) ≈ Co(II) < Cu(II) ≈ Hg(II) < Pb(II). Romera et al. [200] verified that experimental data of biosorption of heavy metals using different types of algae obeyed Langmuir equation very well and that in all cases, the removal capacity was greater for Pb followed by Cd. Moreover, the sorption values for Ni, Cu and Zn were very similar, and the general sequence was: Pb > Cu ≥ Cd > Zn > Ni. Grimm et al. [247] noted that the isotherms of Cu(II) ion onto three biosorbents studied are well described by Langmuir model. According to this equation, the maximum sorption loadings showed that the alga had the best Cu-binding ability (23.4 mg g^{-1}), followed by the moss (11.1 mg g^{-1}) and the sawdust (4.9 mg g^{-1}).

Once more, the Langmuir model (based on monolayer adsorption of solute) and the Freundlich model (developed for heterogeneous surfaces) are the most widely used in literature, providing information on metal uptake loadings at equilibrium. However, these equations empirically utilized with correlation purposes only, do not reflect any mechanisms of sorbate uptake and hardly have a meaningful physical interpretation [217]. A detailed treatment of this subject is given in Chap. 3. Moreover, real wastewaters commonly contain a mixture of metal ions, and a significant gap exists in literature regarding binary and ternary sorption and modelling.

11.6.4 Kinetic Models

As has been mentioned in the previous sections, the rate of metal sorption is an important factor and prerequisite for process optimization and equipment sizing in practical applications [238]. Sahu et al. [239] observed that the sorption of Cr(III) on Indion 790 was fast and followed first-order kinetics. Zhang et al. [238] observed that removals of about 60–70% of Zn(II) by the PVA/EDTA resin occurred in 15 min, and equilibrium was attained at around 30 min. Moreover, the kinetic data were well described by the pseudo-second-order model.

On the study of removal of Cu(II), Co(II), Mn(II) and Zn(II) by large-pore titanosilicate materials, ETS-10 and ETAS-10, Choi et al. [245] observed extremely rapid uptake rates, in less than a few minutes, and reported the pseudo-second-order model provides good results.

Studies conducted on the kinetics of metal ions (II) removal by rice husk revealed that the majority of the ionic species were removed within the first 90–120 min of contact and that the equilibrium state was reached after 120–150 min [212]. Grimm et al. [247] observed good rates of Cu(II) uptake by the tested biosorbents (alga, moss and birch wood sawdust) even for low initial concentrations. The pseudo-second-order rate model properly described the experimental data measured. Lin et al. [236] performed fixed bed experiments at 25°C in a water-jacketed glass column to remove of Cu(II) and Ni(II) from aqueous solutions using Amberlite IRC748 resin. In Fig. 11.13, the measured breakthrough curves at three pH values are shown and evidence dispersive effects decreasing with increasing pH. These researchers applied the Thomas model, containing two parameters, to follow the exchange dynamics and found good agreement between the calculated modelling and data (standard deviation 6%).

As for lead, cadmium and mercury ions (previous section), most of the literature available for the recovery, separation and/or pre-concentration usually adopts the pseudo-first-order and pseudo-second-order models to describe the kinetic of the process.

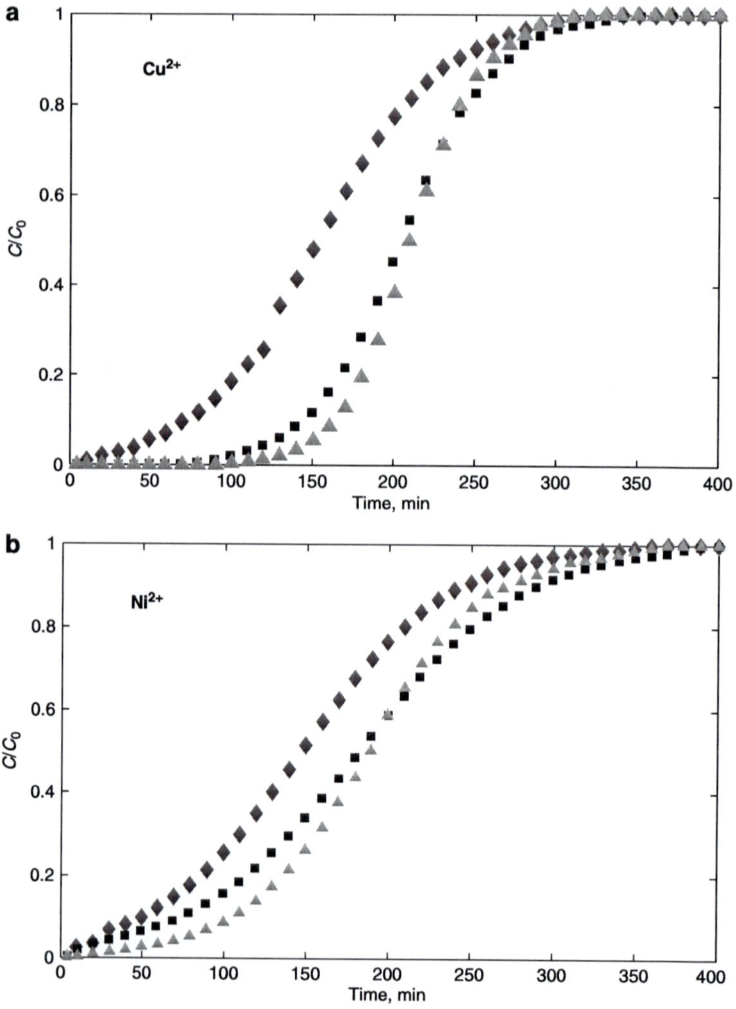

Fig. 11.13 Breakthrough curves of single (**a**) Cu^{2+} and (**b**) Ni^{2+} solution measured in an Amberlite IRC748 resin bed at pH 2 (♦), pH 4 (▲) and pH 5 (■) (Reproduced from Ref. [236] with kind permission of © Elsevier (2008))

References

1. Kadlec RH, Knight RL (1996) Treatment wetlands. CRC Press, Boca Raton
2. Bolton Jr., H. Gorby YA (1995) An overview of the bioremediation of inorganic contaminants. In: Hinchee RE, Means JL, Burris DR (eds) Bioremediation of inorganics. Battelle Press, Columbus
3. Siegel FR (2002) Environmental geochemistry of potentially toxic metals. Springer, Berlin
4. Clark RB, Frid C, Attrill M (2001) Marine pollution, 5th edn. Oxford University Press, Oxford

5. Nriagu JO (2000) Series preface. In: Markert B, Friese K (eds) Trace metals in the environment. Elsevier, New York
6. Dabrowski A, Hubicki Z, Podkoscielny P, Robens E (2004) Selective removal of the heavy metal ions from waters and industrial wastewaters by ion-exchange method. Chemosphere 56(2):91–106
7. Taty-Costodes VC, Fauduet H, Porte C, Ho Y-S (2005) Removal of lead (II) ions from synthetic and real effluents using immobilized Pinus sylvestris sawdust: Adsorption on a fixed-bed column. J Hazard Mater 123(1–3):135–144
8. Baral SS, Das N, Ramulu TS, Sahoo SK, Das SN, Chaudhury GR (2009) Removal of Cr(VI) by thermally activated weed Salvinia cucullata in a fixed-bed column. J Hazard Mater 161(2–3):1427–1435
9. Valderrama C, Arévalo JA, Casas I, Martínez M, Miralles N, Florido A (2010) Modelling of the Ni(II) removal from aqueous solutions onto grape stalk wastes in fixed-bed column. J Hazard Mater 174(1–3):144–150
10. Tchobanoglous G, Burton FL, Stensel HD (2003) In: Eddy M (ed) Wastewater engineering: treatment and reuse. McGraw-Hill, Boston
11. Cruz-Olivares J, Pérez-Alonso C, Barrera-Díaz C et al (2010) Inside the removal of lead(II) from aqueous solutions by De-Oiled Allspice Husk in batch and continuous processes. J Hazard Mater 181:1095–1101
12. Liu P, Liu G-F, Chen D-L, Cheng S-Y, Tang N (2009) Adsorption properties of Ag(I), Au(III), Pd(II) and Pt(IV) ions on commercial 717 anion-exchange resin. Trans Nonferrous Metals Soc China 19(6):1509–1513
13. Hubicki Z, Leszczynska M (2005) Sorption of palladium (II) chloride complexes on weakly, intermediate and strongly basic anion exchangers. Desalination 175:227–236
14. Gelin P, Primet M (2002) Complete oxidation of methane at low temperature over noble metal based catalysts: a review. Appl Catal Environ 39:1–37
15. Farrauto R, Liu Y, Ruettinger W, Ilinich O, Shore L, Giroux T (2007) Precious metal catalysts supported on ceramic and metal monolithic structures for the hydrogen economy. Cat Rev Sci Eng 49(2):141–196
16. Huang XH, El-Sayed IH, Qian W, El-Sayed MA (2006) Cancer cell imaging and photothermal therapy in the near-infrared region by using gold nanorods. J Am Chem Soc 128(6):2115–2120
17. Ramesh A, Hasegawa H, Sugimoto W, Maki T, Ueda K (2008) Adsorption of gold(III), platinum(IV) and palladium(II) onto glycine modified crosslinked chitosan resin. Bioresour Technol 99:3801–3809
18. Atia AA (2005) Studies on the interaction of mercury(II) and uranyl(II) with modified chitosan resins. Hydrometallurgy 80:13–22
19. Qu R, Sun C, Wang M, Ji C, Xu Q, Zhang Y, Wang C, Chen H, Yin P (2009) Adsorption of Au(III) from aqueous solution using cotton fiber/chitosan composite adsorbents. Hydrometallurgy 100:65–71.
20. Patterson JW (1985) Industrial wastewater treatment technology, 2nd edn. Butterworth, Boston
21. Ajiwe VIE, Anyadiegwu IE (2000) Recovery of silver from industrial wastes, cassava solution effects. Sep Purif Technol 18:89–92
22. Çoruh S, Senel G, Ergun ON (2010) A comparison of the properties of natural clinoptilolites and their ion-exchange capacities for silver removal. J Hazard Mater 180:486–492
23. Akgul M, Karabakan A, Acar O, Yurum Y (2006) Removal of silver (I) from aqueous solutions with clinoptilolite. Micropor Mesopor Mater 94:99–124
24. Kononova ON, Leyman TA, Melnikov AM, Kashirin DM, Tselukovskaya MM (2010) Ion exchange recovery of platinum from chloride solutions. Hydrometallurgy 100:161–167
25. Krishna MV, Ranjit M, Chandrasekaran K, Venkateswarlu G, Karunasagar D (2009) On-line preconcentration and recovery of palladium from waters using polyaniline (PANI) loaded in mini-column and determination by ICP-MS; elimination of spectral interferences. Talanta 79:1454–1463
26. Shams K, Goodarzi F (2006) Improved and selective platinum recovery from spent a-alumina supported catalysts using pretreated anionic ion exchange resin. J Hazard Mater B 131:229–237

27. Buslaeva TM (1999) Platinum group metals and their role in contemporary society. Sorosovskiy Obrazovetelny Zhurnal 11:45–49
28. Buyanov RA, Pakhomov NA (2001) Catalysts and processes of dehydrogenation of paraffins and olefins. Kinet Katal 42:72–85
29. Hubickia Z, Wójcika G (2006) Studies of removal of platinum(IV) ion microquantities from the model solutions of aluminium, copper, iron, nickel and zinc chloride macroquantities on the anion exchanger Duolite S 37. J Hazard Mater B 136:770–775
30. Xiong Y, Adhikari CR, Kawakita H, Ohto K, Inoue K, Harada H (2009) Selective recovery of precious metals by persimmon waste chemically modified with dimethylamine. Bioresour Technol 100:4083–4089
31. Neagu V, Paduraru C, Bunia I, Tofan L (2009) Platinum (IV) recovery from chloride solution by functionalized acrylic copolymers. J Environ Manage 91:270–276
32. Cox M, Pichugin AA, El-Shafey EI, Appleton Q (2005) Sorption of precious metals onto chemically prepared carbon from flax shive. Hydrometallurgy 78(1–2):137–144
33. Hubicki Z, Wójcik G (2006) Studies of removal of platinum(IV) ion microquantities from the model solutions of aluminium, copper, iron, nickel and zinc chloride macroquantities on the anion exchanger Duolite S 37. J Hazard Mater B 136:770–775
34. Shams K, Beiggy MR, Shirazi AG (2004) Platinum recovery from a spent industrial dehydrogenation catalyst using cyanide leaching followed by ion exchange. Appl Catal Gen 258:227–234
35. Belyaev AV (2003) Chemical-technological problems of platinum group metals during processing of exhausted nuclear fuel. Zhurnal Strukturnoy Khimii 44:39–47
36. Spektor OV, Ryumin AI, Pochekutova MG (1998) Methods for recovery of platinum group metals from spent catalysts. Tsvet Met 7:31–39
37. Brooks CS (1991) Metal recovery from industrial wastes. Lewis, Chelsea
38. Chang YC, Chen DH (2006) Recovery of gold(III) ions by a chitosan-coated magnetic nano-adsorbent. Gold Bull 39:98–102
39. Gomes CP, Almeida MF, Loureiro JM (2001) Gold recovery with ion exchange used resins. Sep Purif Technol 24:35–57
40. Zhao YZ (2006) The enrichment and separation of race gold, Pt and Pd from the ores based on co-precipitation. Gold 27:42–44
41. Alguacil FJ, Adeva P, Alonso M (2005) Processing of residual gold (III) solutions via ion exchange. Gold Bull 38:9–13
42. Akita S, Yang L, Takeuchi H (1996) Solvent extraction of gold(III) from hydrochloric acid media by nonionic surfactants. Hydrometallurgy 43:37–46
43. Hubicki Z, Leszczynska M, Lodyga B, Lodyga A (2006) Palladium(II) removal from chloride and chloride-nitrate solutions by chelating ion-exchangers containing N-donor atoms. Miner Eng 19(13):1341–1347
44. Weng CH, Huang CP (2004) Adsorption characteristics of Zn (II) from dilute aque- ous solution by fly ash. Colloids Surf A: Physicochem Eng Aspects 247:137–143
45. Annadurai G, Juang RS, Lae DJ (2002) Adsorption heavy metals from water using banana and orange peels. Water Sci Technol 47:185–190
46. Çay S, Uyanık A, Özasık A (2004) Single and binary component adsorption of copper (II) and cadmium (II) from aqueous solutions using tea-industry waste. Sep Purif Technol 38:273–280
47. Srivastava VC, Mall ID, Mishra IM (2007) Adsorption thermodynamics and isosteric heat of adsorption of toxic metal ions onto bagasse fly ash (BFA) and rice husk ash (RHA). Chem Eng J 132:267–278
48. Chassary P, Vincent T, Marcano JS, Macaskiec LE, Guibal E (2005) Palladium and platinum recovery from bicomponent mixtures using chitosan derivatives. Hydrometallurgy 76:131–147
49. Cortina JL, Meinhardt E, Roijals O, Marti V (1998) Modification and preparation of polymeric adsorbents for precious-metal extraction in hydrometallurgical processes. React Funct Polym 36(2):149–165
50. Zhang HG, Dreisinger DB (2002) The adsorption of gold and copper onto ion-exchange resins from ammoniacal thiosulfate solutions. Hydrometallurgy 66:67–76
51. Gallagher NP, Hendrix JL, Milosavljevic EB, Nelson JH, Solujic L (1990) Affinity of activated carbon towards some gold(I) complexes. Hydrometallurgy 25:305–316

52. Zhang HG, Dreisinger DB (2004) The recovery of gold from ammoniacal thiosulfate solutions containing copper using ion exchange resin columns. Hydrometallurgy 72:225–234
53. Sanchéz JM, Hidalgo M, Salvado V (2000) The separation of Au(III) and Pd(II) in hydrochloric acid solutions by strong anion type II exchange resins: the effect of counter ion concentration and temperature. Solvent Extr Ion Exch 18:1199–1217
54. Hubicki Z, Wołowicz A (2009) Adsorption of palladium(II) from chloride solutions on Amberlyst A 29 and Amberlyst A 21 resins. Hydrometallurgy 96:159–165
55. Kovacheva P, Djingova R (2002) Ion-exchange method for separation and concentration of platinum and palladium for analysis of environmental samples by inductively coupled plasma atomic emission spectrometry. Anal Chim Acta 464:7–13
56. Topp K-D, Grote M (1996) Synthesis and characterization of a 1,2,4,5-tetrazine-modified ion-exchange resin. React Funct Polym 31:117–136
57. Parodi A, Vincent T, Pilsniak M et al (2008) Palladium and platinum binding on an imidazol containing resin. Hydrometallurgy 92:1–10
58. Çelik Z, Gülfen M, Aydın AO (2010) Synthesis of a novel dithiooxamide–formaldehyde resin and its application to the adsorption and separation of silver ions. J Hazard Mater 174:556–562
59. Dubois MA, Dozol JF, Nicotra C et al (1995) Pyrolysis and incineration of cationic and anionic ion-exchange resins – identification of volatile degradation compounds. J Anal Appl Pyrolysis 31:129–140
60. Ngah WSW, Hanafiah MAKM (2008) Removal of heavy metal ions from wastewater by chemically modified plant wastes as adsorbents: a review. Bioresour Technol 99:3935
61. Kapoor A, Viraraghavan T (1995) Fungal biosorption – an alternative treatment option for heavy metal bearing wastewaters: a review. Bioresour Technol 53:195
62. Coleman NJ, Bishop AH, Booth SE, Nicholson JW (2009) Ag^+ and Zn^{2+} - exchange kinetics and antimicrobial properties of 11 Å tobermorites. J Eur Ceram Soc 290:1109–1117
63. Rocha J, Anderson MW (2000) Microporous titanosilicates and other novel mixed octahedral-tetrahedral framework oxides. Eur J Inorg Chem 2000:801–818
64. Babel S, Kurniawan TA (2003) Low-cost adsorbents for heavy metals uptake from contaminated water: a review. J Hazard Mater 97:219–243
65. Lihareva N, Dimova L, Petrov O, Tzvetanova Y (2010) Ag+ sorption on natural and Na-exchanged clinoptilolite from Eastern Rhodopes, Bulgaria. Micropor Mesopor Mater 130:32–37
66. Sharma P, Tomar R (2008) Synthesis and application of an analogue of mesolite for the removal of uranium(VI), thorium(IV), and europium(III) from aqueous waste. Micropor Mesopor Mater 116:641–652
67. Al-Attar L, Dyer A, Paajanenb A, Harjulab R (2003) Purification of nuclear wastes by novel inorganic ion exchangers. J Mater Chem 13:2969–2974
68. Riegel M, Tokmachev M, Hoell WH (2008) Kinetics of uranium sorption onto weakly basic anion exchangers. React Funct Polym 68:1072–1080
69. Tykva R, Din KS, Pavel CC, Cecal A, Popa K (2009) Contribution to the external surface of a titanium-rich sand (Abou-Khashaba, Egypt) in the uranium uptake processes. J Radioanal Nucl Chem 279(3):811–816
70. Rao TP, Metilda P, Gladis JM (2006) Preconcentration techniques for uranium (VI) and thorium(IV) prior to analytical determination. Talanta 68:1047–1064
71. Kalina M, Wheelerb WN, Meinrath G (2005) The removal of uranium from mining waste water using algal/microbial biomass. J Environ Radioact 78:151–177
72. Kadous A, Didi MA, Villemin D (2010) A new sorbent for uranium extraction: ethylenediamino tris(methylenephosphonic) acid grafted on polystyrene resin. J Radioanal Nucl Chem 284:431–438
73. Mellah A, Chegrouche S, Barkat M (2006) The removal of uranium(VI) from aqueous solutions onto activated carbon: kinetic and thermodynamic investigations. J Colloid Interface Sci 296:434–441
74. Maheswari MA, Subramanian MS (2004) Selective enrichment of U(VI), Th(IV) and La(III) from high acidic streams using a new chelating ion-exchange polymeric matrix. Talanta 64:202–209

75. Donia AM, Atia AA, Moussa EMM, El-Sherif AM, El-Magied MOA (2009) Removal of uranium(VI) from aqueous solutions using glycidyl methacrylate chelating resins. Hydrometallurgy 95:183–189
76. Eisenbud M, Gesell T (1997) Environmental radioactivity: from natural, industrial, and military sources, 4th edn. Academic, New York
77. Clifford D, Zhang ZH (1994) Modifying ion-exchange for combined removal of uranium and radium. J Am Water Works Assoc 86:214–227
78. Anirudhan TS, Radhakrishnan PG (2009) Improved performance of a biomaterial-based cation exchanger for the adsorption of uranium(VI) from water and nuclear industry wastewater. J Environ Radioact 100:250–257
79. Van Horn JD, Huang H (2006) Uranium(VI) bio-coordination chemistry from biochemical, solution and protein structural data. Coord Chem Rev 250:765–775
80. Anirudhan TS, Bringle CD, Rijith S (2010) Removal of uranium(VI) from aqueous solutions and nuclear industry effluents using humic acid-immobilized zirconium-pillared clay. J Environ Radioact 101:267–276
81. van Leeuwen FWV, Beijleveld H, Miermans CJH, Huskens J, Verboom W, Reinhoudt DN (2005) Ionizable (Thia)calix[4]crowns as highly selective $^{226}Ra^{2+}$ Ionophores. Analytica Chem 77:4611–4617
82. Kathren RL (1998) NORM sources and their origins. Appl Radiat Isotop 49:149–168
83. van Leeuwen FWB, Verboom W, Reinhoudt DN (2005) Selective extraction of naturally occurring radioactive Ra^{2+}. Chem Soc Rev 34(9):753–761
84. Melnikov P, Zanoni LZ (2010) Clinical effects of cesium intake. Biol Trace Elem Res 135:1–9
85. Cornell RM (1993) Adsorption of cesium on minerals: a review. J Radioanal Nucl Chem 171:483–500
86. Cseke JL, Kaufman PB, Polila GK, Tsai C-J (2004) Handbook of molecular and cellular methods in biology and medicine. 2nd edn. CRC Press, Boca Raton.
87. Cortés-Martinéz R, Olguin MT, Solache-Rios M (2010) Cesium sorption by clinoptilolite-rich tuffs in batch and fixed-bed systems. Desalination 258:164–170
88. Wang L, Feng M, Liu CX, Zhao YS, Li SQ, Wang H, Yan L, Tian G, Li SJ (2009) Supporting of potassium copper hexacyanoferrate on porous activated carbon substrate for cesium separation. Sep Sci Technol 44(16):4023–4035
89. Chegrouche S, Mellah A, Barkat M (2009) Removal of strontium from aqueous solutions by adsorption onto activated carbon: kinetic and thermodynamic studies. Desalination 235:306–318
90. Roskill O (1992) The economics of strontium, 6th edn. Roskill Information Service Ltd, London
91. Puziy AM, Kartel NT, Bortun LN, Strelko VV (1996) Adsorbent for strontium adsorption from water solutions. Ukrainian Patene 12352
92. Cho Y, Komarneni S (2009) Cation exchange equilibria of cesium and strontium with K-depleted biotite and muscovite. Appl Clay Sci 44:15–20
93. Ahmadpoura A, Zabihia M, Tahmasbib M et al (2010) Effect of adsorbents and chemical treatments on the removal of strontium from aqueous solutions. J Hazard Mater 182:552–556
94. Behrens EA, Clearfield A (1997) Titanium silicates, $M_3HTi_4O_4(SiO_4)^3.4H_2O$ ($M = Na^+, K^+$), with three-dimensional tunnel structures for the selective removal of strontium and cesium from wastewater solutions. Micropor Mater 11(1–2):65–75
95. Osmanlioglu AE (2006) Treatment of radioactive liquid waste by sorption on natural zeolite in Turkey. J Hazard Mater 137:332–335
96. El-Kamash AM (2008) Evaluation of zeolite A for the sorptive removal of Cs^+ and Sr^{2+} ions from aqueous solutions using batch and fixed bed column operations. J Hazard Mater 151(2–3):432–445
97. Zachara JM, Smith SC, Liu CX, McKinley JP, Serne RJ, Gassman PL (2002) Sorption of Cs^+ to micaceous subsurface sediments from the Hanford site, USA. Geochim Cosmochim Acta 66(2):193–211
98. Luca V, Hanna JV, Smith ME, James M, Mitchell DRG, Bartlett JR (2002) Nb-substitution and Cs^+ ion-exchange in the titanosilicate sitinakite. Micropor Mesopor Mater 55(1):1–13

99. Sylvester P, Clearfield A (1998) The removal of strontium and cesium from simulated hanford groundwater using inorganic ion exchange materials. Solvent Extr Ion Exch 16:1527–1539
100. Ahmadpour A, Zabihi M, Tahmasbi M, Bastami TB (2010) Effect of adsorbents and chemical treatments on the removal of strontium from aqueous solutions. J Hazard Mater 182:552–556
101. Pan J, Zou X, Yan Y, Wang X, Guan W, Han J, Wu X (2010) An ion-imprinted polymer based on palygorskite as a sacrificial support for selective removal of strontium(II). Appl Clay Sci 50:260–265
102. Sureda R, Martínez-Lladó X, Miquel Rovira M, Pablo J, Casas I, Giménez J (2010) Sorption of strontium on uranyl peroxide: Implications for a high-level nuclear waste repository. J Hazard Mater 181:881–885
103. Donat R (2009) The removal of uranium (VI) from aqueous solutions onto natural sepiolite. J Chem Thermodynam 41:829–835
104. Atun G, Ortaboy S (2009) Adsorptive removal of uranium from water by sulfonated phenol-formaldehyde resin. J Appl Polym Sci 114:3793–3801
105. Zou WH, Zhao L, Han RP (2009) Removal of uranium (VI) by fixed bed Ion-exchange column using natural zeolite coated with manganese oxide. Chinese J Chem Eng 17:585–593
106. Mellah A, Silem A, Boualia A, Kada R (1992) Adsorption of organic matter from a wet phosphoric acid using activated carbon: equilibrium study. Chem Eng Process 31(3):191–194
107. Qadeer R, Hanif J, Khan M et al (1995) Uptake of uranium ions by molecular sieve. Radiochimica Acta 68:197–201
108. Camacho LM, Deng S, Parra RR (2010) Uranium removal from groundwater by natural clinoptilolite zeolite: effects of pH and initial feed concentration. J Hazard Mater 175:393–398
109. Misaelides P, Godelitsas A, Filippidis A, Charistos D, Anousis I (1995) Thorium and uranium uptake by natural zeolitic materials. Sci Total Environ 173(1–6):237–246
110. Barton CS, Stewart DI, Morris KS, Bryant DE (2004) Performance of three resin-based materials for treating uranium-contaminated groundwater within a PRB. J Hazar Mater 116(3):191–204
111. Gu BH, Ku YK, Brown GM (2005) Sorption and desorption of perchlorate and U(VI) by strong-base anion-exchange resins. Environ Sci Technol 39:901–907
112. Ladeira ACQ, Morais CA (2005) Uranium recovery from industrial effluent by ion exchange – column experiments. Miner Eng 18:1337–1340
113. Ladeira ACQ, Goncalves CR (2007) Influence of anionic species on uranium separation from acid mine water using strong base resins. J Hazard Mater 148:499–504
114. Hasan S, Ghosh TK, Viswanath DS, Loyalka SK, Sengupta B (2007) Preparation and evaluation of fullers earth beads for removal of cesium from waste streams. Sep Sci Technol 42:717–738
115. Atia AA, Donia AM, El-Enein SA, Yousif AM (2007) Effect of chain length of aliphatic amines immobilized on a magnetic glycidyl methacrylate resin towards the uptake behavior of Hg(II) from aqueous solutions. Sep Sci Technol 42(2):403–420
116. Phillips DH, Gu B, Watson DB, Parmele CS (2008) Uranium removal from contaminated groundwater by synthetic resins. Water Res 42(1–2):260–268
117. Kilislioglu A, Bilgin B (2003) Thermodynamic and kinetic investigations of uranium adsorption on Amberlite IR-118H resin. Appl Radiat Isot 58:155–160
118. Chanda M, Rempel GL (1995) Polyethyleneimine gel-coat on silica. High uranium capacity and fast kinetics of gel-coated resin. React Polym 25:25–36
119. Moon DS, Burnett WC, Nour S, Horwitz P, Bond A (2003) Preconcentration of radium isotopes from natural waters using MnO_2 resin. Appl Radiat Isot 59:255–262
120. Clifford DA, Zhang Z (1995) Removing uranium and radium from groundwater by ion exchange resins. In: Sengupta AK (ed) Ion exchange technology. Technomic, Lancaster
121. Gu B, Ku Y-K, Jardine PM (2004) Sorption and binary exchange of nitrate, sulfate, and uranium on an anion-exchange resin. Environ Sci Technol 38:3184–3188
122. Bortun AI, Bortun LN, Clearfield A (1997) Evaluation of synthetic inorganic ion exchangers for cesium and strontium removal from contaminated groundwater and wastewater. Solvent Extr Ion Exch 15:909–929
123. Clearfield A (1995) Inorganic-ion exchangers – a technology ripe for development. Ind Eng Chem Res 34:2865–2872

124. Harjula R, Lehto J, Paajanen A, Tusa E, Yarnell P (2004) Use inorganic ion exchange materials as precoat filters for nuclear waste effluent treatment. React Funct Polym 60:85–95
125. Tachi Y, Shibutani T, Sato H, Yui M (2001) Experimental and modeling studies on sorption and diffusion of radium in bentonite. J Contam Hydrol 47:171–186
126. Liu CX, Zachara JM, Qafoku O, Smith SC (2003) Effect of temperature on Cs^+ sorption and desorption in subsurface sediments at the Hanford Site, USA. Environ Sci Technol 37(12):2640–2645
127. Guerra DL, Leidens VL, Viana RR, Airoldi C (2010) Amazon kaolinite functionalized with diethylenetriamine moieties for U(VI) removal: Thermodynamic of cation-basic interactions. J Hazard Mater 180(1–3):683–692
128. Clearfield A (2000) Inorganic ion exchangers, past, present, and future. Solvent Extr Ion Exch 18:655–678
129. Koivula R, Harjula R, Lehto J (2006) Selective removal of radionuclides from nuclear waste effluents using inorganic ion exchangers. In: Loureiro JM, Kartel MT (eds) Combined and hybrid adsorbents. Springer, New York
130. Manos MJ, Kanatzidis MG (2009) Highly efficient and rapid Cs^+ uptake by the layered metal Sulfide $K_{2x}Mn_xSn_{3-x}S_6$ (KMS-1). J Am Chem Soc 131(18):6599–6607
131. Al-Haj AA, El-Bishtawi R (1997) Removal of lead and nickel ions using zeolite tuff. J Chem Technol Biotechnol 69:27–34
132. Borai EH, Harjula R, Malinen L, Paajanen A (2009) Efficient removal of cesium from low-level radioactive liquid waste using natural and impregnated zeolite minerals. J Hazard Mater 172(1):416–422
133. Dyer A, Pillinger M, Amin S (1999) Ion exchange of caesium and strontium on a titanosilicate analogue of the mineral pharmacosiderite. J Mater Chem 9:2481–2487
134. Liang Z, Ni J (2009) Improving the ammonium ion uptake onto natural zeolite by using an integrated modification process. J Hazard Mater 166:52–60
135. Goñi S, Guerrero A, Lorenzo MP (2006) Efficiency of fly ash belite cement and zeolite matrices for immobilizing cesium. J Hazard Mater 137:1608–1617
136. Adabbo M, Caputo D, de Gennaro B, Pansini M, Colella C (1999) Ion exchange selectivity of phillipsite for Cs and Sr as a function of framework composition. Micropor Mesopor Mater 28(2):315–324
137. Abusafa A, Yucel H (2002) Removal of Cs-137 from aqueous solutions using different cationic forms of a natural zeolite: clinoptilolite. Sep Purif Technol 28:103–116
138. Han R, Zou W, Wang Y, Zhu L (2007) Removal of uranium(VI) from aqueous solutions by manganese oxide coated zeolite: discussion of adsorption isotherms and pH effect. J Environ Radioact 93:127–143
139. Lopes CB, Otero M, Coimbra J, Pereira E, Rocha J, Lin Z, Duarte A (2007) Removal of low concentration Hg^{2+} from natural waters by microporous and layered titanosilicates. Micropor Mesopor Mater 103(1–3):325–332
140. Jurado-Vargas M, OIguln MT, Ordóñez-Regil E, Jiménez-Reyes M (1997) Ion exchange of radium and barium in zeolites. J Radioanal Nucl Chem 218:153–156
141. Olguín MT, Solache-Ríos M, Acosta D, Bosch P, Bulbulian S (1999) Uranium sorption in zeolite X: the valence effect. Micropor Mesopor Mater 28:377–385
142. Gu BX, Wang LM, Ewing RC (2000) The effect of amorphization on the Cs ion exchange and retention capacity of zeolite-NaY. J Nucl Mater 278:64–72
143. El-Kamash AM, El-Naggar MR, El-Dessouky MI (2006) Immobilization of cesium and strontium radionuclides in zeolite-cement blends. J Hazard Mater 136:310–316
144. Celestian AJ, Clearfield A (2007) The origin of ion exchange selectivity in a porous framework titanium silicate. J Mater Chem 17:4839–4842
145. Moller T, Harjula R, Lehto J (2002) Ion exchange of Sr-85, Cs-134 and Co-57 in sodium titanosilicate and the effect of crystallinity on selectivity. Sep Purif Technol 28:13–23
146. Tripathi A, Medvedev DG, Nyman M, Clearfield A (2003) Selectivity for Cs and Sr in Nb-substituted titanosilicate with sitinakite topology. J Solid State Chem 175(1):72–83

147. Psareva TS, Zakutevskyy OI, Chubar NI, Strelko VV, Shaposhnikova TO, Carvalho JR, Correia MJN (2005) Uranium sorption on cork biomass. Colloids and Surfaces A: Physicochem Eng Aspects 252:231–236
148. Økland TE, Wilhelmsen E, Solevåg Ø (2005) A study of the priority substances of the water framework directive, monitoring and need for screening. Bergfald & Co, Norway
149. WHO (1996) Health criteria and other supporting information. In: Guidelines for drinking-water quality, vol 2, 2nd edn. World Health Organization, Geneva
150. Otero M, Lopes CB, Coimbra J, Ferreira TR, Silva CM, Lin Z, Rocha J, Pereira E, Duarte AC (2009) Priority pollutants (Hg^{2+} and Cd^{2+}) removal from water by ETS-4 titanosilicate. Desalination 249(2):742–747
151. Rivas BL, Maturana HA, Luna M (1999) Selective binding of mercury ions by poly (4-vinylpyridine) hydrochloride resin. J Appl Polym Sci 74:1557–1562
152. Bessbousse H, Rhlalou T, Verchiere J-F, Lebrun L (2009) Novel metal-complexing membrane containing poly(4-vinylpyridine) for removal of Hg(II) from aqueous solution. Vol. 113. American Chemical Society, Washington, DC
153. Monteagudo JM, Ortiz MJ (2000) Removal of inorganic mercury from mine waste water by ion exchange. J Chem Technol Biotechnol 75:767–772
154. Kocaoba S (2007) Comparison of Amberlite IR 120 and dolomite's performances for removal of heavy metals. J Hazard Mater 147:488–496
155. Saha B, Iglesias M, dimming IW, Streat M (2000) Sorption of trace heavy metals by thiol containing chelating resins. Solvent Extr Ion Exch 18(1):133–167
156. Sharaf MA, Sayed SA, Younis AA, Farag AB, Arida HA (2007) Removal of trace contaminants from water using new chelating resins. Anal Lett 40(18):3443–3456
157. Pagano M, Petruzzelli D, Tiravanti G, Passino R (2000) Pb/Fe Separation and recovery from automobile battery wateeasters by selective ion exchange. Solvent Extr Ion Exch 18(2):387–399
158. Valverde JL, de Lucas A, Carmona M, González M, Rodríguez JF (2004) Equilibrium data of the exchange of Cu^{2+}, Cd^{2+} and Zn^{2+} ions for H+ on the cationic exchanger Lewatit TP-207. J Chem Technol Biotechnol 79(12):1371–1375
159. Atia AA, Donia AM, Elwakeel KZ (2005) Selective separation of mercury (II) using a synthetic resin containing amine and mercaptan as chelating groups. React Funct Polym 65:267–275
160. Singh DK, Srivastava M (2005) Selective uptake and recovery of cadmium (II) by microcapsule containing chelating resin. Sep Purif Technol 45:1–7
161. Badawy NA, El-Bayaa AA, Abdel-Aal AY, Garamon SE (2009) Chromatographic separations and recovery of lead ions from a synthetic binary mixtures of some heavy metal using cation exchange resin. J Hazard Mater 166(2–3):1266–1271
162. Pohl P, Prusisz B (2004) Pre-concentration of Cd, Co, Cu, Ni and Zn using different off-line ion exchange procedures followed by the inductively coupled plasma atomic emission spectrometric detection. Anal Chim Acta 502:83–90
163. Liu RX, Zhang BW, Tang HX (1999) Synthesis and characterization of poly(acrylaminophosphonic-carboxyl-hydrazide) chelating fibre. React Funct Polym 39(1):71–81
164. Denizli A, Kesenci K, Arica Y, Piskin E (2000) Dithiocarbamate-incorporated monosize polystyrene microspheres for selective removal of mercury ions. React Funct Polym 44(3):235–243
165. Manju GN, Anoop Krishnan K, Vinod VP, Anirudhan TS (2002) An investigation into the sorption of heavy metals from wastewaters by polyacrylamide-grafted iron(III) oxide. J Hazard Mater 91(1–3):221–238
166. Nam KH, Gomez-Salazar S, Tavlarides LL (2003) Mercury(II) adsorption from wastewaters using a thiol functional adsorbent. Ind Eng Chem Res 42:1955–1964
167. Osman MM, Kholeif SA, Abou Al-Maaty NA, Mahmoud ME (2003) Metal sorption, solid phase extraction and preconcentration properties of two silica gel phases chemically modified with 2-hydroxy-1-naphthaldehyde. Microchimica Acta 143(1):25–31
168. Rether A, Schuster M (2003) Selective separation and recovery of heavy metal ions using water-soluble N-benzoylthiourea modified PAMAM polymers. React Funct Polym 57:13–21

169. Abderrahim O, Didi MA, Moreau B, Villemin D (2006) A new sorbent for selective Separation of metal: polyethylenimine methylenephosphonic acid. Solvent Extract Ion Exch 24(6):943–955
170. Khan A, Mahmood F, Khokhar MY, Ahmed S (2006) Functionalized sol-gel material for extraction of mercury (II). React Funct Polym 66(10):1014–1020
171. Lopes CB, Lito PF, Otero M, Lin Z, Rocha J, Silva CM, Pereira E, Duarte AC (2008) Mercury removal with titanosilicate ETS-4: Batch experiments and modelling. Micropor Mesopor Mater 115(1–2):98–105
172. Barreira LD, Lito PF, Antunes BM, Otero M, Lin Z, Rocha J, Pereira E, Duarte AC, Silva CM (2009) Effect of pH on cadmium (II) removal from aqueous solution using titanosilicate ETS-4. Chem Eng J 155(3):728–735
173. Camarinha ED, Lito PF, Antunes BM, Otero M, Lin Z, Rocha J, Pereira E, Duarte AC, Silva CM (2009) Cadmium(II) removal from aqueous solution using microporous titanosilicate ETS-10. Chem Eng J 155(1–2):108–114
174. Ferreira TR, Lopes CB, Lito PF, Otero M, Lin Z, Rocha J, Pereira E, Silva CM, Duarte A (2009) Cadmium(II) removal from aqueous solution using microporous titanosilicate ETS-4. Chem Eng J 147(2–3):173–179
175. Liu J, Li T, Hu K, Shao G (2009) Preparation and adsorption performances of novel negatively charged hybrid materials. J Appl Polym Sci 112(4):2179–2184
176. Lopes CB, Otero M, Lin Z, Silva CM, Rocha J, Pereira E, Duarte AC (2009) Removal of Hg^{2+} ions from aqueous solution by ETS-4 microporous titanosilicate -Kinetic and equilibrium studies. Chem Eng J 151(1–3):247–254
177. Alam Z, Inamuddin NSA (2010) Synthesis and characterization of a thermally stable strongly acidic Cd(II) ion selective composite cation-exchanger: polyaniline Ce(IV) molybdate. Desalination 250:515–522
178. Lopes CB, Otero M, Lin Z, Silva CM, Pereira E, Rocha J, Duarte AC (2010) Effect of pH and temperature on Hg^{2+} water decontamination using ETS-4 titanosilicate. J Hazard Mater 175(1–3):439–444
179. Mahmoud ME, Hafez OF, Osman MM, Elmelegy E (2010) Implementation of hybrid inorganic/organic adsorbents for removal and preconcentration of heavy metals from industrial waste and drinking waters. Sep Sci Technol 45(9):1302–1312
180. Mandal B, Ghosh N (2010) Extraction chromatographic method of preconcentration and separation of lead (II) with high molecular mass liquid cation exchanger. Desalination 250:506–514
181. Chojnacki A, Chojnacka K, Hoffmann J, Górecki H (2004) The application of natural zeolites for mercury removal: from laboratory tests to industrial scale. Miner Eng 17(7–8):933–937
182. Mondale KD, Carland RM, Aplan FF (1995) The comparative ion exchange capacities of natural sedimentary and synthetic zeolites. Miner Eng 8:535–548
183. Melamed R, da Luz AB (2006) Efficiency of industrial minerals on the removal of mercury species from liquid effluents. Sci Total Environ 368:403–406
184. Payne KB, Abdel-Fattah TM (2005) Adsorption of divalent lead ions by zeolites and activated carbon: effects of pH, temperature, and ionic strength. J Environ Sci Health A Tox Hazard Subst Environ Eng 39:2275–2291
185. Peric J, Trgo M, Vukojevic Medvidovic N (2004) Removal of zinc, copper and lead by natural zeolite–a comparison of adsorption isotherms. Water Res 38:1893–1899
186. Mousavi HZ, Asghari A (2009) Removal of heavy metal in wastewater by Semnan natural zeolites. Asian J Chem 21:2881–2886
187. Lv L, Hor MP, Su F, Zhao XS (2005) Competitive adsorption of Pb^{2+}, Cu^{2+}, and Cd^{2+} ions on microporous titanosilicate ETS-10. J Colloid Interface Science 287(1):178–184
188. Lv L, Wang K, Zhao XS (2007) Effect of operating conditions on the removal of Pb^{2+} by microporous titanosilicate ETS-10 in a fixed-bed column. J Colloid Interface Sci 305(2):218–225
189. Lv L, Zhang Y, Wang K, Ray AK, Zhao XS (2008) Modeling of the adsorption breakthrough behaviors of Pb^{2+} in a fixed bed of ETS-10 adsorbent. J Colloid Interface Sci 325(1):57–63

190. Choi JH, Kim SD, Noh SH, Oh SJ, Kim WJ (2006) Adsorption behaviors of nano-sized ETS-10 and Al-substituted-ETAS-10 in removing heavy metal ions, Pb^{2+} and Cd^{2+}. Micropor Mesopor Mater 87(3):163–169
191. Hristodor C, Copcia V, Lutic D, Popovici E (2010) Thermodynamics and kinetics of Pb(II) and Hg(II) ions removal from aqueous solution by romanian clays. Rev Chim 61(3):285–289
192. Yang J-S, Lee JY, Park Y-T, Baek K, Choi J (2010) Adsorption of As(III), As(V), Cd(II), Cu(II), and Pb(II) from aqueous solutions by natural muscovite. Sep Sci Technol 45(6):814–823
193. Rocha J, Anderson Michael W (2000) Microporous titanosilicates and other novel mixed octahedral-tetrahedral framework oxides. Eur J Inorg Chem 2000:801–818
194. Tuzen M, Sari A, Mendil D, Soylak M (2009) Biosorptive removal of mercury(II) from aqueous solution using lichen (Xanthoparmelia conspersa) biomass: Kinetic and equilibrium studies. J Hazard Mater 169(1–3):263–270
195. Xie JZ, Chang HL, Kilbane JJ (1996) Removal and recovery of metal ions from wastewater using biosorbents and chemically modified biosorbents. Bioresour Technol 57:127–136
196. Zeroual Y, Moutaouakkil A, Zohra Dzairi F, Talbi M, Ung Chung P, Lee K, Blaghen M (2003) Biosorption of mercury from aqueous solution by Ulva lactuca biomass. Bioresour Technol 90(3):349–351
197. Tuzun I, Bayramoglu G, YalçIn E, Basaran G, Çelik G, ArIca MY (2005) Equilibrium and kinetic studies on biosorption of Hg(II), Cd(II) and Pb(II) ions onto microalgae *Chlamydomonas reinhardtii*. J Environ Manage 77(2):85–92
198. Lodeiro P, Herrero R, Sastre de Vicente ME (2006) Batch desorption studies and multiple sorption-regeneration cycles in a fixed-bed column for Cd(II) elimination by protonated *Sargassum muticum*. J Hazard Mater 137:1649–1655
199. Naja G, Volesky B (2006) Multi-metal biosorption in a fixed-bed flow-through column. Colloids Surf A Physicochem Eng Asp 281:194–201
200. Romera E, González F, Ballester A, Blázquez ML, Muñoz JA (2007) Comparative study of biosorption of heavy metals using different types of algae. Bioresour Technol 98(17): 3344–3353
201. Huang L-Z, Zeng G-M, Huang D-L, Li L-F, Du C-Y, Zhang L (2010) Biosorption of cadmium(II) from aqueous solution onto *Hydrilla verticillata*. Environ Earth Sci 60(8):1683–1691
202. Gaballah I, Kilbertus G (1998) Recovery of heavy metal ions through decontamination of synthetic solutions and industrial effluents using modified barks. J Geochem Explor 62:241–286
203. Lloyd-Jones PJ, Rangel-Mendez JR, Streat M (2004) Mercury sorption from aqueous solution by chelating ion exchange resins, activated carbon and a biosorbent. Process Saf Environ Prot 82:301–311
204. Ghodbane I, Hamdaoui O (2008) Removal of mercury(II) from aqueous media using eucalyptus bark: kinetic and equilibrium studies. J Hazard Mater 160:301–309
205. Sari A, Tuzen M (2009) Removal of mercury(II) from aqueous solution using moss (*Drepanocladus revolvens*) biomass: equilibrium, thermodynamic and kinetic studies. J Hazard Mater 171:500–507
206. Reddy DHK, Seshaiah K, Reddy AVR, Rao MM, Wang MC (2010) Biosorption of Pb^{2+} from aqueous solutions by Moringa oleifera bark: Equilibrium and kinetic studies. J Hazard Mater 174(1–3):831–838
207. López A, Lázaro N, Priego JM, Marqués AM (2000) Effect of pH on the biosorption of nickel and other heavy metals by *Pseudomonas fluorescens* 4F39. J Ind Microbiol Biotechnol 24(2):146–151
208. Pagnanelli F, Toro L, Megliò F (2002) Olive mill solid residues as heavy metal sorbent material: a preliminary study. Waste Manage (Oxford) 22:901–907
209. Farajzadeh MA, Monji AB (2004) Adsorption characteristics of wheat bran towards heavy metal cations. Sep Purif Technol 38:197–207
210. Minamisawa M, Minamisawa H, Yoshida S, Takai N (2004) Adsorption behavior of heavy metals on biomaterials. J Agric Food Chem 52(18):5606–5611

211. Anirudhan TS, Divya L, Ramachandran M (2008) Mercury(II) removal from aqueous solutions and wastewaters using a novel cation exchanger derived from coconut coir pith and its recovery. J Hazard Mater 157:620–627
212. Krishnani KK, Meng X, Christodoulatos C, Boddu VM (2008) Biosorption mechanism of nine different heavy metals onto biomatrix from rice husk. J Hazard Mater 153(3):1222–1234
213. El-Shafey EI (2010) Removal of Zn(II) and Hg(II) from aqueous solution on a carbonaceous sorbent chemically prepared from rice husk. J Hazard Mater 175:319–327
214. Sousa FW, Oliveira AG, Ribeiro JP, Rosa MF, Keukeleire D, Nascimento RF (2010) Green coconut shells applied as adsorbent for removal of toxic metal ions using fixed-bed column technology. J Environ Manag 91(8):1634–1640
215. Gamage A, Shahidi F (2007) Use of chitosan for the removal of metal ion contaminants and proteins from water. Food Chem 104:989–996
216. Akkaya R, Ulusoy U (2008) Adsorptive features of chitosan entrapped in polyacrylamide hydrogel for Pb^{2+}, UO_2^{2+}, and Th^{4+}. J Hazard Mater 151(2–3):380–388
217. Wang J, Chen C (2009) Biosorbents for heavy metals removal and their future. Biotechnol Adv 27:195–226
218. Miretzky P, Cirelli AF (2009) Hg(II) removal from water by chitosan and chitosan derivatives: a review. J Hazard Mater 167:10–23
219. Zhai Y, Duan Se, He Q, Yang X, Han Q (2010) Solid phase extraction and preconcentration of trace mercury(II) from aqueous solution using magnetic nanoparticles doped with 1,5-diphenylcarbazide. Microchimica Acta 169(3):353–360
220. Zhang Y, Li Q, Sun L, Tang R, Zhai J (2010) High efficient removal of mercury from aqueous solution by polyaniline/humic acid nanocomposite. J Hazard Mater 175(1–3):404–409
221. Kaur A, Gupta U (2008) Preconcentration procedure using 1-(2-Pyridylazo)-2-napthol anchored to silica nanoparticle for the analysis of cadmium in different samples. E-J Chem 5:930–939
222. Arief VO, Trilestari K, Sunarso J, Ismadjl S (2008) Recent progress on biosorption of heavy metals from liquids using low cost biosorbents: characterization, biosorption parameters and mechanism studies. Clean 36(12):937–962
223. Khambhaty Y, Mody K, Basha S, Jha B (2008) Hg(II) removal from aqueous solution by dead fungal biomass of marine *Aspergillus niger*: Kinetic Studies. Sep Sci Technol 43(5):1221–1238
224. Nieboer E, McBryde AE (1973) Free-energy relationships in coordination chemistry. III. A comprehensive index to complex stability. Can J Chem 51:2512–2524
225. Nieboer E, Richardson DHS (1980) The replacement of the nondescript term "heavy metals" by a biologically and chemically significant classification of metal ions. Environ Pollut Ser B 1:3–26
226. Brady JM, Tobin JM (1995) Binding of hard and soft metal ions to *Rhizopus arrhizus* biomass. Enzyme Microb Technol 17:791–796
227. Pagano M, Petruzzelli D, Tiravanti G, Passino R (2000) Pb/Fe Separation and recovery from automobile battery wateeasters by selective ion exchange. Solvent Extr Ion Exch 18(2):387–399
228. Farooq U, Kozinski JA, Khan MA, Athar M (2010) Biosorption of heavy metal ions using wheat based biosorbents – A review of the recent literature. Bioresour Technol 101(14):5043–5053
229. Bhattacharya AK, Venkobachar C (1984) Removal of cadmium (II) by low cost adsorbents. J Environ Eng 110:110–122
230. Yalçin S, Apak R, Hizal J, Afsar H (2001) Recovery of copper (II) and chromium (III, VI) from electroplating-industry wastewater by ion exchange. Sep Sci technol 36(10):2181–2196
231. Deepatana A, Valix M (2006) Recovery of nickel and cobalt from organic acid complexes: adsorption mechanisms of metal-organic complexes onto aminophosphonate chelating resin. J Hazard Mater 137:925–933
232. Juang R-S, Kao H-C, Liu F-Y (2006) Ion exchange recovery of Ni(II) from simulated electroplating waste solutions containing anionic ligands. J Hazard Mater 128:53–59

233. Pohl P, Prusisz B (2006) Fractionation of calcium and magnesium in honeys, juices and tea infusions by ion exchange and flame atomic absorption spectrometry. Talanta 69:1227–1233
234. Chen J-H, Huang C-E (2007) Selective separation of Cu and Zn in the citric acid leachate of industrial printed wiring board sludge by D2EHPA-modified Amberlite XAD-4 resin. Ind Eng Chem Res 46:7231–7238
235. Noureddine C, Lekhmici A, Mubarak MS (2008) Sorption properties of the iminodiacetate ion exchange resin, Amberlite IRC-718, toward divalent metal ions. J Appl Polym Sci 107: 1316–1319
236. Lin L-C, Li J-K, Juang R-S (2008) Removal of Cu(II) and Ni(II) from aqueous solutions using batch and fixed-bed ion exchange processes. Desalination 225:249–259
237. Fernández-Olmo I, Fernández JL, Irabien A, Höll WH (2009) Removal of Arsenic(III), Chromium(III) and Iron(III) traces from hydrofluoric acid solutions by specialty anion exchangers. Solvent Extr Ion Exch 27(5):727–744
238. Zhang Y, Li Y, Yang L-q, Ma X-j, Wang L-y, Ye Z-F (2010) Characterization and adsorption mechanism of Zn^{2+} removal by PVA/EDTA resin in polluted water. J Hazard Mater 178(1–3):1046–1054
239. Sahu SK, Meshram P, Pandey BD, Kumar V, Mankhand TR (2009) Removal of chromium(III) by cation exchange resin, Indion 790 for tannery waste treatment. Hydrometallurgy 99(3–4):170–174
240. Mustafa S, Shah KH, Naeem A, Waseem M, Ahmad T, Sarfraz S, Irshad M (2010) Kinetic and Equilibrium Studies of Chromium(III) removal by macroporous ion exchanger amberlyst-15 (H^+). Chinese J Chem 28(1):27–32
241. Kir E, Çengeloglu Y, Ersöz M (2005) The effect of chelating agent on the separation of Fe(III) and Ti(IV) from binary mixture solution by cation-exchange membrane. J Colloid Interface Sci 292:498–502
242. Schulte-Bockholt M, Schuster M (2008) Removal enrichment and recovery of Ni(II), Zn(II) and phosphate from phosphation rinsing waters with liquid-phase polymer-based retention technique. Sep Purif Technol 63:172–178
243. Cheng Z, Wu Y, Wang N, Yang W, Xu T (2010) Development of a novel hollow fiber cation-exchange membrane from bromomethylated poly(2,6-dimethyl-1,4-phenylene oxide) for removal of heavy-metal ions. Ind Eng Chem Res 49(7):3079–3087
244. Wang S, Peng Y (2010) Natural zeolites as effective adsorbents in water and wastewater treatment. Chem Eng J 156:11–24
245. Choi JH, Kim SD, Kwon YJ, Kim WJ (2006) Adsorption behaviors of ETS-10 and its variant, ETAS-10 on the removal of heavy metals, Cu^{2+}, Co^{2+}, Mn^{2+} and Zn^{2+} from a waste water. Micropor Mesopor Mater 96(1–3):157–167
246. Davis TA, Volesky B, Mucci A (2003) A review of the biochemistry of heavy metal biosorption by brown algae. Water Res 37:4311–4330
247. Grimm A, Zanzi R, Björnbom E, Cukierman AL (2008) Comparison of different types of biomasses for copper biosorption. Bioresour Technol 99(7):2559–2565
248. Naik R, Wen G, Ms D, Hureau S, Uedono A, Wang X, Liu X, Cookson PG, Smith SV (2010) Metal ion binding properties of novel wool powders. J Appl Polym Sc 115(3):1642–1650
249. Kadous A, Didi MA, Villemin D (2010) A new sorbent for uranium extraction: ethylenediamino tris(methylenephosphonic) acid grafted on polystyrene resin. J Radioanal Nucl Chem 284:431–438
250. Selvakumar R, Kavitha S, Sathishkumar M, Jayavignesh V, Swaminathan K (2010) Liquid phase separation of As(V) from aqueous solution using pretreated *paecilomyces variotii* biomass. Sep Sci Technol 45(6):776–785
251. Yang DJ, Zheng ZF, Yuan Y, Liu HW, Waclawik ER, Ke XB, Xie MX, Zhu HY (2010) Sorption induced structural deformation of sodium hexa-titanate nanofibers and their ability to selectively trap radioactive Ra(II) ions from water. Phys Chem Chem Phys 12(6):1271–1277
252. Ertugay N, Bayhan YK (2010) The removal of copper (II) ion by using mushroom biomass (*Agaricus bisporus*) and kinetic modelling. Desalination 255:137–142

253. Chen C-Y, Lin M-S, Hsu K-R (2008) Recovery of Cu(II) and Cd(II) by a chelating resin containing aspartate groups. J Hazard Mater 152:986–993
254. Galán B, Castañeda D, Ortiz I (2008) Integration of ion exchange and non-dispersive solvent extraction processes for the separation and concentration of Cr(VI) from ground waters. J Hazard Mater 152:795–804
255. Agrawal A, Sahu KK (2006) Separation and recovery of lead from a mixture of some heavy metals using Amberlite IRC 718 chelating resin. J Hazard Mater 133:299–303
256. Magosso HA, Panteleimonov AV, Kholin YV et al (2006) Synthesis, characterization and metal adsorption properties of the new ion exchanger polymer 3-n-propyl(4-methylpyridinium) silsesquioxane chloride. J Colloid Interface Sci 303:18–24
257. Rengaraj S, Joo CK, Kim Y, Yi J (2003) Kinetics of removal of chromium from water and electronic process wastewater by ion exchange resins: 1200 H, 1500 H and IRN97 H. J Hazard Mater 102(2–3):257–275

Chapter 12
Application of Ion Exchange Resins in Selective Separation of Cr(III) from Electroplating Effluents

Licínio M. Gando-Ferreira

Abstract In this study, a chelating resin containing iminodiacetic acid group (Diaion CR11) was investigated to separate trivalent chromium, copper and iron from synthetic and industrial effluents. To determine single-component equilibrium isotherms for those species, batch experiments were performed at 25°C and 50°C. On the basis of these results, it was found that the resin exhibited better selectivity for Cu(II) and Fe(III) over Cr (III). It was demonstrated that the sorption of Cr(III) increases with increasing temperature. The maximum ion exchange capacities were 0.62 mequiv./$g_{\text{dry resin}}$ at 25°C and 0.95 mequiv./$g_{\text{dry resin}}$ at 50°C. The Langmuir model enabled a good description of the ion exchange equilibrium data. A mathematical model considering dispersed plug flow for the liquid phase, intraparticle mass transfer by pore diffusion and external mass-transfer resistance at the pore/wall interface enabled a reasonable description of the experimental breakthrough curves for mono-component, Cr(III)/Diaion, and multicomponent, Cr(III)/Cu(II)/Diaion, systems. The saturation of the Diaion with industrial effluent demonstrated that the breakthrough capacity of the resin is little affected by the presence of other species in solution. High regeneration efficiency of the resin was found by eluting Cr(III) with HCl followed of the mixture of $NaOH/H_2O_2$.

12.1 Introduction

The electroplating is a chemical or electrochemical process in which a thin metallic layer is deposited on a base material. Chromium is a metal widely used in the electroplating industry for both functional and decorative purposes. Historically,

L.M. Gando-Ferreira (✉)
Department of Chemical Engineering, Centre for Chemical Processes Engineering and Forest Products (CIEPQPF), University of Coimbra, Pólo II, Rua Sílvio Lima, 3030-790, Coimbra, Portugal
e-mail: lferreira@eq.uc.pt

Inamuddin and M. Luqman (eds.), *Ion Exchange Technology II: Applications*,
DOI 10.1007/978-94-007-4026-6_12, © Springer Science+Business Media B.V. 2012

chromium plating based on Cr(VI) has been the current process; however, it has many disadvantages, namely the requirement of a high metal concentration, the low efficiency of the electrodeposition process and the toxicity of the baths containing hexavalent chromium. These inconveniences can be avoided using trivalent chromium salt plating solutions. Therefore, the advantages of the electrodeposition from Cr(III) have contributed substantially to successful commercial applications of trivalent chromium systems [1]. The plating baths of these systems use lower concentrations of total chromium in solution, and this leads to less drag out and greater simplification in the treatment of wastewaters, thus implying costs reduction in the electroplating process.

The electroplating industries may create serious environmental problems since their wastewaters contain a large number of heavy metals, including chromium, copper, iron, nickel, zinc, manganese and lead. The conventional processes in use for metal control are based on chemical precipitation coupled to pre- or post-oxidation/reduction and followed by filtration. The main disadvantage of these processes is the production of solid residues (sludges) containing toxic compounds whose final disposal is generally landfilling that as waste treatment method involves high costs. Therefore, the use of other alternative processes based on physical and chemical mechanisms is advisable to protect the environment and at same time recovering at least one valuable metal in a concentrated solution.

In the last decades, the ion exchange has found widespread application in the removal of heavy metals from industrial wastewaters and is a possible process for treating effluents generated by chromium electroplating baths. Use of ion exchange resins enables selectively removing low concentrations of the contaminant metal ions in the presence of moderate concentrations of chromium ions. Several studies focusing the chromium sorption by ion exchange resins have been reported in the literature. Petruzelli and co-workers [2] reported the IERECHROM process based on the use of a macroporous carboxylic resin that allows the removal of almost pure Cr(III) from other interfering metals. Chmielewski et al. [3] implemented a method for recovering chromium, copper and water from electroplating wastewater using a combined process involving electrochemical oxidation and ion exchange. Lin and Kiang [4] proposed a procedure to recover chromic acid from a waste acid solution using a multi-step ion exchange process, where a large amount of regenerant chemicals were spent. Yalçin et al. [5] developed a laboratory-scale separation process for the recovery of Cr(VI) and Cr(III) with a strongly basic resin and a weakly acidic resin, respectively. The chelating resins are commonly employed as ion exchange materials, once their ligands can selectively bind to certain metallic ions. Recent studies [5–14] have shown that these resins could be used for the selective removal and recovery of chromium (III). The pH, metal concentration and temperature are the key variables that influence the performance of ion exchange process [10, 12]. Kabay et al. [11] evaluated chelating resins containing phosphinic, phosphonic and methylenephosphinic groups for removal of Cr(III) against pH change, being observed that almost 100% of removal occurred at pH range between 2 and 2.5. Gode and Pehlivan [9] studied the effect of these variables on sorption of Cr(III) into macroporous resins containing iminodiacetic acid (Lewatit TP 207 and Chelex 100) and observed that the maximum sorption occurs at pH 4.5, and a slight increase in the equilibrium constant as temperature was increased. Pehlivan and Altun [8] reported optimum pH ranges for the sorption capacity of Lewatit CNP 80 and Lewatit TP 207 for Pb, Cu, Zn, Cd and Ni. They concluded that Lewatit CNP 80 was the most efficient for removing those metals. In the work of Kocaoba and Akcin [12], it was demonstrated that Amberlite IRC 86 and Amberlite IRC 718 had better performance for removal of Cr(III) than Amberlite IR 120. They found that the optimum conditions are 10 mg/L of Cr(III), pH 5, stirring time of 20 min and 250 mg of resin.

In the present chapter, the performance of the chelating resin Diaion CR 11 (Mitsubishi Chemical Corporation) for uptaking trivalent chromium, copper and iron was assessed. Equilibrium and dynamic behaviour of the process in column tests, using synthetic and real effluents, were investigated. Moreover, regeneration efficiencies of the pre-saturated resin employing different regeneration agents were also determined. It is worth mentioning that ion exchange data for this resin are very limited.

12.2 Modelling

12.2.1 Adsorption Equilibrium Model

In order to quantitatively describe the adsorption experimental data obtained by the batch technique, the Langmuir equation was used:

$$q_i = \frac{q_{\max i} k_{Li} C_i}{1 + k_{Li} C_i} \tag{12.1}$$

where q_i corresponds to the adsorbed mass of component i (mg/g wet resin) and C_i is the equilibrium concentration of component i (Cr, Cu and Fe) in the liquid phase (mg/L). K_{Li} and $q_{\max i}$ are the Langmuir model parameters.

12.2.2 Dynamic Model for Fixed-Bed Adsorption and Parameter Estimation

A model was developed to describe the dynamic behaviour of the binary system Cr(III)/Cu(II) during the saturation step, namely regarding the species concentrations in the solution at the outlet of the column as a function of time (breakthrough curve). The model includes pore diffusion, mass-transfer resistance in the film and axial dispersion. Dimensionless model equations are:

Mass balance for the fluid phase

$$\frac{\partial X_i(z^*,\theta)}{\partial\theta}=\frac{1}{Pe}\frac{\partial^2 X_i(z^*,\theta)}{\partial z^{*2}}-\frac{\partial X_i(z^*,\theta)}{\partial z^*}-N_{fi}(X_i(z^*,\theta)-Xp_i(1,z^*,\theta)) \quad (12.2)$$

where $i=1$(Cr), 2(Cu).

Mass balance inside particles

$$\frac{\rho_h}{C_{E_i}}\frac{\partial q_i(u^*,z^*,\theta)}{\partial\theta}+\varepsilon_p\frac{\partial Xp_i(u^*,z^*,\theta)}{\partial\theta}=N_{Di}\left(\frac{\partial^2 X_{pi}(u^*,z^*,\theta)}{\partial u^{*2}}+\frac{2}{u^*}\frac{\partial Xp_i(u^*,z^*,\theta)}{\partial u^*}\right) \quad (12.3)$$

Multicomponent equilibrium adsorption pore/wall interface described by the extended Langmuir model

$$q_i=\frac{q_{\max_i}k_{Li}(C_{Ei}X_{Pi}/\eta_i)}{1+\sum_{j=1}^{2}k_{Lj}(C_{Ej}X_{Pi}/\eta_j)} \quad (12.4)$$

where the $q_{\max i}$, k_i is given by the individual Langmuir (Eq. 12.1) and the η_i is the interaction terms to quantify the competitive adsorption in multicomponent systems.

Initial conditions

$$\theta=0 \quad X_i(z^*,0)=Xo_i \text{ and } Xp_i(u^*,z^*,0)=Xo_i \quad (12.5)$$

Boundary conditions

$$u^*=0 \quad \frac{\partial Xp_i(u^*,z^*,\theta)}{\partial u^*}\Big|_{u^*=0}=0 \quad (12.6)$$

$$u^*=1 \quad N_{fi}(X_i(z^*,\theta)-Xp_i(1,z^*,\theta))=3\frac{1-\varepsilon}{\varepsilon}N_{Di}\frac{\partial Xp_i(u^*,z^*,\theta)}{\partial u^*}\Big|_{u^*=1} \quad (12.7)$$

The equations of the model are written in terms of dimensionless variables: $X_i=C_i/C_{Ei}$, $Xp_i=Cp_i/C_{Ei}$ are the dimensionless concentrations of the species i in the bulk liquid phase and in the liquid inside particle pores, respectively; $\theta=t/\tau$ is a reduced time, where τ is the bed space time; $z^*=z/L$ is the reduced axial coordinate, where L is the fixed-bed length; $u^*=r/R$ is the reduced particle radial coordinate, in which R is the adsorbent particle radius.

The dimensionless groups are the Peclet number $Pe=uL/(\varepsilon D_{ax})$ (D_{ax} is the axial dispersion coefficient), the number of mass-transfer units by pore diffusion for species i, $N_{Di}=\tau\varepsilon_p D_{Pi}/R^2$ ($D_{pi}=D_{mi}/\tau_p$ is the effective diffusivity for the species i and τ_p is tortuosity factor) and the number of film mass-transfer units for species i,

$N_{fi} = 3(1-\varepsilon)\ k_{fi}\ \tau/(\varepsilon\ R)$, where k_{fi} is the film mass-transfer coefficient for species i. The axial dispersion was calculated from the Butt correlation [15]: $(u\ d_p/D_{ax}) = (0.2 + 0.011\ Re^{0.48})$, where $Re = u\ \rho\ d_p/\varepsilon\ \mu$.

Numerical solutions of Eqs. 12.2–12.7 were performed by discretizing Eq. 12.3 in the radial direction using orthogonal collocation in NE finite elements with cubic Hermite polynomials as basic functions. This led to a new boundary value problem of N_{comp} (2NE +1) [N_{comp} is the number of components] parabolic PDEs in the independent variables z^* and θ that was numerically integrated with the PDECOL package.

12.3 Materials and Methods

The results of this work were obtained using the materials and experimental procedures described in the next sections.

12.3.1 *Resin and Reagents*

The chelating resin Diaion CR11 was used in this study, and its chemical characteristics are shown in Table 12.1 [16]. All the chemicals used were of analytical grade and were obtained from Riedel-de Haën, Germany.

12.3.2 *Sorption Experiments*

Before the experiments, the resin was pretreated by cyclic washings with 2 M HCl and 2 M NaOH solutions for removing solvents and other preparation chemicals. The last step of the conditioning consisted in percolating a solution of HCl through the column in order to convert the resin to H^+ form. For all the experiments, synthetic effluent samples were prepared by dissolving an appropriate amount of metal nitrate salt in distilled water. Equilibrium isotherms were determined through batch tests, where 40 mL of a synthetic solution of known pH and composition was added into several flasks containing different amounts of preconditioned resin. Regeneration tests of pre-saturated resin were conducted with different solutions, and 0.25 g of saturated resin with Cr(III) was poured into flasks in contact with 80 mL of different regeneration agents. Firstly, several acid solutions (HCl, HNO_3 and H_2SO_4) were tested at three concentration levels (0.5, 1, 2 M). Secondly, a mixture of 2 M NaOH and 0.33 M H_2O_2 was employed. The flasks were sealed and kept in a shaker at constant temperature (25°C or 50°C) for 24 h. At the end of this period, the resin was separated by filtration, and aliquots of liquid were chemically analysed. The metal content was analysed by flame atomic absorption spectrophotometry, Perkin Elmer 3300, Portugal.

Table 12.1 Characteristics of the ion exchange resin

Polymer structure	Polystyrene/divinylbenzene
Functional group	Iminodiacetic acid
Physical form	Spherical beads
Effective size (mm)	0.40–0.60
Moisture content (%)	61.3
Apparent density, ρ_{ap} ($g_{dry\ resin}/cm_{resin}^{3}$)	0.38
Wet particle porosity, ε_p	0.73
Wet density, ρ_h ($g_{dwet\ resin}/cm_{resin}^{3}$)	1.02
Total capacity (mEq Na^+/g_{resin})	1.21

12.3.3 *Column Experiments*

In the column tests, a glass tube with 1.6 cm of internal diameter and 20 cm height packed with approximately 5 g of resin was used. A peristaltic pump was utilized to percolate 10 mL/min of a synthetic or industrial solution containing metal ions through the column. Several samples were collected at the column outlet, and the concentrations of the metals were analysed along the time as described above.

12.4 Results and Discussion

This study shows results that deal with the evaluation of a chelating resin, Diaion CR11, for treating synthetic and industrial effluents containing Cr(III), Cu(III) and Fe(III). The results comprehend mainly the determination of equilibrium ion exchange isotherms and the analysis of fixed-bed saturation/regeneration steps.

12.4.1 *Equilibrium Studies*

The single-isotherm equilibrium data for the ion exchange systems Diaion/Cr(III), Diaion /Cu(II) and Diaion/Fe(III) are illustrated in Fig. 12.1a–c. The data show that the resin exhibits higher sorption capacity for copper and iron than for the trivalent chromium since these two metals form a stronger chelate complex with the functional group (iminodiacetic acid) of the resin. The Langmuir model fits very well the equilibrium experimental data, in which the values found for the parameters are as follows: $K_L = 8.47{\cdot}10^{-3}$ L/mg and $q_{max} = 10.68$ mg/g_{resin} (0.62 mEq/g_{resin}) for Cr(III); $K_L = 5.97{\cdot}10^{-2}$ L/mg and $q_{max} = 31.75$ mg/g_{resin} (1.0 mEq/g_{resin}) for Cu(II); and $K_L = 2.79{\cdot}10^{-1}$ L/mg and $q_{max} = 25.57$ mg/g_{resin} (1.37 mEq/g_{resin}) for Fe(III). The metal sorption of both metals leads to a decrease in the equilibrium pH of the solution with respect to initial pH as shown in Fig. 12.1d, where $\Delta pH = pH_{final} - pH_{initial}$. This pH variation is a consequence of the hydrogen ions release from the resin, in the H^+ form, since those ions are exchanged with the metallic ions in solution.

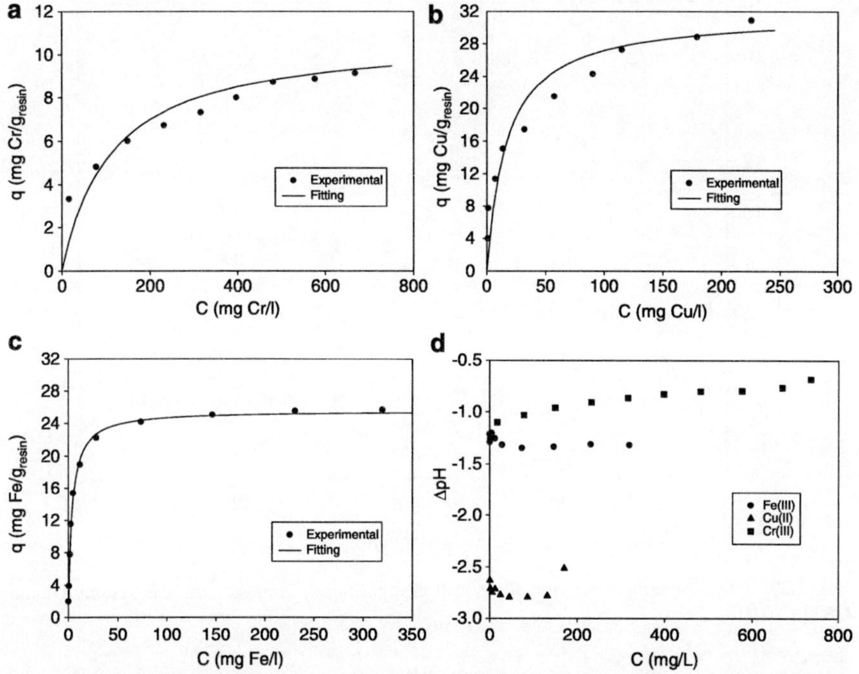

Fig. 12.1 Sorption equilibrium isotherms and pH variation as function of equilibrium metal concentration. Experimental data and model predictions using the Langmuir equation: (**a**) Diaion CR11/Cr(III); (**b**) Diaion CR11/Cu(II); (**c**) Diaion CR11/Fe(III); (**d**) ΔpH vs. C

The effect of temperature on the uptake of Cr(III) was studied by determining equilibrium data at 25°C and 50°C, and the results are shown in Fig. 12.2. It can be observed that the retention of the metal increases when the temperature increases, being that the maximum capacities of the resin(q_{max}) at 25°C and 50°C of 10.68 and 16.47 mg/g_{resin}, respectively. The sorption capacity of conventional resins generally is lower at higher temperatures. However, in the case of chelating resins, the opposite effect can occur depending on how the stability constant of the chelate complex formed between the metal and the resin changes with the temperature [17].

The Langmuir model, where the temperature effect follows the Arrhenius law, $K_L = K_{L0}\ \exp^{(-\Delta H/RT)}$, fits well the experimental data. The parameter values estimated are shown in Table 12.2. The negative value of ΔH (heat adsorption) indicates the exothermic nature of the process.

12.4.2 Regeneration Efficiencies

The ion exchange is a cyclic fixed-bed operation in which the regeneration step plays a crucial role in the global performance of the process. The regeneration of pre-saturated resin samples was studied using different acid solutions followed by

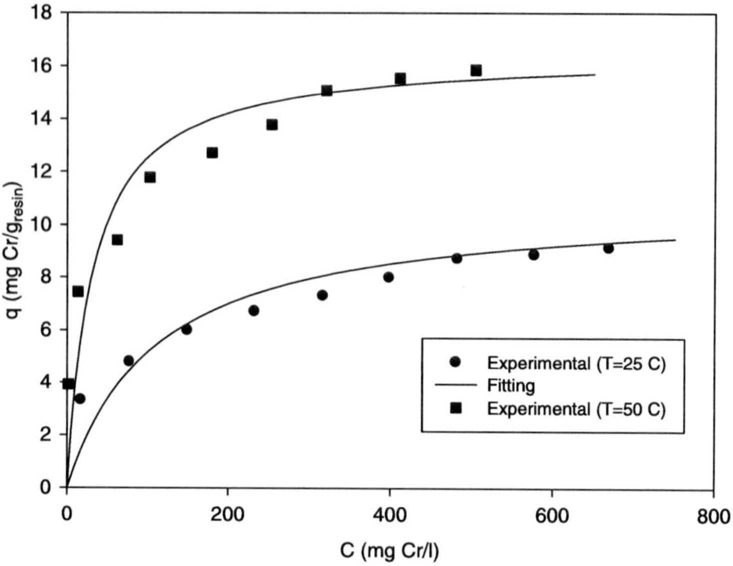

Fig. 12.2 Effect of temperature on the sorption equilibrium isotherm for the system Diaion CR11/Cr(III)

Table 12.2 Sorption isotherm parameters for the system Diaion CR11/Cr(III) at two different temperatures

Temperature (°C)	q_{max} (mg Cr/g_{res})	K_L (L/mg).10^2	K_{L0}.10^5	ΔH (kJ/mol)
25	10.68	5.97	1.89	−19.98
50	16.47	3.2		

a mixture of NaOH and H_2O_2. Figure 12.3 shows that moderate regeneration efficiencies to strip Cr(III) are achieved using solution of HCl. The efficiency is significantly affected by increasing the acid concentration from 0.5 to 2 M, varying from 66.52% to 79.65%. For other acids, low efficiencies (3.8–9%) are obtained. However, it is known that the displacement of Cr(III) from the cation exchangers by using acid regenerants is problematic because of its strong adsorption affinity for the functional groups of the cationic resins. If a prolonged period of contact between the resin and the acid occurs, hydrolyzed and surface-precipitated hydroxo-Cr(III) species are irreversible adsorbed [5]. The use of the regenerant NaOH/H_2O_2 enables to overcome the difficulties of the Cr(III) desorption from the resin. This is achieved by the oxidation in alkaline medium of Cr(III) to Cr(VI) in the anionic form CrO_4^{2-} that is rejected by the cationic resin due to the Donnan effect. The oxidation reaction is as follows:

$$2(RCOO)_3Cr + 3H_2O_2 + 10NaOH \rightarrow 2Na_2CrO_4 + 8H_2O + 6(RCOONa) \quad (12.8)$$

The highest overall efficiency (97.93%) was found for the experiment in which 2 M HCl and the mixture 2 M NaOH/0.33 M H_2O_2 were used.

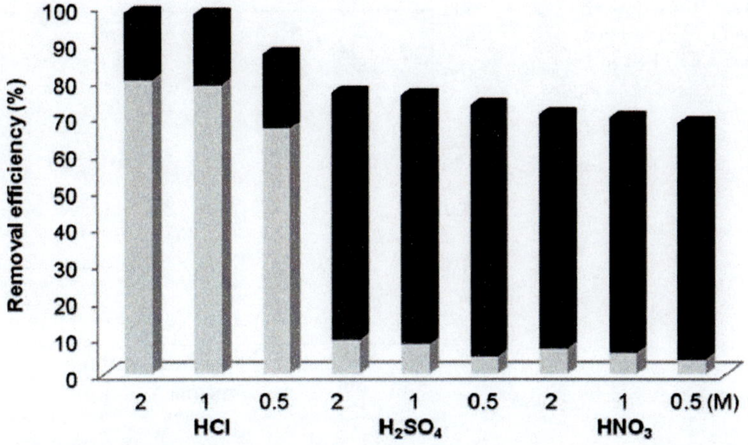

Fig. 12.3 Regeneration of Diaion CR11 resin with different solutions (■ – 2 M NaOH/0.33 M H_2O_2; □ – acid solution)

Table 12.3 Experimental conditions for fixed-bed experiments

	C_{Cr} (mg/L)	C_{Cu} (mg/L)	Feed pH	Bed properties	Flow conditions
Run 1	484	–	3.26	ε=0.35 L=4.0 cm	Q=10 mL/min τ=0.28 min
Run 2	240	160	3.30	ε=0.40 L=5.0 cm	Q=10 mL/min τ=0.40 min
Run 3	230	70	3.11	ε=0.40 L=5.0 cm	Q=10 mL/min τ=0.40 min

12.4.3 Fixed-Bed Studies

Experiments in column were performed in order to study the dynamic behaviour of the sorption process for separating Cr(III) from synthetic and industrial solutions. The first runs involving the metals chromium and copper were conducted with synthetic solutions, and the experimental conditions are shown in Table 12.3. Experimental and simulated breakthrough curves plotted as metal concentration at column outlet against time are depicted in Figs. 12.4–12.6.

A reasonable agreement is observed between experimental and calculated breakthrough curves, being the model solution achieved with the parameter values indicated in the Table 12.4. The Peclet values demonstrate an intermediate effect of axial dispersion on the hydrodynamic conditions of the fixed bed. The $N_{D,i}$ values estimated for the system enables us to conclude that the mass transfer is controlled by the intraparticle resistance.

In the case of multicomponent system, corresponding to the runs 2 and 3, the equilibrium isotherm for each species took into account the following interaction factors adjusted: η_1=1.11 for Cr(III) and η_2=0.5 for Cu(II). The breakthrough

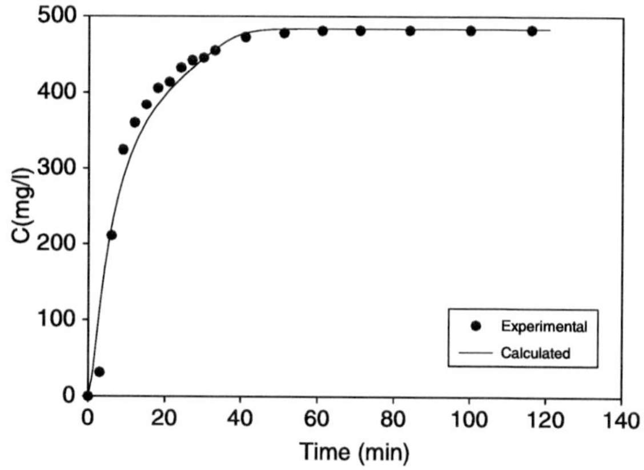

Fig. 12.4 Breakthrough curve of Cr(III) sorption on Diaion CR11 – run 1

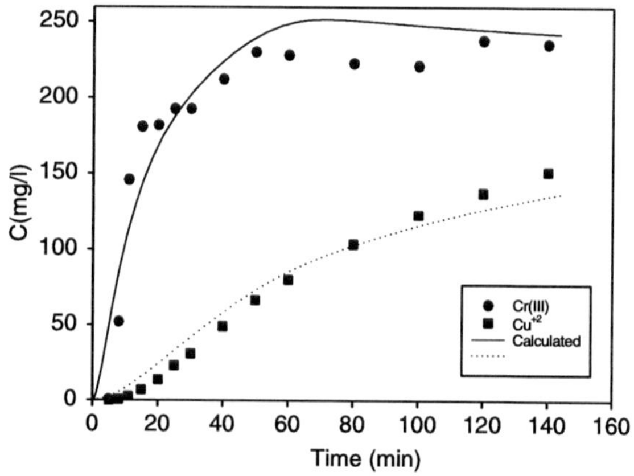

Fig. 12.5 Breakthrough curves for sorption of the mixture Cr(III) and Cu(II) on Diaion CR11 – run 2

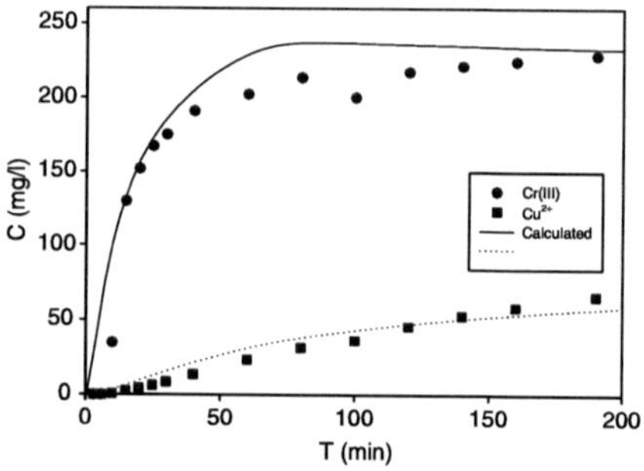

Fig. 12.6 Breakthrough curves for sorption of the mixture Cr(III) and Cu(II) on Diaion CR11 – run 3

Table 12.4 Parameter values used in the simulations of the saturation step

	Pe	$N_{D,cr}$	$N_{D,Cu}$
Run 1	33.0	0.04	–
Run 2	52.3	0.06	0.09
Run 3	49.3	0.06	0.09

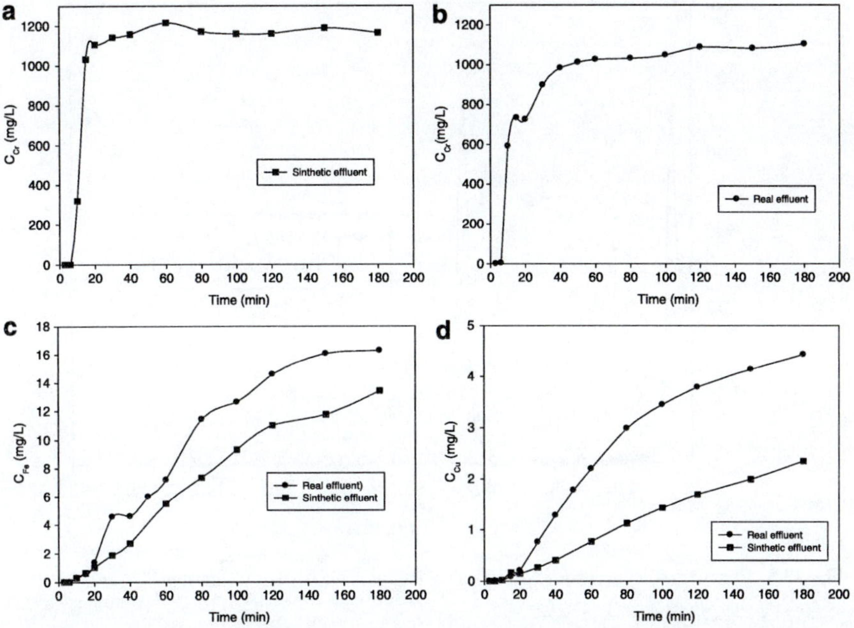

Fig. 12.7 Experimental breakthrough curves for chromium (**a–b**), iron (**c**) and copper (**d**) using synthetic and industrial effluents ($C_{F,Cr}=1195$ ppm, $C_{F,Cu}=5.7$ ppm and $C_{F,Fe}=28.4$ ppm, $pH_0=2.8$, Q=10 mL/min, T=25°C)

curves of these runs (Figs. 12.5 and 12.6), in which the copper concentration in the feed was decreased, show that copper emerges later since this metal is strongly adsorbed by the resin, confirming so the equilibrium studies. It should be noted that the change of the initial copper concentration had effect on the breakthrough time of the saturation curve for this metal. Higher initial concentration leads to faster saturation of the resin, and consequently, the breakthrough curve emerges earlier at column outlet. The chromium behaviour is little affected by the change of copper concentration feed.

The comparison of breakthrough curves for Cr(III), Cu(II) and Fe(III) obtained by percolating synthetic and industrial effluents through the column packed with the resin Diaion CR11 can be seen in Fig. 12.7a–d. Analysing Fig. 12.7a, b, it is visible that the sorption of Cr(III) during the saturation of the resin was little affected by the presence of other metals present in the industrial effluent, such as sodium and calcium. Thus, the treatment of the industrial effluent makes possible the separation of Cr(III) from the other species. A solution of high purity of trivalent, almost free of the metallic contaminants copper and iron, can be recovered for reuse during 20–25 min of saturation of the resin. Regarding the sorption of

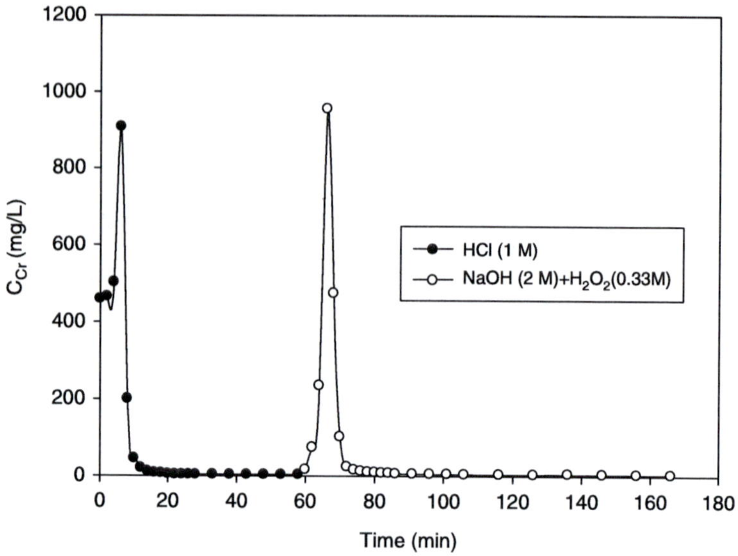

Fig. 12.8 Regeneration curves for Cr(III) using HCl and NaOH/H_2O_2 solutions

Cu(II) and Fe(III), some differences in the kinetics of the ion exchange process occurs for the both effluents, as shown in Fig. 12.7c, d, due to the interference of other species. Also, it can be observed that the saturation of the resin with copper and iron ions is not reached during the time period tested either using synthetic or industrial effluent.

The experimental curve of the Cr(III)-loaded resin regeneration, where the chromium concentration in solution is plotted against the time, is shown in Fig. 12.8. The regeneration process was carried out in a sequential way using 1 M HCl followed of 2 M NaOH /0.33 M H_2O_2. Results indicate that Cr(III) is rapidly desorbed from the resin. The regeneration curve reached a maximum concentration (c_{max}) of 910 mg/L after elution with ≈6 bed volumes (BV) of HCl. In the second regeneration step, a c_{max} = 814 mg/L was achieved after elution with ≈7 BV of NaOH/H_2O_2. Those concentration peaks are equivalent to concentration factors of nearly 2 and 1.8 orders of magnitude, respectively, in comparison with the initial concentration. Regeneration efficiencies of 52.78% and 36.41% were obtained after the percolation of the first and second regeneration agent, respectively. The ion chromium was stripped from the resin with 18 BV of regenerant solution, where an overall efficiency of 89.19% was achieved. The regeneration process can be optimized by varying the operating conditions, such as the concentration of the regenerant agents, flow rate and temperature. It is desirable to find a compromise that enables to maximize the regeneration efficiency and minimize costs and the environmental impact of the wastes generated from the process.

12.5 Conclusions

The chelating ion exchange resin Diaion CR11, in the hydrogenium form, has been tested for uptaking trivalent chromium, copper and iron from synthetic and industrial effluents. The equilibrium isotherms showed that the resin has stronger affinity to the Cu(II) and Fe(II) ions than Cr(III). Higher sorption capacity of the resin for chromium is obtained when the temperature increases. The experimental data were well described by the Langmuir model.

The breakthrough curves resulting from the saturations in fixed bed with synthetic solutions enabled to study the dynamic behaviour of the process for the monocomponent, Cr(III)/Diaion, and multicomponent, Cr(III)/Cu(II)/Diaion, sorption systems. The mathematical model used to simulate the breakthrough curves takes into account the dispersed plug flow for the liquid phase, intraparticle mass transfer by pore diffusion and external mass-transfer resistance at the pore/wall interface. This model simulates well the experimental data, and therefore, it can be used as a suitable approach for the design of industrial systems directed to the treatment of effluents from electroplating processes.

The saturation of the bed with industrial effluent showed that the breakthrough capacity of the resin is little affected by the presence of other ions in solution. Therefore, the treatment of the industrial effluents with this resin enables to separate Cr(III) from other transition metallic ions and simultaneously to recovery for reuse. From regeneration experiments, it can be concluded that a regeneration protocol based on elution with HCl and $NaOH/H_2O_2$ can be used for effectively striping Cr(III) from the resin.

Acknowledgement I would like to thank FCT (Foundation for Technology and Science) for its financial support for this work (Project POCTI/EQU/58149/2004).

References

1. Lei MK, Zhu XP, Xu KW, Xu BS (2008) Study of hard chromium plating from trivalent chromium electrolyte. Key Eng Mat 373–374:200–203
2. Petruzelli D, Passino R, Tiravanti G (1995) Ion exchange process for chromium and recovery from tannery wastes. Ind Eng Chem Res 34:2612–2617
3. Chmielewski AG, Urbański TS, Migdal W (1997) Separation technologies for metals recovery from industrial wastes. Hydrometallurgy 45:333–344
4. Lin SH, Kiang CD (2003) Chromic acid recovery from waste acid solution by an ion exchange process: equilibrium and column ion exchange modeling. Chem Eng J 92:193–199
5. Yalçin S, Apak R, Hizal J, Afşar H (2001) Recovery of copper (II) and chromium (III, VI) from electro-plating industry wastewater by ion exchange. Separ Sci Technol 36:2181–2196
6. Cavaco SA, Fernandes S, Augusto CM, Quina MJ, Gando-Ferreira LM (2009) Evaluation of chelating ion-exchange resins for separating Cr(III) from industrial effluents. J Hazard Mater 169:516–523

7. Janin A, Blais JF, Mercier G, Drogui P (2009) Selective recovery of Cr and Cu in leachate from chromated copper arsenate treated wood using chelating and acidic ion exchange resins. J Hazard Mater 169:1099–1105
8. Pehlivan E, Altun T (2007) Ion-exchange of Pb^{2+}, Cu^{2+}, Zn^{2+}, Cd^{2+}, and Ni^{2+} ions from aqueous solution by Lewatit CNP 80. J Hazard Mater 140:299–307
9. Gode F, Pehlivan E (2003) A comparative study of two chelating ion-exchange resins for the removal of chromium (III) from aqueous solution. J Hazard Mater B100:231–243
10. Gode F, Pehlivan E (2006) Removal of chromium (III) from aqueous solutions using Lewatit S 100: the effect of pH, time, metal concentration and temperature. J Hazard Mater B136:330–337
11. Kabay N, Gizli N, Demircioğlu M, Yŭksel M, Jyo A, Yamabe K, Shuto T (2003) Cr (III) removal by macroreticular chelating ion exchange resins. Chem Eng Commun 190:813–822
12. Kocaoba S, Akcin G (2002) Removal and recovery of chromium and chromium speciation with MINTEQA2. Talanta 57:23–30
13. Kocaoba S, Akcin G (2005) Removal of chromium (III) and cadmium (II) from aqueous solutions. Desalination 180:151–156
14. Kabay N, Demircioğlu M, Ekinci H, Yüksel M, Sağlam M, Akçay M, Streat M (1998) Removal of metal pollutants (Cd(II) and Cr(III)) from phosphoric acid solutions by chelating resins containing phosphonic or diphosphonic groups. Ind Eng Chem Res 37:2541–2547
15. Butt S (1980) Reaction kinetics and reactor design. Prentice-Hall, Englewood Cliffs
16. Cavaco SA, Fernandes SL, Quina MM, Ferreira LM (2007) Removal of chromium from electroplating industry effluents by ion-exchange resins. J Hazard Mater 144:634–638
17. Gode F (2007) Removal of chromium ions from aqueous solutions by adsorption method. In: Lewinsky AA (ed) Hazardous materials and wastewater: treatment, removal and analysis. Nova, New York

Chapter 13
Effect of Temperature, Zinc, and Cadmium Ions on the Removal of Cr(VI) from Aqueous Solution via Ion Exchange with Hydrotalcite

Patricia A. Terry, David M. Dolan, and Kendra Axness

Abstract Experiments were conducted to determine and model the effect of cadmium and zinc ions in solution on the removal of Cr(VI) via ion exchange with hydrotalcite, a clay mineral media. Because many locations that have contaminated groundwater are located in colder climates, the ion exchange process was first characterized as a function of temperature. Binary solutions of Cr(VI) with either cadmium or zinc were investigated followed by a factorial design of ternary systems. Isotherms were modeled and used to compare removal of Cr(VI) in a single ion solution to multi-ion systems. Fixed-effect analysis of variance, PROC GLM (SAS Version 9), was used to analyze main and interactive effects. Results indicated that, while temperature did not significantly affect removal of Cr(VI), both Zn and Cd did reduce Cr(VI) ion exchange, most likely due to the formation of ionic complexes. Also, interactions between all three metal species ion the ternary system negatively affected removal of Cr.

13.1 Introduction

Many heavy metals occur naturally in groundwater supplies, but they may also be introduced by anthropogenic activities. Abandoned mines have resulted in ground and surface water contamination globally, including many areas of the United States.

P.A. Terry (✉) • D.M. Dolan
Department of Natural and Applied Sciences, University of Wisconsin-Green Bay, ES 317 2420 Nicolet Drive, Green Bay, WI 54311, USA
e-mail: terryp@uwgb.edu

K. Axness
Wisonsin Department of Natural Resources, 7921, Madison, WI 53707-7921, USA

Inamuddin and M. Luqman (eds.), *Ion Exchange Technology II: Applications*,
DOI 10.1007/978-94-007-4026-6_13,

The tanning industry, especially in clustered areas of industrial growth, has produced irreversible damage to the water environment [1]. Military installations are also a source of water contamination. Globally, suburban growth is bringing businesses, factories, and industry to once rural areas where people rely on untreated well water for domestic use. Activities such as mining and construction can release large quantities of heavy metals, which pose a health risk, into nearby groundwater sources. Chemical spills at local industrial sites, improper disposal and storage of wastes, and leaking or abandoned underground storage tanks may all contribute to this problem [2]. Studies in urban watersheds in Southeastern Michigan, United States, for example, showed that current and former land use practices have resulted in groundwater and soil contaminated with cadmium, chromium, copper, lead, nickel, and zinc [3]. Sites like this with heavy metal contamination usually involve mixed metals. Plating industries, for example, use anionic chromates in addition to zinc and cadmium cations. Remediation methods must, therefore, consider multiple ions in solution, even if only one is the primary target of removal. In addition, groundwater temperatures in northern climates may be well below the temperature utilized in laboratory testing of techniques, and the effect of temperature on removal should be considered when selecting a technology.

Hexavalent chromium is used in metal finishing, tanning, manufacture of dyes and pigments, and manufacture of circuit boards. The US Environmental Protection Agency lists chromium as one of the top 20 chemicals with the largest total releases. In its hexavalent form, it is highly toxic, carcinogenic, and mutagenic [4]. The maximum acceptable limit in drinking water is 0.1 mg/l. It exists as an anion in aqueous solutions, primarily chromate (CrO_4^{2-}) above a pH of 8 and primarily dichromate ($Cr_2O_7^{2-}$) below a pH of 6. It also appears as bichromate ($HCrO_4^-$) when concentrations are below 1.6 mg/l. Because of its toxicity, effective techniques for Cr(VI) removal are important, not only for remediation of contaminated groundwater but also for treatment of industrial wastewaters. However, because Cr(VI) rarely appears alone in industrial processes, its removal in multi-ion solutions must be considered. For example, plating industry wastes may also include ions of copper, nickel, chromium, zinc, cadmium, and lead.

A number of processes have been used to remove Cr(VI) from water in a single ion solution, including reduction to Cr(III), adsorption, membrane separation, and ion exchange. Reduction to Cr(III) followed by chemical precipitation, an established technology, generates sludges that require costly disposal [5]. Many natural materials have been tested for their adsorption potential including peat, clay, and biomass. Park et al. [6] showed that Cr(VI) was completely biosorbed onto dead *Aspergillus niger.* Aoyama et al. [7] investigated the ability of Japanese cedar (*Cryptomeria japonica*) to absorb Cr(VI) as a function of pH, Cr(VI) concentration, and temperature. Banarjee et al. [8] evaluated impregnated fly ash to remove both Cr(VI) and Hg(II) from solution.

Mass transfer of Cr(VI) through a polyisobutylene liquid-supported microporous polytetrafluoethylene polymer membrane was characterized by Park et al. [9]. Another study using supported liquid membranes showed selective separation of Cr(VI) from Cr(III) [10]. Peng et al. [5] showed that electrodialysis could be used to increase the effectiveness of chemical precipitation processes in high Cr(VI) electroplating wastewaters. Kaplan and Gilmore [11] applied zero-valent iron in reactive well technology to remove Cr(VI). Faghihian and Bowman [12] demonstrated the effectiveness of ion exchange with modified clinoptilolite, a naturally occurring clay mineral media, and Tenorio and Espinosa [13] used Amberlite IR 120 Na, a strong cation ion exchanger, followed by a weak base to remove chromium ions from solution. Another study using clinoptilolite characterized the interaction of Pb(II) (lead), Cd(II) (cadmium), and Cr(VI) (chromium) competing for ion exchange sites [14]. It was shown that Cr(VI) anions interacted with lead and cadmium to form anionic complexes, which reduced ion exchange removal efficiencies.

One promising ion exchange material is a clay mineral hydrotalcite (HTC), $[Mg_2Al(OH)_6]_2CO_3{\cdot}3H_2O$. It belongs to a family of inorganic, double-layered hydroxides that model natural clay minerals. Although rare in nature, hydrotalcites, unlike more traditional polymeric resins, can be easily and inexpensively produced in the laboratory. Hydrotalcite is made by co-precipitation of mixed metal solutions of magnesium and aluminum nitrates, $Mg(NO_3)_2{\cdot}6H_2O$ and $Al(NO_3)_3{\cdot}9H_2O$, in the presence of sodium hydroxide (NaOH) and sodium carbonate (Na_2CO_3) to regulate pH. Depending on pH, carbonate (CO_3^{2-}) or bicarbonate (HCO_3^-) in the interlayer region is the exchangeable ion. Hydrotalcites have higher anionic exchange capacities than traditional resins, and because they are weak base ion exchangers, they can be regenerated without the use of strong bases. Hydrotalcite has also been shown to remove cations from solution and, hence, may present an economic way to treat waters containing a variety of cations and anions without the difficulty of managing sensitive biological units [15]. Terry [16] characterized the removal of Cr(VI) by uncalcined hydrotalcite as a function of pH, temperature, contact time, and both Cr(VI) and hydrotalcite concentration. Lazaridis and Asouhidou [15] studied the kinetics of sorptive removal of Cr(VI) by hydrotalcite, and Martinez-Gallegos et al. [17] found that in addition to Cr(VI), calcined hydrotalcite removed chloride (Cl^-), sulfate (SO_4^{2-}), and bicarbonate (HCO_3^-) from tannery wastes.

This work builds on previous ion exchange studies with hydrotalcite to characterize the effect of zinc and cadmium cations in solution on removal of Cr(VI). First, removal of Cr(VI) by hydrotalcite at cold and warm water temperatures was measured using the established pH, contact time, and hydrotalcite dose established by Terry (2004) for optimal removal. Next, the individual effect of cadmium and zinc cations, respectively, on Cr(VI) removal was performed in binary solutions at metal levels consistent with contaminated groundwater. Finally, a three-way factorial design was used to determine if zinc and cadmium cations in a ternary solution reduced Cr(VI) removal. Isotherms were created for each situation and modeled via Freundlich and Langmuir parameters.

13.2 Materials and Methods

Uncalcined hydrotalcite used in these experiments was purchased from Sigma-Aldrich chemicals. Chromium, zinc, and cadmium were added as research-grade chromium trioxide, zinc chloride, and cadmium chloride, respectively, also from Sigma-Aldrich.

Solution pH adjustments were made with research grade hydrochloric acid and were measured with a Pasco ScienceWorkshop 500 pH meter. All metal concentrations were determined with a Perkin Elmer atomic absorbance spectrophotometer.

Experiments were performed in triplicate in batch mode with average concentrations reported. Samples were prepared in 250 ml acid washed polyethylene flasks. Each sample used 0.5 g/l hydrotalcite and was adjusted to a pH of 2.05 ± 0.05, previously determined optimal for Cr(VI) removal with hydrotalcite. After preparation, samples were placed in an American Scientific products shaking water bath controlled for temperature and set at 200 rpm. Following exposure, hydrotalcite was separated from suspension by vacuum filtration through 0.7 μm Whatman glass fiber filters. Preliminary experiments demonstrated that chromium, zinc, and cadmium ions were not removed by either the polyethylene bottles or the filters.

Initial ion exchange experiments were performed to measure removal of each of Cr(VI), Cd(II), and Zn(II) in single ion solutions under pH conditions optimal for Cr(VI) removal. Each of these was performed at 2°C, 10°C, and 25°C to determine the effect of cold temperatures on removals. Removal of each metal ion was characterized up to a 25 mg/l concentration, consistent with slightly to severely contaminated groundwater. Next, the effect of each of Cd(II) and Zn(II) on removal of Cr(VI) was performed in two ion solutions. For each of these, Cr(VI) removal was characterized as a function of either Cd(II) or Zn(II) concentration for both low, 1 mg/l, and high, 25 mg/l, initial Cr(VI) and compared to removal in the single ion solution. In addition, Cr(VI) isotherms were created for each binary solution comparing the Cr(VI) single ion isotherm to isotherms with low, 1 mg/l, and high, 25 mg/l, initial levels of either Cd(II) or Zn(II). Both the Freundlich, $W\ (mg/g) = b_1 C_{eq}^{\ n}$, and Langmuir, $W(mg/g) = b_2 C_{eq}/(1 + KC_{eq})$, equations were used to model all isotherms, where C_{eq} is the residual Cr(VI) in solution following removal, W is the mass of Cr(VI) removed per unit mass hydrotalcite, and b_1, b_2, n, and K are isotherm constants. Fixed-effect analysis of variance, PROC GLM (SAS Version 9), was used to analyze main and interactive effects. Tukey's studentized range honest significant difference (HSD) at $\alpha = 0.05$ was used as a multiple comparison procedure to determine significant mean differences in treatments.

Finally, the effects of both Zn(II) and Cd(II) on Cr(VI) removal were tested in ternary systems using a factorial design with high and low levels of Zn(II) and Cd(II) over the 1–25 mg/l range of chromium concentrations. Cr(VI) isotherms for each scenario were compared to the Cr(VI) single ion isotherm. These ternary systems were analyzed by an ANOVA model (SAS Version 9). Multiple comparisons were made and significant differences were reported.

13.3 Results and Discussion

Results are presented in the form of residual metal concentrations and percent removal. Except for the single ion solutions, only residual Cr(VI) and isotherms are presented since the goal of the experiments was to determine the effects of Cd(II)

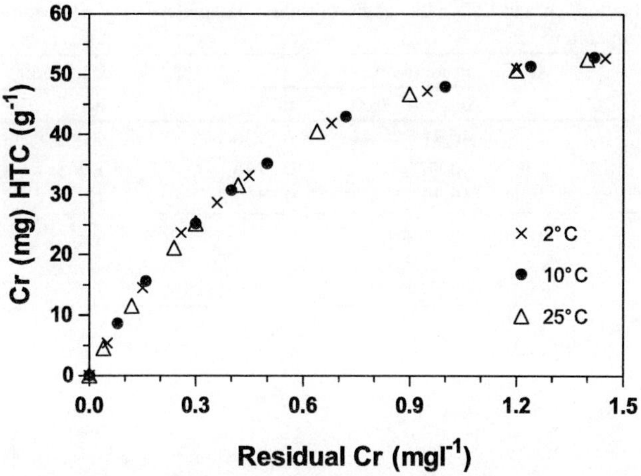

Fig. 13.1 Cr(VI) isotherms as a function of temperature

and Zn(II) on Cr(VI) removal. For each isotherm, the model constants b_1, b_2, n and K and R^2 values describing the model fit are reported.

Samples were prepared by first adding the appropriate metal, adjusting pH to between 2.0 and 2.1, followed by adding hydrotalcite. The alkaline nature of the hydrotalcite caused an immediate shift in pH to 8.2±0.1. At this pH bicarbonate, HCO_3^-, is the exchangeable carbonate ion in solution and chromate, CrO_4^{2-}, is the dominate form of the Cr(VI) ion.

13.3.1 Single Ion Removal

Figure 13.1 shows the isotherms for single ion removal of Cr(VI) as a function of temperature at aqueous solution temperatures of 2°C, 10°C, and 25°C for a range of initial Cr(VI) concentrations from 0 to 25 mg/l. At Cr(VI) concentrations under 10 mg/l, removal was 95%, and at concentrations above 15 mg/l, it was 94%. As the isotherms indicate, temperature in the range considered was not a factor in the efficiency of the ion exchange process. Subsequent experiments were all performed at 10°C. At pH conditions optimal for Cr(VI) ion exchange, neither Cd(II) nor Zn(II) in single ion solutions was greatly removed by hydrotalcite. Zn(II) removal ranged from 27% at an initial concentration of 1 mg/l to less than 10% for 25 mg/l initial Zn(II). Similarly, Cd(II) removal ranged from a maximum of 14% for 25 mg/l initial Cd(II) to less than 5% for 1 mg/l initial Cd(II).

Statistical analysis of Cr(VI) residuals showed that initial Cr(VI) was a significant factor ($P<0.01$) in Cr(VI) removal with residuals varying from 0.05 mg/l for an initial level of 1.0–1.4 mg/l for an initial level of 25 mg/l. Modeling of the Cr(VI) isotherms verified that removal was statistically similar for the three temperatures.

Table 13.1 Freundlich and Langmuir isotherm model parameters for Cr(VI) ion exchange as a function of temperature

	Freundlich		Langmuir	
Temperature (°C)	b_1	m	b_2	K
2	46.71	0.5156	131.2	1.747
10	46.80	0.5146	129.5	1.708
25	46.79	0.5379	126.0	1.638

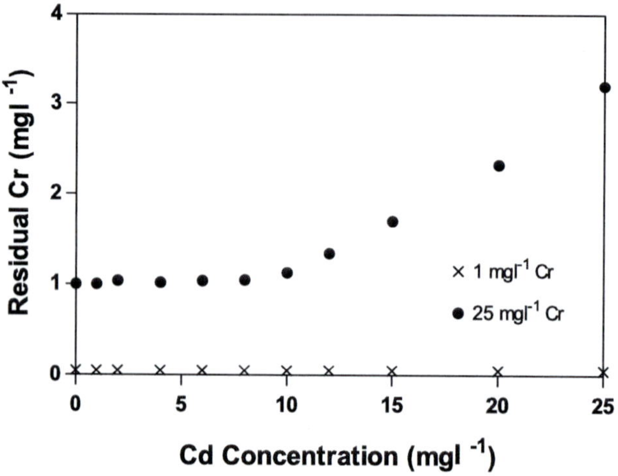

Fig. 13.2 Residual Cr(VI) as a function of cadmium in a binary solution

The Freundlich, b_1 and m, and Langmuir, b_2 and K, parameters for the three temperatures are presented in Table 13.1. For all temperatures, the Langmuir model was a better fit, giving r^2 values greater than 0.998, although r^2 values for the Freundlich model also exceeded 0.98.

13.3.2 *Two Ion Solutions*

Figure 13.2 shows the residual Cr(VI) in solution following ion exchange as a function of Cd(II) in solution for initial Cr(VI) concentrations of 1 and 25 mg/l. At the low Cr(VI) concentration, addition of Cd(II) did not significantly decrease Cr(VI) removal. Residual Cr(VI) was 0.06 mg/l for all Cd(II) concentrations. At Cr(VI) concentrations of 25 mg/l, low levels of Cd(II) did not affect Cr(VI) removal; however, for Cd(II) levels at and above 10 mg/l, Cr(VI) removal was negatively affected. Percent Cr(VI) removal was reduced from 94% in the single ion solution to 87%. This is also demonstrated by the comparative isotherms in Fig. 13.3 for Cr(VI) ranging from 0 to 25 mg/l and either zero, 1 or 25 mg/l Cd(II). The isotherm representing the binary solution with 1 mg/l Cd(II) is similar to that of the single ion Cr(VI)

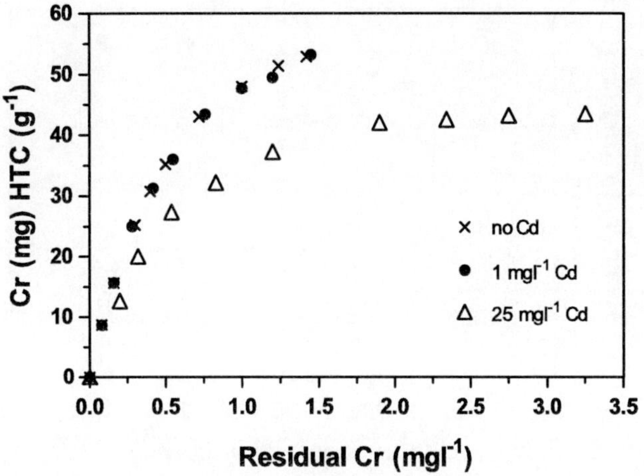

Fig. 13.3 Cr(VI) isotherms in the presence of cadmium

Table 13.2 Freundlich and Langmuir isotherm model parameters for Cr(VI) ion exchange as a function of Cd(II)

Cd(II) concentration	Freundlich		Langmuir	
(mg l^{-1})	b_1	m	b_2	K
0	46.8	0.5146	129.5	1.708
1	46.32	0.512	129.8	1.747
25	31.62	0.3396	101.8	1.966

isotherm, while at higher residual Cr(VI) levels, the isotherm representing the binary solution with 25 mg/l Cd(II) is reduced with higher residual Cr(VI) values.

Fixed-effects analysis of variance was used to analyze the effects of Cr(VI) and Cd(II) on residual Cr(VI). Both Cr(VI) ($P<0.01$) and Cd(II) ($P<0.02$) initial concentrations were shown to significantly affect Cr(VI) removal. In addition, the interaction between Cr(VI) and Cd(II) was also significant ($P<0.01$), but to a lesser degree as only high Cd(II) levels reduced Cr(VI) ion exchange. Table 13.2 shows the Freundlich and Langmuir isotherm model constants for the three isotherms in Fig. 13.3. Again, the parameters for both models are similar for the single ion Cr(VI) and Cr(VI) with low Cd(II) isotherms but are different from the high Cd(II) binary solution isotherm. As before, the Langmuir model is the more descriptive with r^2 values exceeding 0.996 for all three isotherms, but the Freundlich model r^2 values also exceeded 0.96.

Figure 13.4 shows the residual Cr(VI) in solution following ion exchange as a function of Zn(II) in solution for initial Cr(VI) concentrations of 1 and 25 mg/l. At a low Cr(VI) concentration, the addition of Zn(II) decreased the removal from 95% in a Cr(VI) single ion solution to 90% when Zn(II) was present at any level. Residual Cr(VI) in the presence of Zn(II) was only 0.1 mg/l, rather than 0.05 mg/l. At Cr(VI) concentrations of 25 mg/l, the presence of Zn reduced Cr(VI) removal at

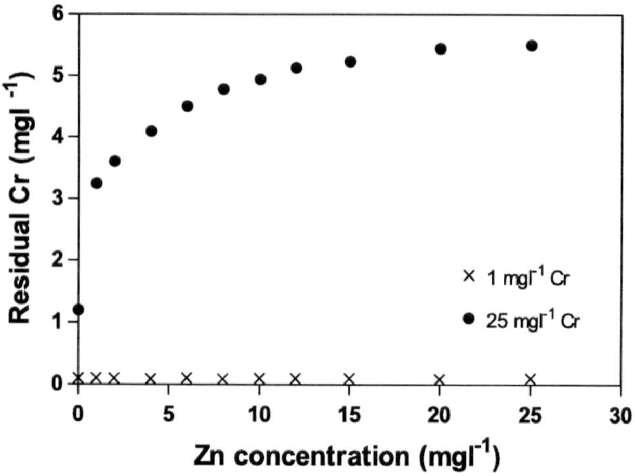

Fig. 13.4 Residual Cr(VI) as a function of Zn(II) in a binary solution

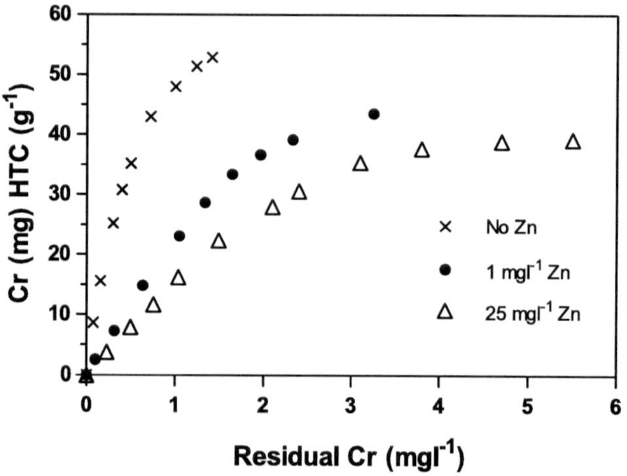

Fig. 13.5 Cr(VI) isotherms in the presence of Zn(II)

all Zn levels. When the Zn was 1 mg/l, Cr(VI) removal decreased from 94% in the single ion solution to 87%, and when Zn was 25 mg/l, the percent Cr(VI) removal was further reduced to 78%. This is also demonstrated by the comparative isotherms in Fig. 13.5 for Cr(VI) ranging from 0 to 25 mg/l and either zero, 1 or 25 mg/l Zn. The isotherm representing the binary solution with 1 mg/l Zn is somewhat reduced from that of the single ion Cr(VI) isotherm, with a maximum capacity of 43 mg Cr/g HTC. The isotherm for the binary system with 25 mg/l Zn is considerably lower than that of either the single ion or low Zn isotherms with a maximum capacity of 37 mg Cr/g HTC.

Table 13.3 Freundlich and Langmuir isotherm model parameters for Cr(VI) ion exchange as a function of Zn(II)

Zn concentration	Freundlich		Langmuir	
(mg l^{-1})	b_1	m	b_2	K
0	46.8	0.5146	129.5	1.708
1	22.12	0.6495	31.98	0.3966
25	16.66	0.5695	22.8	0.3644

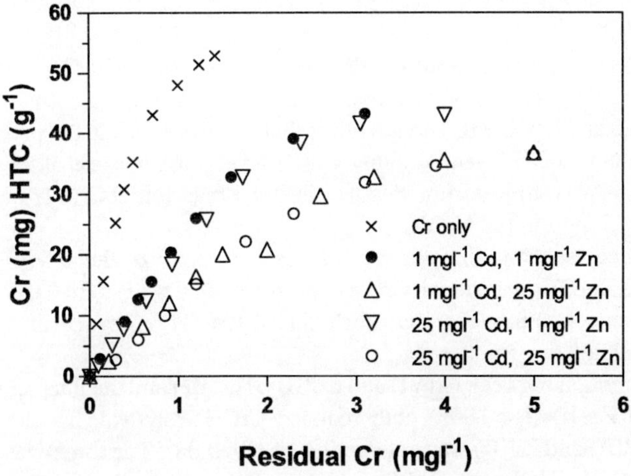

Fig. 13.6 Cr(VI) isotherms in the presence of both Cd(II) and Zn(II)

Fixed-effect analysis of variance was used to analyze the effects of Cr(VI) and Zn on residual Cr(VI). Both Cr(VI) ($P<0.01$) and Zn(II) ($P<0.005$) initial concentrations affected the removal of Cr(VI). The interaction between Cr(VI) and Zn was also significant ($P<0.005$) in decreasing Cr(VI) removal. Table 13.3 shows the Freundlich and Langmuir isotherm model constants for the three isotherms in Fig. 13.3. The parameters for both models indicate reduced Cr(VI) removal in the presence of Zn. The Langmuir isotherms are considered to be favorable when $KC_{eq}>1$, such that the presence of Zn renders the system unfavorable. As before, the Langmuir model is the more descriptive with r^2 values exceeding 0.99 for all three isotherms, but the Freundlich model r^2 values also exceeded 0.96.

13.3.3 Ternary System

Figure 13.6 shows the isotherms for Cr(VI) ion exchange in ternary systems for combinations of high (25 mg/l) and low (1 mg/l) Cd(II) and Zn(II) over the range of 0–25 mg/l Cr(VI). Reductions in the isotherms demonstrate that the presence of

Table 13.4 Freundlich and Langmuir isotherm model parameters for Cr(VI) ion exchange as a function of Zn(II) and Cd(II)

		Freundlich		Langmuir	
Zn (mgl^{-1})	Cd (mgl^{-1})	b_1	m	b_2	K
0	0	46.8	0.5146	129.5	1.708
1	1	21.14	0.6927	29.21	0.3299
1	25	19.64	0.6494	28.11	0.3615
25	1	14.12	0.6513	17.55	0.2541
25	25	13.39	0.6842	16.68	0.2322

both Cd(II) and Zn(II) reduced Cr(VI) ion exchange but that Zn(II) has a greater effect than Cd(II). The Cr(VI) isotherms in the presence of high Zn(II) are similar for both high and low Cd(II) and are lower than the isotherms in the presence of low Zn(II) at either Cd(II) level. At either Cd(II) level, the presence of 25 mg/l Zn(II) decreased Cr(VI) removal from 95% to 75% for 1 mg/l initial Cr(VI) and from 94% to 65% for 25 mg/l initial Cr(VI).

Fixed-effect analysis of variance was used to analyze the effects of Cr(VI), Cd(II), and Zn(II) on residual Cr(VI). Both initial Cr(VI) ($P<0.01$) and Zn(II) ($P<0.01$) concentrations significantly reduced Cr(VI) removal, but initial Cd(II) ($P<0.07$)did not.

The interaction between Cr(VI) and Cd(II) ($P<0.02$) and the interaction between Cr(VI) and Zn(II) ($P<0.005$) both reduced Cr(VI) removal, but the interaction between Cd(II) and Zn(II) was not significant ($P<0.08$). The three-way interaction between Cr(VI), Cd(II), and Zn(II) was ($P<0.04$) statistically significant in reducing Cr(VI) removal. This is demonstrated by the isotherms; low Zn levels reduce the isotherms by similar amounts for both high and low Cr levels, and the presence of high Zn levels further reduce the isotherms, but by similar amounts for both high and low Cr levels.

Table 13.4 shows the Freundlich and Langmuir isotherm model constants for the five isotherms in Fig. 13.6. Again, the parameters for both models are similar for the two isotherms representing 1 mg/l Zn(II) and either 1 or 25 mg/l Cd(II). The same is true for the two isotherms representing 25 mg/l Zn(II) with either 1 or 25 mg/l Cd(II). Presence of either Zn(II) or Cd(II) reduced K values for the Langmiur isotherms sufficiently to render the isotherms unfavorable. As before, the Langmuir model was the more descriptive with r^2 values exceeding 0.985 for all five isotherms, but the Freundlich model r^2 values also exceeded 0.96.

13.4 Conclusions

Experiments performed to characterize ion exchange between the exchangeable carbonate anion in hydrotalcite and the chromate anion in solution demonstrated that the process performs equally well in cold and room temperature water systems.

Experiments performed to characterize the effects of Cd and Zn in aqueous solution on the removal of Cr(VI) via ion exchange at cold water temperatures demonstrated that, while the process is still feasible for treating polluted ground and surface water, the presence of either metal ion reduces its effectiveness, and zinc appears to have a greater deleterious effect than cadmium. Because neither Zn nor Cd is greatly removed in single ion solutions under conditions optimal for Cr(VI) ion exchange, the reduction in Cr(VI) ion exchange due to either of these cations is most likely due more to the formation of ionic complexes between the anionic chromate in solution with the metals than with competition for exchange sites. Statistical analysis of the data showed that Cr(VI), Zn, Cd, and interaction effects between the ions all affected Cr(VI) removal. Lastly, Freundlich and Langmuir isotherms were modeled for all isotherms. While both provided a good fit to the data, the Langmuir model, which is most often used to characterize ion exchange processes, provided the better fit.

References

1. Shankar B (2009) Chromium pollution in the groundwaters of an industrial area in Bangalor, India. Environ Eng Sci 26(2):305–310
2. United States Environmental Protection Agency (2002) Drinking water from household wells. USEPA 16-K-02-003
3. Murray K, Rogers D et al (2006) Dissolved heavy metals in shallow ground water in a southeastern Michigan urban watershed. J Am Water Resour Assoc 42(3):777–791
4. United States Environmental Protection Agency (2003) 2001 toxics release inventory (TRI) public data release executive summary. USEPA 60-S-03-001
5. Peng C, Meng H et al (2004) Elimination of Cr(VI) from electroplating wastewater by electrodialysis following chemical precipitation. Sep Sci Technol 39:1501–1517
6. Park D, Yun Y, Jo S et al (2005) Mechanism of hexavalent chromium removal by dead fungal biomass of *Aspergillus niger*. Water Res 39:533–540
7. Aoyama M, Kishino M, Jo T (2004) Biosprption of Cr(VI) on Japanese cedar bark. Sep Sci Technol 39:1149–1162
8. Banarjee S, Joshi M et al (2004) Removal of Cr(VI) and Hg(II) from aqueous solution using fly ash and impregnated fly ash. Sep Sci Technol 39:1611–1629
9. Park S, Jung H et al (2004) Effect of rheological properties on mass transfer of Cr(VI) through a supported liquid membrane with non-Newtonian liquid. Sep Sci Technol 39:781–797
10. Choi Y, Moon S (2004) A study on supported liquid membranes for selective separation of Cr(VI). Sep Sci Technol 39:1663–1680
11. Kaplan D, Gilmore T (2004) Zero valent iron removal rates of aqueous Cr(VI) ions across polymer inclusion membranes with tridecyl(Pyridine) oxide n-octyllamine. Sep Sci Technol 39:3127–3141
12. Faghihian H, Bowman R (2005) Adsorption of chromate by clinoptilolite exchanged with various metal cations. Water Res 39(6):1099–1104
13. Tenorio J, Espinosa D (2001) Treatment of chromium plating process effluents with ion exchange resins. Waste Manage 21:637–642
14. Vaca Mier M, Lopez Callejas R et al (2001) Heavy metal removal with Mexican clinoptilolite: multi-component ion exchange. Water Res 35(2):373–378
15. Lazaridis N, Asouhidou D (2003) Kinetics of sorptive removal of chromium (VI) from aqueous solutions by calcined Mg-AL-CO_3 hydrotalcite. Water Res 37:2875–2882
16. Terry P (2004) Characterization of Cr ion exchange with hydrotalcite. Chemosphere 57:541–546
17. Martinez-Gallegos S, Martinez V et al (2004) Chromium (VI) separation from tannery wastes utilizing hydrotalcites. Sep Sci Technol 39:667–681

Chapter 14
An Overview of '3d' and '4f' Metal Ions: Sorption Study with Phenolic Resins

J.D. Joshi

Abstract Day by day, society is in need of a variety of materials. Therefore, different types of new materials are required to fulfil the demand of the society. For the synthesis of desired material, various new processes are established, and modifications are made in old processes.

Nowadays, society is very conscious about environment-friendly relationships in atmosphere and climate change; hence, any required change in old processes and for establishing a new one with utmost care is necessary. Chemistry is playing a vital role in daily life. Various chemical industries discharge waste in the form of solid, liquid and gaseous state.

The recovery of the material from the discharge of the waste is a very difficult task, but it is necessary to remove the toxic materials from the waste.

In this chapter, the role of the absorbing material is discussed specially for '3d' and '4f' metal ions. Phenol-based resins can be effectively used as ion exchangers to remove metals ions from the discharge of the industries. The efficiency of the resins has been studied at various pH concentrations, time and with different electrolytes.

14.1 Introduction

Mineral processing and metal finishing industries produce large amounts of waste effluents containing chromium, zinc, cadmium, copper, nickel and other toxic elements [1]. Metal ions are non-biodegradable in nature; therefore, their intake at certain levels is toxic [2]. Copper is both vital and toxic for many biological systems [3]. The increasing strict change in environmental regulations and enforcement of

J.D. Joshi (✉)
Sardar Patel University, Vallabh Vidyanagar-388120, Dist. Anand, Gujarat, India
e-mail: jdjoshi314@gmail.com

Inamuddin and M. Luqman (eds.), *Ion Exchange Technology II: Applications*,
DOI 10.1007/978-94-007-4026-6_14,

discharge limits requires effective decontamination and purification methods. In an analytical point of view, it is known that solid-phase extraction is an attractive technique based on the use of metal sorbent that retains metal ions eluted from the sorbent using a suitable eluent.

Numerous methods have been described for their effective separation and pre-concentration; procedures based on ion exchange [4, 5], solvent extraction [6–8] and solid-phase extraction (SPE) [9–13] are amongst them. Pre-concentration method is a widely accepted technique for monitoring low concentration of metal ions [4, 5]. Chelate-forming selected resins [14–17] have found numerous applications for the separation and monitoring of heavy metals, including lanthanides from aqueous solutions [18–20]. The use of chelating resins for separation and/or removal of metal ions is the method of choice due to its high separation efficiency, good reproducibility of retention parameters and simplicity [21–24]. Considerable efforts have been made for the improvement and economical optimization of the chelating properties of ion exchange resins [25, 26]. Chelating ion exchange resins have specific chelating groups which are extensively useful in separation and pre-concentration of metal ions [27–30]. Vernon has summarized the desirable properties of a chelating polymer which includes high capacity and appropriate selectivity [31]. Polymeric resins which possess chelating properties are found to be more selective by nature [32, 33] as compared to other conventional techniques [34, 35] to remove metal ions. The solid-phase extraction method using molecular imprinted polymers is the most useful method for separation and pre-concentration of trace metals [36, 37]. Molecular imprinting is a methodology for the introduction of selective recognition sites into highly cross-linked polymeric matrices, via the template-directed assembly of functionalized monomers into a polymer network [38–40].

The chelate-forming polymeric ligands, characterized by reactive functional groups containing O, N, S and P donor atoms and capable of coordinating to different metal ions, have been extensively studied [41–47]. Chelate-forming polymers having multi-dentate coordination sites are known to form complexes with metal ions readily [48–50]. In a polymer matrix, they are expected to show affinity and selectivity towards the metal ions at an appropriate pH. This led us to synthesize chelating resins, which have affinity for the metal ions at appropriate pH [51–54]. The phenolic resins are playing a vital role in construction, automobile, electrical and appliance industries [55–57].

14.2 Principle of Ion Exchange Process

Ion exchange is a process in which mobile ions from an external solution are exchanged for ions that are electrostatically bound to the functional groups contained within a solid matrix. When the functional groups are negatively charged, the exchange will involve cations, and when they are positively charged, they involve anions.

Depending on the type of the functional group, ion exchangers can be divided into several types: strongly acidic, strongly basic, weakly acidic and weakly basic.

Table 14.1 pK values for the most common functional groups of some ion exchangers

Cation exchangers		Anion exchangers	
Functional group	pKa	Functional group	pKb
$-SO_3H$ (strongly acidic)	1–2	$\equiv N^+$ (strongly basic)	1–2
$-PO_3H_2$	2–5	=N	4–6
–COOH	4–6	=NH	6–8
–OH (weak acidic)	9–10	$-NH_2$ (weak basic)	8–10

In Table 14.1, the negative logarithm of the dissociation constant (pK) is given for different functional groups. Ion exchangers containing sulpho-, phospho-acidic groups and those containing tetra ammonium basic groups are strong acidic and strong basic exchangers, respectively, whereas those containing phenolic and primary amino groups are weak acidic and weak basic exchangers, respectively. Exchanger with carboxy groups and tertiary amino groups takes a medium position between strong and weak acidic and basic exchangers, respectively. To achieve the removal of both positively and negatively charged ions from solution, a mixture of cation and anion resins in a mixed bed system is often used; for example, in case of NaCl solution, the ion exchange process will be

$$R-H+Na^{+} \rightarrow RNa+H^{+} \tag{14.1}$$

$$R_1-OH+Cl^{-} \rightarrow R_1CL+OH^{-} \tag{14.2}$$

$$H^{+}+OH^{-} \rightarrow H_2O \tag{14.3}$$

Since H_2O is only weakly dissociated, the adsorption reaction is proceeding in the right-hand side of the equation.

14.2.1 Ion Exchange Equilibrium, Selectivity, Sorption and Limitation

Ion exchange equilibrium can be described by any of the following term:

- The ion exchange isotherm
- The separation factor
- The selectivity coefficient
- The thermodynamic equilibrium constant
- The distribution coefficient

These terms are common and well known in ion exchange study [58]. It should be noted that selectivity coefficients are not constant and may vary with the experimental conditions. For general purposes, few 'rules of thumb' can be applied: for cationic organic ion exchange resins at low concentrations and the temperatures normally encountered in waste processing, the affinity increases with (a) an increasing charge on the exchanging cation and (b) an increasing atomic number (decreasing hydrated ionic radii) of the exchanging cation; for example,

(a) $Li^+ < H^+ < Na^+ < K^+ < Cs^+ < Mg^{2+} < Co^{2+} < Ca^{2+} < Sr^{2+} < Ce^{3+} < La^{3+} < Th^{4+}$
(b) $Li^+ < H^+ < Na^+ < K^+ < Cs^+ <$ (Li^+ is an exception, owing to its high hydration energy)

For anions, a typical series is as follows:

$$F^- < CH_3COO^- \text{(acetate)} < Cl^- < Br^- < CrO_4^{2-}$$
$$< NO_3^- < I^- < C_2O_4^{2-} \text{(oxalate)} < SO_4^{2-}$$

Changes in physical parameters as well as in concentration of functional groups affect the distribution coefficient and the driving force for the ion exchange process. High values of the distribution coefficient are always desirable.

Sorption is a separation process involving two phases between which certain components can become differentially distributed. Normally three types of sorption are classified according to the type of bonding involved:

(a) Physical sorption: There is no exchange of electrons in physical sorption; rather, intermolecular attractions occur between 'valency happy' sites and are therefore independent of the electronic properties of the molecules involved. The heat of adsorption or activation energy is low in physical sorption.
(b) Chemical sorption: Chemical sorption or chemisorption involves an exchange of electrons between specific surface sites and solute molecules, which results in the formation of a chemical bond. Such a bond is more stable at higher temperatures; therefore, the chemisorption is a stronger adsorption than physical adsorption.
(c) Electrostatic sorption (ion exchange): This term is reserved for columbic attractive forces between ions and charged functional groups and is more commonly classified as ion exchange.

In addition to being ion exchangers, ion exchange materials can also act as sorbents. When they are in contact with an electrolyte solution, the dissolved ions are absorbed on both the surface and in the pores of the ion exchange media. In a solution of weak electrolytes or non-electrolytes, sorption by ion exchangers is similar to that of non-ionic adsorbents. In a solution of strong electrolytes, sorption equilibrium results, owing to the electrostatic attraction between the ions in solution and the fixed ionic groups on the ion exchange media.

Certain characteristics of ion exchange materials and processes limit their applicability and efficiency. When used as a packed bed in a column, the complete removal of a specific metal ion is not normally possible, owing to leakage or breakthrough. This will cause the metal ions to pass through without being captured. In a properly designed system, this breakthrough may be very low but will still be present.

In batch process, inadequate mixing will also limit the effectiveness of the ion exchange material. The total concentration of dissolved salts in solution must generally be low (<1 g/L). At high concentrations, the exchange potentials for ions diminish, and there is more competition for the available exchange sites between desired ions and the undesired ions. However, for specifically developed ion exchangers (usually inorganic), the tolerance for dissolved salts can be very high (e.g. up to 240 g/L [59]).

14.3 Ion Exchange Materials

A wide range of materials are available for the ion exchange treatment of metal ions in aqueous solution in a variety of forms and have wide difference in their chemical and physical properties. They can be naturally occurring or synthetic materials. Ion exchange materials can be categorized according to their suitability for different applications.

14.3.1 Naturally Occurring Inorganic, Organic and Modified Ion Exchangers

Many natural mineral compounds, such as clays (e.g. bentonite, kaolinite and illite), vermiculite and zeolites (e.g. analcites, chabazite, sodalite and clinoptilolite), exhibit ion exchange properties. Natural zeolites were the first materials to be used in ion exchange processes. Clay materials are often employed as backfill or buffer materials for radioactive waste disposal sites because of their ion exchange properties, low permeability and easy workability. Clays can also be used in batch ion exchange processes but are not generally suited to column operation because their physical properties restrict the flow through the bed [60].

A large number of organic materials exhibit ion exchange properties; these include polysaccharides (such as cellulose, straw and peat), proteins (such as casein, keratin and collagen) and carbonaceous materials (such as charcoals, lignites and coals). Of these, only charcoals, coal, lignite and peat are used commercially. Although they exhibit a very low ion exchange capacity compared with synthetic ion exchangers.

Some naturally occurring organic ion exchangers are modified to improve ion exchange capacity by chemical and/or thermal treatment [52–62].

14.3.2 Synthetic Inorganic and Organic Ion Exchangers

Synthetic ion exchangers are produced by preparing new chemical compounds with desired physical and chemical properties. They can be inorganic (mineral) or organic (generally polymer) based.

Zeolites were largely used for removal of metal ions from solution [63–68]. However, synthetic zeolites can be engineered with a wide variety of chemical properties, pore size and thermally high stable.

Titanates and hydrous titanium oxide are known to be highly selective exchangers for strontium [69–72]. It is a highly effective exchanger in an alkaline solution (i.e. for $pH > 9$) [73]. These exchangers have a high selectivity both for cesium and strontium [74–76].

Insoluble transition metal hexacyanoferrates were found suitable for removing some heavy metals from alkaline reprocessing waste containing a high concentration of sodium salts [77–79]. A freshly precipitated cupric hexacyanoferrate is also used for the removal of cesium from aqueous solution [80, 81].

The largest group of ion exchangers available today are synthetic organic resins in powdered (5–150 μm) or bead (0.5–2 mm diameter) form. The framework or matrix of the resins is a flexible random network of hydrocarbon chains. This matrix carries fixed ionic charges at various locations. The degree of cross-linking of resin determines the mesh width of the matrix, swelling ability, movement of mobile ions, hardness and mechanical durability.

The most common form of ion exchange resin is based on a copolymer of styrene and divinylbenzene. Resin, with low divinylbenzene content, is soft and gelatinous and swells strongly in solvent. Fixed ionic groups are introduced into resin matrix to produce an ion exchanger; these introduced groups then become the mobile or counterions which can be replaced by a treatment with a solution containing another cation. Anion exchangers can be produced by creating $-NH_3^+$ or $-N_2^+$ functional groups on the matrix with OH^-, Cl^- or other anions as the counterion.

Phenol-formaldehyde condensation products, with the phenolic –OH groups as the fixed ionic groups, are very weak acid exchangers. Sulphonation of the phenol prior to polymerization can be used to increase the acid strength. Phenolsulphonic acid resins are bifunctional with both strong acid $-SO_3H$ and weak acid –OH groups. A resorcinol-formaldehyde polycondensation resin was recently developed for removal of heavy metal from alkaline reprocessing waste [82–84]. Incorporation of iminodiacetic acid functional group in the phenolic polymer gives the additional property for uptake of metal ion by chelation [85, 86].

14.4 Ion Exchange Techniques

Ion exchange processes can be implemented in a variety of ways, but batch and column operations are widely accepted techniques.

Batch operation is the simplest method for operating an ion exchange process. It can be used with either organic or inorganic media, does not require sophisticated equipment and can be carried out on any scale at ambient temperatures and pressures. A measured quantity of the ion exchange medium is mixed with the liquid waste in a suitable container. The amount of media required and the rate of exchange can be determined using the following equation:

$$\mathrm{Kd} = (\mathrm{DF} - 1) \times V / m \ldots\ldots \qquad (14.4)$$

where

Kd = measured distribution coefficient
DF = required decontamination factor
V = volume of liquid to be purified
m = amount of the ion exchange medium needed to reach the required decontamination factor

The mixture is allowed to equilibrate for a specified time either with or without stirring. At the end of the reaction time, the ion exchange medium is removed from the liquid using conventional separation techniques such as decantation, filtration or centrifugation. The treatment can be repeated as required by adding further ion exchange media. The powdered or bead-type media can be used in batch process. Bead media is easier to remove by filtration, but powdered media provides larger surface area for contact between the liquid and the media, resulting in faster sorption kinetics, and, therefore, time required for reaction is short.

While in column operation, the most common use of ion exchange media is as packed bed in vessels or columns. The ion exchange medium is enclosed inside a steel pressure vessel, with an engineered inlet, outlet and flow distribution system to allow liquid to percolate through the bed of the medium at a controlled flow rate. Columns are generally used with bead-type ion exchange media and can be constructed in a wide variety of sizes and materials to meet the requirements of the system. The low porosity of bed-powdered material restricts their use to thin layers, usually as a 'pre-coat' on a filter medium.

Column systems can be configured for single-bed operation systems or mixed bed systems, in which cation and anion media are mixed together in a suitable proportion (usually equal portions). However, moving bed and pulsed bed systems can provide a more uniform liquid product and can have a lower media inventory than column systems. Moving bed and pulsed bed systems are also more tolerant to particles in the liquid.

14.5 Synthesized Phenolic Resins and Its Ion Exchange Study

14.5.1 Benzophenone-Based Phenolic Resins and Ion Exchange Characteristic

Benzophenone-based chelating phenolic resins were synthesized by polycondensation of 2,4-dihydroxy benzophenone with different diols, such as ethylene glycol (EG), propylene glycol (PG), 1,3-propane diol (1,3-PD), 1,3-butane diol (1,3-BD) and 1,4-butane diol (1,3- BD), in presence of acid catalyst. Substitution at the fourth position of 2,4-dihydroxy benzophenone has been carried out by replacing –OH with $-OCH_3$, $-OC_4H_9$, etc.

Mixture of 2,4-dihydroxy benzophenone and respective diol was condensed in presence of polyphosphoric acid as a catalyst at suitable temperature for appropriate time. The product obtained was purified by washing with water followed by methanol to remove unreacted acid and monomer [86].

This general procedure was used for the synthesis of 2-hydroxy-4-methoxy benzophenone diol resins, 2-hydroxy-4-butoxy benzophenone diol resins [87, 88].

Here the benzophenone moieties are bonded through third and fifth positions of the aromatic ring with respective diols. The ethylene glycol form $-CH_2-CH_2-$, propylene

Monomer	Bridging groups (OH – R –OH)	Resin
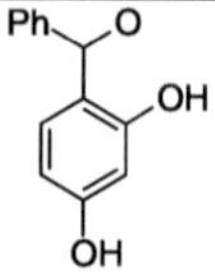 2,4-dihydroxy benzophenone	HO – CH_2 – CH_2 – OH HO – CH_2 – CH_2 – CH_2 – OH HO – CH_2 – CH_2 – CH_2 – CH_2 – OH HO – CH (CH_3) – CH_2 – OH HO – CH (CH_3) – CH_2 – CH_2 – OH Cl – CH_2 – Cl Cl – CH_2 – CH_2 – Cl	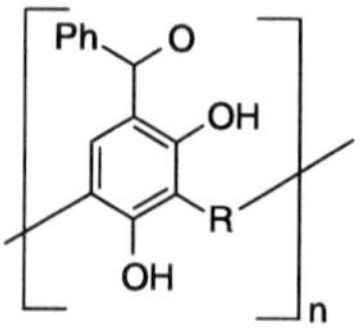
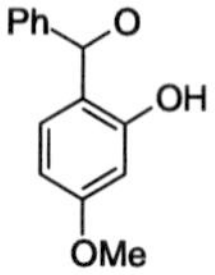 2-hydroxy-4-methoxy benzophenone		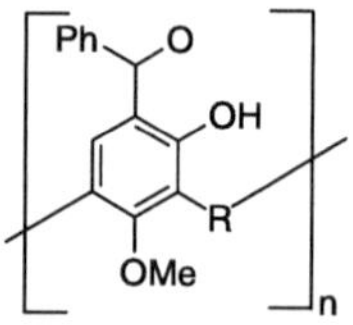
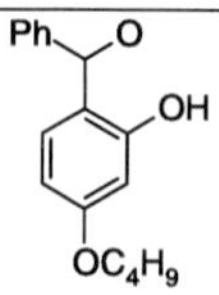 2-hydroxy-4-butoxy benzophenone		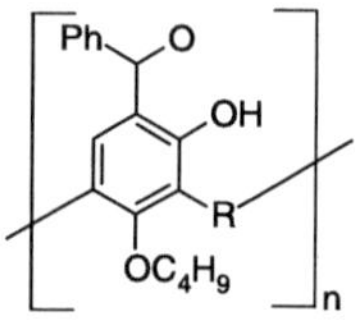

Fig. 14.1 Benzophenone moieties

glycol form –$CH(CH_3)$–CH_2–, 1,3-propane diol form –CH_2–CH_2–CH_2–, 1,3-butane diol and 1,4- butane diol forms –$CH(CH_3)$–CH_2–CH_2– and –CH_2–CH_2–CH_2–CH_2– bridge, respectively (Fig. 14.1).

Benzophenone-based phenolic resins with different bridging groups were synthesized using dichloromethane (DCM), 1,2-dichloroethane (DCE), etc. The resin was synthesized by polycondensation of 2,4-dihydroxy benzophenone and 1,2-dichloromethane. Finely powdered anhydrous aluminium chloride was added in the above mixture. The reaction mixture was refluxed at appropriate temperature and time. The obtained product was purified by washing with hot 10% methanol to remove unreacted monomer.

The resins of 2-hydroxy-4-methoxy benzophenone and 2-hydroxy-4-butoxy benzophenone with dichloromethane, 1, 2-dichloroethane, etc., were synthesized in a similar manner.

The study of '3d' transition and '4f' inner transition metal ion adsorption with different synthesized benzophenone-based phenolic resins was carried out. All the benzophenone-based resins possess ketonic>C=O and phenolic –OH groups as chelating sites, which form coordinated covalent bonds with metal ions. Not only ion exchange techniques and physical parameters but structure and nature of the resin are important factors, which affect the sorption of the metal ions.

2,4-Dihydroxy benzophenone diol resin shows maximum efficiency for binding with transition as well as inner transition metal ions as compared to 2-hydroxy-4-methoxy benzophenone diol and 2-hydroxy-4-butoxy benzophenone diol resins. This is because of bigger size of methoxy–OCH_3 and butoxy–OC_4H_9 than phenolic –OH at fourth position of resins causes steric hindrance in adsorption of metal ions. The acidity of phenolic –OH group is decreasing due to etherification.

The resin possessing –OH group has maximum efficiency than resin possessing methoxy–OCH_3 or butoxy–OC_4H_9 groups. The order of efficiency of the functional groups is as follows:

$$-OH > -OCH_3 > -OC_4H_9$$

The phenolic resin having $-CH_2-CH_2-$ in its backbone shows highest metal-binding capacity for transition as well as inner transition metal ions. The increase in the $-CH_2-$ group is in the bridge; the metal-binding capacity of resin decreases step by step. This is due to the increase in space between two phenolic moieties. This reduces the metal-binding efficiency due to unstable complexation.

Thus amongst all studied phenolic resins, ion exchange capacity and the efficiency of synthesized benzophenone-based phenolic resins have been observed as shown below:

2,4-dihydroxybenzophenone > 2-hydroxy-4-methoxybenzophenone > 2-hydroxy-4-butoxy benzophenone

Metal adsorption capacity of benzophenone-based resins also depends on different diols which form bridge between two substituted benzophenone monomers. The rate at which ions adsorbed under given conditions decreases as one passes from

$$\begin{aligned} &B-DCM > B-DCE \approx B-ED > B-PD > B-1,4-BD \\ &> B-PG > B-1,3-BD \end{aligned}$$

where B = substituted benzophenone moiety

14.5.2 Metal Ion Exchange by Batch Equilibration Method

The chelating behaviour of synthesized phenolic resins for lanthanides and transition metal ions was determined by the batch equilibrium technique. The fine resin powder sample was dried under a vacuum at 60°C for 24 h. The finely powdered dried resin samples were used for metal ion adsorption. Duplicate experiments involving dry, 300 mesh size resin samples were equilibrated with acetate-acetic acid buffer solution of pH 7.0 at ionic known strength using sodium perchlorate for 2 h. To this mixture, desired known concentration of metal ion solution was added. After being shaken for 24 h at 30°C, the mixture was filtered, and metal content remaining in the filtrate was determined by complexometric titration using standard Na_2EDTA solution with appropriate indicator [89].

(a) To investigate the effect of different electrolytes at various concentrations on metal ion uptake, experiments were carried out with a fixed contact time of 24 h, at 30°C and constant pH using $NaNO_3$ (sodium nitrate), NaCl (sodium chloride), Na_2SO_4 (sodium sulphate) and $NaClO_4$ (sodium perchlorate) electrolytes at different concentrations.
(b) Similar experiments were also carried out in various pH range for a fixed contact time of 24 h at constant ionic strength of desired electrolyte solution.
(c) The selectivity of resin for lanthanides and transition metal ions was examined under similar experimental conditions where the contact time was varied from 1 to 24 h at 30°C after being equilibrated with distilled water.
(d) Similar experiments were carried out for different concentrations of metal ion for a fixed contact time at particular ionic strength of desired electrolyte solution.

14.5.3 Influence of an Electrolyte, pH, Time and Metal Concentrations on Uptake of Metal Ions

The effect of nature and concentrations of an electrolyte solution was investigated by determining the metal ion adsorbed by resin at room temperature. The electrolytes $NaNO_3$, NaCl, Na_2SO_4 and $NaClO_4$ were used.

All experiments were carried out with constant weight of the resin samples under specified conditions in presence of various concentrations of electrolytes under study. The constant weight of resin sample was stirred in each of the solutions for 24 h.

Set:	I	II	III	IV
Concentration of an electrolyte:	0.05 M	0.1 M	0.5 M	1.0 M

Desired solution of metal ion (0.1 M, 2 ml) was added to each of the above solutions. The pH of the solution was adjusted to the given value by adding the required amount of aqueous acid or alkali. After 24 h, the content of each filtrate and washings was collected. The metal ion in solution was estimated using complexometric method with standard Na_2EDTA – disodium salt of ethylenediaminetetraacetic acid.

The study reveals that the extent of Ln^{3+} (lanthanide) ion uptake by resin increases with increase in the concentration of NO_3^-, ClO_4^- and Cl^- and decreases with an increase in concentration of SO_4^{2-}. This may be due to the higher charge and bigger size of the sulphate ion, in addition, also responsible for the behaviour of SO_4^{2-} ion. Hence, the influence of NO_3^-, ClO_4^- and Cl^- is less, on the position of metal chelates at equilibrium state that does SO_4^{2-}.

It is observed that in presence of NO_3^-, Cl^- and ClO_4^- ions, good results are obtained for all metal ions. While SO_4^{2-} ion is not found suitable for maintaining high strength of solution in the exchange study, the Cl^- ion has a tendency to form complexes with many metal ions [90]. In the ion exchange study, all the relevant experiments were carried out using $NaNO_3$. The manner of sorption of metal ions

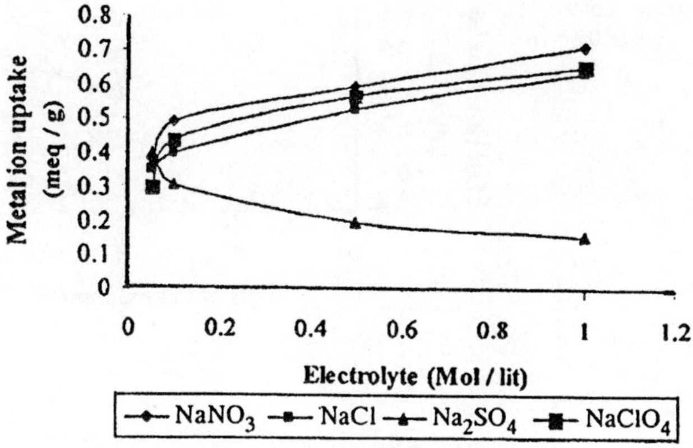

Fig. 14.2 Metal ion uptake as a function of electrolyte concentrations

by benzophenone-based resins in the presence of different electrolyte solutions with different pH is shown in Fig. 14.2.

The binding capacity of resin for lanthanide and transition metal ions was studied in (1.0 M $NaNO_3$ [40 ml]) solution within the pH range of 3.0–6.0 under continuous shaking for a fixed contact time of 24 h at 30°C.

The resin sample (50 mg) was suspended in the electrolyte solution, and pH of the suspension was adjusted to required value by addition of either (0.1 M) HNO_3 or (0.1 M) NaOH solution. To this, known quantity of metal nitrate (0.1 M, 2 ml) was added, and the pH of the medium was adjusted to the desired value. The content was mechanically stirred for 24 h, filtered and washed with distilled water. This experiment was performed using standard known concentration of nitrate solution of transition and lanthanide metal ions at the different pH values of 3.0, 3.5, 4.0, 5.0 and 5.5.

Typical pH-binding capacity profile for the resin indicates that at lower pH values (<pH 6), protonation of chelating group takes place and complexation tendency of ligands is reduced; hence, the percentage uptake decreases. At higher pH values (>pH 6), deprotonation of coordination resin takes place, and hence, the absence of the competition between metal ions and protons for lone pair of electrons is observed. Thus, the percentage of metal ion uptake increases through the formation of stable polymer–metal complexes [91]. It is found that the relative amount of metal ion uptake by the resin increases with increase in pH of the medium [92–94]. Thus maximum uptake of metal ion occurs at pH=5.5. The pH effect on adsorption of metal ion by benzophenone-based resins is shown in Fig. 14.3 as an example.

The distribution ratio K_D of metal ions between the resin phase (solid) and aqueous phase (liquid) is estimated at optimum pH, using 1.0 M $NaNO_3$ or known concentration of electrolyte solution. The experiments were carried out at different pH. Distribution ratio K_D (the concentration of metal ion in the adsorbed form on

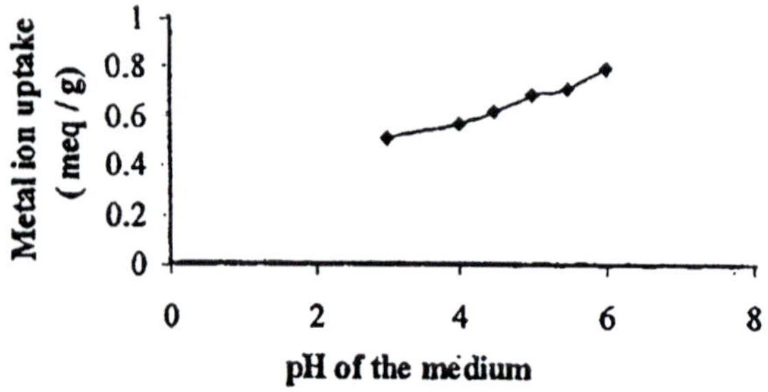

Fig. 14.3 Metal ion uptake as a function of pH of the medium

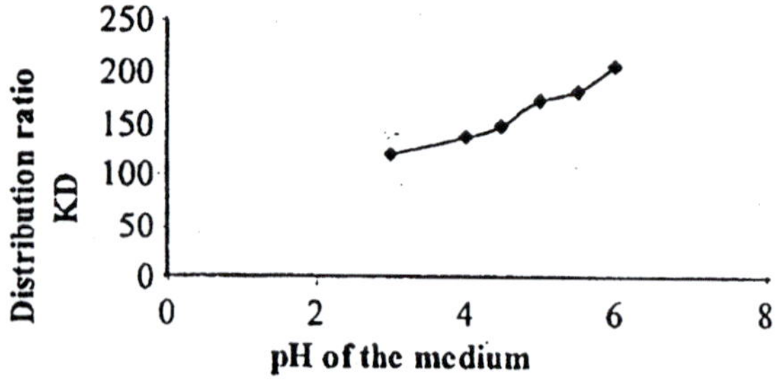

Fig. 14.4 Distribution ratio (K_D) of metal ion as a function of pH

the resin phase divided by concentration of metal ion in solution phase) has been calculated using the following equation:

$$K_D = \frac{\text{Amount of metal ion adsorbed by resin}}{\text{Amount of metal ion in solution}} \times \frac{\text{Volume of solution}}{\text{Weight of resin}} \qquad (14.5)$$

The effect of pH on the amount of metal ions distributed between two phases can be explained. It can be seen that the distribution ratio increases for lanthanide and transition metal ions as the pH of the medium increases. It is found that the value of distribution ratio for a given pH depends upon the nature of the polymeric ligand (resin) as shown in Fig. 14.4.

In this technique, known quantity (0.50 g) of dry 300 mesh sample of the resin was pre-conditioned by allowing the resin to equilibrate for 2 h with the buffer solution at pH 6.0 before the ion exchange experiment. The ion exchange experiments were performed at 30°C under continuous shaking as a function of constant contact time from 1 to 24 h.

The study shows that the uptake of metal ion increases with time until it reaches a steady state. It is assumed that at 30°C and under given conditions, the state of equilibrium is established in 24 h. The rate of metal ion uptake is expressed as percentage of the attainment at equilibrium. The uptake of metal ion by the resin increases with increase in initial metal ion concentration and reaches a plateau at higher concentration [95–97]. The uptake of metal ions by the resin involves either

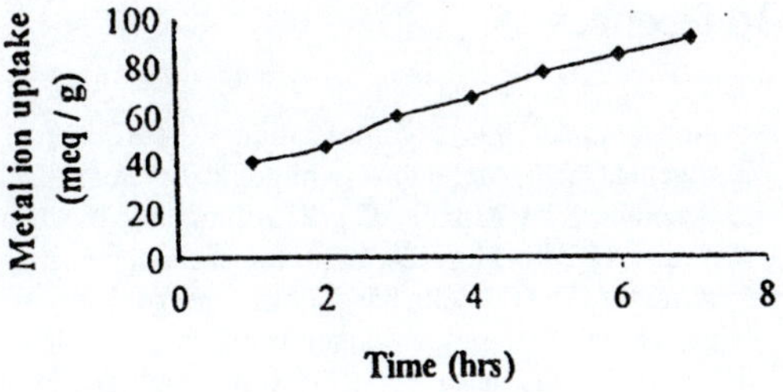

Fig. 14.5 Metal ion uptake as a function of time

chelation or ion exchange or both. Ion exchange behaviour of metal ion by benzophenone-based resin as a function of time is shown in the Fig. 14.5.

The effect of metal ion concentration on uptake behaviour of the different synthesized resins was studied in the desired concentration range (250–2,000 mg per 10 ml) of the metal ion. Increasing the concentration of metal ion enhanced the percentage of loading. However, a levelling effect was noticed at higher concentration because of saturation of available coordination sites. Tikhomirova et al. have observed a similar trend [98]. The adsorption coefficient K_{ad} of the resin for the adsorption of transition and lanthanide metal ions was computed from Freundlich adsorption isotherm:

$$\log(x/m) = \log K_{ad} \times 1/nx \times (\log c) \quad (14.6)$$

where

c = initial concentration of the metal ion in m mole
m = weight of the resin in mg
x = quantity of the metal ion adsorbed by the resin in the *m* mole
n = constant

It is observed that high K_{ad} value indicates that the saturation time for adsorption of metal ion is attained quickly. Luo and co-workers have reported that the slow adsorption rate of metal ions is due to low K_{ad} values [99].

The new benzophenone-based resin with different diols, dichloro alkenes and its chelating properties with lanthanides and transition metal ions has additive values in the present ion exchange material and processes. Chelation properties of the synthesized resins are influenced by the medium, pH of the solution, nature and concentrations of the electrolyte solution.

14.6 Conclusion

The benzophenone-based resins with different diols may be proven effective and efficient ion exchange resins with desired physical parameters.

References

1. Adjemian A (1994) Hydrometallurgy' 94. Chapman & Hall, Cambridge, p 3
2. Sigel H (1998) Metal ions in biological systems. Dekker, New York, p 24
3. Scheinberg IH, Morell AG (1973) Inorganic biochemistry, vol 1. Elsevier, New York, p 306
4. Walton HF, Rocklin RD (1990) Ion exchange in analytical chemistry. CRC Press, Boca Raton
5. Keim R (1983) Gmelin handbook of inorganic chemistry, Uranium Supplement, Catadsorption and chromatography, Springer, Berlin
6. Sekine T, Hasegewa Y (1977) Solvent extraction chemistry, fundamentals and applications. Marcel Dekker, New York
7. Teixeira LSG, Costa ACS, Ferreira SC, Freitas MD, Carvalho MS (1999) J Braz Chem Soc 10(6):519
8. Khuhawar MY, Lanjwani SN (1995) Talanta 42(12):1925
9. Fujikawa Y, Sugahara M, Ikeda E, Fukui M (2002) Radioanal Nucl Chem 252(2):399
10. Kumar M, Rathore DSP, Singh AK (2001) Microchim Acta 137:127
11. Gladis JM, Rao TP (2002) Anal Bioanal Chem 373(8):867
12. Kumar M, Rathore DPS, Singh AK (2000) Analyst 125(6):1221
13. Say R, Ersoz A, Denizli A (2003) Sep Sci Technol 38(14):3431
14. Draye M, Czerwinski KR (2000) Sep Sci Technol 35(8):1117
15. Atia AA, Donia AM, Elwakeel KZ (2005) React Funct Polym 65(3):267
16. Ebraheem KAK, Muharak MS, Al-Gharabli SI (2002) Macromol Sci Pure Appl Chem 39(3):217
17. Soroushian P, Elzafraney M, Chowdhury H, Sarwar G, Aouadi F (2005) Polym Compos 26(5):679
18. Say R, Garipcan B, Emir S, Patir S, Denizli A (2002) Macromol Mater Eng 287(8):539
19. Denizli A, Garipcan B, Karabakan A, Say R, Emir S, Patir S (2003) Sep Pur'f Technol 30(1):3
20. Malachowski L, Stair JL, Holcombe JA (2004) Pure Appl Chem 76(4):777
21. Lee KH, Muraoka Y, Oshima M, Motomizu S (2004) Anal Sci 20:83
22. Lee W, Lee SE, Kim MK, Lee CH, Kim YS (2002) Bull Korean Chem Soc 23:1067
23. Matsumiya H, Masai H, Terazono Y, Iki N, Miyano S (2003) Bull Chem Soc Jpn 76:133
24. Banerjee D, Mondal BC, Das AK (2004) J Indian Chem Soc 81:50
25. Marsh SF, Savitra ZV, Bowen SMJ (1995) Radional Nucl Chem 194:117
26. Kavanaugh MC (1995) Environ Prog 14:117
27. Dingman J Jr, Siggia S, Barton C, Hiscock KB (1972) Anal Chem 44:1351
28. Moyers EM, Fritz JB (1977) Anal Chem 49:418
29. Blount CW, Leyden DE, Thomas TL, Guill SM (1973) Anal Chem 45:1045
30. Bohra S, Mathur R, Mathur NK, Mathur PN (1992) J Polym Mater 9:101
31. Vernon E (1977) Chem Ind 15:634
32. Rivas BL, Pooley SA, Maturana HA, Villegas S (2001) Macromol Chem Phys 202(3):443
33. Rivas BL, Maturana HA, Hauser P (1999) J Appl Polym Sci 73(3):369
34. Shah BA, Shah AV, Shah PM (2008) E-J Chem 5(2):291
35. Al-Rimawi F, Ahmed A, F Khalili, Mubarak MS (2004) Solvent Extr Ion Exc 22(4):721
36. Dai S, Burleig MC, Ju YH, Gao HJ (2000) J Am Chem Soc 122(5):992
37. Buyuktiryaki S, Say R, Ersoz A, Birlik E, Denizli A (2005) Talanta 67(3):640
38. Haupt K (2001) Analyst 126:747
39. Mosbach K, Ramstrom O (1996) Bio Tech 14:163
40. Wulff G (1993) Workshop: Tibtech March 11:85
41. Schmuckler G (1965) Talanta 12:281
42. Blasius E, Brozio B, Flaschka HA, Barnard AJ Jr (1967) Chelates in analytical chemistry, vol 1. Marcel Dekker, New York, p 49
43. Mayasova GV, Savvin SB (1986) Crit Rev Anal Chem 17:1
44. Kantipuly C, Katragadda S, Chow A, Gesser HD (1990) Talanta 37:491
45. Biswas M, Mukherjee A (1994) Adv Polym Sci 115:89
46. Riley JP, Taylor D (1968) Anal Chim Acta 40:479

47. Garg BS, Sharma RK, Mittal S (1999) Microchem 1(61):9
48. Ueno K, Martell AE (1955) J Phys Chem 59:998
49. Ueno K, Martell AE (1956) J Phys Chem 60:1270
50. Che CM, Cheng WK (1986) J Chem Soc Chem Commun 1986:1443
51. Samal S, Das RR, Sahoo D, Acharya S, Panda RL, Rout RC (1996) J Appl Polym Sci 62:1437
52. Samal S, Das RR, Sahoo D, Acharya S (1997) Polym Lot 44:41
53. Samal S, Mohapatra NK, Acharya S, Dey RK (1999) React Polym 42:37
54. Samal S, Das RR, Dey RK, Acharya S (2000) J Appl Polyi'n Sci 77:967
55. Knop A, Pilato LA (1985) Phenolic resins. Springer, Berlin
56. Pasch H, Schrod M (2004) Macromol Rapid Commun 25:224
57. Shafizadeh JE, Guionnet S, Tillman MS, Seferis JC (1999) J Appl Polyrn Sci 73(4):505
58. Heifferich F (1962) Ion exchange. McGraw Hill, New York
59. Harjula R, Lehto J, Brodkin L, Tusa E (1997) Electric Power Research Institute, Palo Alto
60. Hutson GV (1996) Waste treatment. In: Wilson PD (ed) The nuclear fuel cycle. Oxford University Press, Oxford, Ch. 9
61. Chernjatskaja NB (1988) Radiochemistry 27:618
62. Yamasaki N, Kanahara S, Yanagisawa K (1984) Nippon Kagaku Kaishi 12:2015
63. Campbell DA, Collins ED, King LJ (1983) Evaluation of the use of Zeolite mixtures in the submerged demineralizer system. In: Proceedings of the sixth international Zeolite conference, Reno
64. Sinha PK, Panicker PK, Amalraj RV, Krishnasamy V (1995) Waste Manage 15:149
65. Sinha PK, Lal KB, Panicker PK, Krishnasamy V (1996) Radiochim Acta 73:157
66. Sinha PK, Amalraj RV, Krishnasamy V (1994) Radiochim Acta 65:125
67. Sinha PK, Krishnasamy V (1996) J Nuci Sd Technol 33:333
68. Kurath DE, Bray LA, Ross WA, Ploetz DK (1990) In: Mellinger BG (ed) Nuclear waste management III: ceramic transactions, vol 9. American Ceramic Society, Columbus, p 529
69. Samanta SK (1996) J Radioanal Nuci Chem 209:235
70. Heinonen OJ, Lehto J, Miettinen JK (1981) Radiochim Acta 28:93
71. Dosch RG (1981) Report. Sand 80-1212. Sandia Natl Lab, Albuquerque
72. Dosch RG, Brown NE, Stephens HP, Anthony RG (1993) Waste management '93, vol 2. Arizona Board of Regents, Phoenix, p 1751
73. Harjula R, Lehto J, Tusa EH, Paavola A (1994) Nucl Technol 107:272
74. Anthony RG, Philip CV, Dosch RG (1993) Waste Manage 13:503
75. Zheng Z et al (1996) Eng Chem Res 35:4246
76. Braun R, Dangieri TJ, Fennelly DJ (1996) Nuclear and hazardous waste management, vol. 1. In: Proceedings of the international topical meeting, Seattle. American Nuclear Society, La Grange Park p 204
77. Streat M, Jacobi DL (1995) In: Sengupta AK (ed) Ion exchange technology: advances in pollution control. CRC Press, Boca Raton, p 193
78. Samanta SK et al (1995) Nuclear and radiochemistry symposium (NUCAR 95). In: Proceedings of nuclear and radiochemistry symposium, Kalpakkam. Bhabha Atomic Research Centre, Bombay, p 303
79. Sinha PK, Amalraj RV, Krishnasamy V (1993) Waste Manage 13:341
80. Hutson GV (1996) Nucl Energy 35:393
81. Samanta SK, Ramaswamy M, Misra BM (1992) Sep Sci Technol 27:255
82. Samanta SK, Ramaswamy M, Sen P, Varadarajan N, Singh RK (1993) In: National symposium on management of radioactive and toxic wastes (SMART-93), Kalpakkam. Bhabha Atomic Research Centre, Bombay, p 56
83. Samanta SK, Theyyunni TK, Misra BM (1995) J Nucl Sci Technol 32:425
84. Kulkarni Y, Samanta SK, Bakre SY, Raj K, Kurnra MS (1996) Waste management '96. In: Proceedings of the international symposium, Tucson. Arizona Board of Regents, Phoenix
85. Bray LA, Elovich RJ, Carson KJ (1990) Report. PNL-7273. Pacific Northwest Lab, Richiand
86. Joshi JD, Patel NB, Patel SD (2006) Iran Polym J 15(3):219
87. Joshi JD, Patel NB, Patel GP (2006) Int J Polym Mater 55:399
88. Kapadia MA, Patel MM, Patel GP, Joshi JD (2008) J Coord Chem 61(5):677

89. Vogel AI (1978) A text book of quantitative inorganic analysis, 4th edn. Longmans, Green and Co. Ltd, London, p 1978
90. Morris LR, Mock RA, Morshall GA, Hawe JH (1959) J Am Chem Sco 81:377
91. Reddy AR, Reddy KH (2003) Proc Indian Acad Sci (Chem Sci) 115(3):155
92. Patel SD, Patel NB, Joshi JD (2006) J Macromol Sci A: FAQ 43(8):1167
93. Joshi JD, Patel GP, Patel SD (2006) J Macromol Sci A: PAC 44(1):65
94. Kapadia MA, Patel MM, Patel GP, Joshi JD (2007) Int J Polym Mater 56(5):549
95. Patel MM, Kapadia MA, Patel GP, Joshi JD (2007) React Funct Polym 67:746
96. Patel MM, Kapadia MA, Patel GP, Joshi JD (2007) J Appl Polym Sci 106(2):1307
97. Ameta R, Patel V, Joshi J (2007) Iran Polym J 16(9):615
98. Tikhomirova TI, Fadeeva VI, Kubryavtsev GV, Nestorenko PN, Ivanov VM, Savitchev AT, Smirnova NS (1991) Talanta 38:267
99. Luo X-Y, Su Z-X, Gao W-Y, Zhang G-Y, Change X-J (1992) Analyst 117:145

Chapter 15
Inorganic Ion Exchangers in Paper and Thin-Layer Chromatographic Separations

Živoslav Tešić and Dušanka Milojković-Opsenica

Abstract Planar chromatography, i.e., paper and thin-layer chromatography, is a simple and fast method for the separation of various classes of organic and inorganic compounds. Numerous inorganic and organic stationary phases were used in planar chromatography providing effective separations based on different mechanisms. Apart from the most commonly used stationary phases such as cellulose, silica gel, alumina, and modified silica gels, various inorganic ion exchangers were successfully used as sorbents in planar chromatography in a large number of cases. Due to the introduction of ion exchange as an additional separation mechanism, the selectivity of planar chromatography was significantly improved, especially for the separation of inorganic ions and polar organic substances. Numerous literature data related to the usage of inorganic ion exchangers in both paper and thin-layer chromatographic separations are presented through corresponding applications.

15.1 Introduction

Chromatography is a physicochemical separation method based on the distribution of the investigated substances between two phases where one is immobile (stationary) and the other one is mobile. Numerous chromatographic methods used can be classified in different ways by the following [1]:

- Physical nature of the mobile and stationary phases (*gas chromatography*, systems include gas–liquid and gas–solid, and *liquid chromatography*, systems include liquid–liquid and liquid–solid)

Ž. Tešić (✉) • D. Milojković-Opsenica
Faculty of Chemistry, University of Belgrade, Studentski trg 12-16, Belgrade 11158, Serbia
e-mail: ztesic@chem.bg.ac.rs

Inamuddin and M. Luqman (eds.), *Ion Exchange Technology II: Applications*,
DOI 10.1007/978-94-007-4026-6_15, © Springer Science+Business Media B.V. 2012

- Mechanisms of separation, i.e., by the nature of interactions of separated substances with the stationary and mobile phases (*adsorption chromatography, partition chromatography, ion exchange chromatography*, and *gel chromatography*)
- Configuration of stationary phase (*column* and *planar chromatography*)
- Work method (*elution chromatography, displacement chromatography, frontal analysis*).

15.1.1 Planar Chromatography

Planar chromatography (subgroup of liquid chromatography) comprises paper chromatography (PC) and thin-layer chromatography (TLC) [2, 3].

The separation of substances from a mixture using paper chromatography is a complex process mainly based on the distribution of the analyte between two liquid phases, but adsorption and ion exchange should be taken into consideration as well. In paper chromatography, the carrier of the stationary phase is paper (cellulose), while "cellulose–water complex" is the stationary phase [4]. Mixtures of water and one or more organic solvents mixed with water are commonly used as the mobile phase.

Thin-layer chromatography is a simple, fast, sensitive, and economical analytical method of separation, qualitative and quantitative analysis of different organic and inorganic substances, and it can also be used to predict their behavior in the conditions of liquid column chromatography. In addition, TLC can also be utilized as a preparative technique. In TLC, the stationary phase (a sorbent) is in the form of a thin layer on a plane surface (glass, metal, or plastic plate). Solid, porous materials with relatively high specific surface area are most commonly used as chromatographic sorbents. The mobile phase is liquid, comprising one or more solvents.

The movement of substances under conditions of thin-layer chromatography is determined by two opposite forces: the elution force of mobile phase and the ability of the sorbent to retain separated substances on its own surface active centers.

The identification of separated substances (chromatographic zones) in planar chromatography is based on R_F values. The R_F value (retardation factor) represents the ratio of the single zone distance (*a*) and the solvent distance from the starting point to the solvent front (*b*):

$$R_F = a / b; \quad 0 \leq R_F \leq 1 \tag{15.1}$$

Paper chromatography as a part of planar chromatography is slowly losing its relevance, and even though it is still used as a routine laboratory method, there are only sporadic publications in this field of science. Unlike paper chromatography, TLC is still widely used, which is evidenced by a large number of papers published in different analytical and chromatographic journals [5]. The reasons for such popularity may be the availability of various sorbents, diverse possibilities of detection (visualization), and, most recently, the possibility of coupling TLC with mass spectrometry.

Modern thin-layer chromatography is an instrumental technique that is comparable by its accuracy and precision with both gas and high-performance liquid chromatography. It is performed on commercially available precoated plates which enable simultaneous analysis of a great number of samples in a short time, high efficiency, and significant reproducibility.

Apart from most commonly used sorbents such as silica gel, alumina, cellulose, and RP-18 silica gel, in a large number of cases, various ion exchangers (inorganic and organic, natural and synthetic) are successfully used as sorbents in TLC.

15.1.2 Ion Exchange Chromatography

Ion exchange chromatography is one of the chromatographic methods based on reverse (stoichiometric) exchange (replacement) of ions from solutions with mobile ions of the ion exchanger [6]. It is most often used in order to separate ions and ionizable substances, or substances whose ionization depends on the pH value of the solution. The retention of substances in ion exchange chromatography is determined based on the nature of a sample, the type and concentration of other ions present in the mobile phase, and the temperature. In addition to that, the separation can be influenced by the processes of adsorption as well.

Although the ion exchange mechanism has only limited importance in planar chromatography, a great number of suitable stationary phases were used so far [7].

Since ion formation is favored in aqueous solution, the mobile phases in ion exchange chromatography are liquid, aqueous, usually buffered to a particular pH value.

15.2 Inorganic Ion Exchangers in Planar Chromatography

A great variety of inorganic ion exchangers were prepared and used in planar chromatography in the past period [8]. They can be classified in the following categories:

- Synthetic zeolites
- Polybasic acid salts
- Hydrous oxides
- Metal ferrocyanides
- Insoluble ion exchange materials
- Heteropoly acids.

Synthetic zeolites are crystalline aluminosilicate based materials which can be prepared as microcrystalline powders, pellets, or beads. The main advantageous properties of zeolites are their uniform structure and the fact that they are more selective and suitable for the separation of ions based on their different size [9]. On the other hand, the main disadvantage of their use in planar chromatography is the limited chemical stability of zeolites at extreme pH ranges.

Acidic salts of multivalent metals are the most studied group of inorganic ion exchangers. Their composition is nonstoichiometric and depends on the conditions under which they are precipitated. They are extremely insoluble. The well-known materials of this group include phosphates, arsenates, molybdates, and tellurates of zirconium [10, 11], titanium [12], thorium [13], tin [14], cerium [15], chromium [16], iron [17], tantalum [18], etc.

Hydrous oxides of some metal ions, such as hydrous ferric oxide, zinc oxide, are materials which have also been used in ion exchange planar chromatography. Quadrivalent metal oxides, such as SnO_2, SiO_2, ThO_2, and ZrO_2 are also commonly used as inorganic ion exchangers. Consequently, these oxides are usually described as hydrous oxides [19, 20].

Metal ferrocyanides can also be used as inorganic ion exchangers [21]. They are very easily prepared. Molybdenum and tungsten ferrocyanides in particular can be used to separate Cs-137 and Sr-90 from fissionable products in acid media.

Many insoluble ion-exchanging compounds were prepared. These materials were obtained by precipitation from metal salt solutions with Na_2S or H_2S. The ion exchange properties of insoluble sulfides such as Ag_2S, SnS, CuS, PbS, NiS, As_2S_3, and some others were investigated [22]. These compounds are selective toward cations forming insoluble sulfides.

Heteropolyacid salts were also used as inorganic ion exchangers [23]. This group is derived from 12-heteropolyacids of general formula $H_nXY_{12}O_{40}$ nH_2O, where X may be P, As, Si, B, or Ce while Y represents one of the elements such as Mo, W, and V. They are very stable in moderately acidic systems. In comparison to other inorganic exchangers, they are more termically and chemically stable, more selective, and their ion exchange capacity is higher.

A new class of mixed organic-inorganic ion exchange materials was synthesized in the past period [24–30]. The main purpose of these hybrid materials was to combine all the good properties of inorganic and organic exchangers in a new composite material. These days they are applied in a wide variety of industrial, domestic, governmental laboratory operations. These hybrid materials have a good ion exchange capacity, high stability, reproducibility, and selectivity, especially for heavy metal ions [31].

Numerous data related to synthesis, properties, and application of inorganic ion exchangers were reviewed in past period by many authors [8, 32–34].

15.3 Inorganic Ion Exchangers in Paper Chromatography

Chromatography is a widely used method for the separation, identification, and determination of the chemical components in complex mixtures. The separation of the constituents of a mixture in paper chromatography was considered to be due to the difference in their distribution ratios between the water-saturated cellulose and the developing solvent, termed the mobile phase. Cellulose, as present in the form of filter paper, was assumed to play the role of an inert support.

15.3.1 Cellulose

Cellulose is a polar organic stationary phase widely used in planar chromatography. It is a macromolecular polysaccharide composed of long chains of 3-glucopyranose units which are connected to one another at the 1,4 positions. TLC cellulose is not different from PC cellulose except for the shorter length of macromolecule chains. Unlike TLC, where cellulose is on a plane support (glass, aluminum, or plastic plates), in paper chromatography, cellulose is applied as a self-supporting layer. Special papers which are highly purified and have controlled porosity, thickness, and arrangement of cellulose fibers are manufactured for paper chromatography [35].

Numerous oxygen and hydroxyl groups of cellulose are capable of forming hydrogen bonds with each other, with water and with alcohols. Because of that, cellulose is an ideal stationary phase for the separation of hydrophilic substances, mainly by partition chromatography [36, 37].

Different ion exchangers can be obtained by suitable chemical modifications or impregnations of cellulose. The most important cellulose-based ion exchangers contain some of following functional groups: *AE* (aminoethyl), *CM* (carboxymethyl), *DEAE* (diethylaminoethyl), *PEI* (polyethyleneimine), *ECTEOLA* (the product of the reaction of epichlorohydrin, triethanolamine, and alkali cellulose), *P* (phosphate), and *PAB* (4-aminobenzyl).

Ion-exchanging celluloses are particularly important in analytical separations of various classes of inorganic and organic compounds in column, thin-layer, and paper chromatography. Only a few papers deal with the application of paper chromatography impregnated with various ion-exchanging resins and inorganic ion-exchanging salts. The largest number of them deals with the separation of metal ions, although there are papers which deal with the separation of various organic substances.

Papers impregnated with inorganic ion exchangers usually give cleaner and faster cation separations than untreated papers. The separations now depend not only on partition but also on selectivity for various metal ions shown by the ion exchange. A lot of inorganic substances were examined using purely organic and mixed inorganic–organic mobile phases. The papers which deal with this topic will be presented chronologically. *It should be mentioned at this point that one of the most important researchers who has worked on this problem is Mohsin Qureshi, an Indian scientist from Aligarh Muslim University.*

15.3.2 Applications

Alberti and Grassini separated 26 metal ions using chromatography on paper strips impregnated with zirconium phosphate. The strips were developed by ascending method, with hydrochloric acid of various concentrations as a mobile phase. They used mixtures of NH_4Cl–HCl, NaCl–HCl, and KCl–HCl at various concentrations and ratios in order to achieve further separations which were not possible with HCl. The results obtained indicate that Zr phosphate impregnated paper can be successfully employed for the chromatographic separation of numerous cations [38].

Nunes and Jeronimo separated Ag(I) and Cu(II), and Cu(II) and Au(III) using a zirconium selenite as an ion exchanger. They used Whatman No. 1 paper strips and 0.01 M or 0.1 M HCl as a mobile phase. The results were similar to those obtained with zirconium phosphate [39].

During their further investigation of paper impregnated with zirconium phosphate, Alberti and coworkers chromatographed 23 metal ions [40]. They used a very weakly complexing acid ($HClO_4$) and the inorganic acids commonly employed (HCl, HNO_3, and H_2SO_4) at concentration ranging from 0.1 to 6 M as eluents. The R_F values obtained by perchloric acid were not appreciably different from those obtained with other inorganic acids, except in the case of the ions which are well known for their marked tendency to form complexes.

Sastri and Rao found that strips impregnated with zirconium phosphate can be successfully used with aqueous solutions of inorganic acids or their salts for the rapid separation and detection of several elements in different valency states present in admixture [41].

In the next paper [42], Grassini and Padiglione investigated the exchange capacity of papers impregnated with zirconium phosphate. For that purpose, they prepared papers containing various amounts of exchanger and studied the chromatographic behavior of UO_2(II) and Pb(II) using $HClO_4$ as a mobile phase.

Sakodinski and Lederer investigated the chromatographic properties of paper impregnated with titanium hydroxide as a new sorbent [43]. They found that U(VI) is very strongly adsorbed irrespective of the anions present and is not adsorbed even in 10 M LiCl. Some of the cations examined showed that there was usually desorption with increasing electrolyte constriction unless hydrolysis interfered, and for this reason, an ion exchange mechanism could be considered most likely.

In another interesting investigation, Qureshi chromatographed metal ions on stannic phosphate and stannic tungstate papers for specific separations of Au(III), Hg(II), Pt(IV), Mg(II), Mo(VI), and Se(IV). The chromatographic behavior of 44 metal ions in 28 aqueous and nonaqueous solvent systems on stannic phosphate papers and of 46 metal ions in 45 solvent systems on stannic tungstate papers was studied [44]. On stannic phosphate paper, Au(III) was separated from a mixtures of Ag(I), Pb(II), Bi(III), Cd(II), Fe(III), Co(II), Ni(II), Mg(II), Ca(II), Sr(II), Ba(II), Be(II), Al(III), Zr(IV), Mn(II), Cr(III), Th, UO_2(II), and Ru(III) by using the solvent system 7:2:1 butyl alcohol–HCl–10% NH_4Cl. Pt was also separated from the same mixture of cations in 25 min using 1:1 0.05 N HCl–0.05 N NaCl. Sb(III), Hg(II), K(I), Rb(I), and Cs(I) were separated from mixtures containing over 30 cations with various solvent systems. The 32 binary and ternary separations achieved on stannic tungstate papers with 12 solvent systems were tabulated. Mg(II) was separated from 37 metal ions in 20 min on $SnWO_4$ papers. The stannic papers can be used for the chromatographic separations of Mo(VI) from 42 metal ions and Se(IV) from 33 metal ions including Te(IV).

Qureshi used paper strips for a fast separation of Ti(IV), As(III), W(VI), and Au(III) from a large number of metal ions [45]. Ti(IV) was separated from binary mixtures with 32 cations by development in (3:5:2) formic acid–HCl–acetone. The separations were successful in the presence of acetate, tartrate, formate, and citrate. Fe–Cu–Ni–Ti mixtures were resolved with (3:4:3) formic acid–HCl–acetone. Au(II) and W(VI) were separated from numerous metal ions by (6:4) 70% HNO_3–butyl alcohol. In experiments, W(VI) remained at the starting line while Au(III) migrated with the solvent front. W(VI) was separated from binary mixtures with 35 ions. Al(III) was separated from binary mixtures with 25 cations. Fe(III), Al(III), and Ti(IV) mixtures were well resolved using (3:4:3) formic acid–HCl–acetone in the presence of common anions and without special sample preparation.

Paper strips impregnated with Sn(IV) tungstate were employed for the separation of selenium from other metal ions, i.e., Cu(II), Fe(III), Zn(II), Mn(II), Hg(II), Pb(II), and Bi(III), by ascending method using binary solvent 0.1 M ammonium tartrate–4 M ammonia [46].

Continuing their investigation, the same authors used papers impregnated with Ti(IV) tungstate for chromatographic separation of 52 cations in 45 aqueous and mixed solvents [47]. Au(III), Tl(III), Ag(I), Mo(VI), K(I), Rb(I), Cs(I), Ni(II), Sb(III), Bi(III), Be(II), and Tl(I) were separated from numerous metal ions. Several quaternary, ternary, and binary separations were achieved. For that purpose, suitable solvent systems were developed and evaluated.

Finally, Qureshi and coworkers chromatographed metal ions on papers impregnated with the inorganic ion exchangers Sn(IV) tungstate and selenite [48]. They deduced that such sorbents were useful for ion exchange chromatography of several metal ion mixtures.

In their next publication, Qureshi M. and Qureshi S. studied the relative exchange capacity of stannic phosphate papers at temperature range 5–54°C [49]. The R_F values of Ni(II) and Tl(I) in 2 N H_2SO_4 on Whatman No. 1 chromatographic paper impregnated with different amounts of Sn(IV) phosphate expressed as the amount of $SnCl_4$, As = 10–40%, in the hot aqueous solutions into which the chromatographic paper was dipped decreased with increasing As and increased only slightly or were unchanged with increasing temperature. Plots of $[(1/R_F)-1]$ *vs.* As were linear.

In the next publication [50], Qureshi synthesized stannous ferrocyanide as an inorganic ion exchanger having a Sn/Fe ratio of 1:1 with a cation-exchange capacity of 2.03 mequiv/g. Ion distribution of 28 metals on this ion exchanger was studied. Five binary systems (Mg–Ba, Mg–Ca, Mn–Co, Mn–Ni, and Y–Th) were separated on a stannous ferrocyanide column based on the results. Papers impregnated with that sorbent were also used for 18 other separations.

Using the paper strips impregnated with Ti tungstate, Husain separated Cd(II) from 30 cations, Pt(IV) from 38 cations, Se(IV) from 38 cations, Hg(II) from 35 cations, and Pd(II) from 28 cations with the binary solvent systems [51].

Qureshi and Sharma studied the chromatographic behavior of 48 metal ions in butanol–HNO_3 systems on plain and ion exchange papers impregnated with stannic arsenate [52]. Based on the results obtained, they concluded that R_F value depends upon the solubility product of the metal salts, charge of metal ion, and the bare ion radius.

By chromatographing 36 metal ions using the ascending technique on papers impregnated with stannous ferrocyanide ion exchanger and by using 19 different solvents and 23 binary systems, Qureshi separated arsenic from a large number of metal ions using water as the developer [53]. When the H^+ concentration of the solvent was increased, the selectivity of the metal ions for the ion exchanger decreased and their mobility increased.

In the next publication, the same authors investigated chromatographic behavior of 48 cations on stannic and titanium arsenate papers using aqueous nitric acid systems [54]. They found that the selectivity sequence for cations on stannic arsenate is not the same as the one on titanium arsenate, and they defined new quantity R_i ($R_i = R_F$ on untreated papers minus R_F on treated papers) which they proposed as a measure of ion exchange effect.

Chromatographic separations of 49 metal ions on stannic antimonate papers were performed using mixed solvent systems containing dimethyl sulfoxide [55]. Qureshi and Varesshney also used aqueous HNO_3 systems in their publication. The applicability of the papers was demonstrated by achieving various difficult separations. A discussion on the dependence of R_i (R_F on plain paper minus R_F on antimonate papers), log R_F, and R_M on pH of the solvent was presented in order to understand the mechanism of the movement of metal ions on papers impregnated with inorganic materials.

The paper chromatographic behavior of amino acids in various developing solvents was studied on plain papers and papers impregnated with titanium arsenate [56]. The developing solvents were butyl alcohol, acetic acid, butyl alcohol–acetic acid, and methyl alcohol–HCl. The spots were detected using ninhydrin. The R_F values were higher in 10% acetic acid than in butyl alcohol. In butyl alcohol–acetic acid, the R_F values generally increased with the increase in acetic acid content of the solvent. Amino acids were adsorbed more strongly on the ion exchange paper than on the plain paper in most solvents except for butyl alcohol saturated with water where the opposite was true.

Hydrous zirconium oxide paper was prepared by treating Whatman No. 1 chromatographic paper strips with 0.1 M zirconium oxychloride and 6% aqueous ammonia. The R_F values of 28 metal ions in different aqueous solvent systems were reported by Singh and Tandon [57]. Fifty nine binary and/or ternary metal ion separations of analytical and radiochemical interest were successfully achieved.

Paper strips impregnated with zirconium antimonite were successfully employed for rapid separation of alkali metal ions using binary and ternary solvent systems [58].

Impregnation of cellulose with tungstate affects the chromatographic migration rates of a number of polyhydroxy compounds, offering a means, on small and large scales, of achieving the resolution of mixtures which is not easily achieved by other methods [59]. The effects were correlated with the structure of polyhydroxy compounds and their complexes. Pseudostability constants of the complexes were determined.

The paper chromatographic behavior of some pesticides and toxins was also studied using papers impregnated with stannic molybdosilicate in various organic and inorganic solvents [60]. Several binary and ternary separations which were not possible on plain papers were achieved. Methods developed can be successfully applied for certain isomeric differentiation. The minimum detection limit was 25–30 μmol test substance.

Chromatographic separation of anions on papers impregnated with hydrous titanium oxide was investigated using a variety of aqueous and nonaqueous solvent systems [61]. Some anions like F^-, PO_4^{3-}, SO_4^{2-}, VO_3^-, MnO_4^-, AsO_4^{3-}, and WO_4^{2-} were strongly retained on these papers.

Paper strips impregnated with hydrated stannic oxide were also used for the separation of several common anions [62].

Dabral and coworkers conducted a comparative study of the chromatographic behavior of anions, iodide, sulfide, phosphate, arsenate, arsenite, vanadate, chromate, dichromate, thiosulfate, thiocyanate, ferricyanide, and ferrocyanide on papers impregnated with hydrated bismuth oxide and untreated Whatman No. 1 paper strips [63]. The separation of these anions was investigated by employing identical aqueous, nonaqueous, and mixed solvent systems. Sharp and compact spots were obtained with impregnated papers, whereas the opposite was the case with plain papers.

The chromatographic behavior of 15 physiologically important organic acids was studied on papers impregnated with hydrated SnO_2 [64]. Numerous bioanalytically important binary and ternary separations were achieved by using various aqueous and mixed solvent systems.

Ion exchange chromatographic determination on paper strips impregnated with bismuth oxide was employed in the examination of amino acids present in the bulb of *Arisaema curvatum* [65].

The effects of surfactants on the chromatographic behavior of 16 phenols of different nature were observed on Whatman No. 2 paper strips impregnated with hydrated cerium oxide [66]. The movement of phenolic compounds was observed in various mobile phases with or without 4 mM, 8 mM, and 12 mM of sodium dodecyl sulfonate. The use of micellar mobile phase systems improved the selectivity and compactness of the spots, thus offering significant advantages over traditional solvents.

Chromatography of 48 metal ions on stannic antimonate- and stannic arsenate-impregnated papers was carried out using HNO_3–dimethyl sulfoxide solvent systems and on plain paper for comparison [67]. Chromatography was also performed on papers impregnated with stannic antimonate and stannic arsenate in Na^+ form using $NaNO_3$ solutions to verify the modified form of the Lederer's equation ($-n\log a\mathrm{Na}^+ = R_M + \text{constant}$, where n = metal ion charge, $a\mathrm{Na}^+ = \mathrm{Na}^+$ ion activity, and $R_M = \log(1/R_F) - 1$). The R_F values are higher for antimonate-impregnated papers in neat DMSO (apparently due to the slightly higher degree of ionization in the arsenate-impregnated papers). Plots of R_M *vs.* ($-\log a\mathrm{Na}^+$) for K^+, Rb^+, Cs^+, Ni^{2+}, and Co^{2+} were linear with slope depending on cation charge for both stannic antimonate- and arsenate-impregnated papers. Therefore, the modified Lederer's equation holds for both types of papers. The zero R_F values for a number of cations on plain and impregnated papers may be due to strong adsorption or ion exchange.

Planar chromatography of 36 metal ions on Ti(IV) phosphate, Ti(IV) tungstate, and Ti(IV) molybdate impregnated papers in dimethyl sulfoxide–HNO_3 mobile phases was carried out [68]. The ion exchange capacity of papers was detected, and the effects of solvent component, impregnation, and pH on R_F values were studied. The movement of ions was explained based on ion exchange, adsorption, and partition.

The chromatographic behavior of 24 synthetic dyes was studied on zirconium(IV) phosphoantimonate papers with ten pure and mixed mobile phases including tap water and five buffers at different pH values [69]. The effects of mobile phase composition, impregnation, and pH on R_F values were studied. The movement of dyes was explained based on ion exchange, adsorption, intermolecular hydrogen bonding, and the solubility of the dyes in the mobile phases. Differences between R_F values enabled several analytically difficult binary and ternary separations.

15.4 Inorganic Ion Exchangers in Thin-Layer Chromatography

The intensive development of TLC as a method of analysis of various mainly organic substances started in the middle of the twentieth century. Thin-layer chromatography of inorganic substances started around 1960 [70, 71]. Many chromatographic systems were used in inorganic TLC, but silica gel combined with a large number of aqueous, nonaqueous, and mixed aqueous/organic mobile phases was the much-favored stationary phase [72].

Silica gel is a polar inorganic sorbent. It can be prepared by spontaneous polymerization and dehydration of aqueous solution of silicic acid obtained by adding acid to sodium silicate solution [73]. Depending on experimental conditions, the products obtained differ in specific surface area and porosity. Silica gels of medium porosity are suitable for adsorption chromatography. Their characteristics are average pore diameter of 6–10 nm, relatively high specific surface area (over 200 m^2/g), as well as large pore volume (over 0.7 cm^3/g) [74].

Adsorptive properties of silica gel are determined by the presence of silanol groups (Si–O–H) on their surface area. Interactions which determine retention of the analytes include hydrogen bonding, dipole–dipole, and other electrostatic interactions [75]. The intensity of these interactions depends on the number of effective silanol groups, the chemical structure of the separated substances, as well as on the elution strength of the mobile phase used. Since the density of silanol groups (about 8 $\mu mol/m^2$) is constant for all types of silica gels [76], the intensity of adsorptive forces is directly proportional to specific surface area of the sorbent. In adsorption chromatography on silica gel, the analytes containing polar or polarizabile groups are strongly retained due to the stronger interaction with sorbent.

One of the unique characteristics of silica gel is its ion exchange property [77–79]. In the presence of water, silanol groups ionize as weak acids (pK_a is about 7.1) and can act as cation exchanger or form ion pairs with metal ions. Due to that, silica gel was applied in numerous separations of cations. Thin-layer chromatography of metal complexes takes an important place among these separations [80–85].

Numerous applications of silica gel in inorganic thin-layer chromatography are reported in literature and reviewed periodically (biennial reviews in *Analytical Chemistry*; the latest one, Ref. [5]).

The ion exchange became an important method in organic and inorganic chemistry at the time of intensive TLC development [86]. In addition to silica gel and alumina which were mainly used in adsorption chromatography, new ion exchange materials were synthesized and applied as sorbents in TLC [2, 7, 33, 34]. Apart from cellulose-based ion exchangers and synthetic resins, a great number of inorganic ion exchangers were introduced in planar chromatography. Some of them will be presented here through corresponding inorganic and organic applications.

15.4.1 Separations of Inorganic Substances

Zirconium phosphate (ZP in hydrogen form) and hydrous zirconium oxide (HZO in ammonium form) with 3% corn starch as a binder were used for TLC separation of different ions [87]. Ni, Co(II), Pb, Fe(III), Ag, Hg, Cd, and Cu were separated on these cation exchangers using several solvents: 0.01 M hydrochloride acid, 0.1 M and 2.0 M ammonium nitrate. The effect of concentration of the mobile phase was discussed.

The chromatographic behavior of metal ions was studied on thin layer of inorganic ion exchanger obtained by mixing sodium arsenate and silica gel GF [88]. V(V), Ge, Ba, and W(VI) were successfully separated from 28 to 33 metal ions. The following binary mixtures were used as the mobile phase: 1 M NH_4NO_3–0.5 M HNO_3, 1:1; 0.25 M $(NH_4)_2C_2O_4$–0.1 M $H_2C_2O_4$, 1:1, and 2 M HBr–2 M NH_4Br, 1:1. Pb(II)–Cd, Cr(III)–Mn(II), Ag–Ce(III)–Hg, Se–Hg, Pb–Hg, Ag–Bi, As(III)–Mo(VI), Sb(III)–As(III)–Cd, Ag–Pb(II)–Hg, Pb(II)–Mn(II), Ag–Se–Pd(II), Pb(II)–Pd(II), Ag–Pb(II)–Cu(II), Se–Mo(VI)–Au, Cd–Zn, Pb(II)–Zn, and Pb(II)–As(III)–Ag mixtures were resolved on the mentioned ion-exchanger thin layer.

Thin-layer chromatography on silica gel GF_{254} containing stannic arsenate ion exchanger was used for the separation of 57 metal ions [89]. Different mixed solvent systems were used in order to fully explain the usefulness of this cation exchanger. The best results were obtained with mobile phase tributyl phosphate–acetone–50% HNO_3, 2:4:1. The mixtures separated were W(VI)–Rh(III)–Mo(VI); Ti(IV)–V(V)–UO_2(II); W(VI)–Cr(III)–Mo(VI); Pt(IV)–Ni(II)–Pd(II); Ag–Cu(II)–Au(III); Se–Y–La; Be–Mg; Zr(IV)–Y; Cd–Zn; Ce(IV)–Pr; Sm–Nd; Sm–Gd; Tb–Gd; Dy–Ho; Er–Tm; Ga–Al–Zn; and Se–Mo(VI). A satisfactory separation was achieved with developing solvents: butyl acetate–acetic acid–acetone–HCl, 3:1:2:1; ethyl malonate–*iso*-butyl methyl ketone–acetone–50% HNO_3, 1:2:1:0.5; and 2% diacetyl monoxime (in ethanol)–dioxane–10% HNO_3, 1:2:2. The effects of pH, heating plates during preparation, and layer thickness on retention were discussed. It was shown that Lederer's equation is also obeyed by thin layers of stannic arsenate if the activity of Na^+ ion is taken into consideration instead of its concentration.

The chromatographic behavior of platinum-group metals was studied on a thin layer of the semicrystalline stannic arsenate cation exchanger with aqueous and mixed solvents [90]. The developing solvents were *iso*-butanol–acetone–H_2SO_4, 4:3:1; tributyl phosphate–acetone–HNO_3, 2:4:1; 0.5 M ammonia; and 1 M ammonium acetate–1 M NH_3, 2:1. Detection of Pd, Pt, Rh, Ru, and Au was performed using a 3:1 mixture of 5% $SnCl_2$ solution in HCl (4 M) and 5% KI solution. Ir was detected by using 1% alcoholic diphenylcarbazide solution. Os was detected using 2% thiourea solution in 4 M sulfuric acid. The detection of Fe and Cu was performed with 3% potassium hexacyanoferate(II) and the detection of Ni with 1% dimethylglyoxime solution in ethanol. Fast and selective ternary and quaternary separations were listed.

TLC separation and detection of toxic metals were achieved on thin-layer plates prepared by mixing lanthanum tungstate gel with silica gel H [91]. The chromatographic behavior and several binary and ternary separations of the ions Be, V(IV), Cr(III), Mn(II), Fe(III), Ni(II), Cu(II), Zn(II), As(III), Se(IV), Sb(III), Mo(VI), Cd(II), Sn(II), Sn(IV), Ba(II), Hg(II), Tl(I), Pb(II), and Bi(III) were given. Sb was separated from other ions studied by using developing solvent 1 M HBr–1 M KBr, 1:2.

The application of thorium antimonate as a cation-exchange stationary phase in TLC was described [92]. Several important binary and ternary separations on a thin layer of Th(IV) antimonate without binder were performed. Microgram amounts of mercury(II) were quantitatively separated from several metal ions.

Quantitative separation of microgram quantities of Hg(II) from mixtures with several other metal ions was achieved on thin layers of thorium tungstate [93], of zirconium tungstate [94], and of lanthanum antimonite as well [95] by using 1,4-dioxane as solvent.

Thin layers of stannic phosphate without a binder were used in the chromatographic studies of 40 metal ions using 0.1–1.0 M solutions of hydrochloric, nitric, or citric acid and mixed solvent systems as the mobile phase [96]. Some successful binary separations, such as Hg^{2+}–Pb^{2+}, Cd^{2+}–Pb^{2+}, Cd^{2+}–Sb^{3+}, Hg^{2+}–Tl^{+}, and Pb^{2+}–Zn^{2+}, were achieved. The chromatographic behavior of 44 metal ions in dimethyl sulfoxide–nitric acid and dimethyl sulfoxide–water systems was investigated on the same sorbent [97]. The specific DMSO effect was explained and used for some binary mixture separations. Several important separations such as Hg^{2+}–Pb^{2+}, Cd^{2+}–Pb^{2+}, $Hg_2{}^{2+}$–Hg^{2+}, Hg^{2+}–As^{3+}, and Pb^{2+}–Zn^{2+} were achieved.

Thin-layer chromatography of 32 metal ions was performed on titanium(IV) antimonate using aqueous nitric acid systems and mixed systems containing dimethyl formamide [98]. Several important binary and ternary separations achieved are La^{3+}–Pr^{3+}, Gd^{3+}–Nd^{3+}, La^{3+}–Ce^{4+}, Gd^{3+}–Tb^{3+}, and Gd^{3+}–Ho^{3+}; and Cu^{2+}–Hg^{2+}–Pd^{2+} and Au^{3+}–Hg^{2+}–Pt^{4+}. The quantitative separation of Pd(II) from several mixtures containing several other cations was achieved.

The ion exchange behavior of tin(IV) arsenosilicate and arsenophosphate cation exchangers in buffered EDTA (ethylenediaminetetraacetic acid) solution was investigated [99]. The effect of pH values on complex-forming ability of EDTA with different metal ions was discussed. Some important separations, such as Ca^{2+}–Sr^{2+}, Ca^{2+}–Ba^{2+}, as well as the separation of Hg^{2+} from Cu^{2+}, Ni^{2+}, Zn^{2+}, Co^{2+}, and Mn^{2+} were carried out.

The separation of noble metals from commonly present metals on mixed semicrystalline tin phosphate ion exchanger–silica gel G thin layers was studied using different solvents [100]. The ion exchange properties of semicrystalline and amorphous Sn phosphate plates were compared. Better separation of Au, Ru, Rh, Pd, Os, Ir, Pt, Cu, Fe, Co, and Ni ions was achieved on semicrystalline ion exchanger.

The chromatographic behavior of anions on zirconium(IV) molybdate papers and silica gel G thin layers was explored [101]. Several important separations of anions (IO_3^-, BrO_3^-, Br^-, I^-, IO_4^-, NO_2^-, NO_3^-, PO_4^{3-}, F^-) were carried out.

Ion exchange properties of semicrystalline form of stannic phosphate were investigated by centrifugal thin-layer chromatographic separation of different metal ions [102]. Fe, Co, Ni, Cu, Hg, and noble metals were separated on silica gel impregnated with semicrystalline stannic phosphate using nine developing solvents.

Within the scope of the study of inorganic ion exchanger containing phosphorus, chromatographic separation of noble metals on a thin layer of stannic pyrophosphate was investigated [103]. Various binary and ternary mixtures of Ru, Au, Pd, Pt, Rh, Os, Ir, and Ag were separated by thin-layer chromatography on pyrophosphate mixed with silica gel and sodium carboxymethyl cellulose as a binder. Nitric, perchloric, formic, tartaric, or citric acids, ammonia, trimethylamine, ammonium chloride, nitrate, or acetate were used as developing solvents. In addition, the abovementioned solvents were mixed with organic solvents (e.g., ethanol, methanol, *n*-propanol, acetone). The relation between the retention of the investigated metals and the concentration of trimethylamine in the developing solvents was studied. The effect of ethanol content in the mobile phase on the thin-layer chromatographic behavior of noble metals was discussed.

Thin layer of stannic selenopyrophosphate as an inorganic ion exchanger was studied [104]. A series of binary and ternary mixtures of the noble metals was successfully separated by using acidic solutions (nitric, oxalic, or tartaric acid), ammonia–ammonium chloride, trimethylamine, and trimethylamine–ethanol as the mobile phase.

The TLC and ion exchange separation of Fe(III), Co(II), Cu(II), and Ni(II) ions on zeolite molecular sieves as the stationary phase were studied [105]. Separation mechanism was discussed based on the results obtained. Correlation of the kinetic parameters of ion exchange and retention coefficients shows that the migration of ions is determined by ion exchange phenomena.

The chromatographic behavior of 27 cations was examined on thin layers of stannic silicate using 36 solvent systems [106]. Nine typical separations of single metal ions from 15 to 21 other cations using ammonium bromide–containing aqueous solvents were presented. The influence of pH on the R_F values in mobile phases containing lactic acid and citric acid was discussed. The validity of Alberti's relation in ion exchange TLC, as well as the effect of counter-ion concentration, was examined. TLC on the same sorbent was also used for quantitative separation and determination of mercury [107], cadmium, and lead [108]. Mercury was determined in the presence of 11 cations using 1 M ammonium bromide as the mobile phase.

The sorption behavior of heavy metals was examined on layers prepared from mixtures of silica gel and inorganic ion-exchanger gels [109]. Some new sorbents and solvent systems applicable to the separation of cationic species were described.

Thin layers and papers impregnated with Sn(IV) arsenosilicate, Cr(III) arsenosilicate, or Sb(V) arsenophosphate cation exchangers were used for the investigation of chromatographic behavior of heavy metals in buffered nicotine solutions [110]. Fe(III), Cu(II), and Pb(II) were quantitatively resolved from other metals.

The thin-layer chromatographic behavior of 30 cations in 15 solvent systems on cerium(III) silicate ion exchange plates was studied [111]. Separation of Pt from 27 other cations was achieved using 0.5 M ammonia as the mobile phase.

The chromatographic behavior of 28 metal ions in 25 aqueous and mixed solvent systems was investigated on thin layers of lanthanum silicate [112]. As^{3+}, Mo^{6+}, Sb^{3+}, Ti^{4+}, and others were successfully separated from a number of inorganic cations. Fast ternary and binary separations were discussed. The effect of pH on retention in solvent systems containing sorbic acid was studied. Alberti's relation was also obeyed by thin layers of lanthanum silicate.

The application of Ce metaphosphate as inorganic ion exchanger in thin-layer chromatography was demonstrated by the separation of ten precious metal ions [113]. The effects of acids, bases, organic agents, temperature, and thickness of layer on retention behavior were considered. A series of mixed ion solutions was successfully separated and analyzed.

The chromatographic behavior of 26 inorganic cations and 17 anions was studied on plates coated with layered double hydroxides as a new stationary phase with different mobile phases [114]. Separations of one ion from several other ions, as well as numerous binary and ternary separations, were achieved. The separating power of the sorbent was quantified by parameter SR_F.

Thin layers of titanic silicate were used to study the chromatographic behavior of 30 metal ions with aqueous and mixed mobile phases [115]. Separations of the studied ions from numerous other metal ions and many binary and ternary separations were achieved. The relationship between retention and pH in the presence of a complexing anion was studied. The effect of the temperature at which the ion exchange plates were dried on R_F values was also discussed.

The ion-exchanger bismuth silicate was used as a novel sorbent in TLC [116]. The chromatographic behavior of 29 cations and 21 anions in 24 solvent systems was studied. The separations of one cation or anion from a large number of other ions were achieved even in simple aqueous systems. Many binary and ternary separations of toxic ions were carried out on this sorbent as well. The effect of pH of the mobile phase on retention in the presence of EDTA was discussed.

The retention behavior of some inorganic ions was studied on thin layers prepared by mixing one part of a synthetic inorganic ion exchanger (stannic arsenate or tin(IV) molybdosilicate, synthesized under different experimental conditions) with nine parts of silica gel, alumina, or cellulose [117]. Binary system tri-*n*-butyl phosphate–formic acid was used as the mobile phase. Several binary separations of analytical interest were achieved on these blended stationary phases. The effect of pH on the separation of IO_3^- from NO_2^- and BrO_3^- in the presence of other cations was also examined. Quantitative determination of IO_3^- on mixed stannic arsenate–alumina, 1:9, was also performed successfully. It was found that layers containing stannic arsenate with silica gel, alumina, or cellulose are useful for selective separation of cations. Mixed stannic arsenate–alumina layers can be used for a selective separation of Hg^{2+} from multicomponent mixtures of cations. Similarly, stannic arsenate–cellulose layers can be used for selective separation of Ag^+.

Systematic separation of different anions on a thin layer of naturally occurring clay mineral of mixed oxides of silicon, aluminum, titanium, and iron (45.70% SiO_2, 36.60% Al_2O_3, 2.50% TiO_2, and 1.21% Fe_2O_3) was reported [118]. The method was applied to the separation of a number of anions in binary mixtures.

Within the scope of the studies of the applicability of new supports in planar chromatography, a mixed bed of titania and silica was used for TLC separations of toxic metal ions [119]. Some interesting analytical separations were reported and discussed.

The chromatographic behavior of metal ions was investigated on plain and tri-*n*-butylamine impregnated titanium(IV) tungstate layers [120]. Some specific separations were achieved. The selectivity of Ti(IV) tungstate for different metal ions was discussed. The retention behavior of the investigated ions was explained in terms of ion exchange, partition, and adsorption. The effect of pH of the solvent systems used on R_F values of some *d*-block metal ions on both layers was discussed.

A simple, rapid, and selective method of the separation and determination of Cr(VI) from Al(III), Cr(III), Mn(II), Fe(III), Co(II), Ni(II), Cu(II), Zn(II), Se(IV), Sr(II), Zr(IV), Cd(II), La(III), Ce(III), Hg(II), and U(VI) on titanium silicate ion exchange plates with 0.4 M ammonium oxalate–2.2 M aqueous ammonia (1:1, vol./vol.) as the mobile phase was developed [121]. Quantitative measurement of the separated zones of Cr(VI) was achieved by scanning densitometry at $\lambda = 545$ nm.

The same sorbent was used for the separation and determination of Cd in the presence of numerous toxic elements with ammonium buffer solution (pH 10) as the mobile phase [122].

15.4.2 Separations of Organic Substances

In addition to their application in thin-layer chromatographic analysis of different inorganic ions, some inorganic ion exchangers were used as sorbents in the separation of some organic compounds.

The use of thin-layer chromatography in the study the adsorption of cationic dyes on the outer surfaces of zeolites was described [123]. Within the scope of this preliminary investigation, it was found that the R_F values partly depend on ion exchange at the zeolite surface. The method described can be used to distinguish between known zeolites or to estimate the composition of binary mixtures in which one or both components are zeolites.

The thin-layer chromatographic behavior of 28 phenolic compounds was studied on semicrystalline stannic tungstate layers by using various solvent systems [124]. It was found that formic acid and *n*-butanol–formic acid were the most useful for the separation of these compounds. Dimethyl sulfoxide acted as a self-detector for a number of the investigated phenols. A lot of binary, ternary, and quaternary separations of bioanalytical and immunophysiological importance were achieved. Interesting separations of different isomeric compounds were described.

The effect of ionic surfactants on the chromatographic behavior of phenols on bismuth tungstate thin layers was studied using various developing solvents [125]. Numerous binary and ternary separations of phenols were performed. Retention was compared with the results obtained on silica gel layers. It was established that the application of micellar mobile phases improves the selectivity and compactness of the chromatographic zones. The proposed method offers a significant advantage over traditional developing systems.

Thin layers of stannic tungstate were also used in chromatography of 24 amino acids in aqueous and mixed solvent systems [126]. Acetone–formic acid–water and ethyl acetate–formic acid were found to be very useful solvent systems for these separations. DL-3,4-Dihydroxyphenylalanine was selectively separated from a mixture of other amino acids in ethyl acetate–formic acid systems. In addition, specific separations of DL-methionine from a synthetic mixture of numerous amino acids were achieved. *n*-Butanol–acetic acid and acetone–formic acid–water systems were used to successfully separate aspartic acid from glutamic acid.

The selectivity, adsorption, and ion exchange properties of tin(IV) selenoarsenate as a sorbent for TLC separation of amino acids were investigated [127]. Dimethyl sulfoxide was used as the mobile phase.

The chromatographic behavior of amino acids was also investigated on a thin layer of titanium tungstate [128]. In addition to DMSO, butanol was used in order to compare the migration of various amino acids. Furthermore, silica gel G, a weak inorganic ion exchanger, was used to study the retention behavior of these acids in DMSO. Apart from this, the layers prepared by mixing equal amounts of titanium tungstate and silica gel G were employed for the TLC of amino acids. In this way, the separation potential of these layers was increased due to the combined effect of ion exchange behavior of titanium tungstate and the adsorptive behavior of silica gel. Some useful separations of amino acids were realized.

The chromatographic behavior of 19 organic acids was studied on bismuth tungstate thin layers using several developing solvents [129]. It was found that bismuth tungstate was a useful sorbent for thin-layer chromatographic separation of these compounds. Efforts were also made in order to perform binary and ternary separations of bioanalytically important organic acids.

Thin-layer chromatography of 14 organophosphorus, organochlorine, and pyrethroid pesticides was performed on mixed stannic oxide–silica gel G layers [130]. Mixed aqueous and organic mobile phases were utilized for these separations. Several simple, rapid, selective, and reproducible separations of pesticides over a wide range of concentration were achieved. The practical applicability of the method was demonstrated by the quantitative separation of some of the most commonly used organophosphorus pesticides from alcohol extracts of soil samples using butanol–acetic acid–water, 5:1:4, as the mobile phase. The limit of detection was found to be 0.50 μg L^{-1}.

Husain and coworkers developed a simple and rapid method of simultaneous determination of diazinon and ethion residues in the matrix solid phase dispersion extract of pistachio nuts [131]. For this purpose, the investigated organophosphorus pesticides were chromatographed on the lanthanum silicate and lanthanum tungstate ion exchangers with several mobile phases. In the case of lanthanum silicate, methanol–10% ammonia (9.5:0.5) was found to be the most suitable mobile phase for clear separation of diazinon and ethion from the pistachio extract. The most effective mobile phase on thin layers of lanthanum tungstate was methanol–dichloromethane–10% ammonia, 6.5:3.1:0.4. For this purpose, the investigated organophosphorus pesticides were chromatographed on the lanthanum silicate and lanthanum tungstate ion exchangers with several mobile phases. In the case of lanthanum silicate, methanol–10% ammonia (9.5:0.5) was found to be the most suitable mobile phase for clear separation of diazinon and ethion from the pistachio extract. The most effective mobile phase on thin layers of lanthanum tungstate was methanol–dichloromethane–10% ammonia, 6.5:3.1:0.4.

Thin-layer chromatography of seven cephalosporins was studied on stannic oxide using citrate and borate buffers of different pH as mobile phases [132]. Several ternary and quaternary separations were achieved. The TLC procedure described enables a simple and rapid separation and detection of different spontaneous, chemical, and enzymatic degradation products of the cephalosporins. The practical applicability of these separations in estimating cephalosporins in blood serum from patients was demonstrated.

The chromatographic behavior of ten common antibiotics (amoxicillin, ampicillin, cephalexin, cloxacillin, doxycycline, tetracycline, erythromycin, gentamicin, streptomycin, and cotrimoxazole) was studied on thin layers of titanic silicate inorganic ion exchanger with organic, aqueous, and mixed aqueous–organic mobile phases [133]. Fast and selective methods of separating of one antibiotic from others in a single-step process were developed; ternary and binary separations were also achieved. Salting-out TLC using aqueous ammonium sulfate solutions showed the dependence of retention on the concentration of salt in the mobile phase and the existence of a linear relationship between R_M and molarity of ammonium sulfate for some antibiotics. The effect of varying the volume ratio of the binary mobile phase methanol–0.1 M formic acid on the R_F values of the antibiotics was also studied.

The inorganic anion-exchanger bismuth silicate was used to study TLC behavior of 14 performance-enhancing drugs (amphetamine, bemegride, caffeine, chlorphentermine, ephedrine, ethylamphetamine, isoproterenol, methadone, methylenedioxyamphetamine, pentazocine, pethidine, pemoline, strychnine, and salbutamol) using 21 organic, aqueous, and mixed organic–aqueous mobile phases [134]. Rapid separations of one drug from many other drugs and many quaternary, ternary, and binary separations were achieved and discussed. A new variable, SR_F (the difference between the R_F on silica gel and that on bismuth silicate containing a little silica gel as binder), was introduced to quantify the separating power of the bismuth silicate anion exchanger in thin-layer chromatography.

The chromatographic behavior of 24 food additives (ascorbic acid, benzoic acid, butylated hydroxyanisole, butylated hydroxytoluene, butyraldehyde, butyric acid, cinnamaldehyde, citric acid, ethyl acrylate, ethyl benzoate, ethyl *p*-hydroxybenzoate, fumaric acid, lactic acid, lauric acid, maleic acid, methyl *p*-hydroxybenzoate, oleic acid, *p*-hydroxybenzoic acid, propionic acid, propyl gallate, propyl *p*-hydroxybenzoate, salicylic acid, sodium benzoate, and sorbic acid) was studied on thin layers of stannic silicate ion exchanger [135]. Several aqueous, nonaqueous, and mixed mobile phases were used for these separations. Rapid and selective separations of one food additive from many others were achieved. Numerous binary and ternary separations were also performed with simple aqueous mobile phases. The selective adsorption of ionic and neutral species by the exchanger was assumed as the dominant separation mechanism.

Similar results were obtained by TLC of 30 food additives on titanium(IV) silicate ion exchanger with several aqueous, organic, and mixed mobile phases [136]. Fast and selective separations of one food additive from others, as well as some ternary and quaternary separations, were reported. The effect of salt concentration in the mobile phase on retention was studied and discussed.

Rapid separation of methylparaben from seven other food additives was achieved on a thin layer of stannic silicate with *n*-hexane–ethyl methyl ketone–acetic acid, 8:2:0.3 (v/v/v), as the mobile phase [137]. Quantitative analysis was performed by scanning densitometry at $\lambda = 260$ nm. The limits of detection (LOD) and quantitation (LOQ) were 0.29 and 0.50 μg per zone, respectively.

The same ion exchanger was used as a sorbent in TLC separation of *p*-hydroxybenzoic acid from a mixture with ten preservatives [138]. Solvent system n-hexane–ethyl methyl ketone–acetic acid, 8:2:0.3 (v/v/v), was used as the mobile phase. Quantitative analysis was performed by scanning densitometry at 270 nm. The method enables the separation and identification of p-hydroxybenzoic acid which is present as both an impurity and a product of hydrolysis in the parabenes. The limits of detection (LOD) and quantitation (LOQ) for p-hydroxybenzoic acid were 0.05 and 0.51 μg per zone, respectively.

The same authors also used a thin layer of stannic silicate for fast and selective separation of 13 plasticizers [139]. A mixture of toluene–ethyl acetate (10:1, v/v) was used as the mobile phase. In many cases, one plasticizer was separated from others in a single-step process. Fast ternary separations were also achieved. Quantitative determination of di(2-ethylhexyl)phthalate was performed by scanning densitometry at 280 nm with limit of quantitation of 0.50 μg and limit of detection of 0.05 μg per zone.

The effect of ionic surfactants on the thin-layer chromatographic behavior of various amines was investigated [140]. For this purpose, the thin layers of hydrated stannic oxide impregnated silica gel were used in combination with various solvent systems of different nature, with or without surfactant solutions of different concentrations. Some of the important binary separations were reported. The results obtained were compared with chromatographic behavior of amines observed on an unimpregnated thin layer of silica gel.

Finally, to combine the qualities of inorganic and organic ion exchangers, new inorganic–organic ion exchangers were synthesized. Practical utility of the titanium(IV) iodosulfosalicylate [141] and titanium (IV) phosphosulfosalicylate [142] was demonstrated by a novel TLC separations of inorganic cations. Some analytically important binary (Mg^{2+}–Pb^{2+}, Zn^{2+}–Pb^{2+}, Co^{2+}–Pb^{2+}, Ni^{2+}–Pb^{2+}, Ca^{2+}–Mg^{2+}, and Zn^{2+}–Ca^{2+}) and ternary (Mg^{2+}–Pb^{2+}–Ca^{2+}, Zn^{2+}–Ca^{2+}–Pb^{2+}, and Co^{2+}–Ca^{2+}–Pb^{2+}) separations of metal ions were achieved. Due to their high thermal stability and selectivity for lead, these exchangers may be important for the environmentalists.

Acknowledgment The authors are grateful to the Ministry of Science and Technological Development of the Republic of Serbia (Grant No. 172017) for financial support.

References

1. Miller JM (2005) Chromatography: concepts and contrasts. Wiley, New Jersey
2. Sherma J, Fried B (2003) Handbook of thin-layer chromatography. Marcel Dekker, New York
3. Poole CF (2003) The essence of chromatography. Elsevier, Amsterdam
4. Burma DP (1953) Partition mechanism of paper chromatography. Anal Chem 25(4):549–553
5. Sherma J (2010) Planar chromatography. Anal Chem 82:4895–4910
6. Оьшанова КМ, Потапова МА, Копылова ВД, Морозова НМ (1965) Руководство по ионообменной, распределительной и осадочной хроматографии. Химия, Москва
7. Rabel FM (2003) Sorbents and precoated layers in thin-layer chromatography. In: Sherma J, Fried B (eds) Handbook of thin-layer chromatography, 3rd edn. Marcel Dekker, New York
8. Naushad Mu (2009) Inorganic and composite ion-exchange materials and their applications. Ion Exch Lett 2:1–14
9. Townsend RP (1986) Ion-exchange in zeolites: some recent developments in theory and practice. Pure Appl Chem 58(10):1359–1366
10. Troup JM, Clearfield A (1977) Mechanism of ion-exchange in zirconium phosphates. 20. Refinement of the crystal structure of ά-zirconium phosphate. Inorg Chem 16(12):3311–3314
11. Clearified A (1993) Ion-exchange and adsorption in layered phosphates. Mat Chem Phys 35(3–4):257–263
12. Qureshi M, Kumar R, Rathore KS (1972) Separation of cadmium(II) from zinc(II) and other metal ions on a cadmium selective exchangers: titanium selenite. Anal Chem 44(6):1081–1084
13. Alberti GM, Massucci MA (1970) Crystalline insoluble salts of tetravalent metals. IX Thorium arsenate, a new inorganic ion-exchanger specific for lithium. J Inorg Nucl Chem 32(5):1719–1727
14. Inoue Y (1964) Studies on the synthetic inorganic ion-exchangers. Synthesis of stannic phosphate and some of its properties. J Inorg Nucl Chem 26(12):2241–2253
15. Tandon SN (1973) Synthesis and ion-exchange properties of ceric tungstate. Talanta 20(6):585–588
16. Qureshi M, Kumar RM, Rathore HS (1972) Studies on chromium(III) hydroxide, arsenate, antimonate, molybdate and tungstate. Talanta 19(11):1377–1386
17. Rawat JP, Thind PS (1976) Studies on inorganic ion-exchangers. I Preparation, properties and applications of ferric phosphate. Can J Chem 54(12):1892–1898
18. Qureshi M, Gupta JP, Sharma V (1973) Synthesis of a reproducible and chemically stable tantalum antimonate. Anal Chem 45(11):1901–1906
19. Torok G, Diehl JF (1971) Preparation and properties of antimony(V) aquaoxide for separation of sodium in activation analysis. Radiochim Acta 15(2):96–100

20. Nagy LG, Torok G, Foti G (1975) Investigation on the structure and sorption properties of hydrated antimony pentoxide used for the selective removal of radiosodium. Proc Int Conf Coll Surf Sci 1:33–40
21. Gill JS, Tandon SN (1972) Preparation and ion-exchange properties of stannic ferrocyanide. J Inorg Nuc Chem 34(12):3885–3892
22. Gupta AP, Qureshi M, Sunanda Y (1983) Studies on a new adsorbent ion-exchange material: hydrous antimony sulphide. J Chem Sec A 22A(8):721–722
23. Krtil J, Chavko M (1967) Ion-exchange properties of ammonium salts of heteropolyacids: VII sorption of ^{137}Cs and ^{86}Rb on acid and normal ammonium and thallous salts of phosphotungstanic and phosphomolybdic acid. J Chromatogr 27:460–473
24. Wei H, Collinson MM (1998) Functional-group effects on the ion-exchange properties of organically modified silicates. Anal Chim Acta 397:113–121
25. Varshney KG, Tayal N et al (2001) Synthesis, characterization and analytical applications of lead(II) selective polyacrylonitrile thorium(IV) phosphate: a novel fibrous ion-exchanger. Coll Surf 181:123–129
26. Pandit B, Chudasma U (2001) Synthesis, characterization and application of an inorganic organic material: p-chlorophenol anchored onto zirconium tungstate. Bull Mater Sci 24(3):265–271
27. Khan AA, Niwas R, Alam MM (2002) Ion-exchange materials on styrene supported zirconium(IV) tungstate: an organic-inorganic type cation-exchanger. Ind J App Chem 9(3):256–260
28. Inamuddin KSA, Siddiqui WA et al (2007) Synthesis, characterization and ion-exchange properties of a new and novel 'organic–inorganic' hybrid cation-exchanger: Nylon-6,6, Zr(IV) phosphate. Talanta 71:841–847
29. Sairam Sundaram C, Meenakshi S (2009) Fluoride sorption using organic–inorganic hybrid type ion-exchangers. J Colloid Interface Sci 333:58–62
30. Khan AA, Akhtar T (2009) Synthesis, characterization and ion-exchange properties of a fibrous type 'polymeric-inorganic' composite cation-exchanger Nylon-6,6 Sn(IV) phosphate: its application in making Hg(II) selective membrane electrode. Electrochim Acta 54:3320–3329
31. Alam Z, Inamuddin NSA (2010) Synthesis and characterization of a thermally stable strongly acidic Cd(II) ion selective composite cation-exchanger: polyaniline Ce(IV) molybdate. Desalination 250:515–522
32. Amphlett CB (1964) Inorganic ion-exchangers. Elsevier, Amsterdam
33. Walton HF (1968) Ion-exchange chromatography. Anal Chem 40(5):51R–62R
34. Walton HF (1974) Ion-exchange. Anal Chem 46(5):398R–413R
35. Sherma J, Zweig G (1971) Paper chromatography and electrophoresis, vol 2, Paper chromatography. Academic, New York
36. Sherma J, Fried B (1990) Handbook of thin-layer chromatography. Marcel Dekker, New York
37. Gocan S (2002) Stationary phases for thin-layer chromatography. J Chromatogr Sci 40(10):538–549
38. Alberti G, Grassini G (1960) Chromatography on paper impregnated with zirconium phosphate. J Chromatogr 4:83–85
39. Nunes MJ, Jeronimo MAS (1961) Zirconium selenite as ion-exchanger. J Chromatogr 5:546–546
40. Alberti G, Dobici F, Grassini G (1962) Chromatography on paper impregnated with inorganic ion-exchangers. III Chromatography of inorganic ions on zirconium phosphate papers with HCl, HNO_3, $HClO_4$ and CH_3COOH at various concentration. J Chromatogr 8:103–107
41. Sastri MN, Rao AP (1962) Separation of the valency states of some elements on paper impregnated with zirconium phosphate. J Chromatogr 9:250–251
42. Grassini G, Padiglione C (1964) The exchange capacity of papers impregnated with zirconium phosphates. J Chromatogr 13:561–564
43. Sakodinsky K, Lederer M (1965) The chromatographic properties of paper impregnated with titanium hydroxide. J Chromatogr 20(2):358–366
44. Qureshi M, Akhtar I, Mathur KN (1967) Separation of metal ions on stannic phosphate and stannic tungstate papers. Anal Chem 39(14):1766–1770
45. Qureshi M, Fahmida K (1968) Fast separation of titanium(IV), arsenic(III), tungsten(VI), gold(III) from numerous metal ions and of iron-aluminium-titanium by paper chromatography. J Chromatogr 34(2):222–236

46. Qureshi M, Kishore MN (1968) Quantitative separation of selenium from metal ions on tin(IV) tungstate papers. Anal Chim Acta 41(3):560–563
47. Qureshi M, Husain W (1969) Separation of metal ions on titanum(IV) tungstate papers. Sep Sci 4(3):197–207
48. Qureshi M, Mathur KN, Israli AH (1969) Separation of metal ions on tin(IV) tungstanate and selenite papers. Talanta 16(4):503–509
49. Qureshi M, Qureshi ZS (1972) Relative exchange capacities on stannic phosphate papers. Sep Sci 7(2):187–191
50. Qureshi M, Varshney KG, Fahmida K (1972) Synthesis, ion-exchange behavior, and analytical applications of stannous ferrocyanide. Separation of magnesium-calcium, magnesium-barium, manganese-nickel and yttrium-thorium. J Chromatogr 65(3):547–555
51. Husain SW (1972) Chromatographic separation of metal ions on ion-exchange papers. Analusis 1(4):314–316
52. Qureshi M, Sharma SD (1973) Prediction of Ksp from RF values. Chromatography of 48 metal ions on stannic arsenate and plain papers in butanol – nitric acid media. Anal Chem 45(7):1283–1288
53. Qureshi M, Varshney KG, Fahmida K (1973) Chromatographic separation of metal ions on stannous ferrocyanide papers. Sep Sci 8(2):279–284
54. Qureshi M, Sharma SD (1975) Chromatographic behavior of 48 cations on stannic and titanium arsenate papers in aqueous nitric acid system. Talanta 22(2):129–133
55. Qureshi M, Varshney KG, Rajput RPS (1975) Chromatography of 49 metal ions on stannic antimonate papers in dimethylsulfoxide-nitric acid systems. Anal Chem 57(9):1520–1524
56. Qureshi M, Nabi SA, Zehra N (1975) Chromatographic separations of amino acids on titanium arsenate. Sep Sci 10(6):801–808
57. Singh NJ, Tandon SN (1977) Rapid separation of metal ions on hydrous zirconium oxide papers. Chromatographia 10(6):309–311
58. Rajeev JM, Tandon SN (1977) Rapid separation of alkali metal ions on zirconium antimonite papers. Chromatographia 10(1):45–46
59. Angus HJF, Briggs J, Sug NA et al (1978) Chromatography of polyhydroxy compounds on cellulose impregnated with tungstate: determination of pseudostability constants of complexes. Carbohydr Res 66(1):25–32
60. Qureshi M, Anees A, Mohammed SA, Ahmad SN (1987) Separation of certain organic pollutants containing tertiary nitrogen on stannic molybdosilicate papers. Anal Lett 20(8):1157–1169
61. Surendra KVN, Tandon SN (1988) Chromatographic separation of anions on papers impregnated with hydrous titanium oxide. J Liq Chromatogr Related Technol 11(7):1433–1439
62. Dabral SK, Singh MKP, Rawat JR (1988) Ion-exchange chromatographic separations of some anions on hydrate stannic oxide impregnated paper. Ind J Chem Sect A 27A(8):745–746
63. Dabral SK, Muktawat KPS, Rawat JP (1988) Ion-exchange chromatographic separation on hydrated bismuth oxide impregnated papers. Anal Lett 21(4):613–620
64. Dabral SK, Muktawat KPS, Rawat JP (1989) Ion-exchange chromatographic studies of some physiologically important organic acids on impregnated papers. Anal Lett 22(1):61–69
65. Dabral SK, Rawat JP, Muktawat KPS (1989) Ion-exchange chromatographic determination of amino acids on impregnated papers. Anal Lett 22(3):537–543
66. Dabral SK (1989) The effect of surfactants on the chromatographic separation of phenols on paper impregnated with hydrated cerium oxide. Anal Lett 22(6):1623–1629
67. Sharma SD, Misra S, Gupta R, Agarwal A (1995) Increased selectivity for planar chromatography by ion-exchange: ion chromatography on papers impregnated with stannic antimonate and stannic arsenate. J Ind Chem Soc 72(9):657–660
68. Sharma SD, Gupta R (2000) Increased selectivity for planar chromatography by ion-exchange: cation chromatography on papers impregnated with titanium(IV) based inorganic exchangers in DMSO-HNO_3 mobile phase. J Ind Chem Soc 77(4):204–206
69. Misra AK, Pachauree S (2001) Retention behavior of synthetic dyes on zirconium antimonate papers. Acta Chromatogr 11:139–153
70. Lederer M (1999) Fifty years of inorganic chromatography. Chem Intell 5(3):31–34

71. Lederer M (1994) Chromatography for inorganic chemistry. Wiley, Chichester
72. Mohammad A (2003) Inorganic and organometallic compounds. In: Sherma J, Fried B (eds) Handbook of thin-layer chromatography, 3rd edn. Marcel Dekker, New York
73. Geiss F (1987) Fundamentals of thin-layer chromatography (planar chromatography). Huthig, Heidelberg
74. Engelhardt H (1979) High performance liquid chromatography. Springer, Heidelberg
75. Hauck HE, Mack M, Jost W (1991) Chromatogr Sci 55:87–111
76. Boehm HP (1966) Functional groups on the surface of solids. Angew Chemie 78(12):617–628
77. Smith RL, Pietrzyk DJ (1984) Liquid chromatographic separation of metal ions on a silica column. Anal Chem 56:610–614
78. Ohta K, Sando M, Tanaka K et al (1996) Simultaneous determination of common mono- and divalent cations by ion chromatography with an unmodified silica gel column. J Chromatogr A 752:167–172
79. Hirata Sh, Kozaki D, Sakanishi K et al (2010) Simultaneous determinations of Cr(VI) and Cr(III) by ion-exclusion/cation-exchange chromatography with an unmodified silica-gel column. Anal Sci 26:387–390
80. Druding LF, Hagel RB (1966) Separation of *cis-trans*-cobalt(III) amine complexes by thin layer chromatography. Anal Chem 38:478–480
81. Mohammad A (1996) Inorganic and organometallics. In: Sherma J, Fried B (eds) Handbook of thin-layer chromatography, 2nd edn. Marcel Dekker, New York
82. ŽLj Tešić, Grgurić SR, Trifunović SR et al (1997) Planar chromatography of facial and meridional isomers of some aminocarboxylato cobalt(III) complexes. J Planar Chromatogr 10:457–459
83. Milojković-Opsenica DM, Malinar MJ, Tešić ŽLj (1999) Effect of the chelate ring size of diamine-Co(III) complexes on their hR_F values obtained under salting-out thin-layer chromatography conditions. J Chromatogr A 847:291–296
84. Shepherd RE (2003) Chromatographic and related electrophoretic methods in the separation of transition metal complexes or their ligands. Coord Chem Rev 247:147–184
85. Radoičić A, Majstorović H, Sabo T et al (2009) Hydrophilic-interaction planar chromatography of some water-soluble Co(III) complexes on different adsorbents. J Planar Chromatogr 22(4):249–253
86. Lederer M (1999) The story of ion-exchange resin papers. Chem Intell 5(4):32–34
87. Zabin BA, Rollins CB (1964) Inorganic ion-exchangers for thin-layer chromatography. J Chromatogr 14(3):534–535
88. Husain SW, Kazmi SK (1972) Thin layer chromatography of metal ions on a new synthetic inorganic ion-exchanger. Experientia 28(8):988–989
89. Husain SW, Eivazi F (1975) Thin layer chromatography of 57 metal ions on an inorganic ion-exchanger in mixed solvent systems. Chromatographia 8(6):277–282
90. Husain SW, Rasheedzad S (1978) Thin-layer chromatographic separation of platinum metals on a semi-crystalline inorganic ion-exchanger. Mikrochim Acta 1(1–2):11–18
91. Husain SW, Rasheedzad S (1979) Chromatography of toxic metals on thin layers of lanthanum tungstate. Recent Dev Chromatogr Electrophor 1:239–245
92. De AK, Rajput RPS, Das SK (1979) Synthetic inorganic ion-exchangers. XII. Thin-layer chromatography of metals ions on thorium antimonate: quantitative separation of mercury(II) from several metal ions. Sep Sci Technol 14(8):735–739
93. De AK, Pal BK (1979) Synthetic inorganic ion-exchangers. XV. Thin-layer chromatography of metal ions on thorium tungstate: quantitative separation of mercury(II) from several other metal ions. J Liq Chromatogr 2(7):935–941
94. De AK, Pal BK (1980) Synthetic inorganic ion-exchangers. XVI. Thin-layer chromatography of metal ions on zirconium tungstate: quantitative separation of mercury(II) from several other metal ions and from mixtures. Sep Sci Technol 15(5):1271–1275
95. De AK, Chakraborty P (1981) Synthetic inorganic ion-exchangers. XX. Thin layer chromatography of metal ions on lanthanum antimonate. Quantitative separation of mercury(II) from several metal ions. J Liq Chromatogr 4(12):2223–2228

96. Yin B, Lei G, Xie L (1982) Ion-exchange properties of stannic phosphate. II. The thin-layer chromatographic behavior of forty metallic ions. Huaxue Xuebao 40(3):281–286
97. Ying B, Lei G, Liu G (1982) Ion-exchange properties of stannic phosphate. (III). Chromatographic behavior of forty-four cations on stannic phosphate layers. Gaodeng Xuexiao Huaxue Xuebao 3(1):28–34
98. Seth NS, Rajput RPS (1983) Thin-layer chromatography of metal ions on titanium(IV) antimonate in aqueous and mixed solvent systems containing DMF: quantitative separation of palladium(II) from several metal ions. Indian J Chem Sect A 22A(12):1088–1090
99. Varshney KG, Khan AA, Anwar S (1985) Chromatography of alkaline earths and transition metals on tin(IV) arsenosilicate and arsenophosphate thin layers in buffered EDTA solutions. J Liq Chromatogr 8(7):1347–1353
100. Yin B, Liu J (1985) Thin layer chromatographic behavior of noble metal ions on a semicrystalline inorganic ion-exchanger. Huaxue Tongbao 7:25–27
101. Rajput RPS, Agrawal Sh (1985) Chromatographic behavior of anions of zirconium(IV) molybdate papers and silica gel 'G' thin layers. Anal Lett 18(A14):1783–1788
102. Song J, Yin B (1985) Ion-exchange properties of semicrystalline form of stannic phosphate -centrifugal thin-layer chromatographic separation of noble metals. Xibei Daxue Xuebao Ziran Kexueban 15(2):56–60
103. Liu L, Liu J, Huang Zh et al (1987) Inorganic ion-exchanger containing phosphorus. III. Thin-layer chromatographic separation of noble metals on stannic pyrophosphate. Wuhan Daxue Xuebao. Ziran Kexueban 1:64–70
104. Liu Xu, Liu J, Liu L et al (1987) Studies on the new inorganic ion-exchanger containing phosphorus. (IV). Thin-layer chromatography separation of noble metals on stannic selenopyrophosphate. Gaodeng Xuexiao Huaxue Xuebao 8(8):688–692
105. Bold A, Popa A, Vasile A et al (1989) Zeolite molecular sieves as stationary phase in thin layer chromatography. V. On migration mechanism. Rev Roum Chim 34(1):59–63
106. Husain SW, Ghoulipour V (1989) Thin-layer chromatography of metal ions on an ion-exchanger – stannic silicate. J Planar Chromatogr-Mod TLC 2(6):474–476
107. Husain SW, Ishghy Z, Chaloosi M (1990) Quantitative TLC of toxic elements on inorganic ion-exchangers. I. Separation and determination of mercury. J Planar Chromatogr-Mod TLC 3:271–272
108. Husain SW, Ishghy Z (1992) Quantitative TLC of toxic elements on inorganic ion-exchangers. IV. Cadmium and lead. J Liq Chromatogr 15(10):1681–1687
109. Mohammad A, Ajmal M, Fatima N et al (1992) Some observations on the sorption behavior of heavy metals on layers prepared from mixtures of silica and inorganic ion-exchange gels. J Planar Chromatogr-Mod TLC 5(5):368–375
110. Varshney KG, Khan AA, Siddiqui MS (1993) Thin layer and paper chromatography of metal ions on some inorganic ion-exchangers using nicotine as a developer at different pH values. Proc Nat Acad Sci India Sect A Phys Sci 63(3):495–499
111. Husain SW, Avanes A, Ghoulipour V (1996) Thin layer chromatography of metal ions on cerium(III) silicate: a new ion-exchanger. J Planar Chromatogr-Mod TLC 9(1):67–69
112. Husain SW, Mirzaie A (1997) Chromatographic separation of inorganic ions on thin layers of lanthanum silicate ion-exchanger. Chromatographia 45:347–350
113. Su Zh, Feng H, Yin B (1997) Studies on inorganic ion-exchangers. IV. Thin-layer chromatographic application of metaphosphate. Lizi Jiaohuan Yu Xifu 13(5):459–465
114. Ghoulipour V, Husain SW (1999) New sorbents in planar chromatography: I. Separation of inorganic ions on layered double hydroxides. J Planar Chromatogr-Mod TLC 12(5):378–382
115. Ghoulipour V, Husain SW (2000) New adsorbents in planar chromatography: II. Behavior of metal ions on titanium(IV) silicate ion-exchanger. J Planar Chromatogr-Mod TLC 13(5):354–358
116. Ghoulipour V, Husain SW (2000) Ion-exchanger bismuth silicate as a novel sorbent in thin layer chromatography. Anal Sci 16:1079–1081
117. Mohammad A, Yousuf R, Hamid Y (2001) Thin layer chromatography of inorganic ions on blended inorganic ion-exchangers with tributyl phosphate-formic acid as mobile phase. Acta Chromatogr 11:171–182

118. Ghatuary RK, Mukhopadhyay AK, Sarkar CK (2001) Thin layer chromatographic separation of anions by naturally occurring mixed oxides impregnated plates. J Indian Chem Soc 78(9):492–494
119. Ghoulipour V, Husain SW (2001) Thin layer chromatography of toxic metals on mixed bed of titania and silica. Ann Chim (Rome) 91(1–2):111–116
120. Sharma SD, Gupta R (2001) Retention behavior of metal ions on plain and tri-*n*-butylamine impregnated titanium(IV) tungstate layers. J Indian Chem Soc 78(6):301–304
121. Ghoulipour V, Husain SW (2002) Inorganic ion-exchangers for quantitative TLC of toxic elements. V. Separation and determination of chromium(VI). Acta Chromatogr 12:170–176
122. Ghoulipour V, Husain SW (2006) Quantitative TLC of toxic elements on inorganic ion-exchangers. VI. Separation and determination of cadmium. J Planar Chromatogr-Mod TLC 19(109):246–250
123. Lowe BM, Cook HL (1982) Investigation of zeolites by thin-layer chromatography. Zeolites 2(1):29–32
124. Nabi SA, Farooqui WU, Rahman N (1985) A semicrystalline inorganic ion-exchanger for thin-layer chromatographic separation of phenolic compounds. Chromatographia 20(2):109–111
125. Kulshrestha S, Dabral SK, Muktawat KPS et al (2002) The effect of ionic surfactants on the behavior of phenols on bismuth tungstate thin layers. J Indian Chem Soc 79(9):739–742
126. Nabi SA, Farooqui WU, Siddiqui ZM et al (1983) Thin-layer chromatographic separations of amino acids on stannic tungstate. J Liq Chromatogr 6(1):109–122
127. Siddiqi ZM, Rani S (1995) Studies of Tin(IV) selenoarsenate III: thin layer chromatography of α-amino acids. J Planar Chromatogr-Mod TLC 8(2):141–143
128. Sharma SD, Sharma H, Sharma SC (2002) Thin layer chromatography of amino acids on titanium tungstate using dimethyl sulfoxide as the mobile phase. Chem Environ Res 11 (1 & 2):179–187
129. Kulshrestha Sh, Dabral SK, Muktawat KPS (2001) Bismuth tungstate inorganic ion-exchanger – an absorbent for thin layer chromatography of organic acids. J Ind Chem Soc 78(7):374–376
130. Nabi SA, Gupta A, Khan MA et al (2002) Thin-layer chromatographic separations of some common pesticides on mixed stannic oxide–silica gel G layers. Acta Chromatogr 12:201–210
131. Husain SW, Kiarostami V, Morrovati M et al (2003) Multiresidue determination of diazinon and ethion in pistachio nuts by use of matrix solid phase dispersion with a lanthanum silicate co-column and gas chromatography. Acta Chromatogr 13:208–214
132. Nabi SA, Laiq E, Islam A (2004) Selective separation and determination of cephalosporins by TLC on stannic oxide layers. Acta Chromatogr 14:92–101
133. Husain SW, Ghoulipour V, Sepahrian H (2004) Chromatographic behavior of antibiotics on thin layer of an inorganic ion-exchanger. Acta Chromatogr 14:102–109
134. Hassankhani-Majd Z, Ghoulipour V, Husain SW (2006) Chromatographic behavior of performance-enhancing drugs on thin layers of bismuth silicate ion-exchanger. Acta Chromatogr 16:173–180
135. Purghazi K, Ghoulipou V, Husain SW et al (2007) Rapid chromatographic separation of food additives on thin layers of an inorganic ion-exchanger. Acta Chromatogr 18:219–225
136. Ghoulipour V, Amini S, Haghshenas A et al (2010) Chromatographic behaviour of food additives on thin layers of titanium(IV) silicate ion-exchanger. J Planar Chromatogr 23(4):250–254
137. Mirzaie A, Jamshidi A, Husain SW (2007) TLC quantification of methylparaben on an inorganic ion-exchanger in the presence of other food additives. J Planar Chromatogr-Mod TLC 20(2):141–143
138. Mirzaie A, Jamshidi A, Husain SW (2007) Quantitative ion-exchange TLC of p-hydroxybenzoic acid in the presence of preservatives. J Planar Chromatogr-Mod TLC 20(4):303–306
139. Jamshidi A, Husain SW (2007) Fast chromatographic separation of plasticizers on thin layers of an inorganic ion-exchanger: quantitative determination of di(2-ethylhexyl)phthalate. Chromatographia 65(3/4):245–248

140. Kulshrestha Sh, Awasthi A, Dabral SK (2010) Effect of ionic surfactants on the resolution of amines of hydrated stannic oxide impregnated thin layer chromatography. Asian J Chem 22(4):2507–2510
141. Thomas R, Janardanan C (2007) Synthesis, properties and analytical applications of titanium(IV) iodosulfosalicylate – a new hybrid inorganic-organic ion – exchanger. J Ind Counc Chem 24(2):5–10
142. Thomas R (2008) Synthesis, properties and analytical applications of titanium(IV) phosphosulphosalicylate – a new hybrid inorganic-organic ion-exchanger. Orient J Chem 24(1):139–146

Chapter 16
Cation-Exchanged Silica Gel–Based Thin-Layer Chromatography of Organic and Inorganic Compounds

Ali Mohammad, Abdul Moheman, and Gaber E. El-Desoky

Abstract The main goal of this chapter is to encapsulate the data published during last 27 years (1984–2010) on cation-exchanged silica gel–based thin-layer chromatography (TLC) of organic and inorganic species. It includes the types of impregnating cations/salts, mobile phase, and technique involved in the separation, identification, and determination of organic/inorganic substances present either singly or as components of closely related mixtures in a variety of matrices. The use of cation-exchanged silica gel layers in TLC/HPTLC analyses of organic and inorganic species is described by citing several examples. It has been noticed that little attention has been paid on the use of cation-exchanged silica gel as layer material in the analysis of inorganic substances as compared to organic compounds. The alternative features of using cation-exchanged silica gel layers include the better differential migration, increase separation efficiency, and improved resolution of analytes. According to the literature survey, the interest of chromatographers in using the cations/salts for impregnation of silica gel layers has been in the following order: Ag > Cu > Ni > Mn > Fe ≈ Zn > Co.

16.1 Introduction

Thin-layer chromatography (TLC), originally developed by Kirchner et al. [1] and latter standardized by Stahl [2], is an expensive, simple, and versatile analytical technique for qualitative identification and separation of multicomponent mixtures.

A. Mohammad (✉)
Department of Applied Chemistry, Aligarh Muslim University, Aligarh 202 002, India
e-mail: alimohammad08@gmail.com

A. Moheman
Department of Chemistry, Aligarh Muslim University, Aligarh 202 002, India
e-mail: amohemanappchem@gmail.com

G.E. El-Desoky
Department of Chemistry, College of Science, King Saud University, Riyadh, Saudi Arabia

Inamuddin and M. Luqman (eds.), *Ion Exchange Technology II: Applications*, DOI 10.1007/978-94-007-4026-6_16, © Springer Science+Business Media B.V. 2012

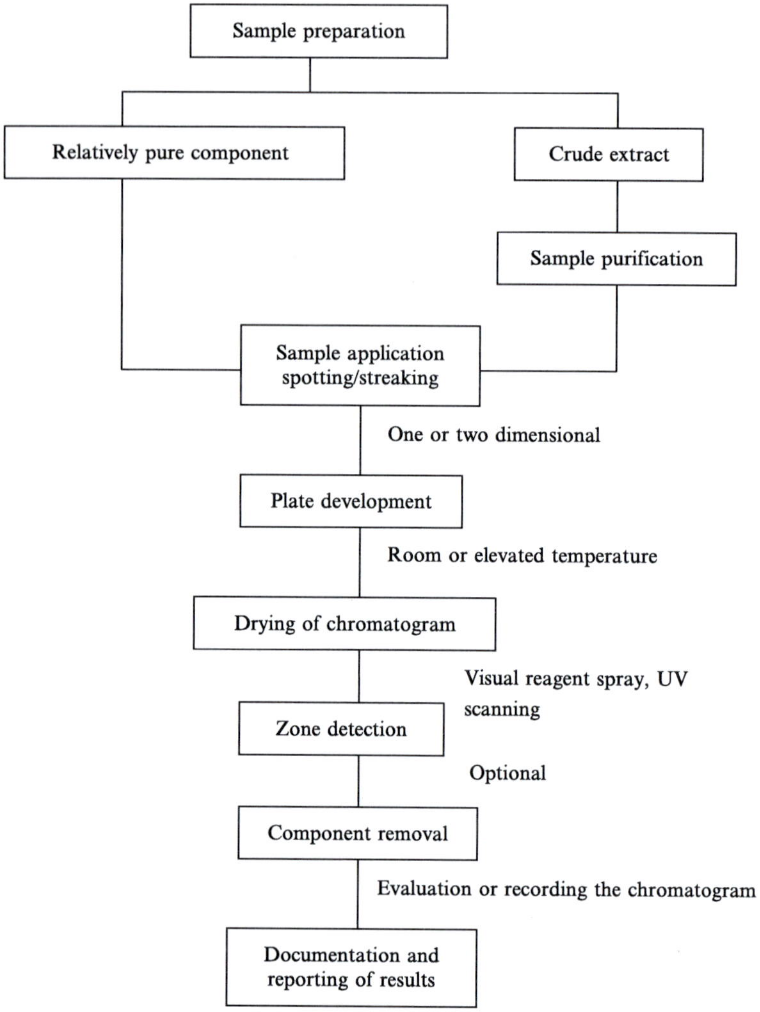

Fig. 16.1 Scheme of typical thin-layer chromatographic process

However, modern TLC which is usually called as high-performance thin-layer chromatography (HPTLC) and originated around 1975 is fast and very useful for rapid and high-resolution separations of chemically closely related substances.

TLC is an off-line process where all the involved steps are carried out independently (Fig. 16.1). Some of the advantages of such arrangement are (a) stationary phase needs not to be reused, (b) carryover of material from one sample to another is very simple, (c) possibility of analyzing large number of samples and standard simultaneously on single plate, (d) ability of constructing calibration curves from standards chromatographed under identical conditions as the samples, (e) static post-chromatographic detection of analytes using various specific as well as universal chromogenic reagent, (f) high selectivity and flexibility because of the availability of great variety of layer materials, (g) minimal sample cleanup and low solvent consumption, and (h) disposal nature of TLC plates and densitometric evaluation of optical properties of all fractions appeared on the chromatogram. Despite the differences in the formate of the stationary phase, kinetic control of the separation process, and use of the development mode (TLC) and the elution mode (HPLC), both techniques (HPLC and HPTLC) are currently being considered as complementary rather than competitive. HPTLC is faster whereas HPLC is more sensitive separation mode.

The majority of TLC separations are carried out on normal phase silica gel, cellulose, and aluminum oxide. However, TLC plates modified with alkyl, amino, cyano, and diol functional groups bonded to the silica have affected the predominant role of normal phase silica gel TLC. Additionally, the chiral layers have been very useful in the resolution of optical isomers.

The layer materials used in TLC analysis of organic and inorganic substances belong to following groups:

(a) Non-surface-modified layers
(b) Impregnated layers
(c) Chemically modified and bonded layers
(d) Inorganic ion exchangers
(e) Mixed layers
(f) Miscellaneous layers

16.2 Silica Gel

Out of layer materials used so far, silica gel has been the most formed material for TLC. All silica gels are silicon dioxides where each silicon atom is surrounded by four oxygen atoms to provide tetrahedral structure (Fig. 16.2). On the surface, the free valencies of the oxygen are connected either with silicon atom as siloxane group (Si–O–Si) or with hydrogen as silanol group (Si–O–H). The Si–O–H groups at the surface of silica gel act as represent sorption-active centers to interact with the analyte. The selective interactions of the species occurring at active sites on the silica gel provide chromatographic separation at the surface. The forces responsible to influence interactions include hydrogen bonding, dipole–dipole, and electrostatic interactions. The intensity of these forces depends upon the magnitude of effective silanol groups.

The silanol group is weakly acidic, and immersion in aqueous salt solution enables cation exchange to take place as follows:

$$M^{n+} + m(-SiOH) \rightleftharpoons M(OSi-)_m{}^{n-m} + mH^+ \qquad (16.1)$$

The exchanged metal cations bring about the change in character of active centers on silica gel surface because their free orbitals are capable of forming coordination complexes with solvent molecules and separated compounds during chromatographic process.

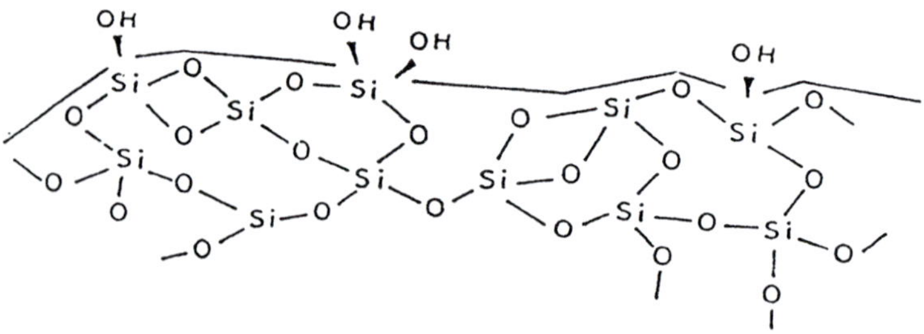

Fig. 16.2 Structure of silica gel

16.3 Cation-Exchanged Silica Gel

The use of silica gel modified with metal salts as stationary phase has been a very attractive area of research. Adsorbents impregnated with metal ions (mainly from transient group) have been successfully used in the analysis of inorganic and organic compounds. The cation exchange and the surface complexation seem to be responsible for differential retention of these compounds on silica gel impregnated with metal ions. The silica gel modified with following metal cations has been frequently used as layer material:

(a) Monovalent cations: Na, K, Ag, and Li
(b) Bivalent cations: Cu, Mn, Ni, Co, Cd, Zn, Sr, Mg, and Hg
(c) Tervalent cations: Al, Cr, Fe, and Eu
(d) Tetravalent cation: Th
(e) Pentavalent cation: V

An important group, among these types of stationary phases, is constituted by the adsorbents impregnated with silver ions (called argentation TLC) which display great selectivity for unsaturated compounds due to the ability of Ag^+ ion to create π -complexes with the unsaturated ligands. In 1962, Barett, Dallas, and Padley described the first application of Ag-TLC to the analysis of triacylglycerol (TAG) [3]. Most of the work [4–21] done afterward follow the general patterns suggested by these authors. Silver ions are incorporated into the layer by adding the salt to the or, better yet, by impregnation of the layer via spraying with or by dipping into a methanolic solution of the salt. Silver content of 10–30% has been considered essential for good resolution [22].

The important applications of metal-cation-modified silica gel as used in the analysis of organic and inorganic substances are listed in Tables 16.1 and 16.2, respectively. Important information about the various combinations of impregnating cations and mobile phases is listed.

Table 16.1 Chromatographic performance of cation-exchanged silica gel in TLC analysis of organic substances

Impregnating cations/salts	Mobile phase	Remarks	Ref.
Ag^+	Benzene–ethyl acetate (4:1)	Improved procedure for thin-layer chromatography of terpenoids on silver ion-silica gel layer. Detection by spraying with methanol–conc. sulfuric acid (1:1) and heating at 110°C for 5 min	[23]
Ag^+	Chloroform–methanol	Argentation TLC was used for the analysis of molecular species of egg lecithin	[24]
Ag^+	Chloroform–methanol–water (70:30:4, and 90:10:1)	TLC of lysophospholipids on 5% silver nitrate-impregnated silica layer developed to 3 cm above the origin with chloroform–methanol–water (70:30:4) followed by drying and redevelopment to 16 cm from the origin using chloroform–methanol–water (90:10:l)	[25]
Ag^+	Chloroform–methanol (95:5 or 98.75:1.25)	Separation of phospholipid molecular species using argentation TLC. Detection by autoradiography	[26]
Ag^+	Chloroform–methanol (99:1 or 95:1)	HPTLC for the separation of monoenoic and dienoic fatty acids and methyl esters and corresponding fatty acid dimethylacetals on silica gel impregnated with 10% $AgNO_3$. Visualization by spraying with 2′,7′-dichlorofluorescein	[27]
Ag^+	Petrol ether – acetone (20:9)	TLC of triglycerides on silica impregnated with 0.5% methanolic silver nitrate solution. After development, dried plates were treated consecutively with bromine and sulfuryl chloride vapors, followed by charring at 180–200°C	[28]
Ag^+	Hexane–ether (65:35)	TLC of isomers of methyl oleate hydroperoxides on plain silica or on silver nitrate-impregnated silica. Detection under UV 254 nm after spraying with a 0.2% ethanolic solution of 2′,7′-dichlorofluorescein sodium salt. Analysis by HPLC and determination by GC–MS	[29]
Ag^+	Petrol ether–acetone (40:3)	TLC of positional isomers of monoenoic fatty acids as their phenylacyl esters on silica impregnated by methanolic silver nitrate (0.5% or 10%). Quantification by densitometry at 450 nm after treatment with sulfuryl chloride vapors for 30 min and, finally, heating at 180–200°C for 30 min	[30]
Ag^+	1: Hexane, hexane–ethyl acetate (3:1 or 2:1) 2: Hexane–diethyl ether (10:1 or 5:1), 3: Hexane–toluene (10:1)	Very convenient and practical procedures for the preparation of TLC plates using silica impregnated with silver nitrate. The application of resultant adsorbent to the separation of steroids and triterpenes	[31]

(continued)

Table 16.1 (continued)

Impregnating cations/salts	Mobile phase	Remarks	Ref.
Ag^+	Petroleum ether–ether–acetic acid (80:30:1)	Argentation TLC of non-methylene-interrupted fatty acids on silica gel impregnated with 5% silver nitrate in acetonitrile. Detection with 0.01% primuline in acetone–water (4:1) under UV radiation	[32]
Ag^+	Chloroform	TLC of triglycerides on silica gel incubated overnight with 20% silver nitrate solution. Visualization of spots by spraying with a 0.15% ethanolic solution of 2′,7′-dichlorofluorescein and observation under UV. Quantitation after elution by GC	[33]
Ag^+	Binary mixture of light petroleum ether and acetone in different proportions and volumes	Examination of the efficiency of silver ion thin-layer chromatography in the analysis of triacylglycerols. Examples are presented to demonstrate the elucidation of triacylglycerol structures of natural and modified lipid samples	[34]
Ag^+	Toluene–acetonitrile (97:3), petroleum ether–ether (1:1), and benzene–petroleum ether (1:1)	Separation of fatty acid methyl esters by argentation thin-layer chromatography. Plates were either impregnated by manual immersion, or homemade plates were prepared from a suspension of silica gel in 10% aq. or ammonical $AgNO_3$ solution	[35]
Ag^+	Petroleum–acetone or chloroform (25:1, 50:1, 50:3, and 40:1)	TLC of fatty acids, butter, oleic type margarines, and linoleic type margarines on silica gel plates impregnated by dipping into a 0.5% or 1% methanolic solution of silver nitrate. Quantitation by densitometry at 450 nm in reflectance mode	[36]
Ag^+	Petroleum ether–acetone (25:1)	Quantitative silver ion TLC for determination of triacylglycerol composition of sesame seeds. Detection by treatment with bromine and sulfuryl chloride vapors and charring at 180–200°C. Quantitation by densitometry at 450 nm	[37]
M-nitrates [M = Cu^{2+}, Ni^{2+}, Mn^{2+}, Cd^{2+}, Zn^{2+}, Ag^+]	Methanol, butanol, isopropanol, isobutanol, $CHCl_3$–C_6H_6–MeOH–AcOH (50:35:10:5), and $CHCl_3$–CCl_4–MeOH (80:50:10) systems	Use of thin layers of silica gel G impregnated with transition metal ions for separation, identification, and estimation of purines. Examination of the influence of transition metal ions and eluting solvents on chromatographic behavior (hR_F) and the application for qualitative and quantitative analysis of purine bases in the mixture as well as in pharmaceutical formulations. The lowering in hR_F on metal ion-impregnated silica gel G layers due to the complex formation between electron acceptor (metal ions) and electron donor (purines) improves the separation efficiency. On the basis of compactness, spherical shape, and easy detection of spots, Cu(II)-impregnated thin layers were selected for detail studied	[38]

Ag^{+}	Toluene, benzene, toluene–hexane (1:1), hexane–diethyl ether (9:1), and hexane–chloroform	TLC of trans- and cis-18:1 isomers of fatty acids on silica gel impregnated with silver nitrate (10% solution of silver nitrate in acetonitrile). The plates were normally developed at room temperature in a dark place. The resolution of some positional 18:1 and 18:2 isomers may be improved by development at about −20°C or −25°C. Detection by spraying with 0.05% rhodamine B in ethanol. Quantification by GC after extraction	[39]
Ag^{+}	Hexane–diethyl ether–acetic acid (85:15:1)	Identification of conjugated linoleic acids and trans-18:1 isomers in dairy fats by using a combination of gas chromatography and silver ion thin-layer chromatography or silver ion liquid chromatography. Visualization after spraying with methanol–sulfuric acid (2:1)	[40]
Ag^{+}	25 Mobile phases	HPTLC and TLC of sterols (cholesterol, cholestanol, beta-sitosterol, stigmasterol, ergosterol, campesterol, desmosterol, and brassicasterol) on silica gel impregnated with 10% silver nitrate. Detection by spraying with ethanolic phosphomolybdic acid and heating at 115°C for 10 min	[41]
Ag^{+}	Hexane–diethyl ether (90:10)	Pre-separation of cis- and trans-18:1 isomers by Ag-TLC. Comparison of different GLC methods suitable to measure the total trans-18:1 isomers, vaccenic acid, and trans-18:1 acid isomeric distribution in milk fat	[42]
Ag^{+}	Borate–phosphate buffer (pH 2.3)	Separation of coexisting tryptophan, alanine, and phenylalanine or tyrosine on silica gel 60 F254 high-performance thin-layer chromatography (HPTLC) plates impregnated with silver nitrate	[43]
Cu^{2+}	Ethyl acetate–benzene (4:2)	TLC separation and quantitative identification of some closely related sulfa drugs on silica layers, impregnated with copper sulfate. Detection of drugs on plain silica plates by spraying with Dragendorff reagent and on impregnated plates spots were self visualized	[44]
Cu^{2+}	Dioxane–water (9:1)	Separation of D- and L-lactic acid enantiomers on silica gel-precoated plates impregnated with Cu^{2+}. Confirmation of method is useful for rapid control of the radiochemical stability of both D- and L-forms of lactic acid	[45]
Cu^{2+}	Mixtures of two solvents consisting of water, propanol-2, ethyl-methyl ketone, acetonitrile, and acetone	TLC separation of carbohydrates on silica gel modified with copper(II) salts. Carbohydrates in water form weak complexes with divalent or trivalent metal ions. The retention mechanism under proposed analytical conditions is quite complex due to the combination of partition, extraction, and ligand exchange processes	[46]

(continued)

Table 16.1 (continued)

Impregnating cations/salts	Mobile phase	Remarks	Ref.
Cu^{2+}	Different mixtures of acetonitrile, methanol, and water	TLC of the enantiomers of the beta-blockers (+/–)-propranolol, (+/–)-metoprolol, and (+/–)-atenolol on silica gel impregnated with a Cu(II)-L-arginine complex in a glass chamber saturated for 20–25 min. The Cu(II)-L-arginine complex was prepared by mixing 1 mM copper(II) acetate and 2 mM L-arginine in water–methanol 9:1 and adjusting the final pH to 6–7 with aqueous ammonia. Detection with iodine vapor. Successful separation of all three racemic drugs was achieved with acetonitrile–methanol–water (15:2:2 and 15:2:1)	[47]
Cu^{2+}	Mixture of acetone and n-hexane in different volume compositions	Comparison of the separation of nicotinic acid and its derivatives on a layer prepared from silica gel 60 plus kieselguhr F254 mixture (nonimpregnated and impregnated with an aqueous solution of $CuSO_4$) using adsorption TLC. Impregnation of the mixture of silica gel and kieselguhr with 2.5% and 5% aqueous solutions of $CuSO_4$ influences separations of nicotinic acid and its derivatives, causing a reduction in their R_F values	[48]
Cu^{2+}	Different binary or ternary mobile phases such as acetone–methanol (1:9), methanol–acetonitrile–dichloromethane (3:1:2)	Resolution of enantiomers of three β-blockers by complexation chiral TLC. The enantiomers of atenolol were best resolved on plates impregnated with Cu(II)–L-proline complex whereas those of propranolol and salbutamol were best resolved on plates impregnated with Cu(II)–*N,N*-dimethyl-L-phenylalanine-L- Phe complex. The used mobile phases enabled successful resolution of the enantiomers of the three racemates on plates impregnated with the Cu(II) complexes of L-proline, L-phenylalanine, or L-histidine	[49]
Cu^{2+}	Binary and ternary mixtures of acetone, methanol, acetonitrile, dichloromethane, and water	Atenolol and propranolol (the ß-blocking agents) and salbutamol (broncho- and vasodilator) were resolved into their enantiomers by adopting different modes of loading/impregnating the Cu(II) complexes of L-proline (L-Pro), L-phenylalanine (L-Phe), L-histidine (L-His), N,N-dimethyl-L-phenylalanine (N,N-Me_2-L-Phe), and L-tryptophan (L-Trp) on commercial precoated normal phase plates. Spots were located using iodine vapor	[50]
Cu^{2+} and Ni^{2+}	Heptane–chloroform (7:3) and heptane–ether (4:1)	Conditions for modification of silica gel and amino- as well as cyanopropyl-bonded silica with metal salts were optimized. The effect of impregnation conditions such as the type of salt and the solvent used was examined. The importance of the modification conditions, especially the role of the impregnation solvent used, was examined by analyzing retention data obtained for unsaturated fatty acid methyl esters in thin-layer chromatography	[51]

Al^{3+} and Cu^{2+}	Chloroform, chloroform–acetone (65:35), chloroform–ethyl acetate (95:5 or 10), toluene–chloroform–acetone (40:25:15 or 40:25:25), hexane–acetone (80:20 or 80:40), benzene, and benzene–acetone (70:30)	Chromatographic behavior of substituted phenolic compounds on silica gel layers impregnated with Al^{3+} and Cu^{2+}. A very good separation of hydroquinone and resorcinol on silica layers impregnated with $CuSO_4$ was achieved. Amino phenols can be successfully separated using silica layers modified by $CuSO_4$. The optimum resolution of *meta* and *para* isomers was achieved on silica gel containing 0.51% Cu^{2+} ions	[52]
Cu^{2+}, Co^{2+}, and Fe^{3+}	Heptane–dioxane–acetic acid (60:40:1)	Identification of chlorophenoxy herbicides on silica gel plates, and silica gel prewashed and impregnated with Cu, Co, and Fe salts. The best results were with silica gel impregnated with 0.2% $CuSO_4$. In addition to efficient separation, such impregnation makes possible visualization without further use of reagents or UV lamps	[53]
Cu^{2+}, Ni^{2+}, Fe^{2+}, and Mn^{2+}	Mixtures of n-hexane–ethyl acetate–acetic acid in the volume compositions: 22:20:5, 25:20:2, 25:20:5, and 25:20:8	Impregnation of silica gel 60 F254 glass plates with aqueous solutions of $CuSO_4$, $MnSO_4$, $NiSO_4$, and $FeSO_4$ improved the separation of GC/GDC and C/GLC, which separated poorly on glass plates precoated with nonimpregnated silica gel 60 F254 at 18°C	[54]
Cu^{2+}, Ni^{2+}, Fe^{2+}, and Mn^{2+}	The mixtures of n-hexane–ethyl acetate–acetic acid in the volume compositions: 22:20:5 and 25:20:2 (v/v/v) for both aluminum plates, 22:22:5 (v/v/v) for silica gel 60 F254 (#1.05554) plates, and 25:20:5 (v/v/v) for silica gel 60 (#1.05553) plates	Adsorption TLC was used to separate seven bile acids: cholic acid (C), glycocholic acid (GC), glycolithocholic acid (GLC), deoxycholic acid (DC), chenodeoxycholic acid (CDC), glycodeoxycholic acid (GDC), and lithocholic acid (LC). Impregnation of silica gel 60 (#1.05553) and silica gel 60 F254 (#1.05554) with aqueous solution of $CuSO_4$, $MnSO_4$, $NiSO_4$, and $FeSO_4$ improved the separation of bile acids, which separated poorly on aluminum plates precoated with nonimpregnated silica gel 60 and 60 F254 The mobile phase n-hexane–ethyl acetate–acetic acid in volume composition 25:20:2 (v/v/v) allowed separating all pairs of bile acids on TLC plates (#1.05553) impregnated with 5% $CuSO_4$	[55]
Fe^{2+}, Cu^{2+}, Mn^{2+}, Zn^{2+}, or Ni^{2+}	Butanol–acetic acid–H_2O (8:2:1 or 4:1:1) or benzene–butanol–acetic acid–H_2O (7:8:5:2)	Separation of antihistamines on silica gel plates impregnated with transition metal cations	[56]

(continued)

Table 16.1 (continued)

Impregnating cations/salts	Mobile phase	Remarks	Ref.
Mn^{2+}, Fe^{2+}, Co^{2+}, Ni^{2+}, and Cu^{2+}	Propanol–H_2O–butanol and butanol–methanol–H_2O in different volume ratio	Cephalosporin antibiotics were separated on thin-layer plates impregnated with transition metal ions at different concentrations. Impregnation reduces the tailing of analytes and improved the resolution. The best conditions of separation have been identified. The hR_F values were affected by the concentration of impregnating reagent in all the solvent systems. The spots were more compact on impregnated layers than on plain silica gel layers	[57]
Cu^{2+}, Co^{2+}, Ni^{2+}, Mn^{2+}, Al^{3+}, Cr^{3+}, and Fe^{3+}	CCl_4, CCl_4–$CHCl_3$ (8:2), C_6H_{14}–$CHCl_3$ or CH_2Cl_2 (8:2), C_6H_{14}–$CHCl_3$ or CCl_4 (9:1), CCl_4–C_6H_{14}–$CHCl_3$ (7:2:1), and CCl_4–C_6H_{14}–$CHCl_3$ or CH_2Cl_2 (7:1:2)	Experiments were performed on silica impregnated with metal cations such as Cu(II), Co(II), Ni(II), Mn(II), Al(III), Cr(III), and Fe(III) – and on unmodified silica as reference material. Comparison of R_F values of monosulfides chromatographed on unmodified and impregnated silica reveals that impregnation of the silica with metal cations enhances the retention performance of the chromatographic systems	[58]
Mn^{2+}, Fe^{2+}, Co^{2+}, Ni^{2+}, Cu^{2+}, Cd^{2+}, Zn^{2+}, or Mg^{2+}	*n*-PrOH–*n*-BuOH–H_2O–NH_3 (7:5:1:2, 7:5:1:1.5 or 7:5:0.75:2)	Separation of vitamin B complex and folic acid on silica gel layers impregnated with some transition metal ions. The spots of vitamins were located by exposing the TLC plates to iodine vapors	[59]
Mn^{2+}, Fe^{2+}, Co^{2+}, Ni^{2+}, Cu^{2+}, Zn^{2+}, and Hg^{2+}	(a) $CHCl_3$–*n*-BuOH–AcOH–NH_3 (4:7:5:1) (b) $CHCl_3$–*n*-BuOH–H_2O–AcOH–NH_3 (3:5:0.5:5:0.5) (c) C_6H_6–BuOAc–*n*-PrOH–AcOH–NH_3 (1:4:1:5:1) (d) CCl_4–BuOAc–propionic acid–NH_3 (3:7:9:3) (e) CCl_4–BuOAc–MeOH–NH_3 (1.5:4.5:7:0.5) (f) CCl_4–BuOAc–propionic acid–MeOH–H_2O (2:3:1:0.5:3)	By using any of these solvent systems, vitamins which were not resolved on the untreated plates were resolved on impregnated layers with most of the impregnating cations. The spots were more compact on impregnated layers than on plain silica gel layer. Constituents of vitamin B complex and folic acid were separated and identified in pharmaceutical and multivitamin preparations	[60]

Cu^{2+}, Co^{2+}, Ni^{2+}, and Mn^{2+}	Dioxane–water mixtures	Separation of the enantiomers of D- and L-lactic acid with transition metal cations used to impregnate the silica gel. For purposes of comparison, D- and L-lactic acid on nonimpregnated silica gel layers were also chromatographed	[61]
Na^{+}, K^{+}, Mg^{2+}, Cu^{2+}, Zn^{2+}, Mn^{2+}, Co^{2+}, and Ni^{2+}	Mixture of *n*-PrOH–$CHCl_3$–H_2O (14:8:2)	Thin-layer chromatography of neutral sugars contained in water-soluble binders (gum and honey) and separation of sugars on silica gel plates impregnated with sulfates, chlorides, and nitrates of various metals. The influence of metal cations can be arranged in order of increasing effect on the retention factor of sugar as follows: $K^{+} < Na^{+} < Ni^{2+} < Co^{2+} < Mg^{2+} < Cu^{2+} < Zn^{2+} < Mn^{2+}$	[62, 63]
Li^{+}, Na^{+}, K^{+}, Rb^{+}, Cs^{+}, Ca^{2+}, Sr^{2+}, and Ba^{2+}	CH_2Cl_2, CH_2Cl_2–$(C_2H_5)_2O$ (98:2 or 95:5), and pentane–$(C_2H_5)_2O$ (6:4)	Chromatographic separation of diastereoisomeric 1,2:3,4-diepoxides on silica gel TLC plates impregnated with various inorganic salts. LiBr proved to give the best results with considerably improved separations	[64]
Li^{+}, Cu^{2+}, Ni^{2+}, Zn^{2+}, Cd^{2+}, Hg^{2+}, and Th^{4+}	Water-in-oil microemulsion	Improved separation of amino acids, but with reduced sensitivity, was realized on silica gel layers impregnated with metal ions such as Li^{+}, Cu^{2+}, Ni^{2+}, Zn^{2+}, Cd^{2+}, Hg^{2+}, and Th^{4+}. Silica gel impregnated with a 1:1 mixture of 2% aqueous $CuSO_4$ and 3% Brij-35 was found to be the best layer material for rapid separation of amino acids using a water-in-oil microemulsion as mobile phase. The experimental conditions have been optimized, and TLC conditions for separation of DL-phenylalanine in the presence of foreign impurities were determined	[65]
Cu^{2+}, Ni^{2+}, Zn^{2+}, or Cd^{2+}	*n*-PrOH–H_2O (8:4, v/v) and i-PrOH–H_2O (8:4, v/v)	TLC separation of glucose, maltose, lactose, sorbitol, and sucrose on silica gel plates impregnated with transition metal ions. The identification is very distinct with $KMnO_4$ (0.5%) in 0.1 M NaOH as spray reagent	[66]
Ni^{2+}, Zn^{2+}, or Cd^{2+}	Butanol–formic acid (1:1, v/v)	TLC separation of five alkaloids (atropine, berberine, brucine, ephedrine, quinine) on nickel chloride, zinc sulfate, or cadmium sulfate impregnated silica gel TLC plates	[67]
Ni^{2+}	Butanol–acetic acid–water or chloroform or ethyl acetate (3:1:1)	TLC separation of 23 amino acids on plain silica and nickel chloride-impregnated silica layers. Detection by spraying with freshly prepared ninhydrin solution. Slight improvement of separation by impregnation	[68]
Zn^{2+}, Mg^{2+}, Mn^{2+}, Fe^{2+}, and Co^{2+}	Three solvent systems	Thin-layer chromatography of 15-component mixtures of phenyl thiohydantoin derivatives of amino acids on silica gel plates impregnated with various Zn salts (chloride, sulfate, phosphate, and acetate) and sulfates of Mg, Mn, Fe, and Co. The effect of various cations on the chromatographic behavior is discussed	[69]

(continued)

Table 16.1 (continued)

Impregnating cations/salts	Mobile phase	Remarks	Ref.
Zn^{2+}	Binary nonaqueous mixtures of 2-propanol or ethyl acetate with heptanes	Effect of impregnation of silica gel with different zinc salts (nitrate, bromide, chloride, acetate, and sulfate) on the TLC behavior of 13 phenols, 15 quinolines, and 9 anilines. Development after pre-saturation in horizontal DS chambers. Detection under UV light at 254 nm. The impregnation with zinc acetate leads to improvement of spot shape and separation selectivity, especially for quinoline derivatives	[70]
Cd^{2+}, Sr^{2+}, Eu^{3+}, and V^{4+}	CCl_4, C_6H_{14}–$CHCl_3$ (9:1, v/v), C_6H_{14}–CCl_4 (9:1, v/v), C_6H_{14}–$C_6H_5CH_3$ (9.5:0.5, v/v), and CCl_4–C_6H_{14}–$CHCl_3$ (7:2:1, v/v)	Experiments were conducted on silica impregnated with metal cations having 2^+, 3^+, and 4^+ oxidation states and also on the plain silica as reference material. Comparison of data obtained for the monosulfides chromatographed on plain and impregnated silica reveals that impregnation of the silica with metal cations results in greater differentiation of R_F values	[71]
Different metal salts	Benzene–dimethyl formamide–AcOH (30:10:7)	Different metal salts have been tried as impregnating reagents for developing TLC separation schemes for some antihistamines on silica gel "G" plates	[72]
Sulfates, oxides, acetates, thiocyanates, chlorides, carbonates, nitrates of transition metals	Butanol–water–acetic acid (4:2:2)	The systems reported were considered as improvements with regard to the number of amino acids resolved from their complex mixtures	[73]

Table 16.2 Performance of cation-exchanged silica gel in the analysis of inorganic substances

Impregnating cations/salts	Mobile phase	Remarks	Ref.
0.3 M Sodium molybdate	Mixtures of aqueous solution of formic acid and sodium formate	Impregnation of silica gel with aqueous sodium molybdate solution provides a new adsorbent phase which is sufficiently stable in the formic acid–sodium formate medium. Cu^{2+} has been selectively separated on impregnated silica gel G thin layers. The semiquantitative determination of nine metal ions on impregnated silica gel G layers has been attempted	[74]
NaCl, KBr, KI, and NH_4Cl	Solutions of formic acid and aqueous salt solutions in various proportions	Study on the chromatographic behavior of 14 heavy metals on thin layers of plain silica gel and silica gel impregnated with aqueous salt solutions of NaCl, KBr, KI, and NH_4Cl (0.1 and 1.0 M). More compact spots of cations with clearer detection of cations were observed on KI impregnated silica gel layers. In addition to a series of effective qualitative separations, quantitative separations of microgram to milligram quantities of Ti^{4+}, Fe^{3+}, and Al^{3+} from each other were achieved	[75]
Th^{4+}	1.0 M Sodium formate	Separation of Zn^{2+} from Cd^{2+}, Hg^{2+}, Ni^{2+}, or Ti^{2+} in environmental samples using thorium nitrate-impregnated silica layers. Zn^{2+} is successfully recovered from soil, seawater, and river water samples. Zinc is separated from nickel and quantitatively determined in industrial wastewater samples produced during the electroplating process. Detection limits for these toxic metals were reported	[76, 77]
K^+, Na^+, Sr^{2+}, Cd^{2+}, Ag^+, uranyl nitrate, or La^{3+}	1 M Aqueous formic acid (pH 1.8), 1 M-aq. formic acid plus 1 M-aq. sodium formate of pH 3.3, and 1 M-aq. sodium formate (pH 7.65)	Chromatographic analysis of some inorganic pollutants on cation-modified silica gel layers. Silica gel G and aq. 0.1% or 1% solution of KCl, NaCl, $SrCl_2$, $CdCl_2$, $AgNO_3$, uranyl nitrate, or La_2O_3 were mixed in the ratio 1–3 for 10 min, and the slurry was coated onto glass plates to give 0.25-mm layers. Ternary separations of metals are achieved. $AgNO_3$, $CdCl_2$, and La_2O_3 impregnated silica proved unsatisfactory for the separation of the metals	[78]

(continued)

Table 16.2 (continued)

Impregnating cations/salts	Mobile phase	Remarks	Ref.
Li^+	Aqueous formic acid (1 M), aqueous sodium formate (IM), and their mixtures	Separation and microgram detection of metal ions on lithium chloride-impregnated silica gel. The effects of mobile phase pH and impregnation level are discussed	[79]
Na^+	Aqueous sodium perchlorate solution	Investigation of TLC behavior of the alkaline earth metal ions (Mg^{2+}, Ca^{2+}, Sr^{2+}, and Ba^{2+}) on silica gel thin layers impregnated with sodium hydroxide using aqueous sodium perchlorate solutions as mobile phase. The best mutual separation of Ca^{2+}, Sr^{2+}, and Ba^{2+} was at salt concentrations above 6 mol L^{-1}	[80]
Na^+	Water or aqueous solution of surfactants	The chromatographic system comprising silica gel impregnated with NaOH (0.05 mM) as stationary phase and water or micellar solution of sodium dodecyl sulfate (0.01 M) has been identified as most favorable for the identification and separation of coexisting Cu^{2+}, Mn^{2+}, and Ni^{2+} ions. The limit of detection and semiquantitative determination of separated metal cations were also determined	[81]

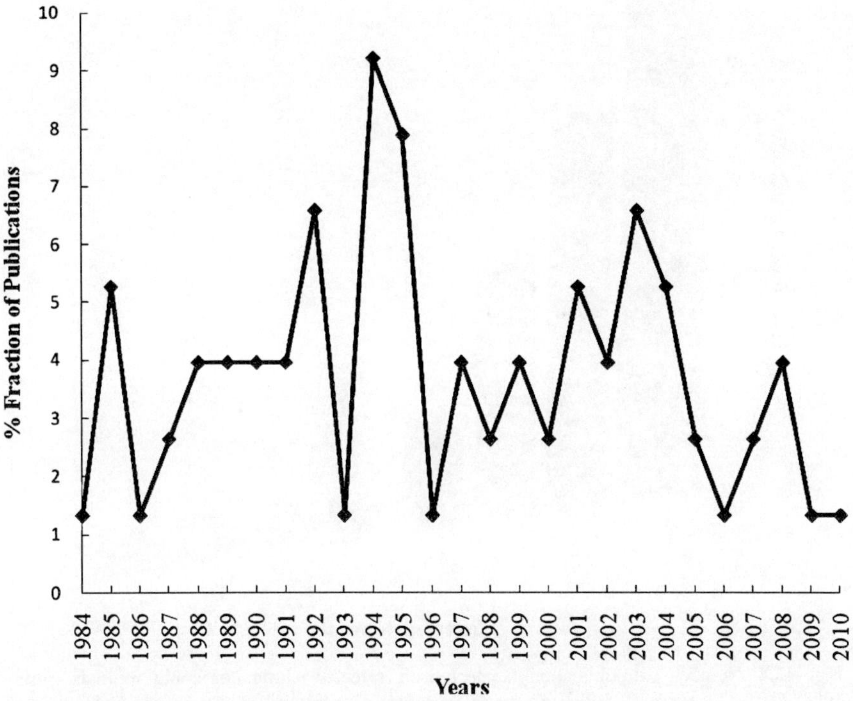

Fig. 16.3 Relative fraction of publications (%) per year appeared on cation-exchanged silica gel layers during the period 1984–2010

The fraction of publications (%) per year appeared in literature during 1984–2010 on cation-exchanged silica gel-based TLC of organic and inorganic species is shown in Fig. 16.3. It is apparent that there is no definite pattern about the magnitude of publications. However, significant publication appeared during 1994–1995.

The number of publications appeared in important chromatographic journals during 1984–2010 are shown in Fig. 16.4. It is clear that the investigators have preferred to publish their papers mainly in three journals dealing with chromatographic studies, e.g., Journal of Planar Chromatography, Journal of Chromatography, and Journal of Liquid Chromatography.

According to literature, the metal ions selected for the impregnation of silica gel follow the following order: Ag > Cu > Ni > Mn > Fe ≈ Zn > Co. It is because of the strong complexing nature of copper and the tendency of Ag^+ to form π – complexes with non-saturated ligands.

From Table 16.3, it is evident that cation-exchanged silica gel has been mostly used for the analysis of organic compounds in comparison to inorganic species.

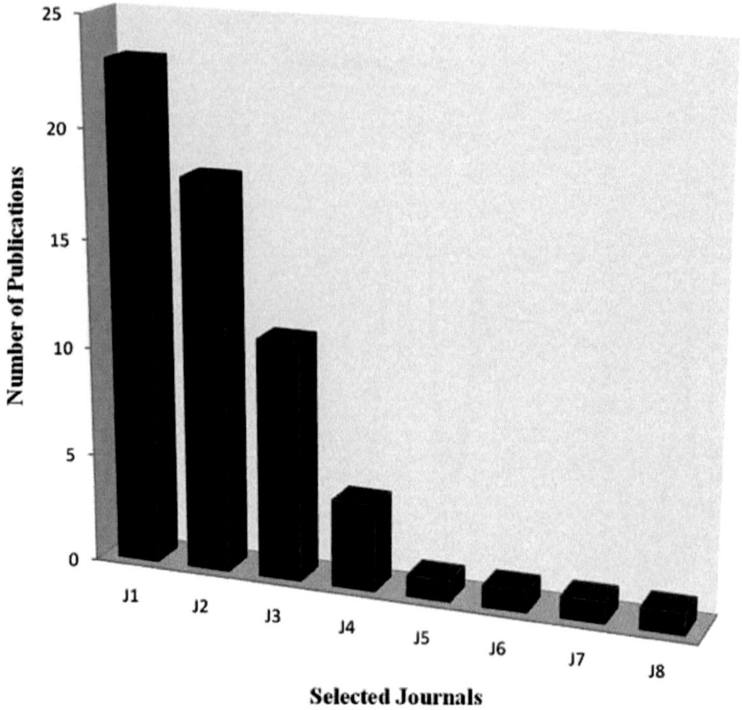

Fig. 16.4 Number of publications appeared in selected chromatographic journals during 1984–2010. J1 = Journal of Planar Chromatography. J2 = Journal of Chromatography A or B. J3 = Journal of Liquid Chromatography & Related Technology. J4 = Biomedical Chromatography. J5 = Chromatographia. J6 = Acta Chromatographica. J7 = Journal of Chromatographic Science. J8 = Separation Science & Technology

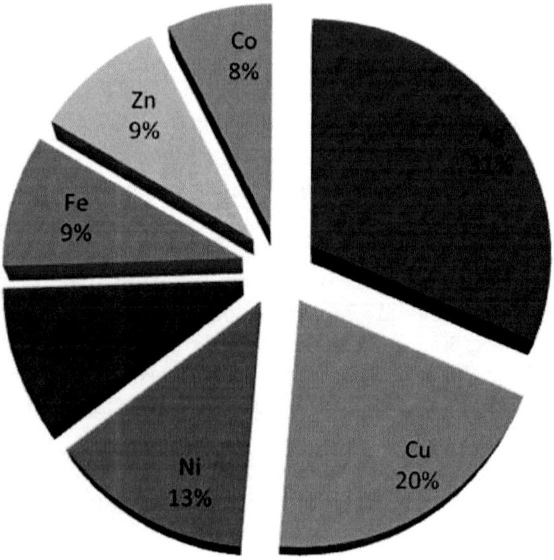

Fig. 16.5 Percentage of publications on selected metal cations (Ag, Cu, Ni, Mn, Zn, Fe, and Co) used for impregnation of silica gel layers during 1984–2010

Table 16.3 Important metal ions used as impregnate of silica gel for the analysis of organic and inorganic substances

Important impregnating cations	Compounds analyses
Ag	Amino acids, cis/trans isomers of capsaicin, fatty acid methyl esters, fatty aldehydes, isomers of methyl oleate hydroperoxide, lipid classes, metal cations, monoacylglycerols, purines, steroids, sterols, terpenoids, triacylglycerols, triglycerides, octadecenoic acids
Cu	Amino acids, antibiotics, bile acids, carbohydrates, D- and L-lactic acid, drugs, enantiomers of D- and L-lactic acid, fatty acid methyl esters, monosulfides, nicotinic acids and its derivatives, pesticides, phenolic compounds, purines, vitamins, sugars
Ni	Alkaloids, amino acids, antibiotics, bile acids, enantiomers of D- and L-lactic acid, monosulfides, vitamins, sugars
Mn	Amino acids, antibiotics, bile acids, enantiomers of D- and L-lactic acid, monosulfides, vitamins, sugars
Fe	Alkaloids, amino acids, antibiotics, bile acids, monosulfides, pesticides, vitamins
Zn	Amino acids, antibiotics, aromatic hydrocarbons, sugars, vitamins
Co	Amino acids, antibiotics, D- and L-lactic acid, monosulfides pesticides, sugars, vitamins

16.4 Conclusion

Modification of silica gel with metal salts produces new selective stationary phases. By changing the kind of metal cations used, these phases have a great chance to become a more popular adsorbent in thin-layer chromatography. The choice of metal cations and its concentration in the adsorbent layer decide the retention mechanisms of analytes. The use of cation-modified silica gel layers in the analysis of inorganic substances seems to be an alternative area of future research in addition to their use in the analysis of pharmaceutical products.

Acknowledgment One of the authors (Abdul Moheman) is thankful to the Council of Scientific and Industrial Research (CSIR) of New Delhi, India, for providing financial assistance. They thank the Deanship of Scientific Research at King Saud University for funding the work through the research group project No. RGP-VPP-130.

References

1. Kirchner JG, Miller JM, Keller GJ (1951) Separation and identification of some terpenes by a new chromatographic technique. Anal Chem 23:420–425
2. Stahl E (1956) Thin-layer chromatography. Pharmazie 11:633–637
3. Barrett CB, Dallas MSJ, Padley FB (1962) Chem Ind (London) 1050–1052
4. Stuhlfauth T, Fock H, Huber H, Klug K (1985) The distribution of fatty acids including petroselinic and tariric acids in the fruit and seed oils of the Pittosporaceae, Araliaceae, Umbelliferae, Simarubaceae and Rutaceae. Biochem Syst Ecol 13:447–453

5. Sventashev VI, Zhukova NV (1985) Analysis of labelled fatty acid methyl esters by argentation and reversed-phase two-dimensional thin layer chromatography. J Chromatogr 330:396–399
6. Xu S, Norton RA, Crumley FG, Nes WD (1988) Comparison of the chromatographic properties of sterols, select additional steroids and triterpenoids: gravity-flow column liquid chromatography, thin layer chromatography, gas liquid chromatography and high performance liquid chromatography. J Chromatogr 452:377–398
7. Michalec C (1990) Argentation TLC and HPTLC of cholesterol and related stanols and stanones. J Planar Chromatogr 3:273–275
8. Mares P, Rezanka T, Novak M (1991) Analysis of human blood plasma triacylglycerols using capillary gas chromatography. Silver ion thin layer chromatographic fractionation and desorption chemical ionization mass spectrometry. J Chromatogr 568:1–10
9. Chobanov D, Tarandjiiska R, Nikolova-Damyanove B (1992) Quantitation of isomeric unsaturated fatty acids by argentation TLC. J Planar Chromatogr 5:157–163
10. Masterson Ch, Fried B, Sherma J (1992) Comparison of mobile phase and detection reagents for the separation of triacylglycerols by silica gel, argentation, and reversed phase thin layer chromatography. J Liq Chromatogr Rel Technol 15:2967–2980
11. Wilson R, Sargent JR (1992) High resolution separation of polyunsaturated fatty acids by argentation thin layer chromatography. J Chromatogr 623:403–407
12. Sempore BG, Bezard JA (1994) Separation of monoacylglycerol enantiomers as urethane derivatives by chiral-phase high performance liquid chromatography. J Liq Chromatogr Rel Technol 17:1679–1694
13. Martinez-lorenzo MJ, Marzo I, Naval J, Pineiro A (1994) Self-staining of polyunsaturated fatty acids in argentation thin layer chromatography. Anal Biochem 220:210–212
14. Wardas W, Pyka A (1994) Visualizing agents for cholesterol in TLC. J Planar Chromatogr 7:440–443
15. Nguyen H, Tarndjiiska R (1995) Lipid classes, fatty acids and triglycerides in papaya seed oil. Fat Sci Technol 97:20–23
16. Molkentin J, Precht D (1995) Optimized analysis of trans-octadecenoic acids in edible fats. Chromatrographia 41:267–272
17. Dobson G, Christie WW, Nikolova-Damyanova B (1995) Silver ion chromatography of lipids and fatty acids. J Chromatogr B 671:197–222
18. Wall PE (1997) Argentation HPTLC as an effective separation technique for the cis/trans isomers of capsaicin. J Planar Chromatogr 10:4–9
19. Wardas W, Pyka A (1998) Visualizing agents for cholesterol derivatives in TLC. J Planar Chromatogr 11:70–73
20. Wilson R, Sargent JR (2000) Chain separation of monounsaturated fatty acid methyl esters by argentation TLC. J Chromatogr A 905:251–257
21. Nikolova-Damyanova B, Momchilova S (2001) Silver ion thin layer chromatography of fatty acids. A survey. J Liq Chromatogr Rel Technol 24:1447–1466
22. Chrisrie WW (1982) Lipid analysis. Pergamon, Oxford
23. Kohli JC, Badaisha KK (1985) Improved procedure for the thin layer chromatography of terpenoids on silver ion-silica gel layer. J Chromatogr 320:455–456
24. Kennerly DA (1986) Improved analysis of species of phospholipids using argentation thin layer chromatography. J Chromatogr 363:462–467
25. Kennerly DA (1987) Molecular species analysis lysophospholipids using high performance liquid chromatography and argentation thin layer chromatography. J Chromatogr 409:291–297
26. Kennerly DA (1988) Two dimensional thin layer chromatographic separation of phospholipid molecular species using plates with both reversed-phase and argentation zones. J Chromatogr 454:425–431
27. Schick PK, Levy C (1990) Reversed phase and argentation HPTLC for the separation of fatty acids and fatty aldehyde. J Planar Chromatogr 3:269–271
28. Nikolova-Damyanove B, Amidzhin B (1991) Densitometric quantitation of triglycerides. J Planar Chromatogr 4:397–401

29. Bortolomeazzi R, Pizzale L, Lercker G (1992) Chromatographic determination of the position and configuration of isomers of methyl oleate hydroperoxides. J Chromatogr 626:109–116
30. Nikolova-Damyanova B, Christie WW, Herslolf B (1994) Improved separation of some positional isomers of monounsaturated fatty acids, as their phenacyl derivatives, by silver ion thin-layer chromatography. J Planar Chromatogr 7:382–385
31. Li T-S, Li Ji-Tai, Li H-Z (1995) Modified and convenient preparation of silica impregnated with silver nitrate and its application to the separation of steroids and triterpenes. J Chromatogr A 715:372–375
32. Tauaka T, Shibata K, Hino H, Murashita T, Kayama M, Satouche K (1997) Purification and gas chromatographic-mass spectrometric characterization of non-methylene interrupted fatty acid incorporated in rat liver. J Chromatogr B 700:1–8
33. Fraga MJ, Fontecha J, Lozada L, Juarez M (1998) Silver ion adsorption thin layer chromatography and capillary gas chromatography in the study of the composition of milk fat triglycerides. J Agric Food Chem 46:1836–1843
34. Nikolova-Damyanova B (1999) Quantitative thin layer chromatography of triacylglycerols. Principles and application B. J Liq Chromatogr Rel Technol 22:1513–1537
35. Flieger J, Szumilo H (2000) Optimizing chromatographic conditions for separation of fatty acid methyl esters by argentation thin layer chromatography. J Planar Chromatogr 13:426–431
36. Marevkov I, Tarandjiiska R, Panayotova S, Nikolova N (2001) Valuation of argentation TLC densitometry for the determination of trans fatty acids in butterfat. J Planar Chromatogr 14:384–390
37. Nikolova-Damyanova B, Velikova R, Kuleva L (2002) Quantitative TLC for determination of triacylglycerol composition of sesame seeds. J Liq Chromatogr Rel Technol 25:1623–1632
38. Singh DK, Srivastava B, Sahu A (2003) Thin layer chromatography of purines on silica gel G impregnated with transition metal ions; assay of caffeine and theophylline in pharmaceutical formulations. J Chinese Chem Soc 50:1031–1036
39. Nimal Ratnayake WM (2004) Overview of methods for the determination of trans fatty acids by gas chromatography, silver-ion thin-layer chromatography, and gas chromatography/mass spectrometry. J Assoc Off Anal Chem 87:523–539
40. Cruz-hernandez C, Deng Z, Zhou J, Hill AR, Yurawecz MP, Delmonte P, Mossoba MM, Dugan MER, Kramer JKG (2004) Methods for the analysis of conjugated linoleic acids and trans-18:1 isomers in dairy fats by using a combination of gas chromatography, silver-ion thin-layer chromatography/gas chromatography, and silver ion liquid chromatography. J Assoc Off Anal Chem 87:545–562
41. Jarusiewicz J, Sherma J, Fried B (2005) Separation of sterols by reversed phase and argentation thin layer chromatography. Their identification in snail bodies. J Liq Chromatogr Rel Technol 28:2607–2617
42. Destaillats F, Golay PA, Joffre F, De Wispelaere M, Hug B, Giuffrida F, Fauconnot L, Dionisi F (2007) Comparison of available analytical methods to measure trans-octadecenoic acid isomeric profile and content by gas-liquid chromatography in milk fat. J Chromatogr A 1145:222–228
43. Mohammad A, Zehra A (2008) Separation of coexisting tryptophan, alanine, and phenylalanine or tyrosine by silver ion high performance thin layer chromatography. J Planar Chromatogr 21:299–304
44. Srivastava SP, Reena (1985) TLC separation of some closely related sulfa drugs on copper sulfate impregnated plates. Anal Lett 18:239–244
45. Cecchi L, Malaspina P (1991) Lactic acid enantiomers: separation by thin-layer chromatography on silica gel plates impregnated with Cu^{2+}. Anal Biochem 192:219–221
46. Szumilo H, Flieger J (2001) TLC separation of carbohydrates on silica gel modified with copper (II) salts. Acta Pol Pharm—Drug Res 58:3–8
47. Bushan R, Gupta D (2006) Ligand-exchange TLC resolution of some beta-adrenergic blocking agents. J Planar Chromatogr 19:241–245
48. Pyka A, Klimczok W (2008) Influence of impregnation of a mixture of silica gel and kieselguhr with copper (II) sulphate (VI) on profile change of the spectrodensitograms and the R_F values of nicotinic acid and its derivatives. J Liq Chromatogr Rel Technol 31:526–542

49. Bhushan R, Tanwar S (2009) Direct TLC resolution of the enantiomers of three β-blockers by ligand exchange with Cu(II)–L-amino acid complex, using four different approaches. Chromatographia 70:1001–1006
50. Bhushan R, Tanwar S (2010) Different approaches of impregnation for resolution of enantiomers of atenolol, propranolol and salbutamol using Cu(II)-l-amino acid complexes for ligand exchange on commercial thin layer chromatographic plates. J Chromatogr A 1217:1395–1398
51. Flieger J, Szumiło H, Giełzak-Koćwin K, Matosiuk D (2002) Effect of impregnation conditions on the structure and chromatographic behavior of TLC adsorbents modified with Cu(II) and Ni(II) salts. J Planar Chromatogr 15:354–360
52. Petrovic' M, Kaštelan-Macan M, Horvat AJM (1992) Thin-layer chromatographic behaviour of substituted phenolic compounds on silica gel layers impregnated with aluminium(III) and copper(II). J Chromatogr A 607:163–167
53. Sober M, Lekic M, Korac B, Imamovic B, Marjanovic A (2003) Identification of chlorophenoxy herbicides by TLC in clinical toxicology. Proc Int Symp Planar Sep Plan Chrom 423–432
54. Pyka A, Dołowy M, Gurak D (2005) Separation of selected bile acids by TLC. VIII. Separation on silica gel $60F_{254}$ glass plates impregnated with Cu(II), Ni(II), Fe(II), and Mn(II) cations. J Liq Chromatogr Rel Technol 28:2273–2284
55. Dołowy M (2007) Separation of selected bile acids by TLC. IX. Separation on silica gel 60 and on silica gel $60F_{254}$ aluminum plates impregnated with Cu(II), Ni(II), Fe(II), and Mn(II) cations. J Liq Chromatogr Rel Technol 30:405–418
56. Bhushan R, Joshi S (1996) TLC separation of antihistamines on silica gel plates impregnated with transition metal ions. J Planar Chromatogr 9:70–72
57. Bhusan R, ThukuThiong O (2002) Separation of cephalosporins on thin silica gel layers impregnated with transition metal ions and by reversed-phase TLC. Biomed Chromatogr 16:165–174
58. Grygierczyk G, Wasilewski J, Witkowska M, Kowalska T (2003) Use of complexation TLC to investigate selected monosulfides. Part I. Silica impregnated with Cu(II), Co(II), Ni(II), Mn(II), Al(III), Cr(III), and Fe(III) cations as stationary phase. J Planar Chromatogr 16:11–14
59. Bhushan R, Parshad V (1994) Separation of vitamin B complex and folic acid using TLC plates impregnated with some transition metal ions. Biomed Chromatogr 8:196–198
60. Bhushan R, Parshad V (1999) Improved separation of vitamin B complex and folic acid using some new solvent systems and impregnated TLC. J Liq Chromatogr Rel Technol 22:1607–1623
61. Sajewicz M, John E, Kronenbach D, Gontarska M, Kowalska T (2008) TLC study of the separation of the enantiomers of lactic acid. Acta Chromatographica 20:367–382
62. Kalinina KB, Litvinova LS (1999) Impregnation of silica gel with inorganic salts from the standpoint of specific adsorption theory. The effect of this impregnation on the separation of sugars by TLC. J Planar Chromatogr 12:190–195
63. Kalinina KB, Litvinova LS (2001) Thin layer chromatography of neutral sugars as influenced by the nature of the cation of impregnating salt. Russ J Appl Chem 74:1343–1347
64. Wyss D, Esser T, Sequin U (1988) Chromatographic separation of diastereoisomeric 1,2:3,4-diepoxides on silica gel TLC plates impregnated with various inorganic salts. Tetrahedron 44:1393–1396
65. Mohammad A, Agrawal V, Kumar S (2003) Use of a water-in-oil microemulsion as mobile phase in complexation TLC of amino acids on silica layers impregnated with metal cations. J Planar Chromatogr 16:220–226
66. Bhushan R, Kaur S (1997) TLC separation of some common sugars on silica gel plates impregnated with transition metal ions. Biomed Chromatogr 11:59–60
67. Bhushan R, Ali I (1989) TLC resolution of alkaloids on silica layers impregnated with metal ions. J Planar Chromatogr 2:397–398
68. Bhushan R, Ali I (1990) Adsorption and partition of amino acids in a batch process compared to their TLC separation on silica gel. J Planar Chromatogr 3:85–87

69. Bhushan R, Reddy GP (1987) Thin-layer chromatography of phenyl thiohydantoin derivatives of amino acids on silica gel thin layers impregnated with zinc salts and some metal sulphates. Analyst 112:1467–1469
70. Flieger J, Szumilo H, Tatarczak M, Matosiuk D (2004) Effect of impregnation of silica gel with different zinc salts on the TLC behaviour of aromatic hydrocarbons with polar groups. J Planar Chromatogr 17:65–71
71. Grygierczyk G, Wasilewski J, Łomankiewicz D, Klimczok W, Kowalska T (2003) Use of complexation TLC to investigate monosulfides. II. Silica impregnated with the Cd(II), Sr(II), Eu(III), and V(IV) cations as stationary phase. J Liq Chromatogr Rel Technol 26:2651–2661
72. Bhushan R, Chauhan R, Chauhan RS (1989) Separation of some antihistamines on impregnated TLC silica plates. Biomed Chromatogr 3:46–47
73. Bhushan R, Parshad V (1994) TLC of amino acids on thin silica layers impregnated with transition metal ions and their anions. J Planar Chromatogr 7:480–484
74. Fatima N, Mohammad A (1984) Thin layer chromatography of metal ions in formic acid medium on impregnated and unimpregnated silica Gel G: semiquantitative determination of Fe^{3+}, Cd^{2+}, Th^{4+}, Al^{3+}, Zn^{2+}, UO_2^{2+}, VO^{2+}, Ce^{4+} and Ni^{2+}. Sep Sci Technol 19:429–443
75. Ajmal M, Mohammad A, Fatima N, Ahmad J (1989) Aqueous alkali halides as impregnant and eluent in TLC of heavy metals: quantitative separation of Al, Ti, Fe, Zn and Cd. J Liq Chromatogr Rel Technol 12:3163–3191
76. Mohammad A, Khan MAM (1993) New surface-modified sorbent layer for the analysis of toxic metals in sea water and industrial wastewater. J Chromatogr A 642:455–458
77. Mohammad A, Khan MAM (1995) Thin layer chromatographic separation of zinc from Cd(ll), Hg(ll), and Ni(ll) in environmental samples using impregnated silica layers. J Chromatogr Sci 33:531–535
78. Mohammad A, Fatima N, Khan MAM (1994) New surface-modified sorbent layers for chromatographic analysis of some inorganic pollutants. J Planar Chromatogr 7:142–146
79. Mohammad A, Khan MAM (1995) TLC separation and microgram detection of metal ions on lithium chloride-impregnated silica gel and alumina layers. J Planar Chromatogr 8:134–140
80. Takedaa Y, Ishidaa K, Hasegawab T, Katohc A (2004) Thin layer chromatographic behavior and separation of alkaline earth metals on silica gel in aqueous sodium perchlorate solution. J Chromatogr A 1049:233–236
81. Mohammad A, Shahab H (2008) Identification and simultaneous separation of manganese(II), copper(II) and nickel(II) on sodium hydroxide impregnated silica layers using water and aqueous surfactants as eluents. Acta Universitatis Cibiniensis Seria F Chemia 11:3–14

Chapter 17
Ion Exchange Technology: A Promising Approach for Anions Removal from Water

Amit Bhatnagar and Eva Kumar

Abstract Water pollution due to increased agricultural, industrial, and domestic activities has become a serious concern worldwide. Various toxic pollutants have been detected in drinking water sources at alarming levels. Anionic pollutants, one of the important classes of aquatic pollutants, need special attention for their removal from water as usually there will be no organoleptic changes in water when anions are present even at low concentrations, thereby increasing the health risks. Ion exchange (IE) technology has been proven as one of the best technologies for water and wastewater treatment. In this review, a compilation of various IE materials reported in the vast literature, used for the removal of various anions (nitrate, fluoride, perchlorate, arsenate, chromate, phosphate, thiocyanate, etc.) from water, has been presented, and their main findings are discussed.

17.1 Introduction

Aquatic environmental systems are of great value to the human civilization as water is an important necessity than any other natural source. Unfortunately, water systems have always been prone to pollution from different anthropogenic and natural sources. The available water is distributed in an uneven manner. Of the total amount of water present on the earth, 97.5% is present in oceans and cannot be used without treatment [1, 2]. The remaining water is generally fresh, but most of it (2.24%) is locked in polar ice caps and glaciers, whereas groundwater accounts for 0.61%. The rivers, lakes, and streams contain only 0.02% of the total, which is immediately available for use.

A. Bhatnagar (✉) • E. Kumar
Chemical Engineering Department, LSRE – Laboratory of Separation and Reaction Engineering, University of Porto (FEUP), Rua Dr. Roberto Frias, Porto 4200-465, Portugal
e-mail: dr.amit10@gmail.com; amit_b10@yahoo.co.in

Inamuddin and M. Luqman (eds.), *Ion Exchange Technology II: Applications*,
DOI 10.1007/978-94-007-4026-6_17,

However, in recent times, the demand for water has risen tremendously to meet the requirements of industrial, agricultural, and domestic activities. Rapid pace of industrial and agricultural activities, human population outburst, deforestation, unplanned urbanization, and rapid exploitation of every bit of natural resources have resulted in destruction of natural water reservoirs and caused severe water pollution at global scale [3–5]. The increasing worldwide contamination of freshwater systems with several toxic in(organic) chemicals discharged from industrial and other anthropogenic activities is one of the key environmental problems [6]. Approximately, one-fifth of the world's population does not have access to the clean drinking water, and two-fifths lack proper sanitation facilities [7]. Due to the adverse affects of polluted waters on human health and animal life, pollution control and management is now becoming a high priority area.

A number of water treatment technologies are available with varying degree of success to control/minimize water pollution. Some of them are coagulation [8], foam flotation [9], filtration [10], ion exchange [11], aerobic and anaerobic treatment [12, 13], advanced oxidation processes [14], solvent extraction [15], adsorption [16], electrolysis [17], microbial reduction [18], and activated sludge [19]. However, most of the above mentioned technologies available for water remediation are found to be expensive, inefficient, and generate additional by-products, and hence, their use is sometimes restricted because of cost factors overriding the importance of pollution control.

17.2 Ion Exchange (IE) Technology for Water Treatment and Purification

Among various water treatment technologies, *ion exchange (IE)* has emerged as one of the most promising technologies for water treatment. The process in which exchange of an ion from a solution with a similarly charged ion present on a stationary solid particle, called IE resin, occurs in a reversible chemical reaction is termed as IE. IE reactions are stoichiometric (i.e., predictable based on chemical relationships) and reversible in nature. In the perspective of water purification, IE reaction occurs by the replacement of ionic impurities in the water with ions released by an IE resin. The resins are periodically regenerated to restore its original ionic form. IE technology has been widely used for various applications on commercial scale with proven success. Various naturally occurring materials such as zeolites, montmorillonite, clay, and soil humus are representative ion exchangers [20].

Since long, cation- and anion-exchange resins have been widely employed for softening the water and for removal of anions, respectively. On the other hand, the combination of cation- and anion-exchange resins is effective for deionization of water. The cation- and anion-exchange resins release positively and negatively charged ions, respectively, in exchange of the cationic and anionic impurities, respectively. The most efficient IE resins, which have been used for water treatment, are prepared from synthetic polymers such as styrene-divinylbenzene and are either sulfonated or aminated for targeted pollutants.

17.3 Application of Ion Exchange (IE) for Anions Removal from Water

Among several aquatic pollutants, inorganic anions are of great importance as there will usually be no organoleptic changes in drinking water when the inorganic anions are present in trace levels and it is rather possible that some of them may remain undetected, thus increasing the possible health risks [21]. These can also be referred as micro-pollutants as the maximum allowed concentrations of some anions are generally relatively low (μg/L to mg/L) as set by the drinking water quality regulatory standards [21]. A number of inorganic anions have been found in potentially harmful concentrations in numerous drinking water sources and can be present at the same time in rather different levels (e.g., nitrate and perchlorate), thus leading to the emerging issue of their control and simultaneous removal from drinking water supplies [21–24]. Intensive research on effective water treatment has resulted in several technologies which can be employed for the removal of inorganic contaminants from water. Coagulation and filtration processes are mainly employed at large-scale water treatment, but many anions such as nitrate do not coordinate or coprecipitate with metal ions. In this perspective, IE technology offers one of the best approaches for the selective removal of trace levels of anions present in drinking water. This review emphasizes the application of IE technology for the removal of important anions from water and wastewater.

17.3.1 Nitrate Removal from Water

Nitrate pollution in groundwater and drinking water is of increasing environmental concern on a global scale [25]. The primary cause of nitrate pollution is excessive use of nitrogenous fertilizers in agricultural sector. Additional sources of nitrate pollution include wastewaters and agricultural and urban runoff, disposal of untreated sanitary and industrial wastes in unsafe manner, leakage in septic systems, landfill leachate, animal manure, and airborne nitrogen compounds given off by industry and automobiles, which are deposited on the land in precipitation and dry particles [26, 27]. Methemoglobinemia also known as blue-baby syndrome and formation of carcinogenic nitrosamines are potential health threats which are caused due to the increased nitrate levels in drinking water [28, 29]. Keeping in view of the serious health problems associated with excess nitrate concentrations in drinking water, US Environmental Protection Agency (US EPA) has set the maximum safe concentration of 10 mg/L of nitrate-N in drinking water [30].

IE has been introduced by World Health Organization (WHO) as a nitrate removal technology [31] and approved as the Best Available Technology (BAT) for nitrate removal by US EPA [32]. Numerous studies have been conducted to test the feasibility of nitrate removal from drinking water by IE, and full-scale IE plants have been built and operated to reduce excess nitrates in groundwater. Different types of IE resins such as, Amberlite® 400, Purolite® A 520E, Amberlite™ IRN-78, and Amberlite® IRA 400 have been examined for their potential to remove/minimize excess nitrate from water [33–36].

Samatya et al. observed that the breakthrough point of nitrate in the groundwater tested was about half that for the synthetic water (deionized water spiked with nitrate) due to the presence of other competitive background ions and higher concentration of nitrate in groundwater (synthetic water: 100 mg NO_3^-/L, groundwater: 195 mg NO_3^-/L) [34]. de Heredia et al. reported that the capacity of Amberlite® IRN-78IE resin decreased for the removal of nitrate as chloride ion concentration in the feed water increased [35]. The efficiency of nitrate removal by IE is found to be influenced by several factors. These factors are classified into four main groups which include operating conditions (e.g., flow rate, hydraulic loading (surface area loading rate), contact time, resin volume, bed depth, and headloss), feed water characteristics, type of resin, and finally regeneration and waste disposal [37]. It has been reported that decreasing the flow rate and increasing the bed depth can increase the removal capacity of the system [37, 38]. However, decreasing the flow rate increases the process time, and increasing the bed depth (resin volume) increases cost [37]. It is also important to emphasize that, in general, higher total dissolved solids (TDS) also influence nitrate removal [33, 39, 40]. Darbi et al. compared the performance of biological denitrification, IE, and reverse osmosis (RO) technologies for nitrate removal from drinking water in a field study and stated that although IE is a promising technology for nitrate removal, the amount of regenerant is decisive to diminish the problems of disposal [40]. Few workers observed that ionic leakage occurred earlier and nitrate removal ability of resins decreased with increasing feed water nitrate concentration [33, 41]. Boumediene and Achour used nitrate-selective resin to study the effect of sulfate on nitrate removal and observed that presence of sulfate could accelerate the time to nitrate leakage and reduced the overall nitrate removal capacity of the nitrate-selective resin [41]. The efficiency of Amberlite® IRA 400 resin for the simultaneous removal of organic pollutants and nitrate was examined by Dore et al. [42]. Authors reported that the resin showed low adsorption capacity for organics and did not alter the organic characteristics of water [42]. Furthermore, the capacity of Amberlite® IRA 400 resin was found to decrease in the presence of high sulfate concentrations during the nitrate removal from water.

Liang et al. have also investigated the effects of volatile organic chemicals (VOCs), strong oxidants such as chlorine, and arsenic adsorption by IE resin [43]. Results exhibited that arsenic was also adsorbed onto the resins, but did not affect the nitrate removal capacity. Moreover, some VOCs were also adsorbed by the resin, and some changes in IE capacity were noticed. The performance also declined to some extent due to the exposure to chlorine. Further tests were recommended by the workers to investigate the long-term effects of VOCs on resins.

As discussed above, the presence of sulfate can reduce nitrate removal due to preferential adsorption of sulfate. To overcome this problem, resins with higher selectivity for nitrate rather than sulfate have been developed by changing the characteristics of matrix and functional groups of the resins. These resins are called nitrate selective [43, 44].

One of the main drawbacks to IE at full scale is the regeneration of resin and the costs associated with preparing the brine and its disposal [44, 45]. Disposing the regenerated brine is one of the main shortcomings of IE. This critical factor needs to be considered when designing a full-scale IE plant [38, 44, 45].

The nitrate concentration of Dutch groundwater (18.4 mg NO_3^--N/L) treated with the nitrate-selective resin (Amberlite IRA 996) at a flow rate of 35 bed volume (BV)/h was found to reach the European guideline level of 5.6 mg NO_3^--N/L after 600 BV (17 h) [46]. Regeneration of Amberlite IRA 996 with 30 g sodium bicarbonate/L was achieved in about 6 h at a flow rate of 11 BV/h.

An IE treatment using strong acid cation and weak basic anion exchangers was employed for fertilizer wastewater treatment in Croatia [47]. The regeneration of cation- and anion-exchange resins was performed by 56% HNO_3 and 16% ammonium hydroxide, respectively, as an external regeneration process. The wastewater in the investigation period contained a high content of nitrogen in ammonium form (with an average of 325 mg/L N-NH_4^+) as well as nitrogen in nitrate form (with an average of 201 mg/L N-NO_3^-). After treatment, the wastewater contained 8.7 mg/L of NH_4^+-N and 7.0 mg/L of NO_3^--N. Four types of anion resin were employed for the wastewater treatment. Decrease in exchange capacity (from the initial 1.19 mol/L to a final 0.71 mol/L) was observed when LEWATIT S 4428 was treated with 2% HNO_3 for 24 days under laboratory conditions. The dynamic contact of 2% HNO_3 with LEWATIT S 4428 in the column resulted in a decrease of the exchange capacity by 32%. LEWATIT S 4428 proved to be the most effective resin for the process under dynamic equilibrium.

17.3.2 Fluoride Removal from Water

Fluoride contamination in groundwater and drinking water due to natural and anthropogenic activities has been recognized as one of the serious health problems worldwide [48]. Excess fluoride intake through drinking water often causes skeletal fluorosis, causing debilitating bone structure disease, as well as discoloration and mottling of teeth, cancer, or adverse effects on the brain and kidney [49]. It was estimated that globally, more than 70 million people are suffering from fluorosis. Realizing such harmful effects of excess fluoride on human health, WHO has set the safe limit of fluoride concentration in drinking water as 1.5 mg/L.

IE resins have also been employed for defluoridation of water. An amberlite-type cation-exchange resin (Amb200CT) chemically modified by trivalent metal ions, such as lanthanum(La(III)), cerium(Ce(III)), yttrium(Y(III)), iron(Fe(III)), and aluminum(Al(III)), was used to remove fluoride from hot spring water in Japan [50].

La(III)-loaded and Ce(III)-loaded Amb200 resins were found to be the most effective for defluoridation process, while Fe(III)- and Al(III)-loaded Amb200 resin exhibited the least fluoride removal. Further, the adsorption capacity of different metal(III)-loaded Amb200 resin for fluoride was in the order

$$\mathrm{La}(\mathrm{III}) \geq \mathrm{Ce}(\mathrm{III}) > \mathrm{Y}(\mathrm{III}) > \mathrm{Fe}(\mathrm{III}) \approx \mathrm{Al}(\mathrm{III}).$$

In order to improve the IE capacity (IEC), mesoporous Ti-oxohydroxide ($TiO_x(OH)_y$) was prepared by Ho et al. using dodecylamine as template [51]. Zirconia and silica were introduced into the mesoporous Ti-oxohydroxide to enhance the IEC. It was found that mesoporous Ti-oxohydroxide containing zirconia exhibited the highest fluoride IEC, as it had the smallest particle size, with high uniformity among the mesoporous materials prepared.

In another study by Solangi et al., Amberlite XAD-4TM resin was modified by introducing amino group onto the aromatic ring, and its performance was further assessed for fluoride removal [52]. The optimum fluoride sorption occurred at pH 9.0. The desorption of fluoride was achieved using 10% HCl. Langmuir, Freundlich, and Dubinin-Radushkevich (D-R) isotherm models were successfully used to model the experimental data. The fluoride uptake by modified Amberlite XAD-4 was found to be 5.04×10^{-3} mol/g. It was concluded that the uptake of fluoride by modified resin occurred through IE mechanism as observed by D-R isotherm parameters.

The defluoridation capacities (DCs) of Indion FR 10 (IND), a chelating resin, and Ceralite IRA 400 (CER), an anion-exchange resin, were compared by Meenakshi and Viswanathan [53]. The chelating resin was found more selective than an anion-exchange resin for fluoride removal. The fluoride sorption data was fitted well by Freundlich and Langmuir isotherm models. The pH of the medium was maintained neutral as the pH showed no significant effect on the DC of either of the resins studied. The efficiency remained the same in both alkaline and acidic medium. The dependence of the DC of the resins on the presence of other co-ions, namely, Cl^-, SO_4^{2-}, HCO_3^-, NO_3^-, Ca^{2+}, and Mg^{2+}, was investigated with varying initial concentrations of these ions, namely, 100–600 mg/L, by keeping 3 mg/L as the initial fluoride concentration. The results showed that the competing ions did not show any significant effect on the DC of the resin IND, which might be attributed to the fact that IND selectively removed fluoride. However, in the case of the resin CER, a significant reduction of DC at higher concentrations of co-ions present in the medium was observed. The maximum Langmuir monolayer capacity was found to be 1.305–1.356 mg/g for IND and 1.504–1.658 mg/g for CER at different temperatures. The sorption process was found to be controlled by pseudo-second-order and particle diffusion models. Various thermodynamic parameters were also calculated to understand the nature of sorption.

Modified Indion FR 10 resin in Na^+ and Al^{3+} forms was tested for fluoride removal by the same workers [54]. The DCs of Na^+ and Al^{3+} forms were found to be 445 and 478 mg/kg, respectively, whereas the DC of H^+ form was 265 mg/kg at 10 mg/L initial fluoride concentration. Fluoride uptake was found independent of pH of the medium and unaltered in the presence of co-anions. Al^{3+} form had higher DC among the sorbents studied. Chemisorption was found to be the controlling mechanism during the process. 0.1 M HCl was identified as the best eluent for the regeneration of the sorbents.

17.3.3 Perchlorate Removal from Water

Perchlorate (ClO_4^-) is becoming a pollutant of serious concern worldwide due to its health hazards caused even in the presence of lower concentrations in groundwater [55]. Perchlorate and its salts are mainly used in missile/rocket propellants and in various industrial applications (e.g., manufacturing of matches, airbag inflators, safety flares, and fireworks) [56, 57]. The major sources of perchlorate contamination in groundwater are due to unsafe disposal of rocket fuel and explosives by the aerospace and chemical industries [58]. The toxic health effects of perchlorate contamination in drinking water are well documented [59]. High dosage of perchlorate causes serious health problems even if the person is exposed to it for a short period [60]. Perchlorate exposures for a longer period can result in the inhibition of iodine uptake in the thyroid gland, affecting/altering the production of thyroid hormones and possibly causing mental retardation in fetuses and infants [61, 62]. The removal of perchlorate from contaminated aquatic systems is a challenging task as perchlorate ions are nonvolatile, highly soluble, and kinetically inert in water [56].

IE technologies currently appear to be relatively economical and technically feasible for both small- and large-scale applications for perchlorate removal from drinking water. Various strong basic anion-exchange (SBA) resins have been developed which have very high selectivity for perchlorate. Gu and Brown developed the bifunctional resins for higher perchlorate selectivity with better reaction kinetics [63]. The bifunctional resins exhibited high selectivity and reaction rates especially at low influent concentrations. A new methodology to regenerate ClO_4^- loaded resins with improved regeneration efficiency, recovery, and waste minimization in comparison with conventional brine regeneration techniques was developed by the same workers [64]. Tetrachloroferrate ($FeCl_4^-$) ions formed by the reaction of ferric chloride ($FeCl_3$) with excess chloride ions which were effective in displacing ClO_4 (sorbed on the resin). Nearly 100% recovery of IE sites was achieved by washing with ~5 BV of the regenerant solution.

Gu et al. successfully demonstrated the degradation of perchlorate by $FeCl_3$-HCl solution in less than 1 h at temperature higher than 200 °C and/or pressure ~ 20 atm [65]. The production of secondary waste was minimized as the properties of regenerant solution remained unchanged making it usable for repeated cycles. Regenerable IE technology for treating ClO_4^- with influent concentration of ca. 450 μg/L was demonstrated by Gu et al. [66]. The bifunctional anion-exchange resin treated approximately 40,000 BV before breakthrough occurred. Relatively high concentrations of competing anions, namely, Cl^- and SO_4^{2-}, did not influence the resin ability to remove ClO_4^-.

A polymeric ligand exchanger (PLE), DOW 3N-Cu, was prepared for perchlorate removal from contaminated water [67]. The equilibrium could be achieved within 20 h as seen by batch kinetic studies. The regeneration of spent DOW 3N-Cu resin was found to be pH dependent. Nearly 90% of the perchlorate capacity of saturated DOW 3N-Cu was recovered using 35 BV of a regenerant containing 4% (w/w) NaCl, 2% NaOH, 5% Tween 80, and 5% ethanol.

Xiong et al. compared perchlorate sorption and desorption behaviors of selected representative standard strong-base anion (SBA) resins, weak-base anion (WBA), bifunctional resin, PLE, and an IE fiber (IEF) [68]. A combination of different factors, namely, hydrophobicity, intraparticle diffusion, surface area, and pore accessibility, was found to be important and needed to be considered before finalizing the suitability of a resin. IEF resins showed rapid kinetics for perchlorate and had significantly higher regeneration efficiencies (~85%) using 38 BV of a brine-only solution (12% NaCl at pH ~7.0). Kim et al. synthesized highly dispersed ammonium mono- or bifunctionalized mesoporous media to study the perchlorate removal kinetics [69]. The bifunctionalized mesoporous media exhibited higher adsorption capacity and faster perchlorate removal kinetics than other mono-functionalized mesoporous media, suggesting that simultaneous existence of both ammonium functional groups could enhance perchlorate removal. Polyethylene glycol (PEG)-grafted aqueous biphasic extraction chromatographic (ABEC) resins were used to remove both mercury and perchlorate from aqueous media by Dilip et al. [70]. The columns exhibited moderate capacity for perchlorate, and it was further noticed by the workers that reuse of resin with little or no loss in capacity was possible.

The mechanisms of perchlorate removal by SR-7 resin, an anion-exchange resin, along with activated carbon (AC) were investigated using Raman, Fourier transform infrared (FTIR), and zeta potential analyses by Yoon et al. [71]. Perchlorate adsorption and the zeta potential of resin were unaffected by the change in solution pH, but in case of AC, the adsorption and zeta potential of the AC were affected by the change in pH. The results of the Raman analyses were helpful in elucidating the perchlorate sorption mechanism onto resin. The electrostatic attraction between the anion and positively charged surface sites was found responsible for the perchlorate exchange from water.

17.3.4 Arsenic Removal from Water

Arsenic is toxic pollutant found in the soil and groundwater globally and requires immediate attention due to its adverse health effects [72, 73]. The high concentrations of arsenic in surface and groundwater have been found in several countries including USA, China, Chile, Bangladesh, Taiwan, Argentina, and India leading to the severe health effects to the community [74]. Arsenic occurs in its inorganic form as trivalent arsenite, As(III), or pentavalent arsenate, As(V), in the aquatic environment. Under typical anoxic groundwater conditions at near neutral pH, As(III) is the predominant form, which is more toxic and mobile than As(V), while in oxic groundwater, As(V) dominates. Aqueous arsenic species in most natural waters (pH ~ 4–10) are the neutral species H_3AsO_3 (arsenious acid) for As^{3+} and $H_2AsO_4^-$ (dihydrogen arsenate) and $HAsO_4^{2-}$ (hydrogen arsenate) for As^{5+}. The rate of oxidation of As(III) to As(V) or reduction of As(V) to As(III) depends on many factors like pH, Eh (activity of electrons), presence of microorganisms, presence of other oxidizing agents like Mn and Fe, and exposure to light and temperature. The removal of arsenic from water by IE has been widely investigated by various researchers.

DOW 3N-Cu, a novel PLE, was prepared by immobilizing Cu^{2+} onto a chelating polymer with high content of pyridine-N donor atoms [75]. The results exhibited that PLE is highly selective for inorganic contaminants such as phosphate, arsenate, and ionizable organic contaminants. The sorption efficiency of DOW 3N-Cu for inorganic contaminants was quite comparable to standard macroporous IE resins. Over 3,000 BV of arsenic contaminated water was treated in each operation cycle using DOW 3N-Cu. Brine efficiently regenerated the arsenate-exhausted DOW 3N-Cu.

Hydrated Fe(III) oxide (HFO) nanoparticles were dispersed in an IE fiber containing quaternary ammonium functional groups to prepare hybrid anion-exchange fiber (HAIX-F) [76]. The HAIX-F was evaluated for selective and simultaneous removal of arsenate and perchlorate in fixed-bed columns. High sorption affinity for perchlorate and arsenate was observed in the presence of high concentrations of sulfate and chloride. Perchlorate was found to bound to the quaternary ammonium functional groups, whereas dispersed HFO nanoparticles selectively captured As(V) through ligand exchange. Desorption of perchlorate and arsenate was done by a two-step regeneration using NaOH and NaCl solutions.

Recently, An et al. prepared six chelating resins by functionalizing three commercially available XAD resins of various matrix properties followed by the loading of pyridinyl functional groups on those sorbents [77]. Consequently, six PLEs were prepared by immobilizing Cu(II) ions onto the chelating resins, and the highest Cu(II) loading was 44 mg/g. The Cu-loaded PLEs showed higher removal rate for arsenate compared to standard SBA resins. Matrix type and accessibility of the functional sites were also found critical for arsenic adsorption. The maximum Langmuir capacity ranged from 15 mg/g for XAD16-2N-Cu to 26 mg/g for XAD1180-3N-Cu. XAD-based PLEs treated 780–920 BV of arsenic laden water before the breakthrough point of 10 μg/L. Complete arsenate recovery was achieved at pH 10 using 20–30 BVs of 8% NaCl.

A novel fibrous anion exchanger (FVA) of non-ligand exchange type was prepared for the selective removal of arsenate from water [78]. Two types of FVA were prepared, one being FVA-c, nonwoven cloth type, for the batch studies, and other was FVA-f, a filamentary type, used for column-mode study. The FVA-c preferred arsenate species to chloride and sulfate in neutral pH regions, which was different from a traditional strong-base anion-exchange resin. The FVA-f was able to remove arsenate very rapidly from the feed containing arsenate. One percent breakthrough point in uptake from the arsenate solution containing 1.0 mg of As/L at the high feed flow rate of 1050/h in SV was found to be 4,670 BV, giving the 1% breakthrough capacity of 0.298 mmol/g of FVA-f. Adsorbed arsenate was able to be quantitatively eluted with 1 M HCl, and FVA-f was simultaneously regenerated.

Shao et al. chemically modified the amberlite resins (200CT resin) using several trivalent metal ions including La(III), Ce(III), Y(III), Fe(III) and Al(III) and evaluated their performance for the arsenic removal from the wastewater [79]. The removal of As(III and V) using metal(III)-loaded 200CT resins was strictly pH dependent. The As(III) adsorption followed the order Y(III)-200CT > Ce(III)-200CT > Fe(III)-200CT > La(III)-200CT > Al(III)-200CT. IE was suggested as the driving mechanism for arsenic removal. The maximum uptake of As(III) by Y(III) and Ce(III)-200CT were reported as 0.4835 and 0.4592 mol/kg, respectively. Further, Fe(III)-200CT resin was found the most effective for the removal of As(V) with the maximum capacity of 1.450 mol/kg. Phosphate had greater influence on As(III and V) than sulfate ions.

Agricultural wastes have also been used to prepare anion exchangers (AE). Anirudhan and Unnithan prepared AE from coconut coir pith by treating it with epichlorohydrin and dimethylamine followed by treatment with HCl [80]. The prepared ion exchanger (CP-AE) was evaluated for its performance for the uptake of As(V) from aqueous solution. The extent of As(V) removal by CP-AE at 1 and 5 mg/L initial arsenic concentration at pH 7.0 was found to be 0.50 mg/g and 2.36 mg/g, respectively. Regeneration studies were performed using 0.1 N HCl.

Selective removal of As(V) from drinking water was carried out with strong-base anion-exchange resins by Korngold et al., and 99% arsenic removal was reported [81]. In the presence of sulfate and chloride, the effectiveness of the process was found to decrease.

Vatutsina et al. also explored the possibility of synthesized hybrid inorganic/polymeric fibrous sorbents for arsenic removal [82]. Nanoparticles of hydrated Fe(III) oxides were dispersed in polymer filaments to prepare the sorbent. The functional groups of weak-base AE allowed high (1.0–1.4 mmol/g) and fairly uniform Fe(III) loading. The prepared hybrid sorbent FIBAN-As exhibited excellent arsenic removal efficiency. The maximum Langmuir sorption capacity for As(III) and As(V) was 1.09 and 1.01 mmol/g, respectively. Arsenic sorption was not suppressed in the presence of sulfate and chloride ions. However, As(V) sorption decreased in the presence of phosphate. The breakthrough capacity (>10 μg/L outlet As concentration) of FIBAN-As was reported 2.9 g of As/kg.

The metal-doped anionic biopolymer was also examined for arsenic(V) removal using alginic acid by Min and Hering [83]. The Ca^{2+} gel beads were treated with Fe(III) for attaining enhanced sorption capacity of arsenate. Nearly 94% removal of As(V) from solution was achieved for an initial As(V) concentration of 400 μg/L at pH 4 in 120 h. Polymeric/inorganic hybrid particles were prepared by the dispersion of agglomerated microparticles of hydrated Fe(III) oxides in gel and pore phases of the spherical polymer beads by DeMarco et al. [84]. The new hybrid ion exchange (HIX) showed selective As(III) and As(V) sorption properties at neutral pH without any pre- or posttreatment. Efficient in situ regeneration with caustic soda and a subsequent short carbon dioxide-sparged water rinse was achieved. Katsoyiannis and Zouboulis modified polystyrene and polyHIPE (PHP) by coating their surface with iron hydroxide [85]. The neutral pH was found optimum for the removal of mixture of As(V) and As(III). PHP was found to be the most effective among the compared media for arsenic removal. Three modified alginates, namely, calcium alginate beads coated with iron oxides and calcium alginate beads doped and coated with iron oxides, were tested. Ca-Fe-doped alginate was found the most efficient.

Monoclinic or cubic hydrous zirconium oxide-loaded porous resin was prepared by Suzuki et al. [86]. They incorporated $ZrOCl.8H_2O$ into porous spherical polymer beads followed by hydrolysis and hydrothermal treatment of the zirconium salt. The prepared resin was able to adsorb As(III) at pH 9–10 and As (V) at pH 4–6. The maximum adsorption capacities of Zr resin for As(III) and As(V) were 1.5 and 1.2 mmol/g. One molar NaOH was required for the regeneration of the resin, followed by conditioning with 0.2 M acetate buffer solution. The leaching of zirconium was negligible during adsorption and regeneration. The same researchers also reported adsorption of Se(VI), Se(IV), As(IIII), As(V), and methyl derivatives of As(V) by porous polymer beads loaded with monoclinic hydrous zirconium oxide [87].

Lysine-N^{α}, N^{α} diacetic acid (LDA) moiety chelating resin was loaded with zirconium(IV), and the performance of Zr-LDA resin was evaluated for the removal of As(V) and As (III) [88]. As(V) was strongly adsorbed in the pH range from 2 to 5, while As(III) was adsorbed between pH 7 and 10.5. Complexation between arsenate or arsenite and Zr complex of LDA leads to the removal of the two arsenic species. The Langmuir sorption capacities of 0.656 mmol/g for As(V) and 1.184 mmol/g for As(III) were reported. One molar NaOH was used to successfully desorb As(V) from the resin. Adsorption/desorption of As(V) was achieved up to six times without any significant loss of Zr in the resin.

In an another study, Zr was loaded on a fibrous phosphoric acid adsorbent, which was synthesized by radiation-induced grafting of 2-hydroxyethyl-methacrylate phosphoric acid onto polyethylene-coated polypropylene nonwoven fabric [89]. The prepared sorbent was evaluated for the removal of As(V), and the maximum breakthrough capacity of the prepared adsorbent, Zr-FPA, was reported 1.5 mmol/g at pH 2. Chloride and nitrate were found to reduce the breakthrough capacity of As(V) adsorption.

Zhu and Jyo studied the uptake of As(V) by a Zr(IV)-loaded phosphoric acid chelating resin (Zr-RGP) prepared from a copolymer of divinylbenzene and 2-hydroxy-3-OPO_3H_2-propyl methacrylate [90]. Minimal Zr(IV) leakage was observed within a wide pH range. The maximum As(V) capacity was 0.2 mmol/mL of wet resin (0.67 mmol/g of dry resin). Adsorbed As(V) was eluted with 0.4 mol/L sodium hydroxide, and repeated use of Zr-RGP was possible.

Egawa et al. prepared chelating resins containing mercapto groups (RES), using 2,3-epithiopropyl methacrylate-divinylbenzene copolymer beads. RES exhibited high affinity for As(III) and high resistance against hot water [91]. Adsorbed As(III) was eluted by NaOH solution containing 5% of sodium hydrogen sulfide. The RES also exhibited high adsorption ability for As in the geothermal power waste solution.

A hydrophilic thiol resin, poly(ethylene mercaptoacetamide), prepared from branched polyethyleneimine and mercaptoacetyl chloride, was examined by Styles et al. for the sorption of As(III) and As(V) [92]. The sorption capacity of the thiol resin was 106 mg As/g dry resin for arsenate at pH 2 and only 30 mg As/g dry resin for arsenite at pH 8. Significant amount of arsenic removal was also attributed to take place via complexation by thiol and anion exchange on protonated amine sites of the branched poly(ethylene mercaptoacetimide) (PEM). NaCl and Na_2SO_4 significantly affected the sorption of arsenate and arsenite. Arsenic was desorbed by 0.2 N NH_4OH. Tatineni and Hideyuki investigated the adsorption of As(III) and As(V) by titanium dioxide loaded onto an Amberlite XAD-7 resin [93]. The resin showed affinity for As(V) from pH 1 to 5 and for As(III) from pH 5 to 10. Langmuir adsorption capacities of 0.063 mmol/g for As(V) at pH 4.0 and 0.13 mmol/g for As(III) at pH 7.0 were achieved. Lin et al. also studied the adsorption of arsenate by iron-based adsorbent (Bayoxide E33 from Bayer) and one AE resin (Arsenex from Purolite) [94].

17.3.5 Chromium Removal from Water

Chromium(VI) compounds are more toxic than Cr(III) due to their high water solubility and mobility. On the other hand, trivalent chromium is insoluble and thus immobile under ambient conditions. The hexavalent form is rapidly reduced to trivalent chromium under aerobic conditions. The health problems associated with chromium toxicity are well documented in literature [95]. Many researchers have examined the removal of chromium by IE.

Amino (RPA) and quaternary amine (RQA) resins were prepared by immobilizing pentamine and quaternary ammonium chloride moieties on glycidyl methacrylate/N,N′-methylene bis-acrylamide resin (GMA/MBA), and the adsorption behavior of chromate by the two resins was evaluated [96]. Resin RPA exhibited high uptake in acidic media as the chromate removal undergo through the protonated amino active sites in acidic media, whereas it dramatically decreased in basic conditions. Resin RQA exhibited uptake at both acidic and basic conditions with slightly improved capacities in the case of the former. This behavior of RQA was attributed to the strong basic nature of the quaternary amine moiety. Regeneration of the resin was achieved using a mixture of 0.05 NaOH and 2 M NaCl.

Two strong-base anion exchangers containing 4-vinyl-1-ethylpyridinium chloride structural units from the quaternization of the gel type 4-vinylpyridine, 8% divinylbenzene copolymer and a commercial type I and gel styrene strong-base anion exchanger, Purolite A 400, were compared for the removal of CrO_4^{2-} from K_2CrO_4 (potassium dichromate) aqueous solutions by Neagu et al. [97]. The two resins exhibited similar isotherms and had similar preference for CrO_4^{2-} over Cl^-. The preference order of the two strong-base anion exchangers for the CrO_4^{2-} was Purolite A 400 > 4-VP:8% DVB + EtCl.

Korngold et al. investigated the removal of Cr(VI) from tap water using strong-base anion-exchange resins [98]. Relite A-490 exhibited higher selectivity for $HCrO_4^-$ and CrO_4^{2-} than the two other resins (Amberlite 416 and Purolite A420). The leakage of Cr(VI) was higher when the concentration of chromate in the feed water was higher, especially at high flow rate. The resin was regenerated with H_2SO_4 or HCl.

Solvent impregnated resins (SIR) were prepared by a wet-impregnation technique using two different polymer matrices, Diaion HP-20 and HP-2MG by Kabay et al. [99]. Aliquat 336 and acetone were employed as the extractant and the solvent, respectively, for impregnation. The maximum uptake was ca. 12 mg Cr(VI)/g SIR for both SIR prepared at 0.5 g Aliquat 336/g polymer. Column study was performed using HP-2MG. The total capacity was calculated as 10.32 mg/g SIR (or 3.83 mg/ml) and the column utilization efficiency was 76.4%. Cr(VI) bound by the SIR was desorbed using a mixture of 0.1 M NaOH and 0.1 M NaCl.

Spinelli et al. chemically modified the surface of chitosan with glycidyl trimethyl ammonium chloride in order to produce a chitosan quaternary ammonium salt that worked as a strongly basic exchanger and studied the adsorption of Cr(VI) onto the prepared quaternary chitosan salt (QCS) [100]. The chitosan salt was cross-linked with glutaraldehyde making it insoluble in aqueous solution. The maximum adsorption capacity of the resin QCS occurred at a pH value of around 3.5–4.5.

17.3.6 Phosphate Removal from Water

The release of phosphate into water bodies, mainly from domestic and municipal wastes and industrial effluents, is one of the prime causes of eutrophication. Removal of phosphate from effluents is therefore necessary to control eutrophication and to meet the discharge limit. Besides other techniques, IE process has also been employed for the removal of phosphate from water. Anirudhan et al. prepared a novel adsorbent system (BS-DMAHP) containing dimethylaminohydroxypropyl (DMAHP) weak-base groups by the reaction of banana stem (BS) with epichlorohydrin and dimethylamine followed by the treatment of HCl [101]. BS-DMAHP was found to be more effective than BS for the removal of phosphate due to increase in the adsorbent stability of the components present in BS and, thereby, the improvement in the access of phosphate ions to the phosphate binding sites of the adsorbent. Ninety-nine percent phosphate uptake was achieved in the pH range of 5.0–7.0. The maximum Langmuir sorption capacity for phosphate was 72.46 mg/g. The removal efficiency of BS-DMAHP was compared with a commercial chloride-form weak-base AE, Duolite A-7. The Langmuir sorption capacity of Duolite A-7 was 31.74 mg/g, which was lower than BS-DMAHP. Adsorbed phosphate on BS-DMAHP was recovered by 0.1 M NaOH solution. After four cycles, the phosphate adsorption capacity of BS-DMAHP decreased ca. 12.0%.

Blaney et al. utilized strong-base AE resin, IRA-900, as parent resin toward the production of hybrid AE (HAIX). HAIX is essentially a polymeric anion exchanger within which HFO nanoparticles have been dispersed irreversibly [102]. Phosphate removal by HAIX continued after 2,000 BV although competing anions, namely, Cl^-, HCO_3^-, and SO_4^{2-}, were present at two orders of magnitude greater concentrations than PO_4^{3-}. Single-step alkaline brine regeneration consistently recovered over 90% PO_4^{3-} in less than 10 BV. HAIX offered the highest phosphate removal capacity at pH slightly above 7.0 where HPO_4^{2-} (mono hydrogen phosphate) is the most predominant species.

17.3.7 Sulfate Removal from Water

Sulfate can be present in almost all natural water. The origin of most sulfate compounds is the oxidation of sulfite ores, the presence of shales, or the industrial wastes. High concentrations of sulfate in the drinking water can have a laxative effect when combined with calcium and magnesium. Lewatit K 6362 anion-exchange resin was used for the removal of sulfate from the industrial wastewater in a fixed-bed column by Haghsheno et al. [103]. The results showed that the maximum removal of sulfate occurred in the resin dosage of 1,000 mg/100 mL. Removal of sulfate using IE resin was found to be dependent on the initial concentration of the adsorbent, flow rate, and the bed height.

17.3.8 Thiocyanate Removal from Water

The interaction of free cyanide with sulfur forms thiocyanate which is generally present in mining wastewater. The toxic effects of thiocyanate include inhibition of halide transport to the thyroid gland, cornea and gill, and stomach and inhibition of a variety of enzymes [104]. Cyanide, cyanate, and thiocyanate compounds are serious hazardous substances, and therefore, the concentration of thiocyanate must be reduced to level that satisfy environmental regulations. The removal of thiocyanate from aqueous solutions by an AE resin (Purolite A-250) was investigated by Dizge et al. [104]. The maximum sorption capacity was observed at pH 8, and the decrease in sorption capacity at pH greater than 8.0 was attributed to the competitiveness of SCN^- and OH^- ions in the bulk. The adsorption capacity of Purolite A-250 for thiocyanate was calculated as 191.20 mg/g at 323 K.

17.3.9 Removal of Multi-anions from Water

Ruixia et al. prepared a new type of IE fiber through the cross-linking reaction of polyacrylonitrile fiber (PANF) by hydrazine hydrate and followed by a functionalized reaction in a mixture of sulfur powder and ethylenediamine [105]. The IE fiber was tested for the removal of fluoride, phosphate, and arsenate ions. The adsorption properties of the IE fiber for fluoride, phosphate, and arsenate ions were found to be pH dependent and on anion concentration. The adsorption of arsenate on the sorbent reached a maximum of 97.9% in the pH range 3.5–7.0, whereas the adsorption percentage of phosphate was more than 99% in the pH range 3.0–5.5. The adsorption of fluoride on the IE fiber was reported 90.4% at pH 3.0. The results of the column experiment showed that dynamic adsorption capacities of fluoride, phosphate, and arsenate ions onto the fiber column were 45.0, 155.5, and 96.0 mg/g, respectively. Five milliliter of 0.50 mol/L NaOH at elution rate of 1 mL/min was required for desorption of fluoride from the column, whereas 30 mL of NaOH was essential for the quantitative recovery of phosphate and arsenate.

A novel IE based on double hydrous oxide ($Fe_2O_3.Al_2O_3.xH_2O$) was prepared by the original sol–gel method and tested for the adsorption of F^-, Cl^-, Br^-, and BrO_3^- from aqueous solutions by Chubar et al. [106]. Fluoride uptake was favorable at pH greater than 5.0 due to the lower concentration of OH^- ions. Bromide also showed similar behavior like fluoride. Sorption of bromate ions was nearly equal through the whole pH interval investigated. The isotherm of F^- sorption fits well the Langmuir model, and the sorption capacity was 90 mg F^-/g. At lower concentrations of the ions in the solution (less than 40 mg/L), bromate dominated, and its sorption was depressed completely by bromide ions.

Table 17.1 summarizes the application of ion exchange technology for some of the important anions removal from water and wastewater.

Table 17.1 Application of ion exchange for the removal of different anions from water and wastewater

Anion	Resin type	Remark/main results	Reference
Nitrate	Amberlite® IRN-78	Resin capacity decreased for the removal of nitrate with an increase in chloride ion concentration	[35]
Nitrate	Amberlite® IRA 400	Capacity of resin decreased in the presence of high sulfate concentration during nitrate removal	[42]
Nitrate	Amberlite IRA 996	600 BV were treated in 17 h to reach the NO_3^- level of 5.6 mg NO_3^--N/L	[46]
Fluoride	Chemically modified amberlite-type cation-exchange resin (Amb200CT)	La(III)- and Ce(III)-loaded Amb200 showed fluoride uptake rapidly and effectively	[50]
Fluoride	Amberlite XAD-4TM resin modified with amino group onto the aromatic ring	Optimum pH 9; sorption capacity = 5.04×10^{-3} mol/g; sorption mechanism: ion exchange	[52]
Fluoride	Indion FR 10 (IND) and Ceralite IRA 400 (CER)	Langmuir monolayer capacity = 1.305–1.356 mg/g for IND and 1.504–1.658 mg/g for CER	[53]
Fluoride	Modified Indion FR 10 resin in the form of Na^+ and Al^{3+} forms	Defluoridation capacity (DC) of Na^+ and Al^{3+} forms was found to be 445 and 478 mg/kg, respectively; chemisorption-controlled process	[54]
Perchlorate	Bifunctional anion-exchange resin	ca. 40,000 bed volumes were treated	[65]

(continued)

Table 17.1 (continued)

Anion	Resin type	Remark/main results	Reference
Perchlorate	DOW 3 N-Cu	Equilibrium achieved within 20 h; ~ 90% of the perchlorate capacity of saturated DOW 3 N-Cu was recovered using 35 bed volumes of a regenerant containing 4% (w/w) NaCl, 2% NaOH, 5% Tween 80, and 5% ethanol	[67]
Arsenic	DOW 3 N-Cu	Over 3,000 bed volumes of arsenic-contaminated water treated	[75]
Arsenic	Modified 200CT resin	The Y(III)- and Ce(III)-200CT resins were better for As(III) adsorption with maximum uptakes as 0.4835 and 0.4592 mol/kg, respectively. Further, Fe(III)-200CT resin was best for the removal of As(V) with the maximum capacity of 1.450 mol/kg	[79]
Arsenic	Coconut coir pith ion exchanger (CP-AE)	Adsorption capacity of 0.50 mg/g (99.5%) and 2.36 mg/g (94.5%) at pH 7	[80]
Arsenic	Synthesized hybrid inorganic/polymeric fibrous sorbents	Sorption capacity for As(III) and As(V) was 1.09 and 1.01 mmol/g, respectively	[82]
Arsenic	Metal-doped anionic biopolymer	Up to 94% removal of As(V) from solution was achieved for an initial As(V) concentration of 400 μg/L at pH 4 in 120 h	[83]
Arsenic	Monoclinic or cubic hydrous zirconium oxide loaded porous resin	Maximum adsorption capacities of Zr resin for As(III) and As(V) were 1.5 and 1.2 mmol/g; optimum pH for As(III) at pH 9–10; optimum pH for As(V) at 4–6	[86]
Arsenic	Zr(IV)-loaded lysine-N^{α}, N^{α} diacetic acid	Sorption capacities of 0.656 mmol/g for As(V) and 1.184 mmol/g for As(III) were reported	[88]
Arsenic	Zr(IV)-loaded phosphoric acid chelating resin (Zr-RGP)	Maximum As(V) capacity was 0.2 mmol/mL of wet resin (0.67 mmol/g of dry resin)	[90]
Arsenic	Amberlite XAD-7	Optimum pH for As(V) was from 1 to 5 and 5 to 10 for As(III); adsorption capacities of 0.063 mmol/g for As(V) at pH 4.0 and 0.13 mmol/g for As(III) at pH 7.0 were achieved	[93]
Chromium	Relite A-490, Amberlite 416 and Purolite A420	Relite A-490 exhibited higher selectivity for $HCrO_4^-$ and CrO_4^{2-} than the two other resins	[98]

(continued)

Table 17.1 (continued)

Anion	Resin type	Remark/main results	Reference
Chromium	Chemically modified chitosan as strongly basic exchanger	Maximum adsorption capacity of the resin QCS occurred at a pH value of around 3.5–4.5 for Cr(VI)	[100]
Phosphate	BS-DMAHP	Maximum sorption capacity was 72.46 mg/g; 99.0% phosphate removal was achieved in the pH range of 5.0–7.0	[101]
Phosphate	HAIX	2,000 bed volumes were treated	[102]
Thiocyanate	Purolite A-250	Sorption capacity was 191.20 mg/g	[104]
Fluoride, phosphate, and arsenate	Modified ion exchange fiber	Dynamic adsorption capacities of fluoride, phosphate, and arsenate ions onto the fiber column were 45.0, 155.5, and 96.0 mg/g, respectively	[105]

17.4 Conclusions and Future Perspectives

It is evident from the literature survey presented in this work that IE technology plays an important role for water and wastewater treatment. It exhibits high performance and selectivity toward the removal of various anions from aqueous solution. However, selectivity and competitive adsorption for a multicomponent system using IE materials vary. Research on surface modification of IE materials has shown that the removal of anions can be enhanced making them ideal candidate for the water and wastewater treatment. However, extensive research is needed using modified forms of these materials on commercial scale. In addition, regeneration of spent IE materials employing simple methods for recycling should also be a focus in future research. Furthermore, design and conducting some pilot-plant scale studies to check their feasibility at industrial level should also be undertaken. Last but not the least, real wastewater should be tested instead of synthetic wastewater.

References

1. Snoeyink VL, Jenkins D (1980) Water chemistry. Wiley, New York
2. World Health Organization, Health in water resources development. http://www.who.int/docstore/water_sanitation_health/vector/water_resources.htm Accessed on 14 July 2010
3. Lehr JH, Gass TE, Pettyjohn WA, DeMarre J (1980) Domestic water treatment. McGraw-Hill, New York
4. Nemerrow NL (1978) Industrial water pollution: origins, characteristics, and treatment. Addison-Wesley, Boston
5. Helmer R, Hespanhol I (1997) Water pollution control-a guide to the use of water quality management principles. E & FN Spon, London

6. Schwarzenbach RP, Escher BI, Fenner K, Hofstetter TB, Johnson CA, Gunten U, Wehrli B (2006) The challenge of micropollutants in aquatic systems. Science 313:1072–1077
7. United Nations Educational, Scientific, and Cultural Organization (2003) World water assessment programme, water for people, water for life-the United Nations world water development report. Berghahn Books, Barcelona
8. Tan BH, Teng TT, Omar AKM (2000) Removal of dyes and industrial dye wastes by magnesium chloride. Water Res 34:597–601
9. Mavros P, Danilidou AC, Lazaridis NK, Stergiou L (1994) Color removal from aqueous solutions. Part I. Flotation. Environ Technol 15:601–616
10. Zouboulis AI, Lazaridis NK, Grohmann A (2002) Toxic metals removal from waste waters by upflow filtration with floating filter medium. I. The case of zinc. Sep Sci Technol 37:403–416
11. SenGupta AK (1995) Ion exchange technology: advances in pollution control. Technomic, Lancaster
12. LaPara TM, Konopka A, Nakatsu CH, Alleman JE (2000) Thermophilic aerobic wastewater treatment in continuous-flow bioreactors. J Environ Eng ASCE 126:739–744
13. Bell J, Plumb JJ, Buckley CA, Stuckey DC (2000) Treatment and decolorization of dyes in an anaerobic baffled reactor. J Environ Eng ASCE 126:1026–1032
14. Esplugas S, Gimenez J, Contreras S, Pascual E, Rodriguez M (2002) Comparison of different advanced oxidation processes for phenol degradation. Water Res 36:1034–1042
15. Lin SH, Juang RS (2002) Removal of free and chelated Cu(II) ions from water by a non-dispersive solvent extraction process. Water Res 36:3611–3619
16. Faust SD, Aly OM (1987) Adsorption process for water treatment. Butterworths, Stoneham
17. Szpyrkowicz L, Naumczyk J, Zilio-Grandi F (1995) Electrochemical treatment of tannery wastewater using Ti/Pt and Ti/Pt/Ir electrodes. Water Res 29:517–524
18. Shen H, Wang YT (1994) Biological reduction of chromium by *E. Coli*. J Environ Eng ASCE 120:560–571
19. Pala A, Tokat E (2002) Color removal from cotton textile industry wastewater in an activated sludge system with various additives. Water Res 36:2920–2925
20. Zagorodni AA (2007) Ion exchange materials: properties and applications. Elsevier, The Netherlands
21. Velizarov S, Crespo JG, Reis MA (2004) Removal of inorganic anions from drinking water supplies by membrane bio/processes. Rev Environ Sci Biotechnol 3:361–380
22. DeZuane J (1997) Handbook of drinking water quality, 2nd edn. Wiley, New York
23. Smith AH, Lopipero PA, Bates MN, Steinmaus CM (2002) Arsenic epidemiology and drinking water standards. Science 296:2145–2146
24. Petrović M, Gonzalez S, Barceló D (2003) Analysis and removal of emerging contaminants in wastewater and drinking water. Trends Anal Chem 22:685–696
25. Liu A, Ming J, Ankumah RO (2005) Nitrate contamination in private wells in rural Alabama, United States. Sci Total Environ 346:112–120
26. Agrawal GD, Lunkad SK, Malkhed T (1999) Diffuse agricultural nitrate pollution of ground waters in India. Water Sci Technol 39:67–75
27. Almasri MN, Kaluarachchi JJ (2004) Assessment and management of long term nitrate pollution of ground water in agriculture-dominated watersheds. J Hydrol 295:225–245
28. Majumdar D, Gupta N (2000) Nitrate pollution of groundwater and associated human health disorders. Indian J Environ Health 42:28–39
29. Tate CH, Arnold KF (1990) Health and aesthetic aspects of water quality. In: Pontius FW (ed) Water quality and treatment. McGraw-Hill, New York
30. US Environmental Protection Agency (2000) Drinking water standards and health advisories. US Environmental Protection Agency, Office of Water, 822-B-00-001
31. World Health Organization (1992) Revisions of the WHO guidelines for drinking-water quality-report on a WHO consultation. WHO Regional Office for Europe, Medmenham, UK

32. U.S. Environmental Protection Agency (2004) National primary drinking water regulations: maximum contaminant levels for inorganic contaminants. Code of Federal Regulations. Title 40, Part 141.62(b)
33. Korngold E (1972) Removal of nitrates from potable water by ion exchange. Water Air Soil Pollut 2:15–22
34. Samatya S, Kabay N, Yuksel U, Arda M, Yuksel M (2006) Removal of nitrate from aqueous solution by nitrate selective ion exchange resins. Rea Funct Polym 66:1206–1214
35. de Heredia JB, Dominguez JR, Cano Y, Jimenez I (2006) Nitrate removal from drinking water using amberlite IRN-78: modelling the system. Appl Surf Sci 252:6031–6035
36. Chabani M, Amrane A, Bensmaili A (2007) Kinetics of nitrates adsorption on Amberlite IRA 400 resin. Desalination 206:560–567
37. Helfferich FG (1995) Ion exchange. Dover, New York
38. Crittenden J (2005) Water treatment principles and design. Wiley, New Jersey
39. Buelow RW, Kropp KL, Withered J, Symons JM (1975) Nitrate removal by anion-exchange resins. J AWWA 67:528–534
40. Darbi A, Viraraghavan T, Butler R, Corkal D (2003) Pilot-scale evaluation of select nitrate removal technologies. J Environ Sci Health A38:1703–1715
41. Boumediene M, Achour D (2004) Denitrification of the underground waters by specific resin exchange of ion. Desalination 168:187–194
42. Dore M, Simon PH, Deguin A, Victot J (1986) Removal of nitrate in drinking water by ion-exchange-impact on the chemical quality of treated water. Water Res 20:221–232
43. Liang S, Mann MA, Guter KPHS, Hardan DL (1999) Nitrate removal from contaminated groundwater. J AWWA 91:79–91
44. Kapoor A, Viraraghavan T (1997) Nitrate removal from drinking water-review. J Environ Eng 123:371–380
45. Kim J, Benjamin MM (2004) Modeling a novel ion exchange process for arsenic and nitrate removal. Water Res 38:2053–2062
46. Hoek JPVD, Hoek WFVD, Klapwijk A (1988) Nitrate removal from ground water – use of a nitrate selective resin and a low concentrated regenerant. Water Air Soil Pollut 37:41–53
47. Leaković S, Mijatović I, Cerjan-Stefanović S, Hodzić E (2000) Nitrogen removal from fertilizer wastewater by ion exchange. Water Res 34:185–190
48. Das DP, Das J, Parida K (2003) Physicochemical characterization and adsorption behavior of calcined Zn/Al hydrotalcite-like compound (HTlc) towards removal of fluoride from aqueous solution. J Colloid Interface Sci 261:213–220
49. Harrison PTC (2005) Fluoride in water: a UK perspective. J Fluor Chem 126:1448–1456
50. Luo F, Inoue K (2004) The removal of fluoride ion by using metal(III)-loaded amberlite resins. Solvent Extr Ion Exch 22:305–322
51. Ho LN, Ishihara T, Ueshima S, Nishiguchi H, Takita Y (2004) Removal of fluoride from water through ion exchange by mesoporous Ti oxohydroxide. J Colloid Interface Sci 272:399–403
52. Solangi IB, Memon S, Bhanger MI (2009) Removal of fluoride from aqueous environment by modified Amberlite resin. J Hazard Mater 171:815–819
53. Meenakshi S, Viswanathan N (2007) Identification of selective ion-exchange resin for fluoride sorption. J Colloid Interface Sci 308:438–450
54. Viswanathan N, Meenakshi S (2009) Role of metal ion incorporation in ion exchange resin on the selectivity of fluoride. J Hazard Mater 162:920–930
55. US Environmental Protection Agency (US EPA) (2002) EPA report. Perchlorate environmental contamination: toxicological review and risk characterization, external review draft
56. Urbansky ET (1998) Perchlorate chemistry: implications for analysis and remediation. Bioremed J 2:81–95
57. ITRC (Interstate Technology Regulatory Council) (2005) Perchlorate: overview of issues, status, and remedial options
58. GAO (U.S. Government Accountability Office) (2005) Perchlorate: a system to track sampling and cleanup results is needed, GAO-05-462

59. NAS (National Research Council of the National Academies) (2005) Health implications of perchlorate ingestion
60. NIOSH (2007) (National Institute for Occupational Safety and Health)
61. Li Z, Li FX, Byrd D, Deyhle GM, Sesser DE, Skeels MR, Lamm SH (2000) Neonatal thyroxine level and perchlorate in drinking water. J Occup Environ Med 42:200–205
62. Crump KS, Gibbs JP (2005) Benchmark calculations for perchlorate from three human cohorts. Environ Health Perspect 113:1001–1008
63. Gu B, Brown GM (2006) Recent advances in ion exchange for perchlorate treatment, recovery and destruction. In: Gu B, Coates JD (eds) Perchlorate: environmental occurrence, interactions and treatment. Springer, New York
64. Gu B, Brown GM, Maya L, Lance MJ, Moyer BA (2001) Regeneration of perchlorate (ClO_4^-)-loaded anion exchange resins by a novel tetrachloroferrate ($FeCl_4^-$) displacement technique. Environ Sci Technol 35:3363–3368
65. Gu B, Dong W, Brown GM, Cole DR (2003) Complete degradation of perchlorate in ferric chloride and hydrochloric acid under controlled temperature and pressure. Environ Sci Technol 37:2291–2295
66. Gu B, Ku YK, Brown GM (2002) Treatment of perchlorate-contaminated groundwater using highly selective, regenerable ion-exchange technology: a pilot-scale demonstration. Remediat J 12:51–68
67. Xiong Z, Dimick P, Zhao D, Kney A, Tavakoli J (2006) Removal of perchlorate from contaminated water using a regenerable polymeric ligand exchanger. Sep Sci Technol 41: 2555–2574
68. Xiong Z, Zhao D, Harper WF (2007) Sorption and desorption of perchlorate with various classes of ion exchangers: a comparative study. Ind Eng Chem Res 46:9213–9222
69. Kim TH, Jang M, Park JK (2008) Bifunctionalized mesoporous molecular sieve for perchlorate removal. Micropor Mesopor Mater 108:22–28
70. Dilip M, Griffin ST, Spear SK, Rijksen C, Rodriguez H, Rogers RD (2008) Dual nature of polyethylene glycol-based aqueous biphasic extraction chromatographic (ABEC) resins: uptakes of perchlorate versus mercury(II). Ind Eng Chem Res 47:7390–7396
71. Yoon IH, Meng X, Wang C, Kim KW, Bang S, Choe E, Lippincott L (2009) Perchlorate adsorption and desorption on activated carbon and anion exchange resin. J Hazard Mater 164:87–94
72. Frankenberger WT Jr (2001) Environmental chemistry of arsenic, 1st edn. CRC Press, New York
73. Bhattacharya P, Arun B, Mukherjee B, Bundschuh J, Zevenhoven R (2007) Arsenic in soil and groundwater environment: biogeochemical interactions, health effects and remediation, vol 9, 1st edn. Elsevier, Oxford
74. Mohan D, Pittman CU Jr (2007) Arsenic removal from water/wastewater using adsorbents-A critical review. J Hazard Mater 142:1–53
75. Henry WD, Zhao D, SenGupta AK, Lange C (2004) Preparation and characterization of a new class of polymeric ligand exchangers for selective removal of trace contaminants from water. React Funct Polym 60:109–120
76. Lin JC, SenGupta AK (2009) Hybrid anion exchange fibers with dual binding sites: simultaneous and reversible sorption of perchlorate and arsenate. Environ Eng Sci 26:1673–1683
77. An B, Fu Z, Xiong Z, Zhao D, SenGupta AK (2010) Synthesis and characterization of a new class of polymeric ligand exchangers for selective removal of arsenate from drinking water. React Funct Polym 70:497–507
78. Awual Md R, Urata S, Jyo A, Tamada M, Katakai A (2008) Arsenate removal from water by a weak-base anion exchange fibrous adsorbent. Water Res 42:689–696
79. Shao W, Li X, Cao Q, Luo F, Li J, Du Y (2008) Adsorption of arsenate and arsenite anions from aqueous medium by using metal(III)-loaded amberlite resins. Hydrometallurgy 91:138–143

80. Anirudhan TS, Unnithan MR (2007) Arsenic(V) removal from aqueous solutions using an anion exchanger derived from coconut coir pith and its recovery. Chemosphere 66:60–66
81. Korngold E, Belayev N, Aronov L (2001) Removal of arsenic from drinking water by anion exchangers. Desalination 141:81–84
82. Vatutsina OM, Soldatov VS, Sokolova VI, Johann J, Bissen M, Weissenbacher A (2007) A new hybrid (polymer/inorganic) fibrous sorbent for arsenic removal from drinking water. React Funct Polym 67:184–201
83. Min JH, Hering JG (1998) Arsenate sorption by Fe(III)-doped alginate gels. Water Res 32:1544–1552
84. DeMarco MJ, SenGupta AK, Greenleaf JE (2003) Arsenic removal using a polymeric/inorganic hybrid sorbent. Water Res 37:164–176
85. Katsoyiannis IA, Zouboulis AI (2002) Removal of arsenic from contaminated water sources by sorption onto iron-oxide coated polymeric materials. Water Res 36:5141–5155
86. Suzuki TM, Bomani JO, Matsunaga H, Yokoyama T (2000) Preparation of porous resin loaded with crystalline hydrous zirconium oxide and its application to the removal of arsenic. React Funct Polym 43:165–172
87. Suzuki TM, Tanco ML, Tanaka DAP, Matsunaga H, Yokoyama T (2001) Adsorption characteristics and removal of oxy-anions of arsenic and selenium on the porous polymers loaded with monoclinic hydrous zirconium oxide. Sep Sci Technol 36:103–111
88. Balaji T, Yokoyama T, Matsunaga H (2005) Adsorption and removal of As(V) and As(III) using Zr-loaded lysine diacetic acid chelating resin. Chemosphere 59:1169–1174
89. Seko N, Basuki F, Tamada M, Yoshii F (2004) Rapid removal of arsenic(V) by zirconium(IV) loaded phosphoric chelate adsorbent synthesized by radiation induced graft polymerization. React Funct Polym 59:235–241
90. Zhu X, Jyo A (2001) Removal of arsenic(V) by zirconium(IV) loaded phosphoric acid chelating resin. Sep Sci Technol 36:3175–3189
91. Egawa H, Nonaka T, Maeda H (1985) Studies of selective adsorption resins. XXII. Removal and recovery of arsenic ion in geothermal power waste solution with chelating resin containing mercapto groups. Sep Sci Technol 20:653–664
92. Styles PM, Chanda M, Rempel GL (1996) Sorption of arsenic anions onto poly(ethylene mercaptoacetimide). React Funct Polym 31:89–102
93. Tatineni B, Hideyuki M (2002) Adsorption characteristics of As(III) and As(V) with titanium dioxide loaded Amberlite XAD-7 resin. Anal Sci 18:1345–1349
94. Lin TF, Liu CC, Hsieh WH (2006) Adsorption kinetics and equilibrium of arsenic onto an iron-based adsorbent and an ion exchange resin. Water Sci Technol Water Supply 6:201–207
95. Mohan D, Pittman CU Jr (2006) Activated carbons and low cost adsorbents for remediation of tri- and hexavalent chromium from water. J Hazard Mater B137:762–811
96. Atia AA (2006) Synthesis of a quaternary amine anion exchange resin and study its adsorption behaviour for chromate oxyanions. J Hazard Mater B137:1049–1055
97. Neagu V, Untea I, Tudorache E, Luca C (2003) Retention of chromate ion by conventional and N-ethylpyridinium strongly basic anion exchange resins. React Funct Polym 57:119–124
98. Korngold E, Belayev N, Aronov L (2003) Removal of chromates from drinking water by anion exchangers. Sep Purif Technol 33:179–187
99. Kabay N, Arda M, Saha B, Streat M (2003) Removal of Cr(VI) by solvent impregnated resins (SIR) containing aliquat 336. React Funct Polym 54:103–115
100. Spinelli VA, Laranjeira MCM, Favere VT (2004) Preparation and characterization of quaternary chitosan salt: adsorption equilibrium of chromium(VI) ion. React Funct Polym 61:347–352
101. Anirudhan TS, Noeline BF, Manohar DM (2006) Phosphate removal from wastewaters using a weak anion exchanger prepared from a lignocellulosic residue. Environ Sci Technol 40:2740–2745
102. Blaney LM, Cinar S, SenGupta AK (2007) Hybrid anion exchanger for trace phosphate removal from water and wastewater. Water Res 41:1603–1613

103. Haghsheno R, Mohebbi A, Hashemipour H, Sarrafi A (2009) Study of kinetic and fixed bed operation of removal of sulfate anions from an industrial wastewater by an anion exchange resin. J Hazard Mater 166:961–966
104. Dizge N, Demirbas E, Kobya M (2009) Removal of thiocyanate from aqueous solutions by ion exchange. J Hazard Mater 166:1367–1376
105. Ruixia L, Jinlong G, Hongxiao T (2002) Adsorption of fluoride, phosphate, and arsenate ions on a new type of ion exchange fiber. J Colloid Interface Sci 248:268–274
106. Chubar NI, Samanidou VF, Kouts VS, Gallios GG, Kanibolotsky VA, Strelko VV, Zhuravlev IZ (2005) Adsorption of fluoride, chloride, bromide, and bromate ions on a novel ion exchanger. J Colloid Interface Sci 291:67–74

Index

Inamuddin and M. Luqman (eds.), *Ion Exchange Technology II: Applications*,
DOI 10.1007/978-94-007-4026-6, © Springer Science+Business Media B.V. 2012